Direct Natural Gas Conversion to Value-Added Chemicals

Direct Natural Gas Conversion to Value-Added Chemicals

Edited by
Jianli Hu
Dushyant Shekhawat

CRC Press
Taylor & Francis Group
Boca Raton London New York

CRC Press is an imprint of the
Taylor & Francis Group, an **informa** business

First edition published 2021
by CRC Press
6000 Broken Sound Parkway NW, Suite 300, Boca Raton, FL 33487-2742

and by CRC Press
2 Park Square, Milton Park, Abingdon, Oxon, OX14 4RN

© 2021 Taylor & Francis Group, LLC
CRC Press is an imprint of Taylor & Francis Group, LLC

ISBN: 978-0-367-07793-8 (hbk)
ISBN: 978-0-367-53949-8 (pbk)
ISBN: 978-0-429-02285-2 (ebk)

Typeset in Times
by Deanta Global Publishing Services, Chennai, India

Contents

Preface

The emergence of shale gas basins in North America and other countries around the world has offered natural gas as an abundant, accessible, and low-cost energy source and feedstock for many value-added chemicals. Generally, it is believed that the long-term price of natural gas, at least in the United States, will remain much lower than equivalent crude oil; therefore, the conversion of natural gas to value-added chemicals presents significant economic benefits. However, commercial natural gas-to-chemical technologies are based on an indirect route via syngas production. Indirect routes are generally energy inefficient and capital intensive. Typically, more than 50% of the capital cost is incurred in syngas production, which is an intermediate. In contrast, direct natural gas conversion eliminates the syngas production step. However, these technologies have not been commercialized because of technical challenges such as low selectivity, coking, heat management in a reactor, catalyst deactivation and regeneration. In spite of the increasing technical and commercial importance of direct natural gas conversion, there are few books on the direct conversion of natural gas.

The objective of this book is to introduce the recent advances in direct natural gas conversion. It provides a broad spectrum of new developments in direct natural gas conversion with the emphasis on catalysis, reactor, and modeling work. This book includes sixteen chapters contributed by experts around the world on natural gas conversion technologies. This book covers nontraditional approaches such as electrochemical, nonthermal plasma, and microwave-assisted natural gas conversion processes. This book also offers state-of-the-art reviews of thermochemical methods such as chemical looping, oxidative coupling, and pyrolysis. Single-atom catalysts, an emerging novel concept that could be highly selective for natural gas conversions reactions due to their unique structure of the active sites are also covered in the book. Homogeneous catalysis and bioconversion of natural gas are also reviewed in the book by experts in those areas. The applicability of additive manufacturing, or 3D printing, in biocatalysis is appealing due to its higher specificity and selectivity for natural gas conversion is also described in the book.

In addition to wide-range of natural gas conversion approaches, the book also covers some product-specific reactions such as dehydroaromatization (DHA) of hydrocarbons for aromatics production, pyrolysis of natural gas for hydrogen and carbon production. The book also discusses membrane reactors, methane and hydrogen, and/or oxygen-permeable membranes for the DHA reaction to remove hydrogen and to increase methane equilibrium conversion. The use of microwave-assisted catalytic reaction for DHA is also discussed in the book. Technoeconomic analysis of microwave-assisted methane DHA is also investigated.

It is a pleasure to thank those who have made this project possible. First of all, we would like to thank all of our coauthors. We are very grateful that they were able to dedicate their valuable time to this project. It has been our pleasure to work with all the contributors involved in this book. Their effort in combining their own research with recent literature in the field of natural gas conversion is highly appreciated.

This effort would not have been possible without their willingness to share their valuable knowledge, insight, and experience. Moreover, we express our gratitude for their responsiveness to deadlines and review comments. We also want to express our sincere gratitude to all reviewers who provided their thoughtful and timely comments. Special thanks are due to Xinwei Bai, a graduate student of Prof. Jianli Hu, for providing editorial help on a number of chapters. We would like to thank Shatkin Allison (Senior Commissioning Editor) and Gabrielle Vernachio (Editorial Assistant) from CRC Taylor and Francis Group for their help in coordinating the publication process. We would also thank CRC Taylor and Francis Group for their commitment to this project.

Editors

Dr. Jianli Hu is a Chair Professor and the Director of the Shale Gas Center at West Virginia University. He leads an interdisciplinary faculty team carrying out cutting-edge research in natural gas conversion as well as renewable energy utilization. He has demonstrated strong leadership in partnering with U.S. national laboratories and industrial companies to undertake a number of research projects funded by U.S. federal agencies. These research projects span across the fields of reaction engineering, surface chemistry, heterogeneous catalysis, plasma, and microwave-enhanced catalytic reactions. Specifically, his research group at WVU is focused on microwave catalysis approaches to produce ammonia from renewable power under low pressure and temperature, chemicals, and carbon nanomaterials from natural gas.

Before joining West Virginia University, Dr. Hu worked as a research leader at Koch Industries, Pacific Northwest National Laboratory and BP Oil. While working in industry, Dr. Hu led efforts in developing technologies for petroleum refineries and future biorefineries using biomass as a feedstock. The efforts had led to a number of acquisitions that facilitated technology commercialization. He has been granted 31 U.S. patents and published more than 150 journal articles, conference proceedings, technical reports and book chapters. Dr. Hu has been serving as an editor and a guest-editor for a number of peer-reviewed technical journals and a book. He has been serving in a number of industrial advisory boards including AIChE RAPID focus area lead. He has been chairing for Advanced Fossil Energy Utilization in AIChE national meetings.

At Pacific Northwest National Laboratory, Dr. Hu was among the key contributors in developing microchannel catalytic reactor technology and received NASA Technology Brief Awards. Dr. Hu received his BS in Chemistry and PhD in Chemical Engineering from Tsinghua University, China, and did his postdoctoral training at the University of Pittsburgh, Pennsylvania.

Dr. Dushyant Shekhawat serves as a team supervisor for the Reaction Engineering team at National Energy Technology Laboratory (NETL), U.S. Department of Energy, Morgantown, West Virginia. Dr. Shekhawat's research interests include fuel processing for fuel cell applications, reaction engineering, surface chemistry, heterogeneous catalysis, energy, fuel cell, plasma and microwave-assisted catalytic reactions, and plasma chemistry. Currently, his team at NETL is leading efforts in developing microwave-assisted technologies and approaches to produce value-added fuels and chemicals from natural gas, higher hydrocarbon fuels, and coal. Dr. Shekhawat was among the key contributors to pyrochlore-based reforming catalyst, which was licensed to a spin-off company, Pyrochem Catalyst Company. He is a corecipient of the Federal Laboratory Consortium 2011 Mid-Atlantic Regional Award for Excellence in Technology Transfer "Novel Pyrochlore Catalysts for Hydrocarbon Reforming". He has published over 200 peer-reviewed journal publications, conference proceedings, and reports, seven book chapters, three books, and fifteen patents (disclosed or full). Dr. Shekhawat has served as an associate editor of the Catalysis book series published by the Royal Society of Chemistry. Dr. Shekhawat has also served as a guest-editor for the special issues (sections) of *Catalysis Today, Fuel, Energy & Fuels*, and *International Journal of Hydrogen Energy*. He was a program chair for a topical "Advanced Fossil Energy Utilization" in AIChE national meetings for the past 8 years. Under the same topical, he has organized multiple sessions of fuel processing for hydrogen production and value-added chemicals from natural gas. He is a registered Professional Engineer (PE) in West Virginia and also serves in the NCEES's PE Chemical Engineering Examination Development committee. Dr. Shekhawat received his BS in Chemical Engineering from the University of Minnesota in Twin Cities and his PhD in Chemical Engineering from the Michigan State University in East Lansing.

List of Contributors

Victor Abdelsayed
Leidos Research Support Team
and
National Energy Technology Laboratory
U.S. Department of Energy
Morgantown, West Virginia

Shabbir Ahmed
Argonne National Laboratory
Argonne, Illinois

Xinwei Bai
Department of Chemical and
  Biomedical Engineering
West Virginia University
Morgantown, West Virginia

Sonit Balyan
Department of Chemical Engineering
Indian Institute of Technology Delhi
New Delhi, India

Hisham Bamufleh
Chemical and Materials Engineering
  Department
King Abdulaziz University
Jeddah, Saudi Arabia

Mark Bearden
Pacific Northwest National Laboratory
Richland, Washington

Madan Mohan Bhasin
Mid-Atlantic Technology
Research & Innovation Center
South Charleston, West Virginia

Debangsu Bhattacharyya
Department of Chemical and
  Biomedical Engineering
West Virginia University
Morgantown, West Virginia

Sichao Cheng
Department of Chemical and
  Biomolecular Engineering
University of Maryland
College Park, Maryland

Robert Dagle
Pacific Northwest National Laboratory
Richland, Washington

Vanessa Dagle
Pacific Northwest National Laboratory
Richland, Washington

Dong Ding
Idaho National Laboratory
Idaho Falls, Idaho

Qiang Fei
School of Chemical Engineering and
  Technology
Xi'an Jiaotong University
Xi'an, China

Rongzhan Fu
School of Chemical Engineering
Northwest University
Xi'an, China

Anne Gaffney
Idaho National Laboratory
Idaho Falls, Idaho

Fausto Gallucci
Department of Chemical Engineering
  and Chemistry
Eindhoven University of Technology
  (TU/e)
Eindhoven, the Netherlands

Consuelo Álvarez Galván
Instituto de Catálisis y Petroleoquímica
    (CSIC)
Madrid, Spain

Yunfei Gao
Department of Chemical and
    Biomolecular Engineering
North Carolina State University
Raleigh, North Carolina

Hamid Reza Godini
Department of Chemical Engineering
    and Chemistry
Eindhoven University of Technology
    (TU/e)
Eindhoven, the Netherlands

Niles Jensen Gunsalus
The Scripps Energy & Materials Center
The Scripps Research Institute
Jupiter, Florida

Puneet Gupta
Department of Chemistry
Indian Institute of Technology Roorkee
Roorkee, India

M. Ali Haider
Department of Chemical Engineering
Indian Institute of Technology Delhi
New Delhi, India

Mahmoud M. El-Halwagi
Chemical and Materials Engineering
    Department
King Abdulaziz University
Jeddah, Saudi Arabia
and
Chemical Engineering Department
Texas A&M University
College Station, Texas

J Holladay
Pacific Northwest National Laboratory
Richland, Washington

Jianli Hu
Department of Chemical and
    Biomedical Engineering
West Virginia University
Morgantown, West Virginia

Tuhin S. Khan
Department of Chemical Engineering
Indian Institute of Technology Delhi
New Delhi, India

Jennifer M. Knipe
Lawrence Livermore National
    Laboratory
Livermore, California

Anjaneyulu Koppaka
The Scripps Energy & Materials Center
The Scripps Research Institute
Jupiter, Florida

Theodore Krause
Argonne National Laboratory
Argonne, Illinois

Fanxing Li
Department of Chemical and
    Biomolecular Engineering
North Carolina State University
Raleigh, North Carolina

Dongxia Liu
Department of Chemical and
    Biomolecular Engineering
University of Maryland
College Park, Maryland

Hua Liu
CAS Key Laboratory of Science and
    Technology on Applied Catalysis
Dalian Institute of Chemical Physics
Chinese Academy of Sciences
Dalian, China
and
University of Chinese Academy of
    Sciences
Beijing, China

Xiaoyan Liu
University of Chinese Academy of
Sciences
Beijing, China

Su Cheun Oh
Department of Chemical and
Biomolecular Engineering
University of Maryland
College Park, Maryland
and
Exponent, Hong Kong Science Park
Shatin, Hong Kong

Haritha Meruvu
School of Chemical Engineering and
Technology
Xi'an Jiaotong University
Xi'an, China

Chirag Mevawala
Department of Chemical and
Biomedical Engineering
West Virginia University
Morgantown, West Virginia

Tomohiro Nozaki
Tokyo Institute of Technology
Department of Mechanical Engineering
Tokyo, Japan

Ying Pan
Department of Chemical and
Biomolecular Engineering
University of Maryland
College Park, Maryland
and
State Key Laboratory of Separation
Membranes and Membrane Processes
Tianjin Polytechnic University
Tianjin, China

K.K. Pant
Department of Chemical Engineering
Indian Institute of Technology Delhi
New Delhi, India

Roy A. Periana
The Scripps Energy & Materials Center
The Scripps Research Institute
Jupiter, Florida

Brandon Robinson
Department of Chemical and
Biomedical Engineering
West Virginia University
Morgantown, West Virginia

Emily Schulman
Department of Chemical and
Biomolecular Engineering
University of Maryland
College Park, Maryland

Dushyant Shekhawat
National Energy Technology Laboratory
U.S. Department of Energy
Morgantown, West Virginia

Sreedevi Upadhyayula
Department of Chemical Engineering
Indian Institute of Technology Delhi
New Delhi, India

Aiqin Wang
University of Chinese Academy of
Sciences
Beijing, China

Hui Wu
State Key Laboratory of Bioreactor
Engineering
East China University of Science and
Technology
Shanghai, China

Wei Wu
Idaho National Laboratory
Idaho Falls, Idaho

Rekha Yadav
Department of Chemical Engineering
Indian Institute of Technology Delhi
New Delhi, India

1 Electrochemical Conversion of Natural Gas to Value Added Chemicals

Wei Wu, Anne Gaffney, and Dong Ding

CONTENTS

1.1 INTRODUCTION

The shale gas revolution results in an increasingly abundant production of natural gas (NG) and light hydrocarbons, most of which is located in remote areas that making transportation through pipelines less economic attractive. In consequence, conversion of NG into value added chemicals and fuels has become increasingly important for petrochemical industry, which has noticed the opportunities for new business and technologies, i.e., how to efficiently produce value added higher hydrocarbons from light alkanes. Since the top three contents of NG are methane (70~98%), ethane and propane (0~18% in total), NG conversion is commonly treated as conversion of light alkanes (i.e., methane conversion and ethane conversion). Methane can be converted to chemicals and fuels in two ways, either through synthesis gas or directly into C_2 hydrocarbons and/or aromatics. Ethane and propane are normally used to produce ethylene and propylene through steam cracking that represents one of the most energy-consuming

1

process in petrochemical industry. Take ethylene production for an example. The ethane steam cracking method is an energy-intensive, nonselective and noncatalytic technology, producing an array of chemicals including hydrogen, ethylene, acetylene, propylene, etc. In addition to the high temperature needed, steam cracking requires high investment in separation and purification equipment (Ren, Patel, and Blok 2006). The ethane steam cracking process consumes 17–21 GJ (specific energy consumption, SEC) of process energy per ton of ethylene, of which 65% is used in high temperature pyrolysis, 15% in fractionation and compression, and 20% in product separation. (Kaiser et al. 1993; Houston 1993; Worrell et al. 2000) Not only is the current steam cracking technology capital and energy intensive, it is a large carbon dioxide producer (1–1.5 t/t of ethylene) (Ramirez-Corredores, Ding, and Gaffney 2019).

However, the current high temperature steam-cracking technology is hard to be replaced due to the maturity of the manufacturing industry, where moderate changes were made over the past 50 years. The energy efficiency was also optimized by reordering of the separation units, improvements in controls, and reactor redesign etc., with reliable operation. In order to reduce both the energy consumption and the CO_2 emission in NG conversion, disruptive methods that feature a lower working temperature with higher energy efficiency, instead of simple process optimization, are preferred, because the working temperature and the energy efficiency are critical in NG and NGLs related manufacturing (Gao et al. 2019).

Electrochemistry, which has been in development for centuries, holds great potential in NG conversion and chemical engineering technology. Electrochemical methods are increasingly attractive in converting electricity power into chemical bonds for energy storage and have obvious advantages over traditional thermochemical methods. First, electrochemical methods use electricity as alternative driving force to enable bond-formation steps, making most electrochemical reactions run at reduced temperatures and pressures with low thermal energy requirement. Second, real-time products separation can be easily achieved with an electrochemical membrane reactor that keeps products generated at the cathode and anode separated. Moreover, the Faradaic efficiency (FE) of most electrochemical processes is close to unity, such as solid oxide fuel cell), resulting in high energy efficiency.

Technically, NG electrochemistry can be divided into two categories: room temperature electrochemistry (RTEC) and elevated temperature electrochemistry (ETEC). Typical cases of RTEC in chemical production are room temperature C_2H_4 production from CO_2 reduction in aqueous solutions (Ren et al. 2015; DeWulf, Tuo, and Bard 1989; Hori et al. 1986) and proton exchange membrane fuel cell (PEMFC) with carbon-supported platinum catalyst.(Li et al. 2005) Since room temperature electrochemistry uses corrosive electrolyzer solution and precious metal as catalyst, the environmental issue and processing cost make it lack economic attractiveness. This topic is not covered in this section.

The solid oxide electrochemical cell (SOC), on the other hand, is suitable to operate at elevated temperatures to incorporate less expensive catalysts and is becoming an alternative for more efficient NG conversion (Wang et al. 2007b; Shi et al. 2008; Guo et al. 1997). Its advantages include: (i) the feedstock conversion rate is no longer subject to the thermodynamic equilibrium if one or more product species can be real-time separated (Champagnie et al. 1992); (ii) the potential to customize the thermodynamics and kinetics of the desired reaction with electrochemical assistance; and (iii) the much higher energy efficiency compared to the conventional high-temperature thermal cracking. We here narrow the discussion of NG electrochemical conversion to direct conversion of light alkanes (CH_4, C_2H_6, and C_3H_8) based on the SOC technology, which is an emerging technology and attracts great technoeconomic interest from petrochemical industry.

1.2   ELECTROCHEMICAL DIRECT CONVERSION OF METHANE

Methane is a very stable, symmetrical molecule. The difference between indirect and direct methane conversion is the intermediate step of producing syngas (mixture of CO and H_2) mainly from steam reforming or dry reforming. The direct conversion of natural gas into higher hydrocarbons has obvious advantage over the indirect way, as it simplifies the production procedure without producing syngas (Lunsford 2000). There are several possible routes for the direct methane conversion process, as shown in Figure 1.1. Methane pyrolysis is an equilibrium and thermal decarbonization reaction in the absence of oxygen to produce hydrogen and carbon black, which begins to produce carbon and hydrogen around 300°C and goes to completion around 1000°C. It is not a new process and has been employed for decades. However, there is no electrochemical method ever reported in this topic, making electrochemical methane decomposition an interesting area to explore for the continuous production of hydrogen and carbon with drastically reduced CO_2 emissions. Electrochemical oxidation of methane into CO_2 and water is typically achieved in methane fed fuel cells to produce electricity. There are several good review papers (Gür 2016; Choudhury, Chandra, and Arora 2013) on this topic and will not be discussed here.

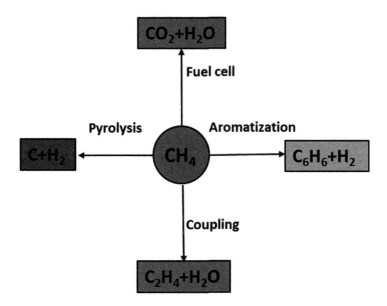

FIGURE 1.1   Different routes of direct methane conversion.

1.2.1 ELECTROCHEMICAL METHANE COUPLING TO PRODUCE ETHYLENE

The electrochemical methane coupling (EMC) process involves the reaction of CH_4 and O_2 over a catalyst at high temperatures to form C_2H_6 and/or C_2H_4. Conventionally, discussions about methane coupling involved co-feeding of reactants oxygen, steam, or carbon dioxide along with methane. In this section and hereafter, we only focus on oxygen as the only oxidative species to react with methane. To avoid confusion, the term "direct" methane conversion here implies that no reactants other than the methane is fed as feedstock.

There are two primary routes for the EMC process, depending on the existence of oxygen containing species in the feedstock: electrochemical oxidative methane coupling (EOMC) and electrochemical nonoxidative methane coupling (ENMC) processes. Figure 1.2 shows the operating principal schematic for both categories.

1.2.1.1 Electrochemical Oxidative Methane Coupling (EOMC)

In EOMC process, the reaction on the methane feedstock side is described in Equation 1.1:

$$CH_4 + O^{2-} \rightarrow \frac{1}{2}C_2H_4 + H_2O + 2e^- \qquad (1.1)$$

Typical oxygen ion conducting electrolyte materials used in the electrochemical membrane methane coupling include oxygen-ion conducting materials such as 8% Y_2O_3– ZrO_2 (8YSZ) (Otsuka, Yokoyama, and Morikawa 1985), gadolinium-doped ceria (GDC)(Liu et al. 2017) and mixed conducting materials such as Yb-doped $SrCeO_3$ (SCY) (Hamakawa, Hibino, and Iwahara 1993). During operation, electricity is either generated or consumed, representing two different approaches. The first approach is direct electro-oxidation of using the methane solid oxide fuel cell technology (SOFC-OMC), where ethylene and electricity are co-generated through an electro generative method. Liu et al. reported a SOFC-OMC tubular membrane reactor design with the use of an optimized Mn–Ce–Na_2WO_4/SiO_2 catalyst (Liu et al. 2017), as shown in Figure 1.3. Optimized performance of 60.7% methane conversion with 41.6% C_{2+} selectivity, 5.8 ethylene to ethane ratio and 19.4% ethylene yield is achieved. The use of the SOFC technology dramatically reduces the cost and safety concerns of conventional fixed

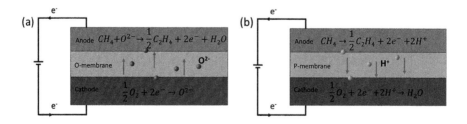

FIGURE 1.2 Operation schematics of SO-EMC through (a) oxidative and (b) nonoxidative approach. The EOMC typically use oxygen ion-conducting electrochemical cells (O-SOCs) while the ENMC uses proton-conducting electrochemical cells (P-SOCs).

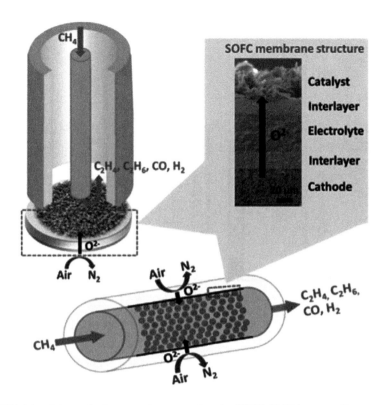

FIGURE 1.3 Schematic diagrams demonstrating the SOFC-OMC button cell reactor. SEM cross-section image of the multilayer SOFC membrane. (Reproduced by permission from *Catal. Commun.*, 96, 24. Copyright 2017 Elsevier.)

bed methane coupling reactions by getting rid of the air separation unit as well as preventing the direct co-feed of explosive methane-oxygen mixtures.

As the methane coupling reaction involves the transfer of multiple electrons, the methane oxidation reaction can only be accomplished in a sequence of elementary steps and has close relationship with electron flux (Park et al. 1999). Tagawa et al. conducted selective oxidation of CH_4 into C_2 and CO using SOFC system (Tagawa et al. 1999). As shown in Figure 1.4, with the increase of voltage, the CO and CO_2 yields increased remarkably while the C_{2+} yield did not remain stable. Other studies with a wide range of catalysts and operation conditions suggested a trade-off between conversion and selectivity,(Guo et al. 1997; Zeng, Lin, and Swartz 1998; Hamakawa, Hibino, and Iwahara 1993; Farrell and Linic 2016) which comes from the intrinsic fact that ethane is intrinsically more active than methane.

The second approach of EOMC can be described as membrane-OMC, where electricity is used to assist OMC reactions via an electrochemical oxygen ion-conducting membrane reactor by driving O^{2-} across the electrolyte. Different from conventional OMC, membrane-OMC is more like an electrochemical oxygen pump using external power to apply either a voltage or a current across the solid electrolytes. In consequence, non-Faradic electrochemical modification of catalytic activity (NEMCA)

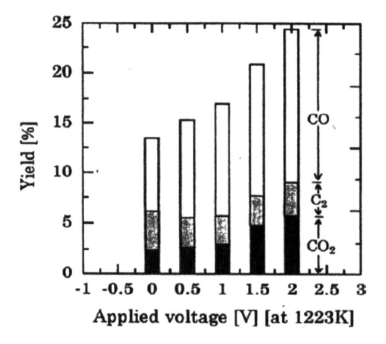

FIGURE 1.4   The effect of applied voltage on products yield in EOMC. (Reproduced by permission from *Chem. Eng. Sci.*, 54, 1553. Copyright 1999 Elsevier.)

is achieved due to the change of both catalysts working function and reaction rate (Eng and Stoukides 1991; Liu et al. 2001). However, the deep oxidation of CH_4 into CO and CO_2 in the presence of oxygen containing species is always a challenge to both membrane-OMC and SOFC-OMC. Figure 1.5 shows the calculated change in the Gibbs free energy at 1073 K and 1 atm for several reactions that can occur in the OCM process. Although the ΔG for the formation of the C_2 products are negative, the formation of CO and CO_2 is even more energetically downhill, indicating that the formation of the undesired products (mainly CO and CO_2) is thermodynamically favorable under high temperature reaction conditions.[6]

1.2.1.2   Electrochemical Nonoxidative Methane Coupling (ENMC)

Like EOMC, the ENMC process also has two approaches, i.e., membrane-ENMC and SOFC-NMC. The ENMC process typically uses proton-conducting materials as electrolyte, by which the deep oxidation of CH_4 to CO and CO_2 is inhibited. The reaction on feedstock side is expressed in Equation 1.2:

$$2CH_4 \rightarrow C_2H_4 + 2H^+ + 2e^- \tag{1.2}$$

Compared with EOMC, the study of the ENMC process is very limited. An electrochemical cell with configuration of Ag/ $SrCe_{0.95}Yb_{0.05}O_{3-\delta}$/Ag was used to conduct ENMC at 1173 K (Hamakawa, Hibino, and Iwahara 1993). On introducing Ar into the cathode side and on passing the applied current across the cell, formation rates

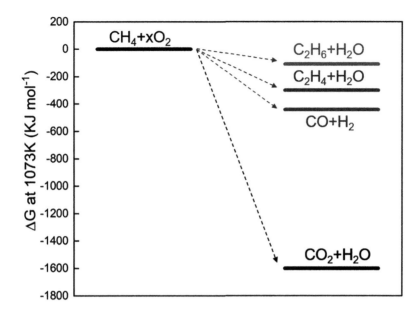

FIGURE 1.5 Change in Gibbs free energy for the reactions that can occur in an OMC reactor at 1073 K and 1 atm.

of C_2-compounds increased from 0.23 to 0.52 μmol min^{-1} cm^{-2}, as shown in Figure 1.6. However, the current efficiency of about 7% for C_2-formation was rather low. The author investigated the current efficiencies of that electrochemical cell using hydrogen pump test. The results show that, at low current densities, the evolution rate of hydrogen agreed with the value calculated from Faraday's law, indicating that the proton transport number in the electrolyte was unity. At high current densities, the evolution rate of hydrogen reached a limiting value, probably due to a diffusion limit of hydrogen through porous silver of the anode. Moreover, during the real ENMC process, the amount of hydrogen formed at the anode was about four times larger than that expected from those of ethane and ethylene, suggesting that coking occurred as a significant reaction. Langguth et al. also studied the ENMC using $SrCe_{0.95}Yb_{0.05}O_{3-\delta}$ as a high-temperature proton-conducting membrane in an electrochemical button cell with methane and air supplied on the anode and cathode side, respectively (Langguth et al. 1997). After analysis of the products composition, CO and CO_2 were also detected besides the target product CH_4. This result revealed a key challenge for ENMC: like SCY material, most of solid oxide proton conducting materials become oxygen-ion and proton mixed conducting at temperatures above 650°C, at which the ENMC processes are typically operated to achieve reasonable yields. Moreover, the development of catalysts for NMC at reduced temperatures is still a big challenge.

1.2.2 ELECTROCHEMICAL METHANE DEHYDRO-AROMATIZATION (EDMA)

Thermodynamically, methane is more favorable to be converted into aromatics than olefins (Xu, Bao, and Lin 2003). Nonoxidative methane dehydro-aromatization

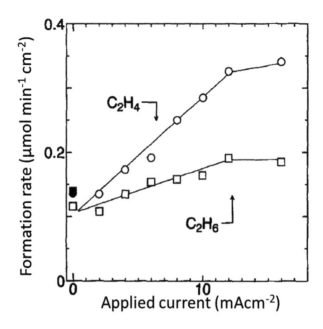

FIGURE 1.6 Formation rates of C-compounds at the anode: closed symbols show formation rates of C2-compounds on cutting off direct current after the experiment. (Reproduced by permission *J. Electrochem. Soc.* 140, 459. Copyright (1993) The Electrochemical Society.)

(MDA) into benzene, as described in Equation 1.3, is a key technology to directly convert methane into value added petrochemicals and is receiving increasing attention (Zhao and Wang 2011; Li, Borry, and Iglesia 2001).

$$6CH_4 \rightarrow C_6H_6 + 9H_2 \tag{1.3}$$

The catalytic MDA process, conventionally operated at 700°C in the presence of bifunctional catalysts, suffers from two major challenges: low single-pass conversion due to the thermodynamics, and the fast catalyst activity degradation because of the accumulation of polyaromatic-type coke on the catalyst surface (Tempelman and Hensen 2015). Recently, Morejudo et al. demonstrated a direct MDA process using a BaZrO3-based, mixed proton and oxygen ion conducting solid oxide electrochemical cell with high aromatic yields and improved catalyst stability (Morejudo et al. 2016). The improved yield and stability originate from the simultaneous removal of hydrogen from products and distributed injection of oxygen ions to the feedstock side to remove the coke. Although aromatics are the target and major products in that research, ethylene was also detected, indicating the possibility of ethylene production through EMDA method (Figure 1.7).

Although remarkable effects have been put on the thermal catalytic direct methane conversion, research on EMDA is limited. Due to highly stable conformation of methane (383 kJ/mol), the electrochemical conversion of methane to higher hydrocarbons is thermodynamically unfavorable and requires relatively high electric energy input (Ngobeni et al. 2009), reducing EMDA's technoeconomic attractiveness.

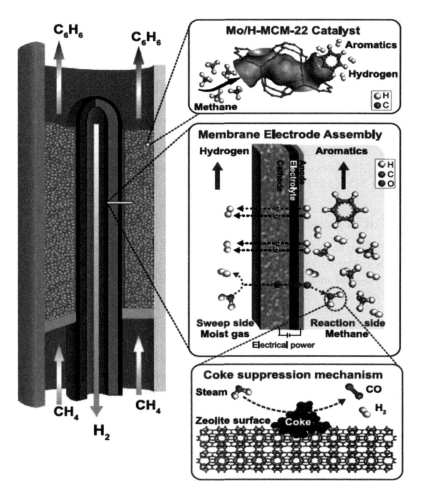

FIGURE 1.7 Current-controlled co-ionic membrane reactor. CH_4 is converted to benzene and hydrogen via a Mo/zeolite catalyst. H_2 is transported as protons to the sweep side. Oxide ions are transported to the reaction medium to react with H_2 and form steam as an intermediate before reacting with coke to form CO and H_2. (Reproduced by permission *Science*, 353, 563. Copyright 2016 American Association for the Advancement of Science.)

Moreover, low conversion rate due to coking issue, and quick catalytic and/or electrochemical performance deactivation remain challenging.

1.3 ELECTROCHEMICAL DIRECT CONVERSION OF ETHANE

Ethane is the second largest component of natural gas. Routes for ethane direct conversion are similar to methane-based approaches: ethane pyrolysis to methane, carbon and hydrogen; ethane dehydrogenation to ethylene and ethane dehydro-aromatization to liquid fuels. Ethylene production through dehydrogenation is the priority use of ethane due to their compositional, structural and chemical similarity.

Ethane pyrolysis and dehydro-aromatization normally appears as side reactions in ethane dehydrogenation process and are not covered in the section.

Similar to SO-EMC, electrochemical dehydrogenation of ethane using SOC technology (SO-EDH) also has two routes: oxidative dehydrogenation (SO-EODH) process and nonoxidative dehydrogenation (SO-ENDH) process, depending on the presence of oxygen-contain species in the feedstock side. Based on the solid oxide electrolyte charge carrier type, the SOCs can be divided into two categories: oxygen ion-conducting electrochemical cell (O-SOC) and proton-conducting electrochemical cell (P-SOC). O-SOCs are typically used in EODH process, while P-SOCs are used in ENDH process. Figure 1.8 shows the operation schematics of SO-EDH processes in O-SOCs and P-SOCs.

1.3.1 Electrochemical Oxidative Dehydrogenation (EODH)

Conversion of ethane to ethylene in a SO-EODH route is a process by which ethane is partially oxidative dehydrogenated to ethylene in an O-SOC, which typically consists of solid-state anode, oxygen ion conducting electrolyte and cathode. The reaction of EODH can be expressed as Equation 1.4:

$$2C_2H_6 + O_2 \rightarrow 2C_2H_4 + 2H_2O \tag{1.4}$$

Typical oxygen ion conducting materials are used in SOFC technology including yttrium-stabilized zirconia (Wu et al. 2014; de Souza, Visco, and De Jonghe 1997) and gadolinium doped ceria (Wu, Guan, and Wang 2015; Wachsman and Lee 2011). During operation, oxygen is reduced to oxygen ions at cathode side and transferred across the electrolyte to the anode. At the anode side, ethane is firstly dehydrogenated to ethylene, electrons and protons are combined with oxygen ions to form water, while electrons go through external circuit to produce electricity. This process can also be described as a chemical/power co-generation procedure in an ethane fueled solid oxide fuel cell. Unfortunately, no oxygen ion conducting fuel cell system was reported to be techonomic viable in conversion of ethane to ethylene. There are two major problems inhibiting further development of the EODH technology: first, the presence of oxygen sources at anode side could significantly lead to undesirable deep oxidation of ethane and/or ethylene to carbon dioxide. Second, O-SOCs typically operate at temperatures above 750°C and results in severe side reactions including thermal cracking (Froment et al. 1976) and decreases the C_2H_4 selectivity.

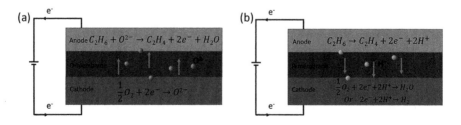

FIGURE 1.8   Operation Schematics of SO-EDH using (a) O-SOCs and (b) P-SOCs.

Hibino et al. constructed three kinds of ionic conductors (protonic, oxide ionic and their mixed) to control the oxidative dehydrogenation of ethane at 700 °C (Hibino, Hamakawa, and Iwahara 1993). After comparison, they found the EDH based on P-SOCs presented the highest current efficiency. When P-SOCs were incorporated into the EDH process, the case was shift to nonoxidative mode, namely SO-ENDH.

1.3.2 ELECTROCHEMICAL NONOXIDATIVE DEHYDROGENATION (ENDH)

As shown in Figure 1.8b, the ENDH process can be achieved either with or without oxygen-contain-species on the cathode side. When oxygen species were involved, the ENDH process is a proton conducting ethane fuel cell (SOFC-ENDH). During that operation, ethane is dehydrogenated to ethylene, electrons and protons at the feedstock side. Different from SOFC-EODH, protons, instead of oxygen ions, move across electrolyte and react with oxidant species to form water. When oxygen acts as the oxidant, the ideal overall reaction would be identical to ethane ODH.

Power/chemical co-production through ethane fuel cell is not a new approach. Remarkable progress has made on this topic, especially using P-SOCs (Gao et al. 2019). An ethane/oxygen fuel cell with a configuration of $Pt/BaCe_{0.85}Y_{0.15}O_{3-\delta}$ (BCY15)/Pt was reported to co-produce ethylene and electricity (Wang et al. 2007a). The proton source was found to be either from conversion of hydrogen generated from dehydrogenation of ethane or from electrochemical conversion of ethane. Shi et al. optimized the operation parameters and achieved 34% ethane conversion with 96% ethylene selectivity, using the same C_2H_6, Pt/BCY15/Pt, O_2 proton conducting fuel cell system (Shi et al. 2008). Relevant follow-on activities were conducted by the same group using different anode materials like, Co_2CrO_4(Lin et al. 2018), Co–Fe nano particles (Liu, Chuang, and Luo 2015) and nano-Cr_2O_3 (Fu et al. 2011) with reasonable performance achieved at intermediate temperatures ($\geq$ 650°C). Recently, perovskite of $SrMo_{0.8}Co_{0.1}Fe_{0.1}O_{3-\delta}$ with in situ exsolved Co nanoparticles was synthesized as C_2H_6 dehydrogenation catalyst (Liu et al. 2018), as shown in Figure 1.9a. The corresponding C_2H_4 yield of the cell with Co-SMCFO at peak current load is

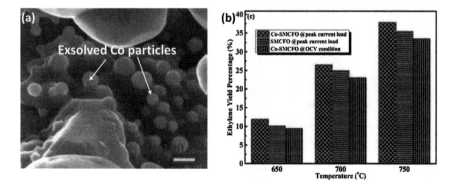

FIGURE 1.9 (a) SMCFO catalyst and (b) C_2H_4 yields of SMCFO and Co-SMCFO as a function of temperature at different conditions. (Reproduced by permission *Appl. Catal*, B. 220, 288. Copyright 2018 Elsevier.)

37.8% with the C_2H_4 selectivity exceeding 91% at 750°C, as demonstrated in Figure 1.9b. The ex-solution method is an emerging method to fabricate nanoscale electrocatalyst for NG conversion reactions. When perovskite is utilized as supporting backbone, the incorporated B-site cation in the perovskite lattice can be partially in situ exsolved as functional metal nanoparticle under reducing flow, which can not only serve as co-catalyst for certain reactions but also improve the electronic conductivity of the parent perovskite (Neagu et al. 2013).

However, it should be noted that $BaCe_{0.7}Zr_{0.1}Y_{0.2}O_{3-\delta}$ (BCZY) material is actually a mixed oxygen-ion and proton conductor at temperature above 600°C (Kreuer 2003, 1999; Kreuer, Schönherr, and Maier 1994), making it highly possible that EODH and ENDH exist simultaneously in BCZY based electrochemical system. For large band gap perovskites (ABO$_3$), the most important reaction leading to the formation of protonic defects at moderate temperatures is hydration, which requires the presence of oxide ion vacancies, as expressed in Equation 1.5:

$$H_2O + V_{\ddot{O}} + O_O^x \leftrightarrows 2OH_O^{\cdot}$$ (1.5)

A hydroxide ion and a proton are formed by the dissociation of water molecule; the hydroxide ion fills an oxide ion vacancy, and the proton forms a covalent bond with a lattice oxygen. Therefore, the proton conductivity depends on the degree of hydration. Figure 1.10 show the hydration degree of different proton conductors at different temperatures. As shown in the figure, most proton conductors demonstrated

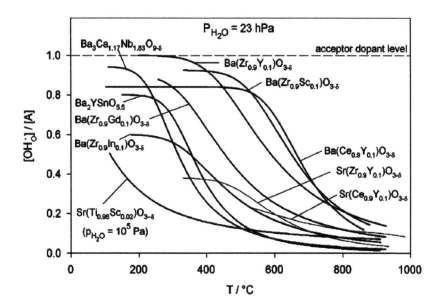

FIGURE 1.10 Normalized hydration isobars (pH$_2$O = 23 hPa) for different perovskites (data for cubic perovskites are shown in bold). For BaZrO$_3$-based compositions, data for different acceptor dopants (Y, Sc, Gd, In) are included (Reproduced by permission *Annual Review of Materials Research*, 33, 340. Copyright (2003) Annual Reviews.)

coexistence of protonic defects and oxide ion vacancies at elevated temperatures above 600°C, making doped $BaZrO_3$ materials proton/oxygen ion mixed conductor. In consequence, due to the fact that the product subjected to catalyst surfaces is often oxidized more easily than the feedstock (Rostrup-Nielsen and Trimm 1977; Lu et al. 2012), as well as the mixed ion conducting characters of proton conductor at elevated temperatures, ENDH process must be operated at low conversions in order to reach high selectivity (Lercher and Naraschewski 2011). Moreover, side reactions (e.g., ethane thermal cracking) and coking on catalysts remain challenging at high operating temperatures (McIntosh and Gorte 2004; Wachsman and Lee 2011).

ENDH to co-produce ethylene and hydrogen with no oxygen species involved, as expressed as Equation 1.6, is an emerging area that was first reported by Idaho National Laboratory:

$$C_2H_6 \rightarrow C_2H_4 + H_2 \qquad (1.6)$$

Compare with SOFC-ENDH, the absence of oxygen in cathode side overcome the drawbacks of oxidative hydrogenation processes, especially when the protonic solid oxide electrolyte was used at temperatures above 600°C. In addition, the strictly non-oxidative environment on both electrodes sides eliminates the competitive reaction between feedstock and product, ensuring the high selectivity while alleviating safety consideration and reducing the carbon footprint. Ding et al. reported a P-SOC with configuration of $PrBa_{0.5}Sr_{0.5}Co_{1.5}Fe_{0.5}O_{5+\delta}$ (PBSCF) $/BaZr_{0.1}Ce_{0.7}Y_{0.1}Yb_{0.1}O_{3-\delta}$ (BZCYYb)/Ni-BZCYYb, which was used to co-produce ethylene and hydrogen from ethane through an ENDH process at 400°C (Ding et al. 2018). Figure 1.11a and 1.11b show the operation mechanism and cell configuration used in Ding's work. Ethane was dehydrogenated at anode, and the generated protons passed across the 15-μm electrolyte to the cathode and reacted with electrons to produce hydrogen. It should be noted that the H_2 produced in ENDH process can be readily used without separation. Considering the heating value of generated H_2, the ENDH technology would be even more attractive.

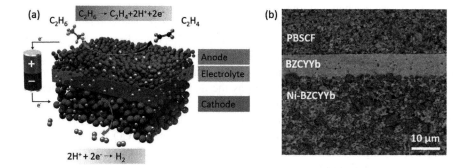

FIGURE 1.11 (a) Schematic of the co-production of ethylene and hydrogen via an ENDH process of ethane in an electrode supported P-SOC. (b) A cross-sectional SEM image of electrochemical cell after testing at 400°C. (Reproduced by permission from *Energy Environ. Sci.*, 11, 1711. Copyright 2018 Royal Society of Chemistry.)

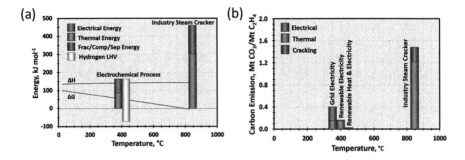

FIGURE 1.12 (a) A comparison of the process energies for ethylene production from ethane and (b) A comparison of the carbon footprint for ethylene production from ethane. The ENDH was carried out at 400°C, whereas the steam cracking was performed at 850°C. (Reproduced by permission from *Energy Environ. Sci.*, 11, 1716. Copyright 2018 Royal Society of Chemistry.)

Figure 1.12 compares the energy consumption and CO$_2$ emission between the ENDH process and industrial steam cracking. Reduced temperature has significant positive impacts on both low thermal budget and thermal coking suppression. The proof-of-concept work presented a 65% savings in energy consumption and CO$_2$ reduction by 72% or more if renewable electricity and heat are used, compared with the industrial steam cracking (Ding et al. 2018). The same team also indicated the feasibility of ethylene production using methane as the feedstock through nonoxidative process at 350°C (Ding et al. 2019). Besides the ready-to-capture hydrogen, such a low-thermal-budget, nonoxidative, electrochemical process enabled the production of value-added chemicals with the additional benefit of CO$_2$ emission reduction, resulting in enhanced process intensification.

1.4 ELECTROCHEMICAL DIRECT CONVERSION OF PROPANE

Propane is basically used to produce propylene through dehydrogenation process. Start-of-art electrochemical routes for propane dehydrogenation process is based on propane fueled fuel cell technology, i.e., solid oxide propane fuel cell (Lo Faro et al. 2009; Zhan, Liu, and Barnett 2004; Zhang et al. 2017). Depending on the electrolyte charge carrier type, the propane dehydrogenation process can also be divided into either an oxidative (EODP) or a nonoxidative route (ENDP). A cermet of sliver and gadolinium-doped ceria is investigated as the anode material for SOFCs operated directly on propane. As described in Figure 1.13, that SOFC are operated with dry propane as the fuel and ambient air as the oxidant. The cell was stably operated for 150 hours at 800°C, with a series of gas products including H$_2$, CO, CH$_4$, C$_2$H$_4$, C$_2$H$_6$, and C$_3$H$_6$ (Zhang et al. 2017). A carbon layer is built on surface of Ag–GDC anode of SOFCs operated on propane, indicating propane pyrolysis was the dominating reaction during SOFC operation at that temperature.

When a propane-air mixture is used as feedstock, propane internal reforming can be achieved in SOFCs (Zhan, Liu, and Barnett 2004). For example, the addition of a Ru–CeO$_2$ catalyst layer promoted propane partial oxidation at temperatures ≥ 500°C

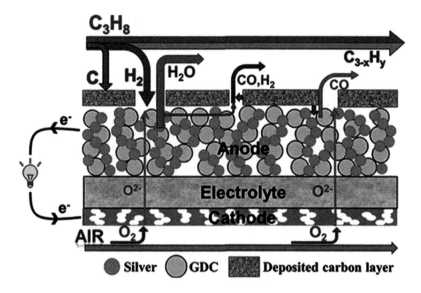

FIGURE 1.13 Working principles of propane fed oxygen ion conducting SOFC. (Reproduced by permission from *J. Power Sources*, 336, 56. Copyright 2017 Elsevier.)

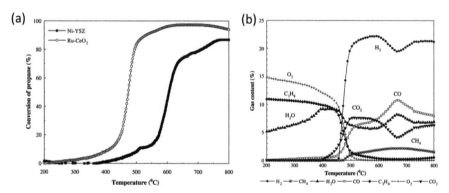

FIGURE 1.14 (a) Propane conversion percentage versus T for 10.7% C_3H_8, 18.7% O_2, 70.6% Ar fuel mixtures flowed over a Ru–CeO$_2$ catalyst layer at open circuit. Shown for comparison is the result for a bulk Ni–YSZ anode. (b) The exhaust gas composition versus temperature at 50 ml min^{-1}. (Reproduced by permission from Solid State Ionics. Copyright 2005 Elsevier.)

without carbon formation, using conventional Ni–YSZ-based oxygen-ion conducting SOFC (Zhan and Barnett 2005). As shown in Figure 1.14, the propane conversion over a Ru–CeO$_2$ functional layer reached 80% at ∼ 475°C, 150 °C lower than the Ni–YSZ anode, indicating that the Ru–CeO$_2$ layer was a more effective low-temperature catalyst, for partial oxidation of propane.

The anode reaction of nonoxidative propane dehydrogenation has no oxygen containing species involved, as expressed in Equation 1.7:

$$C_3H_8 \rightarrow C_3H_6 + H_2 \tag{1.7}$$

Feng et al. carried out the ENDP process using a BCY-based proton conducting fuel cell reactor at 700°C (Feng, Luo, and Chuang 2007). The overall propane conversion was ~ 20% with low C_3H_6 selectivity. The high reaction temperature negatively affects propene selectivity because of parallel side thermal cracking reactions. Alkenes, hydrogen can be co-generated from dehydrogenation of alkanes in a proton-conducting membrane fuel cell reactor when run without an oxygen feed. However, there is very limited work dedicated to this area. Compared with ENDH, this nonoxidative dehydrogenation mode is less appropriate for propane dehydrogenation because C_3H_6 selectivity is more sensitive to reaction temperature than C_2H_4 selectivity.

1.5 CHALLENGES AND OPPORTUNITIES

The shale gas revolution in the United States has changed the natural gas (NG) markets, and the oversupply of light alkanes may continue for decades. Due to the challenges associated with nonoxidatively breaking the C–H bond of light alkanes and the greenhouse gas production, NG utilizations are currently limited to power generation from combustion, syngas production from reforming and partial oxidation and directly usage as transportation fuel and residential heating sources (Ramirez-Corredores, Ding, and Gaffney 2019).

Clearly, conversion of NG through electrochemical methods has the following advantages over the industrial steam/thermal cracking: (1) it can circumvent the thermodynamic limitation, allowing operation at reduced temperatures to mitigate challenges associated with side reactions, coking and catalyst deactivation (Ding, Liu, et al. 2008); (2) proper material selection and efficient integration into a full device to enable good performance, leading to small electrical energy input; (3) separation of products can further overcome thermodynamic limitation, pushing the reaction forward to maximize the conversion; (4) potential to incorporate highly efficient catalyst to maximize the conversion rate; and (5) modularity of solid oxide membrane reactors benefits facile-integration with renewable-based (e.g., nuclear energy) heat and electricity for an integrated/hybrid energy system, resulting in more efficient process intensification (Kim, Boardman, and Bragg-Sitton 2018). Table 1.1 summarizes primary characteristics of different electrochemical converiosn methods using methane, ethane and propane as feedstocks. These emerging electrochemical NG conversion approaches address one or more of the limitations of the commercial steam/thermal cracking technologies, e.g., high reaction endothermicity, limited equilibrium conversion and complex product separation, holding promises in thermal budget reduction by lowering the operating temperature and controlling the reaction kinetics and product types.

Ideally, the electrochemical NG converion should be purely an electrochemical process, which can inherently overcome the thermodynamic limitations associated with its thermochemical counterpart. However, it is realistically a combination of electrochemistry and thermochemistry and becomes much more complicated with the remaining challenges. First and most important, the development and integration of highly selective catalysts/electrocatalysts toward conversion of light alkanes, especially under nonoxidative conditions, is limited. Very recently, Wang

TABLE 1.1

Comparisons of the Various Approaches for NG Conversion

- Spontaneous
- Non-spontaneous

			Electrolyte type	Oxygen involved	Anode/negative electrode reaction	Cathode/positive electrode reaction	Power consumption	Co-product of hydrogen	Coke formation	Main side products
CH_4	EOMC	Membrane-EOCM	O^{2-}	Yes	$CH_4 + O^{2-} \to \frac{1}{2}C_2H_4 + H_2O + 2e^-$	$\frac{1}{2}O_2 + 2e^- \to O^{2-}$	Yes	No	high	CO_2, CO
		SOFC-EOCM	O^{2-}	Yes	$CH_4 + O^{2-} \to \frac{1}{2}C_2H_4 + H_2O + 2e^-$	$\frac{1}{2}O_2 + 2e^- \to O^{2-}$	No	No	high	CO_2, CO
	ENMC	Membrane-ENCM	H^+	Yes	$CH_4 \to \frac{1}{2}C_2H_4 + 2H^+ + 2e^-$	$2H^+ + \frac{1}{2}O_2 + 2e^- \to H_2O$	Yes	No	medium	CO, C_2H_6
				No	$CH_4 \to \frac{1}{2}C_2H_4 + 2H^+ + 2e^-$	$\frac{1}{2}H^+ + 2e^- \to H_2$	Yes	Yes	medium	C_2H_6
		SOFC-ENCM	H^+	Yes	$CH_4 \to \frac{1}{2}C_2H_4 + 2H^+ + 2e^-$	$2H^+ + \frac{1}{2}O_2 + 2e^- \to H_2O$	No	No	medium	CO, C_2H_6
	EDMA	Membrane-EDMA	H^+	No	$6CH_4 \to C_6H_6 + 18H^+ + 18e^-$	$18H^+ + 18e^- \to 9H_2$	Yes	No	medium	C_2H_6, C_3H_8, C_{6+}
C_2H_6	EODH	Membrane-EODH	O^{2-}	Yes	$C_2H_6 + O^{2-} \to C_2H_4 + H_2O + 2e^-$	$\frac{1}{2}O_2 + 2e^- \to O^{2-}$	Yes	No	high	CO_2, CO, CH_4
		SOFC-EODH	O^{2-}	Yes	$C_2H_6 + O^{2-} \to C_2H_4 + H_2O + 2e^-$	$\frac{1}{2}O_2 + 2e^- \to O^{2-}$	No	No	high	CO_2, CO, CH_4

(Continued)

TABLE 1.1 (CONTINUED)
Comparisons of the Various Approaches for NG Conversion

- Spontaneous
- Non-spontaneous

		Electrolyte type	Oxygen involved	Anode/negative electrode reaction	Cathode/positive electrode reaction	Power consumption	Co-product of hydrogen	Coke formation	Main side products
ENDH	Membrane-ENDH	H+	Yes	$C_2H_6 \rightarrow C_2H_4 + 2e^- + 2H^+$	$2H^+ + \frac{1}{2}O_2 + 2e^- \rightarrow H_2O$	Yes	No	medium	CO, CH_4
			No	$C_2H_6 \rightarrow C_2H_4 + 2e^- + 2H^+$	$\frac{1}{2}H^+ + 2e^- \rightarrow H_2$	Yes	Yes	medium	CH_4
	SOFC-ENDH	H+	Yes	$C_2H_6 \rightarrow C_2H_4 + 2e^- + 2H^+$	$2H^+ + \frac{1}{2}O_2 + 2e^- \rightarrow H_2O$	No	No	medium	CO_2, CO, CH_4
C_3H_8 EODP	Membrane-EODP	O^{2-}	Yes	$C_3H_8 + O^{2-} \rightarrow C_3H_6 + H_2O + 2e^-$	$\frac{1}{2}O_2 + 2e^- \rightarrow O^{2-}$	Yes	No	high	CO_2, CO, CH_4, C_2H_6
	SOFC-EODP	O^{2-}	Yes	$C_3H_8 + O^{2-} \rightarrow C_3H_6 + H_2O + 2e^-$	$\frac{1}{2}O_2 + 2e^- \rightarrow O^{2-}$	No	No	high	CO_2, CO, CH_4, C_2H_6
ENDP	Membrane-ENDP	H+	Yes	$C_3H_8 \rightarrow C_3H_6 + 2e^- + 2H^+$	$2H^+ + \frac{1}{2}O_2 + 2e^- \rightarrow H_2O$	Yes	No	medium	CO, CH_4, C_2H_6
			No	$C_3H_8 \rightarrow C_3H_6 + 2e^- + 2H^+$	$\frac{1}{2}H^+ + 2e^- \rightarrow H_2$	Yes	Yes	medium	CH_4, C_2H_6
	SOFC-ENDP	H+	Yes	$C_3H_8 \rightarrow C_3H_6 + 2e^- + 2H^+$	$2H^+ + \frac{1}{2}O_2 + 2e^- \rightarrow H_2O$	No	No	medium	CO, CH_4, C_2H_6

et al. reported that iron supported on ZSM-5 zeolite (Fe/ZSM-5) was a highly efficient catalyst for nonoxidative dehydrogenation of ethane to produce ethylene below 600°C,(Wang et al. 2019) which is promising in integration into P-SOCs for ENDH at lower tempertures to reduce thermal budgets with sufficient conversion and products yield. But it is not enough. it is ultimately required to develop electrocatalysts that can maximize the effect of electrochemistry by efficiently enabling electrons for the entire process. By doing so, it will realize the "true" electrocatalysis and maintain the high conversion of the ethane at reduced temperatures. Additionally, electrode engineering is important to leverage the catalytic performance of well-developed catalysts and electrode backbone by enabling effective coating and surface modification techniques (Ding et al. 2014, 2013; Ding, Zhu, et al. 2008). A representative case is an ultraporous 3D electrode that is suitable for the direct integration of catalysts into electrochemcial cells. Recent advances in achieving direct electro-oxidation of solid carbon at intermediate temperature steam electrolysis using 3D ceramic textile electrodes, have demonstrated that pursuit of this approach has great potential (Wu, Ding et al. 2018; Wu, Zhang et al. 2018) (Figure 1.15).

Second, reduced processing temperature results in active demand for superior solid electrolytes with higher ionic conductivity and lower activation energy than the ones currently used. An easy approach is making the electrolyte layer thinner to lower the working overpotential at reduced temperatures, but a breakthrough in the inherent conducting properties of materials is the ultimate solution. Rational design of super protonic conductors for high quality solid oxide proton conducting electrochemical reactors may offer new opportunities in this direction for further lowering the electrochemical process temperature and reducing the thermal budget (Wu, Ding, and He 2017; Ding, Wu, and Ding 2019). Third, optimal operating conditions (e.g., temperatures, ethane concentration, flow rate, etc.) are required to balance the thermal energy input *vs* ethane conversion as well as ethane conversion *vs* ethylene selectity.

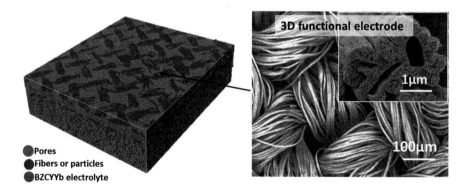

FIGURE 1.15 Solid oxide electrolysis cell with 3D ultraporous functional electrode, which structure is promising in electrode design for NG conversion (Wu, Ding et al. 2018). (Reproduced by permission from *Adv. Sci.* 11. Copyright 2018 WILEY-VCH Verlag GmbH & Co.)

Finally, scale-up/modularity is one undisputed step to push electrochemical NG conversion technology forward into commercialization. Conversely, this step could also bring uncertainties in performance reproducilibity. Despite the increase in the research of NG electrochemical conversion, data related to the the different steps of scale-up is still missing. A best practice is to adopt the innovations of materials and structures from academia as well as to provide the fundamental interpretations and recommendations back to the industrial developers. It is also expected that an effective combination of experiment, modeling, and technoeconomic analyses is crucial for the further development and implementation of the electrochemical approaches.

ACKNOWLEDGMENTS

The authors would like to thank the support of U.S. Department of Energy (USDOE), Office of Energy Efficiency and Renewable Energy (EERE), Advanced Manufacturing Office (AMO) R&D Projects Emerging Research Exploration under DOE Idaho Operations Office with contract no. DE-AC07-05ID14517.

REFERENCES

Champagnie, A. M., Theodore, T. T., Ronald, G. M., and E. Wagner. 1992. The study of ethane dehydrogenation in a catalytic membrane reactor. *Journal of Catalysis* 134(2):713–730. doi:10.1016/0021-9517(92)90355-L.

Choudhury, A., Chandra, H., and A. Arora. 2013. Application of solid oxide fuel cell technology for power generation—a review. *Renewable and Sustainable Energy Reviews* 20:430–442. doi:10.1016/j.rser.2012.11.031.

de Souza, S., Steven, J. V., and L. C. De Jonghe. 1997. Reduced-temperature solid oxide fuel cell based on YSZ thin-film electrolyte. *Journal of the Electrochemical Society* 144(3):L35–L37.

DeWulf, D. W., Tuo, J., and A. J. Bard. 1989. Electrochemical and surface studies of carbon dioxide reduction to methane and ethylene at copper electrodes in aqueous solutions. *Journal of the Electrochemical Society* 136(6):1686–1691.

Ding, D., He, T., Wu, W., and Y. Zhang. 2019. *Methods for Producing Hydrocarbon Products and Hydrogen Gas Through Electrochemical Activation of Methane, and Related Systems and Electrochemical Cells.* Battelle Energy Alliance LLC.

Ding, D., Li, X. X., Lai, S. Y., Gerdes, K., and M. L. Liu. 2014. Enhancing SOFC cathode performance by surface modification through infiltration. *Energy & Environmental Science* 7(2):552–575. doi:10.1039/c3ee42926a.

Ding, D., Liu, M. F., Liu, Z. B., Li, X. X., Blinn, K., Zhu, X. B., and M. L. Liu. 2013. Efficient electro-catalysts for enhancing surface activity and stability of SOFC cathodes. *Advanced Energy Materials* 3(9):1149–1154. doi:10.1002/aenm.201200984.

Ding, D., Liu, Z. B., Li, L., and C. R. Xia. 2008. An octane-fueled low temperature solid oxide fuel cell with Ru-free anodes. *Electrochemistry Communications* 10(9):1295–1298. doi:10.1016/j.elecom.2008.06.026.

Ding, D., Zhang, Y., Wu, W., Chen, D., Liu, M., and T. He. 2018. A novel low-thermal-budget approach for the co-production of ethylene and hydrogen via the electrochemical non-oxidative deprotonation of ethane. *Energy & Environmental Science* 11(7):1710–1716.

Ding, D., Zhu, W., Gao, J. F., and C. R. Xia. 2008. High performance electrolyte-coated anodes for low-temperature solid oxide fuel cells: model and experiments. *Journal of Power Sources* 179(1):177–185. doi:10.1016/j.jpowsour.2007.12.059.

Ding, H., Wu, W., and D. Ding. 2019. Advancement of proton-conducting solid oxide fuel cells and solid oxide electrolysis cells at Idaho National Laboratory (INL). *ECS Transactions* 91(1):1029–1034.

Eng, D., and M. Stoukides. 1991. Catalytic and electrocatalytic methane oxidation with solid gxide membranes. *Catalysis Reviews* 33(3–4):375–412.

Farrell, B. L., and S. Linic. 2016. Oxidative coupling of methane over mixed oxide catalysts designed for solid oxide membrane reactors. *Catalysis Science & Technology* 6(12):4370–4376.

Feng, Y., Luo, J., and K. T. Chuang. 2007. Conversion of propane to propylene in a proton-conducting solid oxide fuel cell. *Fuel* 86(1–2):123–128. doi:10.1016/j.fuel.2006.06.012.

Froment, G. P., Van de Steene, B. O., Van Damme, P. S., Narayanan, S., and A. G. Goossens. 1976. Thermal cracking of ethane and ethane-propane mixtures. *Industrial & Engineering Chemistry Process Design and Development* 15(4):495–504. doi:10.1021/i260060a004.

Fu, X. Z., Luo, X. X., Luo, J. L., Chuang, K. T., Sanger, A. R., and A. Krzywicki. 2011. Ethane dehydrogenation over nano-Cr_2O_3 anode catalyst in proton ceramic fuel cell reactors to co-produce ethylene and electricity. *Journal of Power Sources* 196(3):1036–1041. doi:10.1016/j.jpowsour.2010.08.043.

Gao, Y., Neal, L., Ding, D., Wu, W., Baroi, C., Gaffney, A. M., and F. Li. 2019. Recent advances in intensified ethylene production—a review. *ACS Catalysis* 9(9):8592–8621. doi:10.1021/acscatal.9b02922.

Guo, X. M., Hidajat, K., Ching, C. B., and H. F. Chen. 1997. Oxidative coupling of methane in a solid oxide membrane reactor. *Industrial & Engineering Chemistry Research* 36(9):3576–3582. doi:10.1021/ie9607006.

Gür, T. M. 2016. Comprehensive review of methane conversion in solid oxide fuel cells: prospects for efficient electricity generation from natural gas. *Progress in Energy and Combustion Science* 54:1–64. doi:10.1016/j.pecs.2015.10.004.

Hamakawa, S., Hibino, T., and H. Iwahara. 1993. Electrochemical methane coupling using protonic conductors. *Journal of the Electrochemical Society* 140(2):459–462.

Hibino, T., Hamakawa, S., and H. Iwahara. 1993. Dehydrogenation of ethane using solid electrolytes as membrane reactor. *Nippon Kagaku Kaishi* 1993(3):238–242.

Hori, Y., Kikuchi, K., Murata, A., and S. Suzuki. 1986. Production of methane and ethylene in electrochemical reduction of carbon dioxide at copper electrode in aqueous hydrogencarbonate solution. *Chemistry Letters* 15(6):897–898.

Jens, R. N., and D. L. Trimm. 1977. Mechanisms of carbon formation on nickel-containing catalysts. *Journal of Catalysis* 48(1):155–165. doi:10.1016/0021-9517(77)90087-2.

Kaiser, V., Ruiz-Martinez, J., Santillan-Jimenez, E., and B. M. Weckhuysen. 1993. *Ethylene Plant Energy Analysis. National Meeting of AIChE*, Houston, TX.

Kim, J. S., Boardman, R. D., and S. M. Bragg-Sitton. 2018. Dynamic performance analysis of a high-temperature steam electrolysis plant integrated within nuclear-renewable hybrid energy systems. *Applied Energy* 228:2090–2110.

Kreuer, K. D. 1999. Aspects of the formation and mobility of protonic charge carriers and the stability of perovskite-type oxides. *Solid State Ionics* 125(1):285–302. doi:10.1016/S0167-2738(99)00188-5.

Kreuer, K. D. 2003. Proton-conducting oxides. *Annual Review of Materials Research* 33(1):333–359. doi:10.1146/annurev.matsci.33.022802.091825.

Kreuer, K. D., Schönherr, E., and J. Maier. 1994. Proton and oxygen diffusion in BaCeO3 based compounds: a combined thermal gravimetric analysis and conductivity study. *Solid State Ionics* 70:278–284.

Langguth, J., Dittmeyer, R., Hofmann, H., and G. Tomandl. 1997. Studies on oxidative coupling of methane using high-temperature proton-conducting membranes. *Applied Catalysis A: General* 158(1–2):287–305.

Lercher, J. A., and F. N. Naraschewski. 2011. C–H activation of alkanes in selective oxidation reactions on solid oxide catalysts. In *Nanostructured Catalysts*, Royal Society of Chemistry, 5–32.

Li, L., Borry, R. W., and E. Iglesia. 2001. Reaction-transport simulations of non-oxidative methane conversion with continuous hydrogen removal—homogeneous–heterogeneous reaction pathways. *Chemical Engineering Science* 56(5):1869–1881.

Li, W. S., Lu, D. S., Luo, J. L., and K. T. Chuang. 2005. Chemicals and energy co-generation from direct hydrocarbons/oxygen proton exchange membrane fuel cell. *Journal of Power Sources* 145(2):376–382. doi:10.1016/j.jpowsour.2005.01.077.

Lin, J. Y., Shao, L., Si, F. Z., Liu, S. B., Fu, X. X., and J. L. Luo. 2018. Co_2CrO_4 nanopowders as an anode catalyst for simultaneous conversion of ethane to ethylene and power in proton-conducting fuel cell reactors. *The Journal of Physical Chemistry C* 122(8):4165–4171. doi:10.1021/acs.jpcc.7b11680.

Liu, K., Zhao, J., Zhu, D., Meng, F., Kong, F., and Y. Tang. 2017. Oxidative coupling of methane in solid oxide fuel cell tubular membrane reactor with high ethylene yield. *Catalysis Communications* 96:23–27. doi:10.1016/j.catcom.2017.03.010.

Liu, S., Chuang, K. T., and J. L. Luo. 2015. Double-layered perovskite anode with in situ exsolution of a Co–Fe alloy to cogenerate ethylene and electricity in a proton-conducting ethane fuel cell. *ACS Catalysis* 6(2):760–768. doi:10.1021/acscatal.5b02296.

Liu, S., Liu, Q., Fu, X. Z., and J. L. Luo. 2018. Cogeneration of ethylene and energy in protonic fuel cell with an efficient and stable anode anchored with in-situ exsolved functional metal nanoparticles. *Applied Catalysis B: Environmental* 220:283–289. doi:10.1016/j.apcatb.2017.08.051.

Liu, S., Tan, X., Li, K., and R. Hughes. 2001. Methane coupling using catalytic membrane reactors. *Catalysis Reviews* 43(1–2):147–198.

Lo Faro, M., Rosa, D. L., Nicotera, I., Antonucci, V., and A. S. Aricò. 2009. Electrochemical investigation of a propane-fed solid oxide fuel cell based on a composite Ni–perovskite anode catalyst. *Applied Catalysis B: Environmental* 89(1):49–57. doi:10.1016/j.apcatb.2008.11.019.

Lu, J., Fu, B., Kung, M. C., Xiao, G., Elam, J. W., Kung, H. H., and P. C. Stair. 2012. Coking- and sintering-resistant palladium catalysts achieved through atomic layer deposition. *Science* 335(6073):1205–1208.

Lunsford, J. H. 2000. Catalytic conversion of methane to more useful chemicals and fuels: a challenge for the 21st century. *Catalysis Today* 63(2–4):165–174.

McIntosh, S., and R. J. Gorte. 2004. Direct hydrocarbon solid oxide fuel cells. *Chemical Reviews* 104(10):4845–4866.

Morejudo, S. H., Zanón, R., Escolástico, S., Yuste-Tirados, I., Malerød-Fjeld, H., Vestre, P. K., Coors, W. F. G., Martínez, A., Norby, T., and J. M. Serra. 2016. Direct conversion of methane to aromatics in a catalytic co-ionic membrane reactor. *Science* 353(6299):563–566.

Neagu, D., Tsekouras, G., Miller, D. N., Ménard, H., and J. T. S. Irvine. 2013. In situ growth of nanoparticles through control of non-stoichiometry. *Nature Chemistry* 5(11):916.

Ngobeni, M. W., Carley, A. F., Scurrell, M. S., and C. P. Nicolaides. 2009. The effects of boron and silver on the oxygen-free conversion of methane over Mo/H-ZSM-5 catalysts. *Journal of Molecular Catalysis A: Chemical* 305(1–2):40–46.

Otsuka, K., Yokoyama, S., and A. Morikawa. 1985. Catalytic activity-and selectivity-control for oxidative coupling of methane by oxygen-pumping through yttria-stabilized zirconia. *Chemistry Letters* 14(3):319–322.

Park, S., Craciun, R., Vohs, J. M., and R. J. Gorte. 1999. Direct oxidation of hydrocarbons in a solid oxide fuel cell: I. Methane oxidation. *Journal of the Electrochemical Society* 146(10):3603–3605.

Ramirez-Corredores, M. M., Ding, D., and A. M. Gaffney. 2019. Idaho National Laboratory's advanced design and manufacturing initiative. *Catalysis Today.* doi:10.1016/j.cattod.2019.08.022.

Ren, D., Deng, Y., Handoko, A. D., Chen, C. S., Malkhandi, S., and B. S. Yeo. 2015. Selective electrochemical reduction of carbon dioxide to ethylene and ethanol on copper (I) oxide catalysts. *ACS Catalysis* 5(5):2814–2821.

Ren, T., Patel, M., and K. Blok. 2006. Olefins from conventional and heavy feedstocks: energy use in steam cracking and alternative processes. *Energy* 31(4):425–451.

Shi, Z., Luo, J. L., Wang, S., Sanger, A. R., and K. T. Chuang. 2008. Protonic membrane for fuel cell for co-generation of power and ethylene. *Journal of Power Sources* 176(1):122–127.

Tagawa, T., Moe, K. K., Ito, M., and S. Goto. 1999. Fuel cell type reactor for chemicals-energy co-generation. *Chemical Engineering Science* 54(10):1553–1557.

Tempelman, C. H. L., and E. J. M. Hensen. 2015. On the deactivation of Mo/HZSM-5 in the methane dehydroaromatization reaction. *Applied Catalysis B: Environmental* 176:731–739.

Wachsman, E. D., and K. T. Lee. 2011. Lowering the temperature of solid oxide fuel cells. *Science* 334(6058):935–939.

Wang, L. C., Zhang, Y., Xu, J., Diao, W., Karakalos, S., Liu, B., Song, X., Wu, W., He, T., and D. Ding. 2019. Non-oxidative dehydrogenation of ethane to ethylene over ZSM-5 zeolite supported iron catalysts. *Applied Catalysis B: Environmental* 256. doi:10.1016/j.apcatb.2019.117816.

Wang, S., Luo, J. L., Sanger, A. R., and K. T. Chuang. 2007a. Performance of ethane/oxygen fuel cells using yttrium-doped barium cerate as electrolyte at intermediate temperatures. *The Journal of Physical Chemistry C* 111(13):5069–5074.

Wang, S. Y., Luo, J. L., Sanger, A. R., and K. T. Chuang. 2007b. Performance of ethane/oxygen fuel cells using yttrium-doped barium cerate as electrolyte at intermediate temperatures. *Journal of Physical Chemistry C* 111(13):5069–5074. doi:10.1021/Jp066690w.

Worrell, E., Phylipsen, D., Einstein, D., and N. Martin. 2000. *Energy Use and Energy Intensity of the U.S. Chemical Industry*, Lawrence Berkeley National Laboratory.

Wu, W., Ding, D., and T. He. 2017. Development of high performance intermediate temperature proton-conducting solid oxide electrolysis cells. *ECS Transactions* 80(9):167–173.

Wu, W., Ding, H., Zhang, Y., Ding, Y., Katiyar, P., Majumdar, P. K., He, T., and D. Ding. 2018. 3D self-architectured steam electrode enabled efficient and durable hydrogen production in a proton-conducting solid oxide electrolysis cell at temperatures lower than 600° C. *Advanced Science* 5(11):1800360.

Wu, W., Guan, W., and W. Wang. 2015. Contribution of properties of composite cathode and cathode/electrolyte interface to cell performance in a planar solid oxide fuel cell stack. *Journal of Power Sources* 279:540–548.

Wu, W., Guan, W. B., Wang, G. L., Wang, F., and W. G. Wang. 2014. In-situ investigation of quantitative contributions of the anode, cathode, and electrolyte to the cell performance in anode-supported planar SOFCs. *Advanced Energy Materials* 4(10):1400120.

Wu, W., Zhang, Y., Ding, D., and T. He. 2018. A high-performing direct carbon fuel cell with a 3D architectured anode operated below 600° C. *Advanced Materials* 30(4):1704745.

Xu, Y., Bao, X., and L. Lin. 2003. Direct conversion of methane under nonoxidative conditions. *Journal of Catalysis* 216(1–2):386–395.

Zeng, Y., Lin, Y. S., and S. L. Swartz. 1998. Perovskite-type ceramic membrane: synthesis, oxygen permeation and membrane reactor performance for oxidative coupling of methane. *Journal of Membrane Science* 150(1):87–98. doi:10.1016/S0376-7388(98)00182-3.

Zhan, Z., and S. A. Barnett. 2005. Use of a catalyst layer for propane partial oxidation in solid oxide fuel cells. *Solid State Ionics* 176(9):871–879. doi:10.1016/j.ssi.2004.12.005.

Zhan, Z., Liu, J., and S. A. Barnett. 2004. Operation of anode-supported solid oxide fuel cells on propane–air fuel mixtures. *Applied Catalysis A: General* 262(2):255–259. doi:10.1016/j.apcata.2003.11.033.

Zhang, Y., Yu, F., Wang, X., Zhou, Q., Liu, J., and M. Liu. 2017. Direct operation of Ag-based anode solid oxide fuel cells on propane. *Journal of Power Sources* 366:56–64. doi:10.1016/j.jpowsour.2017.08.111.

Zhao, T., and H. Wang. 2011. Methane dehydro-aromatization over Mo/HZSM-5 catalysts in the absence of oxygen: effect of steam-treatment on catalyst stability. *Journal of Natural Gas Chemistry* 20(5):547–552.

2 Microwaves in Nonoxidative Conversion of Natural Gas to Value-Added Products

Xinwei Bai, Brandon Robinson, Dushyant Shekhawat, Victor Abdelsayed, and Jianli Hu

CONTENTS

2.1 INTRODUCTION

Microwaves are electromagnetic waves with a frequency between 300 MHz to 300 GHz that with a wavelength from 1 mm to 1 m. Since Gedye et al. (1986) observed that using microwave energy could substantially save time in organic synthesis, microwave technology was attracting attentions by chemists and engineers. Today, microwave technology is used primarily in organic compounds synthesis as well as in inorganic material synthesis (metal and metal-oxide nanoparticles) using liquid-based chemical reactions, where the difference in dielectric heating between the solvent and reactant precursors results in selective heating and enhanced conversion rates. Microwaves were also used in preparing materials from gaseous reactants, Silicon carbide films were prepared using microwave plasma under vacuum conditions as reported by Weinreich and Ribner (1968). In recent decades, thanks to the development of different microwave reactor designs, microwave-assisted chemical reaction has gained significant attraction and microwaves have started playing a role in gas–solid heterogeneous catalysis to convert natural gas into value-added chemicals and carbon products, such as methane decomposition and methane-reforming reactions. Recently, microwave-based reactions have been reported to significantly enhances the productivity and conversion of methane (Dąbrowska et al., 2018; Domínguez et al., 2007; Fidalgo et al., 2008; Hamzehlouia et al., 2018; Nguyen et al., 2018).

Microwave-induced catalytic reactions require material that is catalytically active as well as microwave sensitive to achieve chemical conversion reaction. Due to the complexity of microwave-catalyst interaction, it is necessary for microwave-catalysis researchers to equip themselves with basic understanding of electromagnetism and combine it with principles of catalysis. Therefore, the purpose of this chapter is to give an overview of latest available microwave technology applied in catalytic natural gas conversion. In this chapter we address fundamental theory of microwave catalysis, including wave functions of electric and magnetic fields, mechanisms of microwave heating, dielectric property characterization and microwave reactor design and components. In addition, few microwave-assisted direct/indirect natural gas conversion examples and numerical modeling of microwave–material interaction are reviewed.

2.2 BASIC MICROWAVE HEATING THEORY

2.2.1 ELECTROMAGNETIC WAVES

The relationship between electricity and magnetism was first developed by James Clerk Maxwell. By correlating four differential equations, also known as Maxwell's Equations, mathematical models describing electric and magnetic fields could

TABLE 2.1

Maxwell's Equations in Derivative Form (COMSOL, 2018)

Equation name	Differential equation	Note
Gauss's Law	$\nabla \cdot (\epsilon \mathbf{E}) = 0$	ϵ: electric permittivity E: electric field
Gauss's Law for Magnetism	$\nabla \cdot (\mu \mathbf{H}) = 0$	μ: magnetic permeability H: magnetic field
Maxwell–Faraday's Equation	$\nabla \times \mathbf{E} = -\dfrac{\partial(\mu \mathbf{H})}{\partial t}$	
Maxwell–Ampere's Law	$\nabla \times \mathbf{H} = \dfrac{\partial(\epsilon \mathbf{E})}{\partial t} + \sigma \mathbf{E}$	σ: electroconductivity

be derived, including those used in developing microwave-based simulations. By assuming a medium with no free charges, the electric and magnetic fields can be given below by those four equations given in Table 2.1.

Gauss's Law equation in Table 2.1 describes the relationship between electric field and the distribution of electric charges in the space: the static electric field lines begin at positive charges and end at negative charges, and the total charge enclosed by a closed surface can be calculated by the electric flux passing through that closed surface. Also, the Gauss's Law for Magnetism given in Table 2.1 indicates that the magnetic monopoles are not exist. Therefore, the magnetic field lines do not have a start nor an end point. Instead, the magnetic field lines form a loop and the net magnetic flux through any Gaussian surface equals zero. The third equation is Maxwell–Faraday's Equation which states that a changing magnetic field induces an electric field. Finally, the fourth equation which is Maxwell–Ampere's Law states that the magnetic fields can be generated from changing electric field (first term on the right-hand side of the equation, Maxwell's addition) and electric current (second term on the right-hand side of the equation, the original Ampere's Law).

Based on the above equations, one can derive two explicit wave equations (one for electric field and the other for magnetic field) to describe electromagnetic waves by combining Maxwell–Faraday's Equation and Maxwell–Ampere's Law, described in Table 2.1. It is reasonable to assume the medium is time independent and isotropic (i.e. $\mu \neq f(t)$, μ and ϵ do not vary in space). By taking the curl ($\nabla \times$) of Maxwell–Faraday's Equation, combining with Maxwell–Ampere's Law and simplifying, the following equation can be obtained:

$$\nabla \times \left(\mu^{-1} \nabla \times \mathbf{E} \right) + \sigma \frac{\partial \mathbf{E}}{\partial t} + \epsilon \frac{\partial^2 \mathbf{E}}{\partial t^2} = 0 \tag{2.1}$$

Similarly, one can also derive an equation to represent a magnetic field:

$$\nabla \times \left(\nabla \times \mathbf{H} \right) + \sigma \mu \frac{\partial \mathbf{H}}{\partial t} + \epsilon \mu \frac{\partial^2 \mathbf{H}}{\partial t^2} = 0 \tag{2.2}$$

To solve these differential equations, we can assume the electroconductivity (σ) is zero, since most of the supports of metal-doped catalysts are insulators (e.g. γ-Al_2O_3, ZSM-5, etc.) and we assume the metal loading is low. Therefore, the wave equation of an electric field becomes:

$$\nabla^2 \mathbf{E} + \mu\epsilon \frac{\partial^2 \mathbf{E}}{\partial t^2} = 0 \qquad (2.3)$$

Assuming the electric field is sinusoidal, and the wave propagates in positive z direction and has a constant value in x and y directions (i.e. E_0), therefore, one of the possible solutions of Equation 2.3 is:

$$E(x,t) = E_0 \exp(i\omega t)\exp(-ikz) \qquad (2.4)$$

and

$$k = \omega(\epsilon\mu)^{\frac{1}{2}} \qquad (2.5)$$

$$\omega = 2\pi f \qquad (2.6)$$

where f is wave frequency.

Similarly, the mathematical form to represent magnetic field can be derived as Equation (2.7). It is noticed that the electric field, magnetic field and wave propagation direction are perpendicular to each other (Figure 2.1).

$$H(y,t) = H_0 \exp(i\omega t)\exp(-ikz) \qquad (2.7)$$

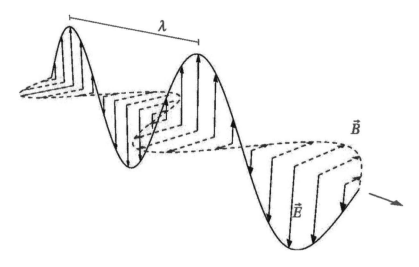

FIGURE 2.1 Electromagnetic wave. (From Kudling and Deknop, 2019.)

2.2.2 INTERACTION BETWEEN ELECTROMAGNETIC FIELDS AND DIELECTRIC MATERIALS

When the material is placed in electromagnetic field, if electric dipoles exist in the material, those dipoles will rotate or oscillate in order to respond to the change in the direction of electric and magnetic fields. It is noticed that the oscillation is inelastic, hence the kinetic energy lost (as the dipoles rotate or oscillate) is transformed to thermal energy, and subsequently the heat is dissipated into bulk material which causes temperature raise of the material (Figura and Teixeira, 2007). Also, for conductive material, alternating microwave electric field causes electrons and ions move through the material (Sun et al., 2016). This movement creates electric current. Meanwhile, changing magnetic field can also induce eddy current within conductive material. Therefore, the material can be heated due to the electric resistant of the material. The abilities of a material to absorb microwave energy and to dissipate some of this energy via inelastic rotation or oscillation are mathematically expressed by its electromagnetic properties, which include complex electric permittivity (ϵ) and complex magnetic permeability (μ). Electromagnetic properties of material are essential of describe how well a material response to microwave irradiation. The definition of permittivity and permeability are given below as:

$$\epsilon = \epsilon' - i\epsilon'' \tag{2.8}$$

$$\mu = \mu' - i\mu'' \tag{2.9}$$

where

$$i = \sqrt{-1}$$

where ϵ' is known as the dielectric constant*, which is a function of temperature and frequency. According to Figura and Teixeira (2007), the real part indicates the elastic component that how much microwave energy can be absorbed by dipoles, and the imaginary part represents the inelastic component that how much energy is loss and transformed into heat. In general, a lossy material is defined when $\epsilon'' \gg \epsilon'$, and the lossy material has a better response to microwave irradiation. In a microwave-catalysis system, it is desired to heat a lossy catalyst material. The "lossiness" of a material can be characterized by dielectric loss tangent:

$$\tan \delta_e = \frac{\epsilon''}{\epsilon'} \tag{2.10}$$

where δ is known as loss angle. The higher value of loss tangent indicates the more "lossy" of the material. Similarly, the magnetic loss tangent is the ratio of imagine part and real part of complex magnetic permeability.

---

* Sometimes dielectric constant is defined as $\epsilon_r = \epsilon'/\epsilon_0$, where ϵ_0 is a constant represents the permittivity of free space (8.854×10^{-12} F/m). Similarly, for real part of magnetic permeability $\mu_r = \mu'/\mu_0$ where μ_0 is a constant represents the permeability of free space ($4\pi \times 10^{-7}$ H/m).

If the material is highly electric conductive, as a result, the fields diminish with greater depths of the material due to skin effect (National Research Council, 1994). Depth of penetration is defined as the depth where the field strength falls to $1/e = 0.368$ of its surface values, given by (von Hippel 1994):

$$d = \frac{1}{\sqrt{\pi f \mu' \sigma}} \qquad (2.11)$$

where f is microwave frequency. As the relative permeability increases, the less microwave will penetrate into the material.

2.2.3 MEASUREMENT OF ELECTROMAGNETIC PROPERTIES

Electromagnetic properties measurement is an important characterization of materials that determine how well the materials interact with microwave irradiation during experiments. In numerical modeling and simulation, these properties could largely affect the accuracy and reliability of calculation results.

von Hippel (1994) provided a detailed introduction and derived complex permittivity and permeability from fundamental electromagnetism. Thanks to the development of the computer, information communication and lab equipment technologies, we can perform electromagnetic characterizations using fully automatic devices that improves data accuracy and work efficiency. In this section, two popular and common methods are further introduced and discussed.

2.2.4 TRANSMISSION-REFLECTION LINE METHOD

Transmission-reflection line method is popular nowadays for complex permittivity and permeability measurement. A vector network analyzer is usually used to determine complex scattering parameters, known as S-parameters. S-parameters indicate the performance of material under different RF/microwave frequencies in terms of incident and reflected waves. It is noticed that the S-parameters changes with frequency. By obtaining S-parameters, computer calculates dielectric constant by using appropriate model (Nicolson–Ross–Weir method, NIST iterative method, etc.). Figure 2.2 shows simplified block diagram of S-parameters determination in a vector network analyzer. Kuek (2012) listed few advantages and disadvantages of transmission/reflection line method:

Advantages:
- Coaxial lines and waveguides are commonly used to measure samples with medium to high loss;
- Can determine both permittivity and permeability.

Disadvantages:
- Measurement accuracy is limited by the air-gap effects;
- It is limited to low accuracy when the sample length is the multiple of one-half wavelength in the material.

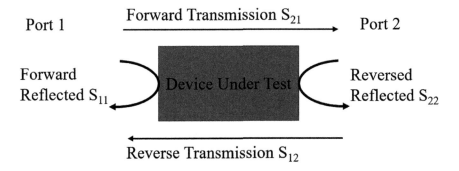

FIGURE 2.2   Schematic diagram of S-parameters determination of a vector network analyzer.

2.2.5   CAVITY PERTURBATION METHOD

Cavity perturbation method is widely applied in dielectric constant measurement at microwave frequencies (Vyas et al., 2008). While a small piece of sample is inserted into the maximum electric or magnetic field of an empty cavity, the resonant frequency of the cavity slightly shifts (perturbed) (Altschuler, 1963; National Research Council, 1994). Perturbation method can be very accurate in determining material with small loss tangents. In this method, the resonant frequency of empty cavity and perturbed cavity (f_c and f_s, respectively); quality factors of empty and perturbed cavity (Q_c and Q_s) are monitored. According to Mathew (2005), if the waveguide cavity is rectangular (TE_{10} mode; refer to Section 2.3.3 for waveguide mode) and the sample is inserted at the maximum electric field, the complex dielectric permittivity can be calculated as:

$$\epsilon'_r - 1 = \frac{f_c - f_s}{2f_s}\left(\frac{V_c}{V_s}\right) \tag{2.12}$$

$$\epsilon''_r = \left(\frac{V_c}{4V_s}\right)\frac{Q_c - Q_s}{Q_c Q_s} \tag{2.13}$$

where V_c and V_s are the volumes of the cavity resonator and the sample, respectively. Similarly, if the sample is inserted at the maximum magnetic field, the complex permeability can be calculated as (Mathew, 2005):

$$\mu'_r - 1 = \frac{\lambda_g^2 + 4a^2}{8a^2}\frac{f_c - f_s}{f_s}\left(\frac{V_c}{V_s}\right) \tag{2.14}$$

$$\mu''_r = \frac{\lambda_g^2 + 4a^2}{16a^2}\left(\frac{V_c}{V_s}\right)\frac{Q_c - Q_s}{Q_c Q_s} \tag{2.15}$$

where a is the broad dimension of the waveguide; $\lambda_g = 2d/n$. λ_g is the guided wavelength; d denotes for cavity length; n = 1, 2, 3, ...

2.2.6   Advantages of Microwave over Conventional Heating

Based on the theory mentioned, microwave heating has its own features that are not available in conventional thermal heating of the materials, including internal heating of the materials, selective heating, rapid heating, and controllable field distributions.

2.2.6.1   Internal Heating, Rapid Heating, and Selective Heating

As shown in Equation (2.12), microwave can penetrate inside the dielectric material. The penetration depth depends on the microwave frequency and magnetic permeability of the materials. Figure 2.3 presents a 2D temperature profile of microwave-heated dielectric material (a potato). At the surface of the dielectric material, the thermal energy obtained from dipoles rotation/oscillation is dissipated into the ambient, while gas cannot be heated by microwave. The center of the material is also heated, and the heat transfer is from internal to external surface. As a result, the materials are internally heated by microwave, which is opposite to conventional thermal heating. Since the microwave directly convey energy to the dipole molecules of the materials, microwave heating of dielectric materials is much faster than conventional thermal heating, and the processing time of a catalytic reaction can be significantly reduced.

Selective heating is another important feature of microwave heating on material. There are two major phenomena due to selective heating: hot region and hot spot*. Hot region is caused by uneven distributed electric field along with catalyst bed. Hot spot is caused by the differences in dielectric properties of metal and catalyst

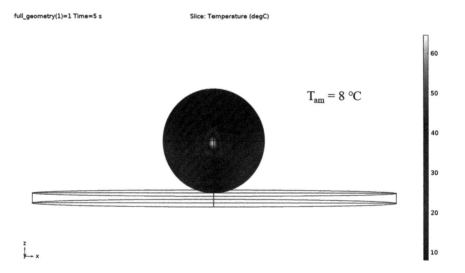

full_geometry(1)=1 Time=5 s   Slice: Temperature (degC)

$T_{am} = 8\ °C$

FIGURE 2.3   Cross-sectional temperature profile of a microwave-heated potato (f = 2.45 GHz, t = 5 s, P = 1 kW). (From COMSOL, 2019).

---

* Hot region and hot spot are the nomenclatures used to distinguish two different selective heating scenarios mentioned. In this chapter, hot region implies macroscopic uneven temperature distribution and hot spot implies microscopic uneven temperature distribution.

support and/or the morphology of the microscopic particles. These two phenomena are very common in microwave-assisted catalysis and it is widely acknowledged that they potentially affect yield and product distribution. Several recent publications mentioned selective heating in catalysis, and this feature will be discussed in Sections 2.4 to 2.7.

2.2.6.2 Controllable Field Distributions

In the single-mode applicators, the electric field can be focused on the tested material by adjusting E-H tuners and sliding short (see Section 2.3.2 and 2.3.4). In this way, it improves the heating of the material with low "lossiness". If the electric field is strong enough, the microwave plasma can also be induced that can greatly affect yield and production distribution in microwave-assisted catalysis.

2.3   MICROWAVE REACTOR DESIGNS

Unlike domestic microwave oven, there are several different laboratory microwave reactors designated for different applications. Based on the applicator cavity, microwave reactors have two major types: single-mode (or monomode) and multimode. The major difference between these two types of microwave reactors is the size of cavity and how the electromagnetic field is distributed in the cavity. In general, single-mode microwave cavity is smaller than the multimode cavity, and the electromagnetic field in multimode cavity is distributed non-uniformly. A flow-type single-mode cavity can form uniform electromagnetic field along the tubular reactor inside the waveguide (Nishioka et al., 2013). Therefore, under the same power, microwave heating is faster in single mode than in multimode cavity reactors. However, multimode microwave reactors can be operated in a much larger scale due to larger cavities. Thus, multimode microwave reactors are widely applied in batch organic compound synthesis, catalyst preparation and sample acid digestion for ICP analysis. In this chapter, we are focusing on the single-mode flow-type microwave reactor, which can be applied in gas-solid heterogenous catalysis science. The microwave reactor design and components are discussed below.

2.3.1   MICROWAVE SOURCE

There are two major types of microwave generator: magnetron generator and solid-state generator. In magnetron microwave generator, shown in Figure 2.4, a cylindrical cathode is surrounded by anode with resonance cavities. While the cathode rotates counterclockwise, the electron (or cyclotron) induces voltage across the resonance cavities that creates electric field, E, and a perpendicular magnetic field, H. The produced electromagnetic wave (microwave radiation) is delivered to a waveguide, which further channels the microwave to the material sample. Figure 2.5 shows the setup of a magnetron microwave reactor. A magnetron generator operating at 2.45 GHz is the most popular microwave source nowadays because of high energy efficiency and relatively low cost.

Figure 2.6 shows a variable-frequency solid-state microwave reactor. Instead of using magnetron, solid-state microwave generator uses semiconductor technology. By the interaction between semiconductor device and electrical resonance circuit, direct current power can be transformed to radio frequency/microwave power. Therefore,

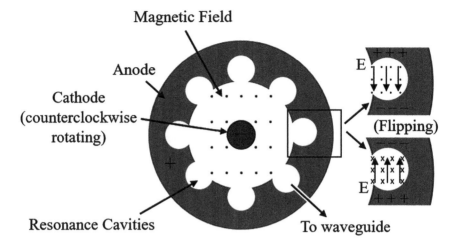

FIGURE 2.4 Scheme of a magnetron in microwave generator.

FIGURE 2.5 Sairem 3 kW magnetron microwave reactor unit (2.45 GHz).

solid-state microwave generator is compact in size, and has a longer life than magnetron devices. In addition, variable frequency can be achieved by using solid-state generator, hence the microwave frequency effect on chemical reaction can be investigated.

2.3.2 E-H Tuners

According to Bogdanovich et al. (2002), an E-H tuner is a Magic Tee with a variable shorting plunger in each of the E- and H-arms that allows arbitrary transformation of load impedance within the Smith-chart area. In laboratory operation, an E-H tuner is used to minimize the reflection factor to 0, keep the Voltage Standing Wave Ratio (VSWR) equal to one and to shift the RF phase over a ±180° range. In this way, the total microwave power required to reach or to maintain at a specific temperature can be minimized and the energy efficiency can be maximized.

2.3.3 WAVEGUIDE CAVITY

A waveguide cavity is used to guide the electromagnetic wave (microwave in this case) propagates at a designated direction. The electromagnetic field pattern of the electromagnetic radiation inside a waveguide is called mode. The modes can be classified as transverse electromagnetic (TEM) modes, transverse electric (TE) modes, transverse magnetic (TM) modes and hybrid modes. TEM modes suggest neither electric nor magnetic field in the direction of wave propagation; TE modes mean only the magnetic field exist in the direction of wave propagation; TM modes mean only the electric field exist in the direction of wave propagation. Rectangular waveguide is one of the most common waveguide shapes for microwave reactor and it only supports TE and TM modes. It is important to identify the mode number (TE_{mn} or TM_{mn}) for finite-element simulation of the microwave system (m, n are the numbers of half-wave patterns across the longer and the shorter transverse of the waveguide, respectively), and the mode number is depended on the size of the waveguide and supported frequency range. Figure 2.7 shows the electric and magnetic field distribution of a TE_{10} waveguide which is the waveguide type of the microwave reactor showing in Figure 2.6. As shown in Figure 2.7, the E-field direction in TE_{10} mode is perpendicular with the direction of wave propagation (−x direction); has only one

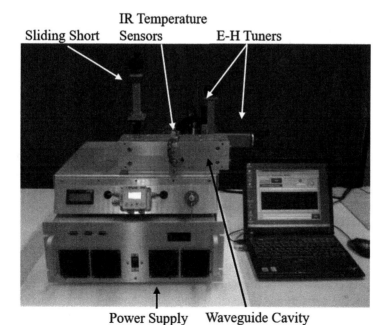

FIGURE 2.6 Lambda MC-1330 180 W variable-frequency solid-state microwave reactor unit (5.85–6.65 GHz).

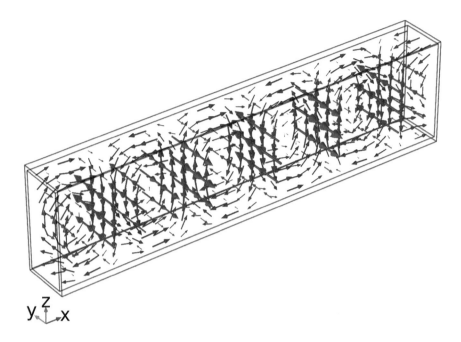

FIGURE 2.7   Field distribution of a TE_{10} waveguide (red arrows: electric field; blue arrows: magnetic field; the direction of wave propagation is $-x$; a larger arrow indicates a stronger field).

lobe in the longer transverse direction (z direction); and has 0 lobes in the shorter transverse direction (y direction).

2.3.4   SLIDING SHORT

Sliding short is used to reflect all electromagnetic energy so that the spatial distribution electromagnetic field can be maximized on the catalyst material (controllable field distribution) (Liao et al., 2016). In most of the cases, the material of sliding short piece is metal with an infinity VSWR. In operation, adjusting sliding short will affect heating pattern under the same microwave frequency, thus, in general, the sliding short should not be adjusted during a non-phase-shifting fixed-frequency experiment.

2.3.5   TEMPERATURE MEASUREMENT

Temperature measurement for catalyst in microwave reactors is difficult but very important to control chemical reactions. Traditionally, thermocouples are used to measure the temperature of the material, and the probe must be in contact with the material to have a more accurate reading. However, according to National Research Council (1994), 'the presence of thermocouple probe in a microwave environment can cause electromagnetic interference problems, causing distortion of the electrical field or affecting the electronics used for temperature measurement, as well as errors due to self-heating; heat conduction; shielding; and excessive localized heating, particularly at the tip of the probe'. Therefore, optical temperature measurement methods (e.g. infrared pyrometer) are commonly applied due to following advantages:

- Non-contact temperature measurement, which will not cause electromagnetic interference problems
- Fast response and compact in design
- Temperature mapping, which is able to visualize the selective heating of microwave

2.4 MICROWAVE-ASSISTED DRY METHANE REFORMING (DMR)

2.4.1 BACKGROUND

Synthetic gas (H_2 and CO) is important feedstock in industry for energy source and higher-valued chemicals synthesis such as methanol. Methane reforming is a key reaction to produce synthetic gas in industry (Ashcroft et al., 1990; Choudhary et al., 1992). Specifically, dry methane reforming uses CO_2 as feedstock that can potentially reduce greenhouse gases in the atmosphere while producing fuels. Many literatures report DMR reactions on noble metals catalysts (Rh, Pt, etc.) (Hou et al., 2006; Nakagawa et al., 1998; Richardson et al., 2003). To lower the cost of the process, several literatures reported that using activated carbon at high temperature (>900°C) was very promising as catalyst (Song et al., 2008). By adding non-noble metals on activated carbon catalyst, the methane and CO_2 conversion can be further enhanced (Zhang et al., 2014; Zhang et al., 2015).

In industrial thermal reactor, operating a chemical reaction that requires a high temperature, a large portion of this heat is lost to the surroundings which lower the total energy efficiency. Therefore, microwave irradiation was recently applied where it has the ability to selectively heat the catalyst to targeted temperature and keeps the reactant gas medium at a lower temperature during the reaction. In this and the following sections, several microwave-assisted chemical reactions are introduced and discussed.

2.4.2 REACTION CHEMISTRY

The overall reaction formula of DMR is shown as Equation (2.16):

$$CO_2 + CH_4 \rightarrow 2CO + 2H_2 \quad \Delta H_0 = 247.3 \text{ kJ/mol} \quad (2.16)$$

which is widely accepted to be the combination of methane decomposition (Equation 2.17) and Boudouard (Equation 2.18) reaction (Arora and Prasad, 2016):

$$CH_4 \rightarrow C + 2H_2 \quad \Delta H_0 = 74.9 \text{ kJ/mol} \quad (2.17)$$

$$CO_2 + C \rightarrow 2CO \quad \Delta H_0 = 171 \text{ kJ/mol} \quad (2.18)$$

While substantial researches focus on catalyst development, very few studies utilize state-of-the-art reactor designs such as microwave reactors. The first study comparing DMR in microwave reactors and in conventional-heated fixed-bed reactors was completed by Fidalgo et al. (2008). In this work, 11 experiments using different space velocity, feed, temperature and heating methods were conducted, and the results were compared. The results showed that by using microwave irradiation as heating method, both CO_2 and methane conversion can be maintained at ~100% at 700°C for at least 3 hours (Fidalgo et al. 2008). By comparing catalyst performance at 800°C, microwave irradiation significantly improved the conversion compared to conventional-heated fixed-bed reactor, showing in Figure 2.8.

For microwave-catalyst interaction in DMR reaction, Domínguez et al. (2007) came up with a concept of "microplasma" within the catalyst bed where "hot spots" are formation. The existence of "microplasma" potentially favors the methane conversion. Based on the discovery of Domínguez et al. (2007), Fidalgo et al. (2008)

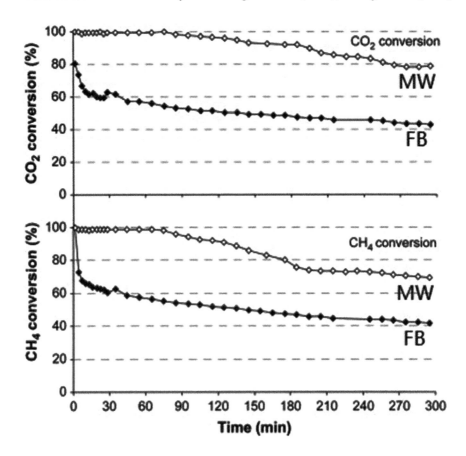

FIGURE 2.8 Reaction performance of DMR comparison between microwave-heated (MW) and thermally heated fixed-bed (FB) reactor, evaluated by CH_4 and CO_2 conversion. The reaction conditions are: T = 800°C, 8 g catalyst, 50 vol% CO_2 with balanced CH_4; GHSV = 0.32 L•h^{-1}•g$_{cat}$$^{-1}$. (Reproduced from Fidalgo et al. 2008, with permission from Elsevier.)

stated that the DMR reaction was catalyzed by char that contains several metal-oxide "impurities"; and the dielectric properties differences of the impurities and char create "hot spots". However, the microplasma formation could be due to the electron distribution of the material, which may be related to the material morphology, polarizability and electric field distribution. Musho et al. (2018) shows an example of how morphology affects electric field distribution near the particle under microwave irradiation. One of the possible ways to solve this type of mechanistic problem is numerical modeling, that theoretically predict electromagnetic field distribution (see Section 2.7 in this chapter). With this information, more accurate mechanistic postulation can be inspired. In summary, microwave irradiation enables higher methane and CO_2 conversion at lower temperatures in DMR reaction by using inexpensive carbon catalysts by enhancing methane decomposition and Boudouard reactions.

2.5 MICROWAVE-ASSISTED DEHYDROAROMATIZATION OF NATURAL GAS

Aromatic products are important ingredients for synthetic material production and polymers. Recently, research on direct conversion of natural gas into higher-value products has gained considerable because indirect natural gas conversion pathway nowadays requires higher capital investment. Natural gas dehydroaromatization is a process that directly convert natural gas (mainly composed of methane and ethane) into aromatics (mainly benzene and toluene) by using catalysts. Many literatures show good results on this reaction using metal-loaded zeolites as catalyst (Chetina et al., 1995; Guisnet et al., 1986; Krogh et al., 2003; Samanta et al., 2017; Skutil and Taniewski, 2006). The challenge of this process is the requirement of high temperature to activate stable C–H bonds in methane and ethane molecules to achieve a reasonable single-pass conversion toward higher-valued chemicals. Figure 2.9 shows the reaction pathway of methane and ethane dehydroaromatization that widely accepted among research community. Both methane dimerization and ethane dehydrogenation reactions are endothermic, which requires high temperature. Therefore, 600–650°C is generally required for ethane dehydroaromatization; while methane dehydroaromatization requires at least 700°C. However, the oligomerization of ethylene to benzene is an exothermic reaction, so lowering the temperature will benefit the benzene yield. It is noticed that pathway 1 and 2 showing in the Figure 2.9 are taking place on metal sites (Mo (Mo_2C), Ga, etc.), and reaction 3 generally takes place on Bronsted acid sites of the zeolite (Ismagilov et al., 2008). Thus, the selective heating feature of microwave heating can play an important role in dehydroaromatization of natural gas.

The pioneering work by Gui et al. (2008) first reported the synergistic effect of microwave irradiation on n-hexane (n-C_6H_{14}) aromatization reaction. Specifically, as shown in Table 2.2, by using Zn-Ni/HZSM-5 catalyst and microwave irradiation, the single-ring aromatics production was significantly higher at lower temperature under microwave heating. They concluded that microwave selective heating of catalyst's acid site is the key reason for better single-ring aromatics yield.

To further investigate the microwave effect on dehydroaromatization, Bai et al. (2019) conducted a series of ethane dehydroaromatization experiments in microwave reactor and fixed-bed reactor under low temperature (400°C). By applying microwave irradiation,

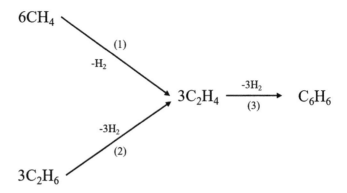

$\Delta H_{(1)} = 605.94$ kJ/mol; $\Delta H_{(2)} = 136.96$ kJ/mol; $\Delta H_{(3)} = -73.91$ kJ/mol.

FIGURE 2.9 Simplified reaction pathway of methane and ethane dehydroaromatization.

TABLE 2.2

Effect of Temperature on the Yields of Single-Ring Aromatics by Feeding n-Hexane (Reaction Conditions: 10 mL 3wt%Zn-1wt%Ni/HZSM-5, GHSV = 1.1 hr⁻¹, 8-hour Reaction (Gui et al., 2008))

Temperature (°C)	Single-ring aromatics yield (wt%), microwave heating	Single-ring aromatics yield (wt%), conventional heating
360	22.7	19.0
400	28.1	18.5
440	27.3	23.0
480	24.3	25.5
520	27.6	26.2

---

a maximum of 80% ethane conversion can be achieved, while only 3% of ethane can be converted in a conventional-heated fixed-bed under same temperature (Figure 2.10). Selective heating effect was also noticed by Bai et al. (2019) by observing metal particle agglomeration in spent catalysts from microwave reactor. Particle agglomeration usually takes place at much higher temperature (> 600°C) and it is one of the key reasons for catalyst deactivation. The reaction pathways for major products observed were presented in the literature showing as Equations (2.19)–(2.22). In this specific case, the metal site temperature was claimed to be much higher than zeolite since microwave-transparent SiO₂ is the main component of ZSM-5 (Si/Al = 32). Therefore, reaction shown as Equation (2.19) was accelerated and lower zeolite temperature favors Equations (2.20) and (2.21). In addition, compared to conventional-heated fixed-bed ethane dehydroaromatization, higher methane production was observed at the early stage of the reaction in microwave reactor. As indicated by Bai et al. (2019), ethylene hydrogenolysis (Equation 2.22) was also accelerated which takes place on hot metal sites. In summary, microwave irradiation showed

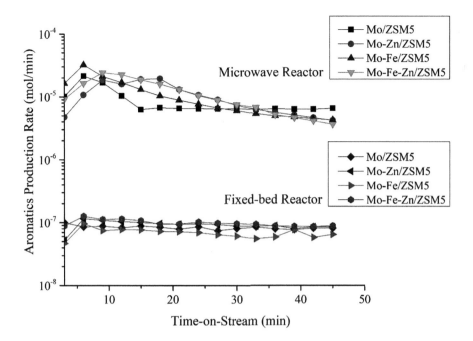

FIGURE 2.10 Catalytic performance comparison between microwave-heated (6.65 GHz) and thermally heated fixed-bed reactor, evaluated for production rate of aromatics (0.8g catalyst; 400°C; 1 atm; feedstock flowrate = 50 mL/min with 36 vol% of ethane in nitrogen). (Reproduced from Bai et al. 2019 with permission from Elsevier.)

promising synergistic effect on accelerating reaction on metal sites that including ethane activation, dehydrogenation (Equation 2.19) and ethylene hydrogenolysis (Equation 2.22) due to the selective heating feature. Microwave irradiation does not alter the catalyst and promoter functions; thus, microwave irradiation is capable for other catalytic reactions involving C–H bond activation:

$$C_2H_6 \rightarrow C_2H_4 + H_2 \tag{2.19}$$

$$3C_2H_4 \rightarrow C_6H_6 + 3H_2 \tag{2.20}$$

$$7C_2H_4 \rightarrow 2C_7H_8 + 6H_2 \tag{2.21}$$

$$C_2H_4 + 2H_2 \rightarrow 2CH_4 \tag{2.22}$$

A recent publication by Julian et al. designed a novel catalyst-monolith system that has excellent performance in microwave reactor. In the literature, Mo/HZSM-5 catalyst was used to study direct methane nonoxidative conversion, and the microwave was utilized to rapidly heat catalyst, to selectively heat active phase that enhance the molybdenum carbide formation, and to create gas-solid temperature gradient that inhibits coke formation (Julian et al., 2019). Instead of directly heated HZSM-5 catalyst, Julian et al. (2019) impregnate Mo/HZSM-5 onto SiC-β monolith, which is an excellent microwave absorber (high loss tangent). The results showed that by coating zeolite catalyst on SiC monolith, the "hot region" effect of the catalyst, due to field distribution, was greatly attenuated (Figure 2.11). Also using SiC monolith can also facilitate temperature control during the reaction, while coking changes the dielectric properties of the catalyst. With large gas–solid temperature gradient and high space velocity (6 $L \cdot g^{-1} \cdot h^{-1}$), polyaromatic coke formation was greatly depressed in microwave reactor, and the benzene selectivity was also decreased. Meanwhile, at the same temperature, C2 production in microwave reactor, especially acetylene, was significantly higher than that in conventional-heated reactor. It was hypothesized that the microplasma hot spots affect product distribution (Julian et al., 2019). As mentioned earlier in this chapter, highly endothermic methane dimerization is the key step in methane hydroaromatization, and it is widely accepted that methane dimerization takes place on the molybdenum carbide sites. Therefore, hot Mo_2C spots can be the reason for high C2 selectivity in microwave heating. Low aromatics selectivity can be also due to the gas–solid temperature gradient, which affects the pore diffusion process of intermediate C2. In summary, Julian et al. (2019) provided a novel approach of microwave-designated catalyst for better temperature control;

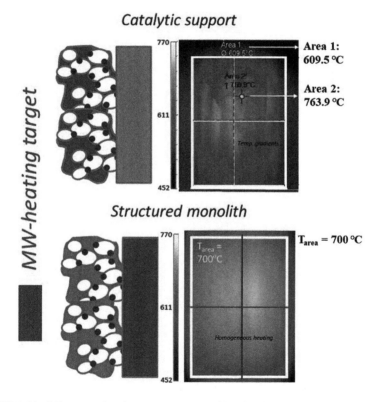

FIGURE 2.11 Microwave heating temperature profiles of (1) Catalyst support: Mo/HZSM-5 at microwave-transparent monolith; (2) Structure monolith: Mo/HZSM-5 at lossy SiC-β monolith. (From Julian et al., 2019.)

and their discoveries of microwave heating effect provided a new solution for low-coking methane conversion to higher hydrocarbons.

2.6 MICROWAVE-ASSISTED AMMONIA SYNTHESIS FROM METHANE AND NITROGEN

2.6.1 BACKGROUND

Developed by Fritz Haber and Carl Bosch, Haber–Bosch ammonia synthesis has been commercialized for more than 100 years and this process has achieved a single-pass nitrogen conversion of 15% (Renner et al., 2015). Ammonia not only is an important feedstock for fertilizer but also has been regarded as an excellent hydrogen storage for further transportation and transformation. For this traditional process, high temperature (~500°C) and pressure (> 100 atm) are required for such a high conversion, which dramatically increase capital and operating costs. It is noticed that in industry nowadays, natural gas reforming (or coal gasification) is the main source of hydrogen. For instance, about 50% cost in ammonia plant is on the hydrogen production from methane steam reforming. Therefore, it is desirable and attractive to synthesis ammonia directly from methane and nitrogen under atmospheric pressure. However, this process is not thermodynamically favorable due to high chemical stability of N_2 and CH_4 molecules. Therefore, microwave irradiation can be considered to facilitate this reaction.

2.6.2 REACTION CHEMISTRY OF AMMONIA SYNTHESIS FROM METHANE AND NITROGEN

2.6.2.1 Reaction in Thermally Heated Fixed-Bed Reactor

The first and the only systematic study of methane–nitrogen ammonia synthesis in thermally heated fixed-bed reactor was finished by Xu (Xu, 2012). In this study, CH_4 and N_2 react directly with the presence of cobalt or nickel catalyst, under atmospheric pressure. High temperature is required to activate methane and nitrogen molecules. Theoretically, the overall reaction formula is shown as Equation (2.23):

$$2CH_4 + N_2 \leftrightarrow 2NH_3 + H_2 + 2C \qquad (2.23)$$

In the literature, it was emphasized that the presence of cobalt is crucial to activate the triple bond in nitrogen molecule. In the presence of cobalt–carbon catalyst, the ammonia production rate can reach 0.0104 $\mu mol \cdot g_{cat}^{-1} \cdot s^{-1}$ with a methane conversion of 10% (600°C, 1 atm). It is also noticed that by impregnating nickel on the cobalt catalyst, while ammonia production rate held almost the same, ethane and ethylene were also observed in the exiting gas.

2.6.2.2 Reaction under Microwave Plasma without Catalyst

The production of ammonia under methane–nitrogen microwave plasma was discovered by Oumghar et al. (1994). By investigating methane conversion under microwave plasma with nitrogen inert, trace amount of ammonia was found in silica-gel trap.

In the literature, the overall formula for ammonia production was shown as (2.45 GHz microwave plasma, < 40 mbar):

$$CH_4 + N_2 \leftrightarrow NH_3 + HCN \qquad (2.24)$$

In the presence of microwave discharge, methane is dissociated into CH_x species and activated hydrogen species. In the same time, dinitrogen is excited, followed by the formation of activated nitrogen and hydrogen species. The simplified ammonia formation steps were given as (* stands for wall, the hydrogen and dihydrogen are obtained from the dissociation of electronically excited methane):

$$N_2 + e \rightarrow 2N + e \qquad (2.25)$$

$$N + * \rightarrow N\left(ads\right) \qquad (2.26)$$

$$N\left(ads\right) + H \rightarrow NH\left(ads\right) \qquad (2.27)$$

$$NH\left(ads\right) + H_2 \rightarrow NH_3\left(ads\right) \qquad (2.28)$$

$$NH\left(ads\right) + 2H \rightarrow NH_3\left(ads\right) \qquad (2.29)$$

$$NH_3\left(ads\right) \rightarrow NH_3 + * \qquad (2.30)$$

2.6.2.3 Reaction in Microwave-Heated Fixed-Bed Reactor

With substantial microwave reactor technologies have been developed, researchers are able to integrate microwave irradiation with catalytic reaction systems. Bai et al. (2018) combined microwave irradiation with cobalt-based catalyst to synthesize ammonia from methane and nitrogen directly. The overall reaction follows Equation (2.23). It is noticed that the experiments were performed with a frequency of 6.65 GHz and this is the first study to show this reaction at higher frequency. The result in the literature shows that the ammonia production was reported as 0.40 µmol·gcat^{-1}·s^{-1} (600°C, 1 atm, Co/γAl$_2$O$_3$), which is 50 times as much as the ammonia production in thermally heated fixed-bed reactor under same temperature and pressure. With the addition of iron promoter, the ammonia yield was further boosted to 0.54 µmol·gcat^{-1}·s^{-1} (600°C, 1 atm, Co-Fe/γAl$_2$O$_3$). Table 2.3 summarizes the ammonia production and methane conversion reported by Bai et al. (2018).

High hydrogen production was reported in this study. It has been noticed that the deactivation of the catalyst was severe due to high coking rate and metal particle agglomeration. By adding iron promoter, part of the amorphous carbon deposit was replaced by higher-value carbon nanotubes (CNTs, Figure 2.12). Extensive literature proved that iron promoter facilitates CNT growing, hence microwave irradiation does not alter the functions of catalysts. The reaction pathway was postulated as (s indicates site):

$$CH_4 + s \rightarrow CH_4\left(ads\right) \qquad (2.31)$$

TABLE 2.3
Catalyst Performance Summary (Co/γ-Al$_2$O$_3$ Has 5 wt% Co Loading, and Co–Fe/γ-Al$_2$O$_3$ Has 5 wt% Co Loading and 0.5 wt% Fe Loading; 600°C; 1 atm; WHSV = 120 mL/g cat.) (Bai et al., 2018)

Catalyst	Reaction condition	Maximum NH$_3$ production rate (10^{-8} mol/g/s)	Total NH$_3$ production for 30 minutes (10^{-8} mol)	Maximum methane conversion
Co/γ-Al$_2$O$_3$	Traditional Heating	1.7	3.0	19.7%
Co/γ-Al$_2$O$_3$	Microwave Irradiation Only	40.8	157.3	73.8%
Co/γ-Al$_2$O$_3$	Microwave Irradiation and Microwave Plasma	39.9	383.1	80.7%
Co–Fe/γ-Al$_2$O$_3$	Microwave Irradiation Only	53.9	441.5	80.6%

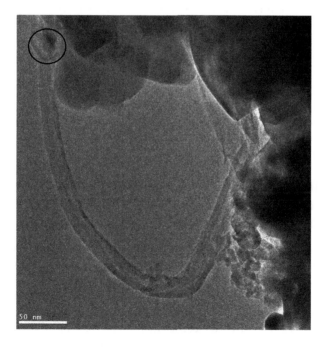

FIGURE 2.12 TEM images of CNTs on spent Co–Fe/γAl$_2$O$_3$ (600°C, 1 atm, microwave irradiation only). The circled particle is iron promoter. (From Bai et al. 2019, author reuse of their own work published by the Royal Society of Chemistry.)

$$CH_4(ads) \rightarrow C(ads) + 4H(ads) \tag{2.32}$$

$$C(ads) \rightarrow C \ (amorphous \ or \ CNT) \tag{2.33}$$

$$2H(ads) \rightarrow H_2 \tag{2.34}$$

$$N_2 + s \rightarrow N_2(ads) \tag{2.35}$$

$$N_2(ads) + 2H(ads) \rightarrow 2NH(ads) \tag{2.36}$$

$$NH(ads) + H_2 \left(or \ 2H(ads)\right) \rightarrow NH_3(ads) \tag{2.37}$$

$$NH_3(ads) \rightarrow NH_3 \tag{2.38}$$

In the mechanism shown above, adsorbed methane decomposition (Equation (2.32)) is one of the key steps that provides hydrogen source of ammonia formation. Nitrogen activation and adsorption (Equation (2.35)) is another key step and it is believed to be the rate determine step. The literature postulated that microwave irradiation facilitates methane decomposition step and nitrogen activation and adsorption (Equations (2.32) and (2.35)) in this reaction by delivering electromagnetic energy onto active sites that can lower the activation energy in these reaction steps when compared to conventional heating. Although the microwave irradiation does not directly activate gas, it changes the electron/dipole distribution on the catalyst surface that enhances the interaction between gas and catalyst surface (Equations (2.32) and (2.35)) (Zhou et al., 2016). As a result, the overall reaction is accelerated. In summary, it is proved that microwave can accelerate methane activation by providing electromagnetic energy to the active sites. The future study can be focused on attenuating the catalyst deactivation and increasing the selectivity of higher-value carbon allotrope or C2 olefins to increase the economic value of this reaction.

2.7 NUMERICAL MODELING OF MICROWAVE–MATERIAL INTERACTION: MICROWAVE-ASSISTED METHANE DECOMPOSITION EXAMPLE

Thermocatalytic decomposition (TCD) of methane in oxygen-free reactor can produce high purity hydrogen without producing carbon oxides. In many literatures, methane TCD requires at least 700°C to carry out in a conventional fixed-bed reactor because it is an endothermic reaction (Asai et al., 2008; Ayillath Kutteri et al., 2018). Carbon-based catalysts are commonly applied in methane TCD and generally, carbon materials are lossy dielectrics that can be used under microwave conditions. Domínguez et al. (2007) observed that microwave irradiation can significantly enhance methane conversion at lower temperatures. However, the mechanism and kinetics of this simple microwave-assisted reaction are unclear, and the microwave reactor becomes a "black box". As mentioned in previous sections, infrared (IR) pyrometers are commonly used to monitor the temperature change in catalyst bed.

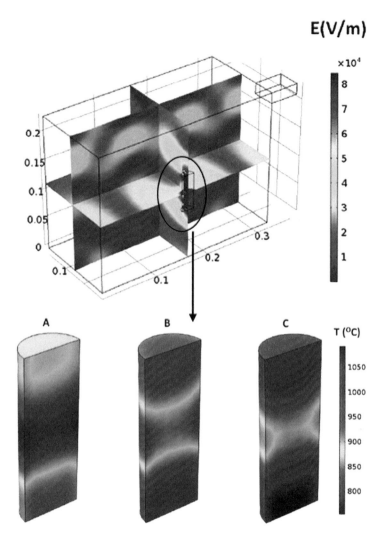

FIGURE 2.13 Electric field distribution of the microwave reactor and temperature profile of cylindrical catalyst bed (A: 800 W; B: 1000 W; C: 1200 W). (Gadkari et al., 2017.)

However, the IR sensors only have a band width in micrometer, and sometimes catalyst bed is not heated evenly (hot regions). As a result, the explanation of experiment data, especially kinetic models, may not be accurate. Under this setting, numerical finite-difference modeling provides a viable approach to combine multiple physical fields and theoretically calculate some important physical properties during a microwave-assisted chemical reaction, such as temperature profile through material and electric/magnetic field distribution.

Gadkari et al. (2017) provides a comprehensive numerical simulation study on a continuous-flow microwave catalytic reaction which includes electromagnetic waves, heat transfer, mass transfer (species transport and reaction) and momentum

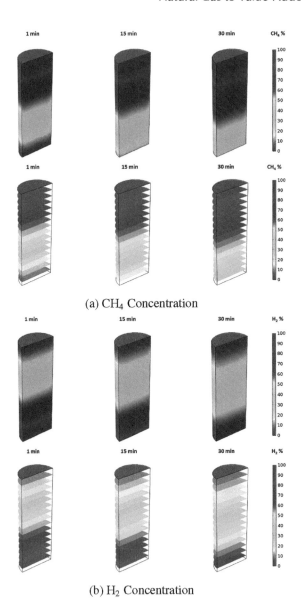

(a) CH$_4$ Concentration

(b) H$_2$ Concentration

FIGURE 2.14   (a) CH$_4$ and (b) H$_2$ concentration profiles in terms of volume and slice plots in the reactor tube for average (bulk) temperature of 800°C at different times, t = 1 min, t = 15 min and t = 30 min. (From Gadkari et al., 2017.)

transfer (fluid flow). For each of the transport phenomenon, one corresponding partial differential equation (governing equation) can be derived. However, it is very complicated and nearly impossible to solve those partial differential equations. Modern finite-element modeling software allow us to couple those governing equations to find a converged solution by meshing the complex 3D or 2D geometries into small elements and doing iterative calculations. The detail of establishing governing equations, boundary conditions, initial values and rate-law setup can be found in literatures (Chen et al., 2013; Gadkari et al., 2017). Through numerical modeling, lots of interesting phenomena can be observed. According to Gadkari et al. (2017), the catalyst bed was not heated evenly due to uneven distributed electric field along catalyst bed (Figure 2.13). The maximum E-field was more likely to be on top and bottom parts of the bed, and this is due to the periodic feature of E-field distribution inside the cavity, as shown in Equation (2.4) earlier in this chapter.

According to Gadkari et al. (2017), Reaction rate profile through catalyst bed is another interesting feature of numerical modeling that provides a straightforward view of catalyst bed usage and catalyst activation/deactivation (Figure 2.14). In methane TCD, the simulation result can be verified by examining the coking extent of catalyst bed. Reported by Gadkari et al. (2017), there is a good agreement between simulation result and experimental data at lower space velocities. In summary, numeric modeling of microwave reactor enables researchers to investigate how microwave assists a chemical reaction, and physical phenomenon of the material under microwave irradiation. With the development of computer technologies, future research of numerical modeling can focus on more complicated reaction system of natural gas conversion such as dehydroaromatization reaction mentioned in Section 2.5 of this chapter. Also, microscopic numerical modeling is another possible direction to simulate metal-support interaction under microwave irradiation.

2.8 CONCLUSIONS

This chapter has provided a brief introduction of the interaction between electromagnetic field and catalyst material. Additionally, several applications of microwave in natural gas conversion have also been discussed, and a possible approach to investigate electromagnetism in catalysis, i.e. numerical modeling, has been elaborated. According to the studies reviewed in this chapter, microwave has been proven to have synergistic effect that assisting several natural gas catalytic conversion reactions due to its thermal and non-thermal effects. Other than assisting heterogeneous catalysis directly, microwave is also able to enhance the performance of catalysts using in natural gas conversion during catalyst synthesis process. Therefore, there are many research topics regarding microwave-assisted natural gas conversion that are worthy to focus on. As illustrated in this chapter, microwave catalysis can activate two stable molecules methane and nitrogen to synthesis ammonia under ambient pressure. In the presence microwave, catalytic ethane dehydroaromatization results in benzene formation rate two order of magnitude higher than conventional thermally heated fixed-bed reactor. All of these applications indicate the unique feature of microwave catalysis in selective bond activation.

REFERENCES

Altschuler, H. 1963. Dielectric constant. In *Handbook of Microwave Measurements* (vol. 2), eds. Sucher, M., Fox, J., and M. Wind, 518–548. New York, NY: Polytechnic Press.

Arora, S. and R. Prasad. 2016. An overview on dry reforming of methane: Strategies to reduce carbonaceous deactivation of catalysts. *RSC Adv.* 6:108668–108688.

Asai, K., Nagayasu, Y., Takane, K., et al. 2008. Mechanisms of methane decomposition over Ni catalysts at high temperatures. *J. Jpn. Pet. Inst.* 51:42–49.

Ashcroft, A. T., Cheetham, A. L., Foord, J. S., et al. 1990. Selective oxidation of methane to synthesis gas using transition metal catalysts. *Nature* 344:319–321.

Ayillath Kutteri, D., Wang, I. W., Samanta, A., Li, L., and J. Hu. 2018. Methane decomposition to tip and base grown carbon nanotubes and COx-free H_2 over mono- and bimetallic 3d transition metal catalysts. *Catal. Sci. Technol.* 8:858–869.

Bai, X., Robinson, B., Killmer, C., Wang, Y., Li, L., and J. Hu. 2019. Microwave catalytic reactor for upgrading stranded shale gas to aromatics. *Fuel* 243:485–492.

Bai, X., Tiwari, S., Robinson, B., Killmer, C., Li, L., and J. Hu. 2018. Microwave catalytic synthesis of ammonia from methane and nitrogen. *Catal. Sci. Technol.* 8:6302–6305.

Bogdanovich, B., Ebert, M., Egorov, M., et al. 2002. Design of an E-H tuner and an adjustable directional coupler for high-power waveguide systems. *Proceed. EPAC* 2002, *Paris, France*, 506–508.

Chen, W. H., Liou, H. J., and C. I. Hung. 2013. A numerical approach of interaction of methane thermocatalytic decomposition and microwave irradiation. *Int. J. Hydrogen Energy* 38:13260–13271.

Chetina, O. V., Vasina, T. V., and V. V. Lunin. 1995. Aromatization of ethane over Pt,Ga/HZSM-5 catalyst and the effect of intermetallic hydrogen acceptor on the reaction. *Appl. Catal. A* 131:7–14.

Choudhary, V. R., Mammon, A. S., and S. D. Sansare. 1992. Selective oxidation of methane to CO and H_2 over Ni/MgO at low temperatures. *Angew. Chem. Int. Ed.* 31:1189–1190.

COMSOL Multiphysics. 2018. *RF Module User's Guide.* https://doc.comsol.com/5.4/doc/com.comsol.help.rf/RFModuleUsersGuide.pdf (accessed Dec 31, 2019).

COMSOL Multiphysics. 2019. *Microwave Oven (Application).*

Dąbrowska, S., Chudoba, T., Wojnarowicz, J., and W. Łojkowski. 2018. Current trends in the development of microwave reactors for the synthesis of nanomaterials in laboratories and industries: a review. *Crystals* 8:379–404.

Domínguez, A., Fidalgo, B., Fernández, Y., Pis, J. J., and J. A. Menéndez. 2007. Microwave-assisted catalytic decomposition of methane over activated carbon for CO_2-free hydrogen production. *Int. J. Hydrogen Energy* 32:4792–4799.

Fidalgo, B., Domínguez, A., Pis, J. J., and J. A. Menéndez. 2008. Microwave-assisted dry reforming of methane. *Int. J. Hydrogen Energy* 33:4337–4344.

Figura, L. O. and A. A. Teixeira. 2007. *Food Physics: Physical Properties – Measurement and Applications.* Leipzig, Germany: Springer.

Gadkari, S., Fidalgo, B., and S. Gu. 2017. Numerical analysis of microwave assisted thermocatalytic decomposition of methane. *Int. J. Hydrogen Energy* 42:4061–4068.

Gedye, R., Smith, F., Westaway, K., et al. 1986. The use of microwave ovens for rapid organic synthesis. *Tetrahedron Lett.* 27:279–282.

Gui, J., Liu, D., Zhang, X., Song, L., and Z. Sun. 2008. Aromatization of n-hexane under microwave irradiation. *Petrol. Sci. Technol.* 26:506–513.

Guisnet, M., Magnoux, P., and C. Canaff. 1986. Formation and nature of coke deposits on zeolites HY and HZSM-5. *Stud. Surf. Sci. Catal.* 28:701–707.

Hamzehlouia, S., Jaffer, S. A., and J. Chaouki. 2018. Microwave heating-assisted catalytic dry reforming of methane to syngas. *Sci. Rep.* 8:Article number 8940.

Hou, Z., Chen, P., Fang, H., Zheng, X., and T. Yashima. 2006. Production of synthesis gas via methane reforming with CO_2 on noble metals and small amount of noble-(Rh-) promoted Ni catalysts. *Int. J. Hydrogen Energy* 31:555–561.

Ismagilov, Z. R., Matus, E. V., and L. T. Tsikoza. 2008. Direct conversion of methane on Mo/ZSM-5 catalysts to produce benzene and hydrogen: achievements and perspectives. *Energy Environ. Sci.* 1:526–541.

Julian, I., Ramirez, H., Hueso, J. L., Reyes, M., and J. Santamaria. 2019. Non-oxidative methane conversion in microwave-assisted structured reactors. *Chem. Eng. J.* 377:119764.

Krogh, A., Hagen, A., Hansen, T. W., Christensen, C. H., and I. Schmidt. 2003. Re/HZSM-5: a new catalyst for ethane aromatization with improved stability. *Catal. Commun.* 4(12):627–630.

Kudling, L. and R. Deknop. 2019. *Electromagnetic Wave in Color.pdf.* Source: https://commons.wikimedia.org/wiki/File:Electromagnetic_wave_color.pdf (accessed Dec 31, 2019).

Kuek, C. Y. 2012. *Measurement of Dielectric Material Properties. Application Note, Rohde & Schwarz.* https://cdn.rohde-schwarz.com/pws/dl_downloads/dl_application/00aps_undefined/RAC-0607-0019_1_5E.pdf (accessed Dec 31, 2019).

Liao, Y., Lan, J., Zhang, C., et al. 2016. A phase-shifting method for improving the heating uniformity of microwave processing materials. *Materials* 9(5):309.

Mathew, K. T. 2005. Perturbation theory. In *Encyclopedia of RF and Microwave Engineering*, ed. Chang, K., 3725–3735. Hoboken, NJ: Wiley-Interscience.

Musho, T. D., Wildfire, C., Houlihan, N. M., Sabolsky, E. M., and D. Shekhawat. 2018. Study of Cu_2O particle morphology on microwave field enhancement. *Mater. Chem. Phys.* 216:278–284.

Nakagawa, K., Anzai, K., Matsui, N., et al. 1998. Effect of support on the conversion of methane to synthesis gas over supported iridium catalysts. *Catal. Lett.* 51:163–167.

National Research Council. 1994. *Microwave Processing of Materials.* National Academies Press. https://www.nap.edu/catalog/2266/microwave-processing-of-materials.

Nguyen, H. M., Pham, G. H., Ran, R., Vagnoni, R., Pareek, V., and S. Liu. 2018. Dry reforming of methane over $Co-Mo/Al_2O_3$ catalyst under low microwave power irradiation. *Catal. Sci. Technol.* 8:5315–5324.

Nishioka, M., Miyakawa, M., Daino, Y., et al. 2013. Single-mode microwave reactor used for continuous flow reactions under elevated pressure. *Ind. Eng. Chem. Res.* 52:4683–4687.

Oumghar, A., Legrand, J. C., Diamy, A. M., Turillon, N., and R. I. Ben-Aïm. 1994. A kinetic study of methane conversion by a dinitrogen microwave plasma. *Plasma Chem. Plasma Process.* 14:229–249.

Renner, J. N., Greenlee, L. F., Herring, A. M., Ayers, K. E., and A. M. Herring. 2015. Electrochemical synthesis of ammonia: a low pressure, low temperature approach. *Electrochem. Soc. Interface.* 24:51–57.

Richardson, J. T., Garrait, M., and J. K. Hung. 2003. Carbon dioxide reforming with Rh and Pt-Re catalysts dispersed on ceramic foam supports. *Appl. Catal. A Gen.* 255:69–82.

Samanta, A., Bai, X., Robinson, B., Chen, H., and J. Hu. 2017. Conversion of light alkane to value-added chemicals over ZSM-5/metal promoted catalysts. *Ind. Eng. Chem. Res.* 56:11006–11012.

Skutil, K. and M. Taniewski. 2006. Some technological aspects of methane aromatization (direct and via oxidative coupling). *Fuel Process. Technol.* 87:511–521.

Song, Q., Xiao, R., Li, Y., and L. Shen. 2008. Catalytic carbon dioxide reforming of methane to synthesis gas over activated carbon catalyst. *Ind. Eng. Chem. Res.* 47:4349–4357.

Sun, J., Wang, W., and Q. Yue. 2016. Review on microwave-matter interaction fundamentals and efficient microwave-associated heating strategies. *Materials* 9:231.

von Hippel, A. 1994. *Dielectrics and Waves* (2nd Ed.). Artech House Print.

Vyas, A. D., Rana, V. A., Gadani, D. H., and A. N. Prajapati. 2008. Cavity perturbation technique for complex permittivity measurement of dielectric materials at X-band microwave frequency. *Proceed. Int. Conf. Microwave*, 836–838.

Weinreich, O. A. and A. Ribner. 1968. Optical and electrical properties of sic films prepared in a microwave discharge. *J. Electrochem. Soc.* 115:1090–1092.

Xu, Q. 2012. *Study on the Catalyst of Co and Ni and their Performance in Ammonia Synthesis Directly from Methane and Nitrogen* (Engl. Transl.), M. S. Thesis, Xinjiang University, China.

Zhang, G., Hao, L., Jia, Y., Du, Y., and Y. Zhang. 2015. CO_2 reforming of CH_4 over efficient bimetallic Co-Zr/AC catalyst for H_2 production. *Int. J. Hydrogen Energy* 40:12868–12879.

Zhang, G., Su, A., Du, Y., Qu, J., and Y. Xu. 2014. Catalytic performance of activated carbon supported cobalt catalyst for CO_2 reforming of CH_4. *J. Colloid Interface Sci.* 433:149–155.

Zhou, J., Xu, W., You, Z., et al. 2016. A new type of power energy for accelerating chemical reactions: the nature of a microwave-driving force for accelerating chemical reactions. *Sci. Rep.* 6:25149.

3 Nonthermal Plasma Conversion of Natural Gas to Oxygenates

Tomohiro Nozaki

CONTENTS

3.1 INTRODUCTION

Energy and resource conservation, as well as reduction of environmental impacts for systems using fossil fuels, call for urgent attention. Energy and materials conversion based on thermochemical processes play a major role, and it is expected that this will continue to be valued in the future. In the meantime, further improvement of energy and materials conversion efficiency of existing thermochemical processes is demanded toward sustainable use of energy resources and the necessity of innovative technology is being highlighted. Many so-called innovative reaction processes employ biochemical reaction, photocatalysis, microwave chemistry, plasma chemistry, and so on. These emerging technologies are anticipated to engender a breakthrough in the medium and long term, in spite of unresolved issues related to energy efficiency and the productivity of desired products. Given this background, we have been developing a novel reaction process to convert methane as a principal component of natural gas and biogas with high efficiency using nonthermal plasma formed at atmospheric pressure. Methane releases the least amount of CO_2 per energy input when combusted and could also be considered a renewable energy obtainable via fermentation of organic substances.

Atmospheric pressure nonthermal plasma is able to activate methane independently of the reaction temperature, triggering a chemical reaction using an electron impact reaction. The chemical reaction might be controlled at a temperature too low to be obtained using conventional thermochemical methods; thus, an overall improved energy and materials conversion efficiency would be implemented together with a significant simplification of processes as well as the flexibility of reactor design. The high flexibility in reactor design provides a lightweight, compact, and highly responsive reactor. This chapter focuses especially on dielectric barrier discharge (DBD) as a source of nonthermal plasma, which is integrated into a microreactor, providing a highly nonequilibrium reaction condition. Other plasma sources are reviewed briefly in Section 3.3. A unique reaction control technique is applied successfully to the direct methane conversion to methanol as described in this chapter. Other applications, such as methane steam/dry reforming by DBD and heterogeneous catalyst combination, are described elsewhere (Nozaki and Okazaki 2011, 2013).

3.2   METHANE REFORMING

3.2.1   DIRECT SYNTHESIS OF METHANOL FROM METHANE

A reaction to insert an oxygen atom into a methane molecule and synthesize methanol directly is considered to be the most ideal methanol synthesis reaction (Reaction 3.1). This has been studied in relation to noncatalytic high-pressure oxidation, gas phase catalytic oxidation, and liquid phase catalytic oxidation (Rasmussen and Glarborg 2008):

$$CH_4 + 0.5O_2 = CH_3OH \qquad (3.1)$$

A direct methane conversion to methanol eliminates the energy-intensive syngas (H_2 and CO) production step and thus simplifies the methanol synthesis process, making energy saving more easily realized (Kondratenko et al. 2017; Taifan and Baltrusaitis 2016). Moreover, onsite conversion of methane into liquid fuels brings tremendous benefits because this technology diminishes the capital investment for storage and transportation such as pipelines and gas liquefaction facilities. In addition, the development of unused energy resources, such as remote gas and biogas resource, is anticipated. However, suppressing decomposition of highly reactive methanol while activating chemically stable methane is considered difficult to implement within a single reactor. In fact, a one-pass yield of methanol has yet to exceed 5% (Rasmussen and Glarborg 2008).

3.2.2 Conventional Methane Reforming

Figure 3.1 presents equilibrium compositions in partial oxidation of methane (Nozaki et al. 2011a). Chemical species considered in the equilibrium calculation were CH_4, O_2, CO, CO_2, H_2, H_2O, $HCHO$, CH_3OH, and $HCOOH$, based on experimental observation described later in this chapter. The O_2/CH_4 ratio of 0.5 and total pressure of 101 kPa were assumed as computational conditions. Data shown in Figure 3.1 suggest complete consumption of oxygen at room temperature and the generation of CO_2 and H_2O by the complete oxidation of methane (Reaction 3.2):

$$CH_4 + 2O_2 = CO_2 + 2H_2O \quad \Delta H = -802 \text{ kJ} \tag{3.2}$$

It is interesting to mention that a low-temperature condition is not desirable thermodynamically, but instead it is kinetically favored for the efficient methane partial oxidation to methanol. CO and H_2 increase gradually over 300°C as unreacted CH_4 reacts with H_2O and CO_2. Although the oxygen is consumed completely, CO_2 and H_2O react to CH_4 as oxidizers at high temperatures:

$$CH_4 + CO_2 = 2CO + 2H_2 \quad \Delta H = 247 \text{ kJ} \tag{3.3}$$

$$2CH_4 + 2H_2O = 2CO + 6H_2 \quad \Delta H = 412 \text{ kJ} \tag{3.4}$$

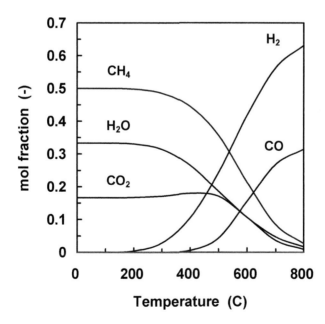

FIGURE 3.1 Equilibrium product compositions for methane reforming at different temperatures at 101 kPa. The initial composition is $CH_4 = 2/3$ mol%, $O_2 = 1/3$ mol% ($O_2/CH_4 = 1/2$), $CO_2 = H_2O = 0$ mol%. (Reproduced with permission from Nozaki et al. 2011a. Copyright 2011 Elsevier.)

Reaction 3.3 plus Reaction 3.4 yields Reaction 3.5:

$$3CH_4 + 2H_2O + CO_2 = 4CO + 8H_2 \quad \Delta H = 659\,kJ \tag{3.5}$$

The H_2/CO ratio approaches 2 when the temperature exceeds 800°C (Reaction 3.5). Reactions 3.2 and 3.5 express a typical methane reforming reaction (syngas production) by a multistage process, indicating that 25% of initial CH_4 must be fully oxidized (Reaction 3.2) to provide heat, which is necessary for the endothermic reforming reaction (Reaction 3.5). Combustion of initial CH_4 leads inevitably to the unavoidable generation of CO_2, NOx, and particulate matters. Moreover, combustion and steam generation processes are known to be the major exergy loss processes. Equilibrium calculations show the selectivity of liquid components as below $10^{-4}\%$ at room temperature and up to 800°C. Liquid oxygenated products such as CH_3OH and HCHO are generated transitionally during full combustion of CH_4 to CO_2 and H_2O. To synthesize liquid components with high selectivity, intermediate products should be quenched and taken out of a reaction field before the complete oxidation of intermediate species occurs.

3.3 PLASMAS FOR FUEL CONVERSION

3.3.1 GENERAL

A type of plasma used for fuel reforming is categorized into three types. The most commonly used plasma is nonthermal plasma where the gas temperature remains at room temperature but electron temperature is elevated up to several tens of thousands of Kelvin (1 eV = 11300 K). Dielectric barrier discharge (DBD), corona discharge, and spark discharge (Kado et al. 2004) are used frequently for this purpose because these plasma sources are readily generated using a simple electrode configuration and an inexpensive power source. Reactive species produced by electron impact can initiate chemical reaction at room temperature without increasing the temperature of the system. The second is the arc plasma where high-temperature heat plays a key role in decomposing carbon containing materials into hydrogen and carbon monoxide. Arc plasma historically has been used for acetylene synthesis (Fincke et al. 2002a) and hydrogen and carbon black production (Fincke et al. 2002b). Recent applications of arc plasma focus on the gasification of biomass, coal, and municipal solid wastes to produce syngas due to the growing concern of energy and environmental issues. The last example of a plasma source, known as "warm plasma" (Gutsol 2010), is relatively new. The precise definition of warm plasma is not clear, but it could be characterized as nonthermal plasma at a relatively high gas temperature (e.g., 1000–2000 K). A peculiarity of warm plasma is that the combination of radical species produced by an electron impact reaction and thermal energy are combined to improve chemical conversion efficiency. Gliding arc plasma (Kalra et al. 2005) and rotating arc plasma (Lee et al. 2007), both characterized as a transient arc plasma, are frequently used for this purpose. This chapter focuses on DBD exclusively as one of the viable plasma sources for fuel reforming due to its flexibility and better controllability of operation parameters over other plasma sources.

3.3.2 DIELECTRIC BARRIER DISCHARGE

Dielectric barrier discharge (DBD) has a long history originating in ozone synthesis. Industrial applications as well as fundamental physics and chemistry are well established (Kogelschatz 2003). A peculiarity of DBD is the presence of dielectric insulator on one or both metallic electrodes. Dielectric material terminates discharge current within 1–10 ns, leadings to formation of a large number of filamentary microdischarges of nanosecond duration. Figure 3.2 shows typical voltage and current waveforms observed in methane-fed DBD (Nozaki et al. 2002). It also shows a snapshot of microdischarge taken with a high-speed intensified CCD camera with 10 ns exposure time (i-Star DH712; Andor Technology). A number of nanosecond current pulses are observed at every half cycle of applied voltage. The immediate termination of developing discharge channels as a result of charge built up on the dielectric barrier creates highly transient nonequilibrium condition at atmospheric pressure. Although DBD in air, nitrogen, and oxygen has been studied extensively (Kogelschatz 2003), limited information is available for methane-fed DBD. Detailed gas phase plasma chemistry of methane-fed DBD is provided in Nozaki et al. (2004a, b). A combination of nonthermal plasma and heterogeneous catalysts is a rapidly growing technology for fuel reforming, including CO_2 hydrogenation to hydrocarbon fuels, CH_4 reforming to hydrogen (syngas), and ammonia synthesis (Tu et al. 2019). The main technological obstacles to be overcome are the low energy efficiency and selectivity of desired products. A key research subject is not limited to the optimization of the plasma process but also to the development of catalysts suitable for the progression of plasma processes.

3.3.3 THERMAL CATALYSIS VERSUS NONTHERMAL PLASMA CATALYSIS OF METHANE

The rate-determining step of methane conversion is the dissociation of a strong C–H bond. Three different approaches are compared for the better understanding of this plasma process: a homogenous gas phase reaction, known as pyrolysis (Reaction 3.6);

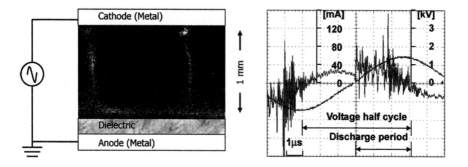

FIGURE 3.2 A snapshot of DBD in pure methane at 20°C and atmospheric pressure (10 ns exposure time), showing filamentary discharge channel between the gas gap (left). Correspondingly, a number of spike-like current pulses are formed (right). (Reproduced with permission from Nozaki et al. 2002. Copyright 2002 IOP Publishing.)

a heterogeneous reaction (Reaction 3.7); and a nonthermal plasma reaction (Reaction 3.8):

$$CH_4(+\text{heat}) = CH_3 + H \quad E_1 = 434 \, \text{kJ/mol} \tag{3.6}$$

$$CH_4\left(+Ni\right) = CH_3{}^* + H^* \quad E_2 = 100 \, \text{kJ/mol} \tag{3.7}$$

$$CH_4 + e = CH_3 + H + e \quad E_3 = 868 \, \text{kJ/mol} \tag{3.8}$$

Here, E_1–E_3 express activation energy and the superscript "*" denotes adsorbed surface species. The activation energy for methane pyrolysis (Reaction 3.6) is 434 kJ/mol (Tsang and Hampson 1986), requiring high-temperature thermal energy (> 1000°C). The activation energy and, therefore, the reaction temperature is greatly lowered by the use of heterogeneous catalysts (Reaction 3.7) (Wei and Iglesia 2004). A peculiarity of nonthermal plasma is that the inert methane molecule is dissociated by electron impact independently of reaction temperature; however, an inelastic electron collision requires a large activation energy (Reaction 3.8) (Davies et al. 1986). Compared to the endothermic enthalpy of methane dissociation, approximately 50% of the electronic energy transferred into methane molecule is wasted via molecular collision, which is eventually dissipated as heat. Consequently, the overall energy efficiency of nonthermal plasma methane conversion is generally low.

In order to increase energy conversion efficiency of nonthermal plasma CH_4 reforming, there are two approaches: the first example is the low-temperature (300–500°C) methane reforming using a DBD-catalyst hybrid reaction. Reactive species produced by DBD enhances the interaction with heterogeneous catalyst, resulting in an increased amount of CH_4 conversion efficiency. In the case of CH_4 steam reforming (an endothermic reaction), radical production by DBD is not sufficient to promote reforming reaction: thermal energy must be added to the reaction system simultaneously in order to satisfy the conservation of energy, which may arise another technological issue: overall heat transfer performance of plasma reactor would become the rate-limiting step. The second example is direct methane conversion to oxygenates via an exothermic reaction system. Plasma-generated radical species enable to trigger the CH_4 partial reaction at room temperature that suppresses the excess CH_4 oxidation to CO_2 and H_2O and a highly nonequilibrium product distribution is obtainable. In this chapter, noncatalytic homogeneous CH_4 partial oxidation to methanol is presented in detail. For the plasma-catalyst hybrid reaction systems, please refer to the published literature (Nozaki and Okazaki 2013; Sheng et al. 2018).

3.4   METHANE CONVERSION USING MICROPLASMA

As presented in Section 3.3, nonthermal plasma activation of methane include dielectric barrier discharge, corona discharge as well as a hybrid reaction of nonthermal plasma and heterogeneous catalysts. Although nonthermal plasma activates methane at room temperature, the synthesized methanol is also decomposed indiscriminately, so that the one-pass yield of methanol has not exceeded 5%. We have succeeded in implementing "separation of methanol," sustaining a highly nonthermal condition by integrating a microreactor and DBD, and have attained improved yield of methanol (Nozaki et al. 2011a, 2004c, d, Agiral et al. 2011).

Figure 3.3 depicts a schematic diagram of the experimental system. The reactor consists of a quartz glass tube with inner diameter of 1.5 mm, outer diameter of 2.7 mm, and length of 50 mm. Sinusoidal high voltage was applied to the stainless wire electrode of 0.2 mm diameter (10 kHz or 80 kHz). Electrolyte (270 mS/cm) at 5°C was circulated around the reactor and was used for a ground electrode. Microplasma reactor was cooled from the outside by the circulating electrolyte. DBD was formed inside a thin glass tube. DBD consists of a large number of transient filamentary discharges (designated as a streamer) with nanosecond duration. Because a lifetime of microdischarge is 1–10 ns, nonthermal plasma can be formed at atmospheric pressure where electron temperature is elevated while gas temperature remains near room temperature.

Distilled water was supplied to the reactor along with a methane and oxygen mixture. Distilled water supplied to the top of the reactor (0.1 cm³/min), forming a liquid film on the reactor inner wall. Gas and liquid components were analyzed quantitatively using a gas chromatograph (GC-8A; Shimadzu Corp.) and a gas chromatograph and mass spectrometer (QP2010 Plus; Shimadzu Corp.), respectively. Generally, plasma produced in a mixture of methane and oxygen leads to CO_2 and H_2O by the complete oxidation. However, methane partial oxidation can be implemented without combustion in a microplasma reactor because the temperature rise due to exothermic reaction is suppressed. Products such as methanol are condensed and dissolve into the liquid water film at room temperature and atmospheric pressure. As a result, liquid products are separated from an plasma reaction field and decomposition of the liquid components are suppressed. Consequently, reactant (gas) and products (liquid) are separable only with the temperature control of a reactor, enabling high methanol selectivity without an expensive separation membrane materials.

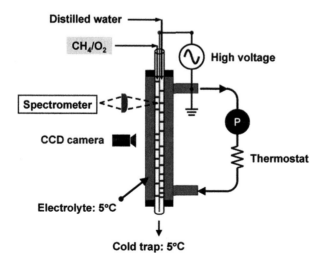

FIGURE 3.3 Schematic diagram of a microplasma reactor.

3.5 RESULTS AND DISCUSSION

3.5.1 EFFECT OF REACTION TEMPERATURE ON METHANOL SELECTIVITY

Figure 3.4(a) shows the ^{1}H-NMR spectrum of liquid products synthesized at a reaction temperature of 5°C (400 MHz; Bruker Analytik) (Nozaki et al. 2011b). The electrolyte was circulated via thermostat so that the temperature of the electrolyte was maintained at 5°C. The major products were methanol (CH_3OH), formic acid (HCOOH), formaldehyde (HCHO), and methyl hydroperoxide (CH_3OOH). Ethanol (C_2H_5OH), acetone (CH_3COH) (Oshima et al. 1988; Shigapov et al. 1998), and acetic acid (CH_3COOH) (Oshima et al. 1988; Larkin et al. 2001) were also detected, although only in a trace amount. Production of various oxygenates is a characteristic of nonthermal plasma. CH_3OOH is uniquely synthesized in this study, and has not been reported in a noncatalytic high-pressure thermal reaction (30–100 atm, 300–500°C) (Rasmussen and Glarborg 2008). It is important to understand the production and decomposition pathways of CH_3OOH from the viewpoint of yielding CH_3OH at high selectivity because CH_3OOH is a key intermediate product of CH_3OH. HOO is an important precursor of H_2O_2 rather than OH at low-temperatures, but OH formation by the pyrolysis of H_2O_2 is not expected at room temperature (Goujard et al. 2011). Generation and decomposition behavior of reactive oxygen species such as OH, HOO, and H_2O_2 by microplasma are described in detail in Section 3.6.3.

In order to emphasize the importance of low-temperature synthesis, the reactor temperature was raised to about 300°C via the heat generated by partial oxidation of methane as well as DBD: such temperature is readily attainable if the reactor is not cooled by electrolyte. In that case, metallic mesh was used for ground electrode. Figure 3.4(b) presents the ^{1}H-NMR spectrum of liquid products synthesized on these conditions. Although the methanol peak intensity decreases only slightly, intensity of other peaks dropped considerably. Presumably, liquid components were decomposed in the gas phase simply because these components did not condense to the water film at 300°C: in other words, water film does not work for product separation at elevated temperature. CH_3OH was not decomposed at 300°C, but chemically unstable species such as HCHO and the important peroxides including CH_3OOH were decomposed

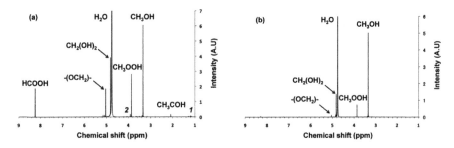

FIGURE 3.4 ^{1}H-NMR spectra of liquid products obtained at different reaction temperatures: (a) 5°C and (b) 300°C. Fine peaks of 1; C_2H_5OH and 2; CH_3OCOH were detected in (a). Experimental conditions: $CH_4 = O_2 = 20$ cm^3/min at STP (101 kPa and 25°C); distilled water flow rate, 0.1 cm^3/min. (Reproduced with permission from Nozaki et al. 2011b. Copyright 2011 IOP Publishing.)

by heat. Similarly, H_2O_2 yield dropped drastically at 300°C. Partial oxidation of methane at low-temperature condition is necessary for their direct synthesis with high selectivity.

3.5.2 SYNGAS GENERATION VIA DIRECT ROUTE

Figure 3.5(a) presents the selectivity of CO and CO_2. CO, a principal component of syngas, increased with methane conversion and reached 40–50%. The H_2/CO ratio was ascertained from the O_2/CH_4 ratio irrespective of the methane conversion (Figure 3.5(b)). A smaller O_2/CH_4 ratio yielded hydrogen-rich syngas, but the methane conversion dropped. Small amounts of ethane, ethylene, and acetylene were also produced aside from the liquid components and syngas (Nozaki et al. 2004c). Equilibrium calculations presented in Figure 3.1 suggest negligible amount of syngas is possible thermodynamically at 10–300°C, i.e., the microplasma reactor resulted in a highly nonequilibrium syngas production at considerably low-temperatures.

3.5.3 METHANE CONVERSION AND PRODUCT SELECTIVITY

Figure 3.6 presents selectivity of liquid components to methane conversion. In the figure, ● represents the total selectivity of three liquid components obtained in the microplasma reactor.

$$S_{liq} = \left[CH_3OH\right]_{sel} + \left[CHOH\right]_{sel} + \left[HCOOH\right]_{sel} \tag{3.9}$$

In addition to liquid products, syngas is produced as a main byproduct as shown in Figure 3.5. Assume one-step synthesis of dimethyl ether (DME) using the syngas (Reaction 3.10), total selectivity of liquid fuels S_{total} was evaluated (Reactions 3.11 and 3.12) which is expressed by ■ in Figure 3.6. Figure 3.6 also displays the

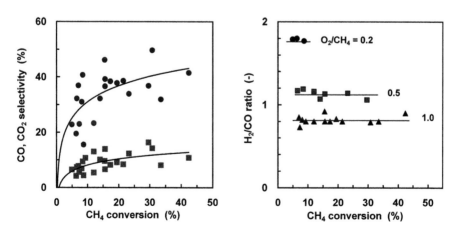

FIGURE 3.5 (a) Selectivity of CO (●) and CO_2 (■): (b) the relation between the H_2/CO ratio and the methane conversion (●, O_2/CH_4 = 0.2; ■, 0.5; ▲, 1.0) (Reproduced with permission from Nozaki et al. 2011a. Copyright 2011 Elsevier.)

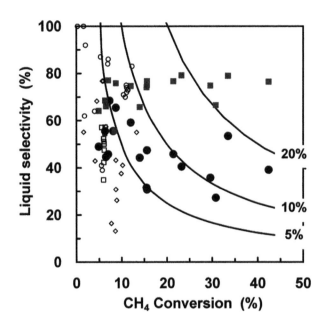

FIGURE 3.6 Selectivity of liquid components and methane conversion (●, S_{liq}; ■, S_{total}). $O_2/CH_4 = 0.5$, 1.0, reaction time = 130–530 ms, three curves represent yields (□, Bjorklund and Carr (2002); ○, Feng et al. (1994); ◇, Yarlagadda et al. (1988)). (Reproduced with permission from Nozaki et al. 2011a. Copyright 2011 Elsevier.)

earlier reports by the thermochemical processes (Bjorklund and Carr 2002; Feng et al. 1994; Yarlagadda et al. 1988), providing better comparison between thermal and nonthermal plasma processes. Three curves crossing the figure represent the yields of liquid components.

$$3CO + 3H_2 = CH_3OCH_3 + CO_2 \tag{3.10}$$

$$S_{DME} = (2/3) \times [CO]_{sel} (H_2/CO > 1) \tag{3.11}$$

$$S_{total} = S_{liq} + S_{DME} \tag{3.12}$$

DBD reactor was used for the direct conversion of methane to methanol without heterogeneous catalysts (Larkin et al. 2001; Zhou et al. 1998; Okazaki et al. 2002); however, the one-pass yield for useful oxygenates was smaller than 5% without taking advantage of the microreactor principle.

A salient shortcoming of conventional methods is that the reaction temperature should be raised to several hundred degrees Celsius to trigger a partial oxidation reaction. Because the oxidation reaction advances explosively once a reaction is initiated, heat flow and reaction temperature are too difficult to control. Consequently, the complete oxidation of methane is reached. The initial oxygen concentration is therefore restricted to 10% or less in almost all studies to avoid this problem. Nonthermal plasma allows a high initial oxygen concentration because the oxidation reaction is triggered at room temperature. Moreover, efficient removal of reaction heat by a microreactor suppresses the excessive rise of the reaction temperature. Liquid components can be condensed on the reactor wall and be separated from the oxygen-rich plasma reaction field. Combined selectivity of liquid products and DME would improve total selectivity up to ca. 80% (■, S_{total}).

3.6 REACTION MECHANISM

3.6.1 INTERACTION OF DBD AND LIQUID

Figure 3.7(a) presents a picture of DBD when distilled water is not supplied, showing almost uniform discharge in the microplasma reactor. A large number of streamer type discharge filaments are produced locally and temporally: time- and space-averaged image show quasi-uniform discharge in the reactor. The distilled water formed a liquid film on the reactor inner wall. The liquid surface was pulled up by the electrohydrodynamic effect: a protruding lump of liquid was formed (Figure 3.7(b)–(d)). Localized discharge was formed because the liquid protrusion enhances local electric field that initiates intense discharge channels: the discharge behavior of DBD and the liquid film influenced each other. The liquid film was unstably rippling by the electric field. Presumably, local overheating and evaporation of distilled water take place or OH radicals are generated at the place where filamentary discharges contact the liquid film. However, this phenomenon has yet to be clarified.

3.6.2 MECHANISM OF LOW-TEMPERATURE METHANE OXIDATION

The direct synthesis of methanol from methane has a long history of research. Details of reaction mechanisms have been investigated intensively. Oshima et al.

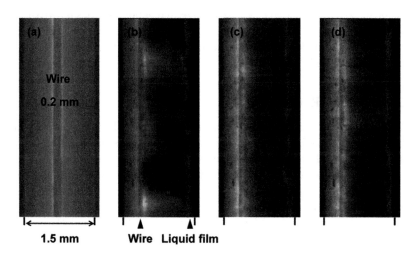

FIGURE 3.7 Magnified view of microplasma: 80 kHz, 30 W, $CH_4 = O_2 = 20$ cm³/min, (a) without distilled water, (b)–(d) with distilled water. (Reproduced with permission from Nozaki and Okazaki 2012. Copyright 2012 de Gruyter.)

(1988) irradiated $CH_4/O_2/N_2O$ mixed gas with an ArF excimer laser and generated oxygen atoms selectively to investigate the mechanism of methane partial oxidation at room temperature and up to 400°C:

$$N_2O + h\nu \rightarrow N_2 + O \tag{3.13}$$

$$O + CH_4 \rightarrow CH_3 + OH \tag{3.14}$$

$$CH_3 + O_2 \rightarrow CH_3OO \tag{3.15}$$

$$CH_3OO + CH_4 \rightarrow CH_3 + CH_3OOH \tag{3.16}$$

$$CH_3OOH \rightarrow CH_3OH \text{ or } HCHO \tag{3.17}$$

Reactions 3.13 and 3.14 correspond to an initiation reaction. Methane partial oxidation continues if CH_3 radicals associated with ground state oxygen molecules to generate CH_3OO (Reaction 3.15), and if CH_3OO abstracts hydrogen from methane (Reaction 3.16). CH_3OOH formed by Reaction 3.16 generates either CH_3OH or HCHO (Reaction 3.17). In the case of nonthermal plasma, CH_3 and O are generated by electron collision reaction (Reactions 3.18 and 3.19):

$$CH_4 + e^- \rightarrow CH_3 + H + e^- \tag{3.18}$$

$$O_2 + e^- \rightarrow O + O + e^- \tag{3.19}$$

A continuous methane oxidation is not anticipated without O_2 even if CH_3 is generated by plasma (Reactions 3.15 and 3.16). Atomic oxygen generated by DBD would abstract hydrogen from methane to assist the chain reaction. Various reactive oxygen species are known to be generated under nonthermal plasma (Eliasson et al. 1987). Accordingly, better understanding of electron collisional processes with methane and oxygen is required.

3.6.3 GAS PHASE REACTION MODEL

Oxidation reaction of methane at low-temperatures (5°C, 300°C) was analyzed using CHEMKIN 4.0; ANSYS. First, radical generation by the collision of methane and oxygen molecules with electrons was analyzed separately (Reactions 3.18 and 3.19). The Boltzmann equation solver, BOLSIG+ (Hagelaar and Pitchford 2005), was used to estimate radical generation rate constants (ki) in a methane and oxygen mixture, assuming DC electric field of 600 Td (1 Td = 10^{-17} V cm^2), pressure of 101 kPa, temperature of 300 K, and O_2/CH_4 ratio of 1/1. The electron density in a streamer and its propagation time were assumed respectively as $Ne = 10^{13}$ cm^{-3} and $\Delta t = 10$ ns. The radical density produced during 10 ns streamer propagation was estimated with the following equation:

$$\Delta Ni = ki \times \left[CH_4 \right] \times Ne \times \Delta t \tag{3.20}$$

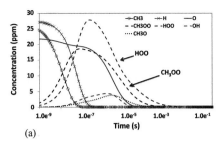 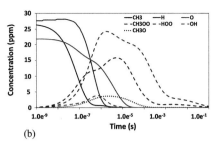

(a)   (b)

FIGURE 3.8 Time evolution of radical species. Conditions: $O_2/CH_4 = 1$, Pressure = 101 kPa, Temperature = (a) 5°C and (b) 300°C. Electron density $(N_e) = 10^{13}$ cm^{-3}. (Reproduced with permission from Goujard et al. 2011. Copyright 2011 IOP Publishing.)

Streamer propagation (< 10 ns) is much shorter than the characteristics time of the chemical reaction (ca. 1 μs). Accordingly, the radical generation by a single streamer was analyzed independently and the results were used as the initial condition for CHEMKIN analysis. Radical species considered in BOLSIG+ were CH_3, CH_2, CH, C, H, O(^{1}D), and O(^{3}P). A set of elementary reactions for methane combustion was obtained from GRI-MECH 3.0 (Smith et al. 2020). Additional elementary reactions were considered (Coppens et al. 2007) because some chemical species and their elementary reactions, such as CH_3O, CH_3OO, and CH_3OOH were needed to be implemented at low-temperature methane oxidation. Consequently, 36 chemical species and 236 elementary reactions were considered in the model. No elementary reactions involving ions were examined.

About 30 ppm of methane was dissociated into fragments directly by electron collision in a single streamer. However, the methane conversion did not increase further with time: no hydrogen abstraction of CH_4 by O and CH_3OO took place. Initial concentration of atomic oxygen was varied intentionally between 10 ppm and 30 ppm, but methane conversion was kept unchanged, implying that the atomic oxygen does not have a capability of methane activation at room temperature. The activation energy for Reaction 3.14 is small ($E = 35$ kJ/mol); however, Reaction 3.14 is an endothermic reaction ($\Delta H = 12$ kJ/mol) thus Reaction 3.14 reaction may not proceed at room temperature without a supply of thermal energy. The hydrogen abstraction from CH_4 by CH_3 or H is similarly slow. According to the model calculation, about 95% of CH_3 is generated by electron collision (Reaction 3.18), suggesting that the electron impact dissociation of methane is the dominant reaction pathway at room temperature, but the atomic oxygen plays a minor role. A time evolution of the key species at 5°C is presented in Figure 3.8(a) (Goujard et al. 2011). The initial concentration of CH_3 and H decreases rather abruptly. Consequently, CH_3 and H were almost fully consumed by 10^{-7} s. CH_3 generates CH_3OO and HOO via coupling with O_2. The initial concentration of atomic oxygen was sustained until 10^{-6} s because the reactivity of atomic oxygen is unexpectedly low. The hydrogen abstraction by CH_3OO is so slow that no chain reaction such as Reaction 3.16 was induced. Instead, CH_3OO reacts to HOO and generates CH_3OOH:

$$CH_3OO + HOO = CH_3OOH + O_2 \tag{3.21}$$

Reaction 3.21 is a termination reaction because no radical species, or chain promoter, is generated. CH_3OOH reacts with OH and OOH to generate CH_3O and CH_3OH. Electron impact dissociation of CH_4 to CH_3 and H is the key initiation reaction; however, the large activation energy of this reaction hinders efficient methane conversion (Nozaki and Okazaki 2013). Figure 3.8(b) shows the result of similar analysis at an elevated temperature of 300°C. About 35% of CH_3 was produced by plasma and 65% was produced by chain reaction induced by OH. However, the selectivity of liquid products decreased considerably and syngas (H_2 and CO) became a principal product. This results shows good agreement with the ^{1}H-NMR analysis shown in Figure 3.4 (Nozaki et al. 2011b).

H_2 concentration in the reaction field increases as methane conversion increases. Due to the higher reactivity of H_2 than of CH_4, partial oxidation of H_2 occurs dominantly at higher H_2 concentration. Abundant reactive oxygen species, such as HOO and H_2O_2, are generated which further accelerates the partial oxidation of H_2 so that the selectivity of H_2 drops sharply. Moreover, liquid products with high reactivity such as CH_3OOH and HCHO are oxidized to CO by reactive oxygen species originated from H_2 partial oxidation: it is quite evident that H_2 is detrimental for increasing the selectivity of liquid components (Nozaki et al. 2014). Removing H_2 from the plasma reaction field and separating liquid components using a liquid film are necessary for maintaining the methane conversion and reactant selectivity high.

3.7   CONCLUSIONS

This chapter discussed direct methanol synthesis from methane by gas–liquid two-phase flow microplasma reactor. Although atmospheric pressure nonthermal plasma is expected to create "innovative process technology", low energy efficiency and product selectivity are the main obstacle. The successful development of plasma gas cleaning technology has encouraged the development of a compact and inexpensive plasma reactor and related application studies in various fields.

Estimation of energy balance and efficiency for a plasma reformer alone is not sufficient for evaluating the economic value of plasma fuel reformer. Because the syngas production needs the highest temperature in the process, the securing of the heat source (combustion of the raw material) and the heat exchange at various temperature levels inevitably lead to a large energy penalty that governs the overall efficiency of the methanol synthesis process. If high-temperature and energy-intensive syngas manufacturing becomes unnecessary by nonthermal plasma technology, overall efficiency and thus the economic impact of plasma methanol synthesis can be viable. Furthermore, nonthermal plasma becomes possible to design a chemical reaction process in a different parameter regime from the conventional thermal catalysis (Nozaki and Okazaki 2012). This design implements innovative simplification and high efficiency material and energy use. Considering scaling-up of microplasma reactor as well as the use of renewable energy source (wind and solar), application to a relatively small scale and distributed energy system is desired rather than a gigantic natural gas field. Once the superiority of plasma fuel reforming is verified, this technology might affect various chemosynthesis processes.

REFERENCES

Ağıral, A., Nozaki, T., Yuzawa, S., Okazaki, K., and Han Gardeniers, J. G. E. 2011. Gas-to-liquids process using multi-phase flow, non-thermal plasma microreactor. *Chem. Eng. J.* 167:560–566.

Bjorklund, M. C. and Carr, R. W. 2002. Enhanced methanol yields from the direct partial oxidation of methane in a simulated countercurrent moving bed chromatographic reactor. *Ind. Eng. Chem. Res.* 41:6528–6536.

Coppens, F. H. V., De Ruyck, J., and Konnov, A. A. 2007. The effects of composition on burning velocity and nitric oxide formation in laminar premixed flames of $CH_4 + H_2 + O_2 + N_2$. *Combust. Flame* 149:409–417.

Davies, D. K., Kline, L. E., and Bies, W. E. 1986. Measurements of swarm parameters and derived electron collision cross sections in methane. *J. Appl. Phys.* 65:3311–3323.

Eliasson, B., Hirth, M., and Kogelschatz, U. 1987. Ozone synthesis from oxygen in dielectric barrier discharges. *J. Phys. D Appl. Phys.* 20:1421–1437.

Feng, W., Knopf, F. C., and Dooley, K. M. 1994. Effects of pressure, third bodies, and temperature profiling on the noncatalytic partial oxidation of methane. *Energy Fuels* 8:815–822.

Fincke, J. R., Anderson, R. P., Hyde, T. A., and Detering, B. A. 2002b. Plasma pyrolysis of methane to hydrogen and carbon black. *Ind. Eng. Chem. Res.* 41:1425–1435.

Fincke, J. R., Anderson, R. P., Hyde, T., Detering, B. A., Wright, R., Bewley, R. L., Haggard, D. C., and Swank, W. D. 2002a. Plasma thermal conversion of methane to acetylene. *Plasma Chem. Plasma Proc.* 22:105–136.

Goujard, V., Nozaki, T., Yuzawa, S., Ağıral, A., and Okazaki, K. 2011. Plasma-assisted partial oxidation of methane at low temperatures: numerical analysis of gas-phase chemical mechanism. *J. Phys. D Appl. Phys.* 44:274011(13pp).

Gutsol, A. 2010. Warm discharges for fuel conversion. In *Handbook of Combustion: Part 5 New Technologies*, ed. M. Lackner, F. Winter, A. K. Agarwal, and J. S. Lighty, 323–353. Wiley-VCH.

Hagelaar, G. J. M. and Pitchford, L. C. 2005. Solving the Boltzmann equation to obtain electron transport coefficients and rate coefficients for fluid models. *Plasma Sour. Sci. Technol.* 14:722–733.

Kado, S., Sekine, Y., Nozaki, T., and Okazaki, K. 2004. Diagnosis of atmospheric pressure low temperature plasma and application to high efficient methane conversion. *Catal. Today* 89:47–55.

Kalra, C. S., Gutsol, A. F., and Fridman, A. A. 2005. Gliding arc discharges as a source of intermediate plasma for methane partial oxidation. *IEEE Trans. Plasma Sci.* 33:32–41.

Kogelschatz, U. 2003. Dielectric-barrier discharges: their history, discharge physics, and industrial applications. *Plasma Chem. Plasma Proc.* 23:1–46.

Kondratenko, E. V., Peppel, T., Seeburg, D., Kondratenko, V. A., Kalevaru, N., Martin, A., and Wohlrab, S. 2017. Methane conversion into different hydrocarbons or oxygenates: current status and future perspectives in catalyst development and reactor operation. *Catal. Sci. Technol.* 7:366–381.

Larkin, D. W., Lobban, L. L., and Mallinson, R. G. 2001. The direct partial oxidation of methane to organic oxygenates using a dielectric barrier discharge reactor as a catalytic reactor analog. *Catal. Today* 71:199–210.

Lee, D. H., Kim, K. T., Cha, M. S., and Song, Y. H. 2007. Optimization scheme of a rotating gliding arc reactor for partial oxidation of methane. *Proc. Combust. Inst.* 31:3343–3351.

Nozaki, T., Abe, S., Moriyama, S., Kameshima, S., Okazaki, K., Goujard, V., and Ağıral, A. 2014. One step methane conversion to syngas by dielectric barrier discharge. *Jpn. J. Appl. Phys.* 54(1S):01AG01(6pp).

Nozaki, T., Ağıral, A., Yuzawa, S., Han Gardeniers, J. G. E., and Okazaki, K. 2011a. A single step methane conversion into synthetic fuels using microplasma reactor. *Chem. Eng. J.* 166:288–293.

Nozaki, T. and Okazaki, K. 2011. Innovative methane conversion technology using atmospheric pressure non-thermal plasma. *J. Jpn. Petrol. Inst.* 54(3):146–158.

Nozaki, T. and Okazaki, K. 2012. Plasma enhanced C1-chemistry: towards greener methane conversion. *Green Process. Synth.* 1:517–523.

Nozaki, T. and Okazaki, K. 2013. Non-thermal plasma catalysis of methane: principles, energy efficiency, and applications. *Catal. Today* 211:29–38.

Nozaki, T., Goujard, V., Yuzawa, S., Ağıral, A., and Okazaki, K. 2011b. Selective conversion of methane to synthetic fuels using dielectric barrier discharge contacting liquid film. *J. Phys. D Appl. Phys.* 44:274010(6pp).

Nozaki, T., Hattori, A., and Okazaki, K. 2004c. Partial oxidation of methane using a microscale non-equilibrium plasma reactor. *Catal. Today* 98:607–616.

Nozaki, T., Kado, S., Hattori, A., Okazaki, K., and Muto, N. 2004d. Micro-plasma technology: direct methane-to-m ethanol in extremely confined environment. *Stud. Surf. Sci. Catal.* 147:505–510.

Nozaki, T., Muto, N., Kado, S., and Okazaki, K. 2004a. Dissociation of vibrationally excited methane on Ni catalyst: Part 1. Application to methane steam reforming. *Catal. Today* 89:57–65.

Nozaki, T., Muto, N., Kado, S., and Okazaki, K. 2004b. Dissociation of vibrationally excited methane on Ni catalyst: Part 2. Process diagnostics by emission spectroscopy. *Catal. Today* 89:67–74.

Nozaki, T., Unno, Y., and Okazaki, K. 2002. Thermal structure of atmospheric pressure non-equilibrium plasmas. *Plasma Sour. Sci. Technol.* 11:431–438.

Okazaki, K., Kishida, T., Ogawa, K., and Nozaki, T. 2002. Direct conversion from methane to methanol for high efficiency energy system with exergy regeneration. *Energy Convers. Manage.* 43:1459–1468.

Oshima, Y., Saito, M., Koda, S., and Tominaga, H. 1988. Partial oxidation of methane by laser-initiated chain reaction. *Chem. Lett.* 17:203–206.

Rasmussen, C. L. and Glarborg, P. 2008. Direct partial oxidation of natural gas to liquid chemicals: chemical kinetic modeling and global optimization. *Ind. Eng. Chem. Res.* 47:6579–6588.

Sheng, Z., Kameshima, S., Sakata, K., and Nozaki, T. 2018. Plasma-enabled dry methane reforming. In *Plasma Chemistry and Gas Conversion*, ed. N. Britun and T. Silva. InTechOpen. doi:10.5772/intechopen.80523.

Shigapov, A. N., Hunter, N. R., and Gesser, H. D. 1998. The direct oxidation of ethane to alcohols at high pressures. *Catal. Today* 42:311–314.

Smith, G. P., Golden, D. M., Frenklach, M., et al. 2020. *What's is in GRI-Mech 3.0*. http://combustion.berkeley.edu/gri-mech/version30/text30.html (accessed on Jan 10, 2020).

Taifan, W. and Baltrusaitis, J. 2016. CH_4 conversion to value added products: potential, limitations and extensions of a single step heterogeneous catalysis. *Appl. Catal. B Environ.* 198:525–547.

Tsang, W. and Hampson, R. F. 1986. Chemical kinetic data base for combustion chemistry. Part I. methane and related compounds. *J. Phys. Chem. Ref. Data* 15:1087–1279.

Tu, X., Whitehead, J. C., and Nozaki, T. 2019. *Plasma Catalysis: Fundamentals and Applications*. Springer.

Wei, J. and Iglesia, E. 2004. Isotopic and kinetic assessment of the mechanism of reactions of CH_4 with CO_2 or H_2O to form synthesis gas and carbon on nickel catalysts. *J. Catal.* 224:370–383.

Yarlagadda, P. S., Morton, L. A., Hunter, N. R., and Gesser, H. D. 1988. Direct conversion of methane to methanol in a flow reactor. *Ind. Eng. Chem. Res.* 27:252–256.

Zhou, L. M., Xue, B., Kogelschatz, U., and Eliasson, B. 1998. Partial oxidation of methane to methanol with oxygen or air in a nonequilibrium discharge plasma. *Plasma Chem. Plasma Proc.* 18:375–393.

4 Natural Gas Conversion to Olefins via Chemical Looping

Yunfei Gao and Fanxing Li

CONTENTS

4.1 INTRODUCTION

Light olefins, such as ethylene and propylene, are essential building blocks of the modern chemical industry, with worldwide consumption exceeding 200 million tons per year. Presently, the primary route for ethylene production is steam thermal cracking of ethane and naphtha as feedstocks. In these processes, the hydrocarbon feedstocks are cracked into hydrogen, ethylene and other light olefins as well at high temperatures (> 750 °C) in the presence of cofeed steam to prevent coke deposition (Tao et al. 2006). The use of high-temperature steam, the high endothermicity of cracking reactions, and the demanding cryogenic product separation make these processes highly energy and carbon-intensive, with up to 2 tons of carbon dioxide emitted per ton of ethylene (Tao et al. 2006). Further optimization of conventional steam cracking is challenging, since the current approach has already achieved over 90% thermal efficiency. However, existing processes still lead to significant exergy loss

(> 13 GJ/ton of ethylene) (Neal et al. 2019a). This indicates the need for new technologies as well as significant process intensifications (Gao et al. 2019). Meanwhile, propylene is a byproduct from cracking processes but is also produced on purpose by endothermic dehydrogenation processes, such as Oleflex and Catofin (Bhasin et al. 2001). These processes are also subject to high amounts of energy consumption. Moreover, catalyst performance such as activity, selectivity, coke-resistance, and lifetime also affect the efficacy of on-purpose propylene production.

In addition to the need to develop robust, economical, and environmentally friendly light olefin production technologies, the increasing availability and lowered costs of light alkane feedstocks, resulting largely from the US shale gas revolution, have also created a significant driving force for technological advancement in olefin production. Due to the large fraction of ethane in many shale formations, the market has yet to bridge the increasing gap between ethane supply and demand. As a result, a significant amount of ethane is rejected through reinjection, flaring, or as a fuel. This circumstance is worsened with ethane price fluctuations versus transportation and fractionation costs. In 2014–2015 alone, approximately 210 million barrels of ethane were rejected in the lower 48 states. Thus, intensified ethane conversion into ethylene with a more economical and efficient approach is highly desirable. In addition to methane and ethane, naphtha, a mixture of C_5–C_9 hydrocarbons, is another abundant and important feedstock for olefin production, especially in Europe and the Asia-Pacific region (Zimmermann and Walzl 2009). Steam cracking of naphtha accounts for about 45% of ethylene produced worldwide (Tao et al. 2006). As discussed earlier, steam cracking has its intrinsic limitations due to thermodynamic equilibrium constraints, energy intensity, and process complexity.

This chapter discusses chemical looping as a potentially efficient strategy to intensify light olefin production from shale gas. In the following sections, the general concept of chemical looping is introduced and described. Then, approaches for light olefin production via chemical looping from different components in shale gas including methane, ethane, and propane, as well as natural gas liquids, are discussed. The specific topics covered include chemical looping – oxidative coupling of methane (CL–OCM), chemical looping – oxidative dehydrogenation (CL–ODH), and redox oxidative cracking (ROC). We also briefly cover the production of syngas via chemical looping reforming and partial oxidation of methane.

4.2 CHEMICAL LOOPING – THE GENERAL APPROACH

The core concept of chemical looping is to divide an overall chemical reaction into multiple sub-reactions with the use of solid reaction intermediates. This idea was investigated as early as the 19th century for oxygen separation using a BaO/BaO$_2$ redox pair (Hepworth 1892; Hardie and Pratt 1966). The term chemical looping combustion (CLC), the most widely investigated chemical looping topic to date, was introduced by Richter and Knoche in 1983 with the aim of improving the thermodynamic second law efficiency of fossil fuel combustion for power generation. This approach integrates metal oxide-based reaction intermediates, also referred to as oxygen carriers or redox catalysts, to fully oxidize carbonaceous fuels into steam and CO_2. Subsequently, the oxygen carriers are reoxidized with air in a separate reactor (Zeng et al. 2018). Since the 1990s, CLC was extensively investigated as a novel technology to combine efficient fuel conversion with integrated CO_2 separation (Ishida et al. 1987, 1994; Ishida and Jin 1994, 1996; Jin and Ishida 2001; He et al. 2007; Li and Fan 2008; Fan and Li 2010; Chung et al. 2017). Partial oxidation, or gasification of fossil fuels into syngas (H_2 and CO mixture), was also investigated in chemical looping partial oxidation (CL–POx) schemes. In comparison to conventional fossil fuel gasification and reforming processes, CL–POx has the potential to generate high-quality syngas with desirable H_2/CO ratios, while avoiding energy- and cost-intensive air separation units and syngas conditioning units (Mihai et al. 2012; He et al. 2013; Bhavsar et al. 2014; Luo et al. 2014; Chen et al. 2017; Mishra et al. 2019a, b). In addition to syngas production, CL–POx can also be applied to thermochemical CO_2 and H_2O splitting (He and Li 2015; Haribal et al. 2017a, 2019; Zhang et al. 2017). With an improved understanding of the redox catalyst design and surface modifications, the concept of chemical looping has been expanded into chemical looping selective oxidation (CLSO). Examples include chemical looping–based ethylene oxidation to ethylene oxide (Chan et al. 2018), ethane oxidative dehydrogenation to ethylene (Gao et al. 2016; Yusuf et al. 2017), methane oxidative coupling to ethylene (Gaffney et al. 1988), redox oxidative cracking (Dudek et al. 2019), etc.

4.3 CHEMICAL LOOPING CONVERSION OF SHALE GAS TO LIGHT OLEFINS

4.3.1 METHANE-BASED OLEFIN PRODUCTION

4.3.1.1 Chemical Looping – Oxidative Coupling of Methane

Given the abundance of methane, oxidative coupling of methane (OCM) to heavier hydrocarbons has become one of the most important research topics in C_1 chemistry. OCM uses gaseous oxygen as a cofeed reactant with methane, resulting in a net exothermic reaction. A schematic drawing of an O_2-cofeed OCM over a heterogeneous catalyst is shown in Figure 4.1(a). In the past 40 years, many catalyst systems have been demonstrated for OCM; the most effective ones include Li/MgO, $Na_2WO4/Mn/SiO_2$, and nanostructure catalyst. First reported in 1985, Li/MgO demonstrates around 20% C_{2+} yield at 800 °C (Ito and Lunsford 1985; Ito et al. 1985). But this catalyst suffers from stability issues due to loss of lithium through volatilization (Korf et al. 1987). $Na_2WO_4/Mn/SiO_2$ is another well-reported OCM catalyst and a ~24% C_{2+} yield was obtained at 800 °C (Fang et al. 1992). Extensive optimization efforts have been applied to both catalyst systems, including alkali-metal/dopant variation (Wang et al. 1995; Palermo et al. 2000; Wu et al. 2007; Yildiz et al. 2014), mechanistic studies (Ito et al. 1985; Driscoll et al. 1985; Korf et al. 1987; Palermo et al. 1998; Hou et al. 2006; Takanabe and Iglesia 2009; Lee et al. 2012) (37; 40; 46; 50; 55; 58; 62), and reactor configuration studies (Pak and Lunsford 1998; Simon et al. 2011), yet the C_{2+} yields are nevertheless limited to less than 30%, as shown in Figure 4.2. More recently, nanostructured catalysts have received more attention for low-temperature OCM (Huang et al. 2013; Yunarti et al. 2014; Scher et al. 2017). For instance, a

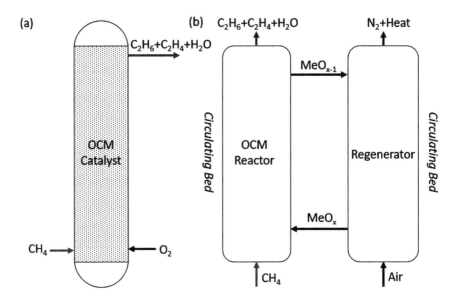

FIGURE 4.1 Schematic drawing of oxidative coupling of methane in (a) gaseous O_2 cofeed mode and (b) chemical-looping mode. (Reproduced with permission from *ACS Catalysis, 9,* Copyright 2019, *ACS publications* Gao et al. 2019.)

La_2O_3–CeO_2 nanofiber catalyst can achieve 22% C_{2+} yield with 55% C_{2+} selectivity at 520 °C (Noon et al. 2013). The low temperature operating range makes this type of catalyst potentially attractive, but C_{2+} yields are still limited.

Techno-economic analysis suggests that in order for an OCM process to be commercially and economically viable, single-pass C_{2+} yield and selectivity need to be greater than 30 and 90%, respectively (Farrell et al. 2016). Given that most, if not all, of the catalytic OCM processes achieved notably less than 30% yields, it is of critical importance to understand whether an upper bond for C_{2+} yield exists. William H. Green and coworkers investigated O_2-cofeed OCM assuming perfect catalyst surface chemistry (i.e. perfectly selective surface for radical imitation) (Su et al. 2003). A C_{2+} yield upper bound of 28% was projected. Such an upper bound is not subjected to a specific O_2-cofeed OCM catalyst. This upper bound in C_{2+} yield, however, has the potential to be overcome by chemical looping – OCM. Rather than reacting with gaseous oxygen, in CL–OCM, methane is oxidized by oxygen donated from the lattice sites of the redox catalyst. As a result, the nonselective gas-phase oxidation reactions can potentially be completely avoided. Labinger (1988) investigated kinetic models using Mg–Mn oxides as a redox catalyst in 1988. A 30% C_{2+} yield upper bound was projected due to CO_2 formation either from deep oxidation of methyl radical or secondary oxidation of the ethylene product. It is noted, however, that the properties of the redox catalyst would significantly affect the CO_2 selectivity. Unlike conventional OCM, the C_{2+} yield upper bound could potentially be overcome in CL–OCM with an optimized redox catalyst.

CL–OCM utilizes two reactors to complete an OCM reaction cycle. First, the oxidized catalyst is reacted with methane in the absence of gaseous oxygen to form

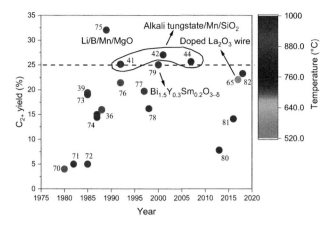

Number	Reference
1	Gaffney et al. 1988
2	Ito et al. 1985
3	Fang et al. 1992
4	Palermo et al. 2000
5	Wu et al. 2007
6	Scher et al. 2017
7	Mitchell III and Waghorne 1980
8	Keller and Bhasin 1982
9	Gaffney 1985
10	Jones et al. 1985
11	Jones et al. 1987
12	Gaffney et al. 1989
13	Gaffney 1992
14	Choudhary et al. 1997
15	Choudhary et al. 1998
16	Zeng et al. 2001
17	Elkins et al. 2013
18	Elkins et al. 2016
19	Cheng et al. 2018

FIGURE 4.2 OCM catalyst performance in "Year vs. C_{2+} yield".

coupling products. The catalyst particles are then reoxidized with air in a separate step to complete the loop (Figure 4.1b). This approach can offer better OCM selectivity control. As has been reported in a number of studies, lattice oxygen species in a redox catalyst have the ability to achieve higher selectivity in selective oxidation reactions (Gao et al. 2016, 2018). This trend holds true for CL–OCM. For example, sodium promoted praseodymium oxide (4% Na on Pr_6O_{11}) demonstrated higher C_{2+} selectivities/yields in CL–OCM than O_2-cofeed (conventional) OCM (Gaffney 1985; Gaffney et al. 1988). As shown in Figure 4.3, a $Pr^{3+} \leftrightarrow Pr^{4+}$ redox cycle was attributed to the formation of OCM-active peroxide species (O_2^{2-}). Another class of active chemical-looping OCM catalysts are Li/B/Mn/MgO mixed oxides developed by Atlantic Richfield Company (ARCO) (Gaffney 1989). The over-oxidation of hydrocarbon products is minimized in the chemical-looping scheme, and about 10% higher C_{2+} selectivity was achieved compared to O_2-cofeed OCM. The addition

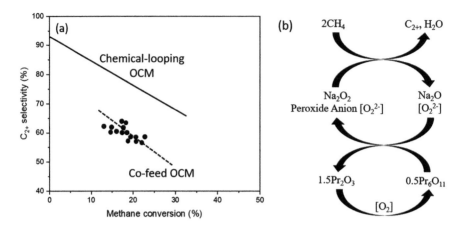

FIGURE 4.3 (a) Redox vs cofeed over 4% Na on Pr_6O_{11} (b) OCM reaction scheme in 4% Na on Pr_6O_{11}. (Reproduced with permission from *ACS Catalysis, 9,* Copyright 2019, *ACS publications* Gao et al. 2019.)

of alkali (e.g. Li, Na) and alkaline earth (e.g. Ca, Mg) metals were used to further inhibit the overoxidation of C_{2+} products (Jones et al. 1987).

More recently, Fan and coworkers revisited the Li-promoted Mn/MgO using a chemical-looping OCM scheme with more detailed characterizations and process simulations. A 63.2% C_{2+} selectivity and a 23.2% yield were obtained on Li-promoted Mn/MgO, whereas unpromoted Mn/MgO exhibited much lower selectivity. Density functional theory simulation indicated that Li doping induces oxygen vacancy formation, thus reducing the adsorption energy of methyl radicals and increasing the C–H activation barrier (Cheng et al. 2018). Process simulations also showed that chemical-looping OCM allows for better heat integration and saves energy from negating air separation units as compared to O_2-cofeed OCM (Chung et al. 2016). Schömäcker and coworkers also investigated chemical-looping OCM with the classical Na_2WO_4-Mn/SiO_2 catalyst. They identified that strongly absorbed or lattice oxygen in Na_2WO_4-Mn/SiO_2 is more intrinsically selective and can further improve C_{2+} selectivity in a redox scheme (Fleischer et al. 2016, 2018).

4.3.1.2 Chemical Looping – Partial Oxidation or Reforming of Methane

Although CL–OCM can produce ethylene in a single step, the limited methane conversion and ethylene yield due to C–H bond activation severely limits its potential for practical implementation. As an alternative, olefins can be indirectly produced from methane via a two-step process with syngas as an intermediate. CL–POx of methane for syngas production has been extensively studied (Bhavsar and Veser 2014; Bhavsar et al. 2014; Neal et al. 2014; Shafiefarhood et al. 2015; Mishra et al. 2016, 2019b). Readers interested in this topic are referred to a few comprehensive reviews and perspectives (Adanez et al. 2012; Fan et al. 2012; Tang et al. 2015; Mishra et al. 2018; Zhu et al. 2018; Zeng et al. 2018). As discussed earlier, CL–POx holds the potential to be more efficient than conventional reforming approaches due to its capacity for autothermal operation without the need for an air separation unit. The potential ability for water or CO_2 splitting can further be applied to adjust the H_2:CO ratio in the syngas and/or to upgrade downstream byproducts (He and Li 2015; Haribal et al. 2017a, 2019; Zhang et al. 2017). The syngas products can subsequently be converted into light olefins via the high-temperature Fischer–Tropsch process using an iron- or cobalt-based catalyst (Van Der Laan and Beenackers 1999; Dry 2002). A few recent studies have also led to highly promising syngas to olefin catalysts with exceptional selectivity and performance (Kang et al. 2011; Peng et al. 2015; Jiao et al. 2016). While CL–POx followed by syngas conversion can lead to light olefin products, this approach is less desirable than OCM due to its high complexity. In addition, most of the CL–POx studies carried out to date were operated under low methane partial pressures (< 1 atm). Downstream conversion of the syngas, however, requires much higher pressures (e.g. > 20 atm). Therefore, demonstration of CL–POx at high pressures is necessary when considering the high capital and energy costs for syngas compression.

4.3.2 ETHANE- AND PROPANE-BASED OLEFIN PRODUCTION VIA CHEMICAL LOOPING

Among the alkane feedstocks, ethane and propane are the most suitable for ethylene and propylene production due to their compositional, structural, and chemical similarities. In comparison with propane to propylene, ethane to ethylene is less challenging due to the higher energy barrier for C–C bond cleavage and higher stability of ethylene compared to propylene. Due to this reason, CL–ODH of ethane was extensively investigated in the past few years for ethylene production. In comparison, studies on chemical looping conversion of propane and higher alkanes is relatively limited but represents an important area for future research.

4.3.2.1 Chemical Looping Oxidative Dehydrogenation of Ethane

Oxidative dehydrogenation of ethane (ODH) is a thermodynamically favorable exothermic reaction that selectively oxidizes ethane into ethylene and water ($\Delta H = \sim 105.5$ kJ/mol at 100–1000 °C). During the past 20 years, many catalysts have been developed for ethane ODH, including supported V and Mo oxide (23–32% yield) (Chao and Ruckenstein 2003; Solsona et al. 2006a), supported Ni oxide (~35–38% yield) (Zhang et al. 2003; Nakamura et al. 2006), Sn doped Pt (> 55% yield) (Huff and Schmidt 1993; Henning and Schmidt 2002; Donsi et al. 2005), La/Sr/Nd and Cl doped La/Sr/Fe oxides (~55% yields) (Dai et al. 2000; Mulla et al. 2001) , Mo/V/Te/Nb/O (Up to 78% yield) (Botella et al. 2004; Xie et al. 2005), carbon nanotube and modified boron nitride (up to 50% yield) (Frank et al. 2010; Shi et al. 2017a) and many others (Cavani et al. 2007). These catalyst systems all require the cofeed of gaseous oxygen. This leads to increased cost from air separation units and safety concerns from mixing ethane with O_2. Similar to CL–OCM, CL–ODH of ethane utilizes the lattice oxygen from a redox catalyst via a cyclic redox scheme. As shown in Figure 4.4, the redox catalyst selectively oxidizes ethane into ethylene and water in an ODH reactor. The reduced redox catalyst is transferred into a second regeneration reactor where its lattice sites are replenished with oxygen to complete the redox loop. In an alternative configuration, a series of packed bed reactors can be used with ethane/air gas switching similar to the Houdry process.

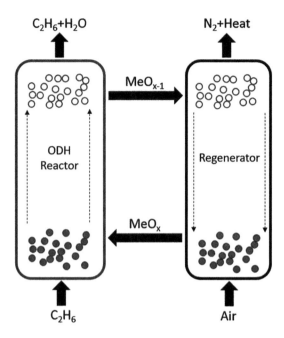

FIGURE 4.4 Schematic drawing of two different routes for CL–ODH of ethane. (Reproduced with permission from *ACS Catalysis, 9,* Copyright 2019, *ACS publications.*)[4]

Recent studies have indicated that ethane CL–ODH can occur via both catalytic and noncatalytic routes. In the catalytic route, ethane is converted to ethylene with the use of a redox catalyst. The temperature is usually limited to below 750°C since ethane thermal cracking becomes significant at higher temperatures. Vanadium oxide (VO_x) and molybdenum oxide (MoO_x) are typical catalysts for O_2-cofeed catalysts and they can also be used as redox catalysts in CL–ODH. Hossain and de Lasa reported γ-Al_2O_3 supported VO_x or MoO_x for chemical-looping ODH (Al-Ghamdi et al. 2013; Bakare et al. 2015; Elbadawi et al. 2016a, b). A circulating fluidized bed simulator was used to evaluate these redox catalysts. Ethylene selectivities of 55%–85% were achieved at 500–650°C but ethane conversion was limited to 7%–28% due partly to the small active lattice oxygen carrying capacity. Active metal oxide supports such as α-Fe_2O_3 were also used with the aim to increase the oxygen capacity of the redox catalysts. Novotny et al. reported that α-Fe_2O_3 supported MoO_3 (0.62 wt% oxygen capacity) doubles the oxygen capacity of un-supported MoO_3 (0.28 wt%). MoO_3/α-Fe_2O_3 achieved 62% ethylene selectivity at 600 °C, however, with low ethane conversions (< 10%) (Novotný et al. 2018). In addition to VO_x and MoO_x, other (mixed) metal oxides can also be used as redox catalysts in ODH. Gao et al. (2016, 2018) demonstrated that alkali-metal promotion can modify the selectivity of promoted $La_xSr_{2-x}FeO_{4-\delta}$ (LSF), where 90 % ethylene selectivity and 61 % ethane conversion was obtained in a fixed bed reactor. The redox catalyst is core-shell structured, where the core material is LSF and the shell material is enriched with lithium oxide and ferrite. It was determined that the surface layer decreased the lattice oxygen conduction rate from the bulk to the surface and suppressed the unselective

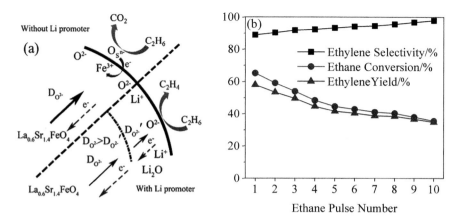

FIGURE 4.5 (a) CL–ODH mechanistic drawing with and without Li promotion; (b) Dynamic change in CL–ODH with consecutive ethane pulses on Li-promoted LSF. (Reproduced with permission from *ACS Catalysis, 6,* Copyright 2016, *ACS publications* Gao et al. 2016.)

surface electrophilic oxygen species (Figure 4.5a). In typical CL–ODH systems, ethane conversion/ethylene selectivity change dynamically as the redox catalysts get reduced (Figure 4.5b). As such, ethylene yields also vary with the duration of the ODH step (for packed bed operation) or solids residence time (for circulating fluidized bed design) (Gao et al. 2016, 2018). Therefore, balancing the degree of reduction for the redox catalyst and its activity is important to ensure optimal CL–ODH performance.

In the noncatalytic route, ethane is thermally cracked into ethylene and H_2. The H_2 byproduct is selectively combusted in situ by the redox catalyst, providing favorable heat of reaction and facilitating higher ethylene yield. These CL–ODH processes are typically carried out above 800°C to achieve substantial ethane conversion into ethylene and C_{3+}. To ensure high olefin selectivity, the ability of the redox catalyst to selectively oxidize H_2 in the presence of C_{2+} hydrocarbons is critical. Dudek et al. examined Na_2WO_4 promoted $CaMnO_3$, $SrMnO_3$ and Mg_6MnO_8 mixed oxide for selective oxidation of hydrogen with ethylene cofeed in the temperature range between 550 and 850 °C (Dudek et al. 2018). Using a similar principle, Neal and Yusuf et al. developed Na_2WO_4 promoted Mn–Mg, Mn–Si and Mn–Fe oxides for CL–ODH under a typical operating temperature of 850 °C (Yusuf et al. 2017, 2018). Na_2WO_4 was shown to be an effective promoter to inhibit hydrocarbon combustion while allowing facile oxidation of H_2. It was proposed that Na_2WO_4 form a molten shell covering the mixed oxides, blocking C–H bond surface activation and conduct lattice oxygen through a $WO_4^{2-} \leftrightarrow WO_3^-$ transition (Figure 4.6) (Yusuf et al. 2019). The molten Na_2WO_4 was also shown to be an effective electronic conductor. Up to 68% ethylene yields were obtained with CO_2 selectivity as low as 1.9% were achieved. Meanwhile, more than 70% of H_2 co-product was combusted, leading to an autothermal operation for ethylene production. A process simulation study by Haribal et al. (2017b) indicated that 82% reduction in energy consumption and CO_2 emission reduction can be achieved via CL–ODH. Moreover, the volumetric flow

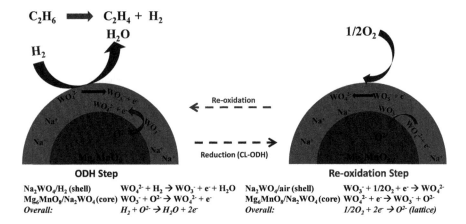

$$C_2H_6 \Longrightarrow C_2H_4 + H_2$$
$$H_2O$$
$$H_2$$
$$1/2O_2$$

Re-oxidation

Reduction (CL-ODH)

ODH Step		**Re-oxidation Step**
Na$_2$WO$_4$/H$_2$ (shell) WO$_4^{2-}$ + H$_2$ → WO$_3^-$ + e$^-$ + H$_2$O	Na$_2$WO$_4$/air (shell) WO$_3^-$ + 1/2O$_2$ + e$^-$ → WO$_4^{2-}$	
Mg$_6$MnO$_8$/Na$_2$WO$_4$ (core) WO$_3^-$ + O^{2-} → WO$_4^{2-}$ + e$^-$	Mg$_6$MnO$_8$/Na$_2$WO$_4$ (core) WO$_4^{2-}$ + e$^-$ → WO$_3^-$ + O^{2-}	
Overall: *H$_2$ + O^{2-} → H$_2$O + 2e$^-$*	*Overall:* *1/2O$_2$ + 2e$^-$ → O^{2-} (lattice)*	

FIGURE 4.6 Mechanistic drawing of CL–ODH on core-shell structured Na$_2$WO$_4$ promoted Mg$_6$MnO$_8$. (Reproduced with permission from *ACS Catalysis, 9,* Copyright 2019, *ACS publications* Yusuf et al. 2019.)

rates of the gaseous products were reduced by nearly 40%, rendering significantly decreased compression and separation costs (Haribal et al. 2017b). The feasibility of CL–ODH via this noncatalytic approach is further supported by a robust redox catalyst that demonstrates stable activity and selectivity for over 1,400 redox cycles in a laboratory-scale fluidized bed reactor (Neal et al. 2019b). Besides thermal cracking, catalytic dehydrogenation of ethane in parallel (or followed with) selective hydrogen oxidation has also been investigated and CL–ODH operating temperatures can be significantly lowered compared to the typical operating temperatures in the noncatalytic route. However, compatibility challenges between the dehydrogenation catalyst and the selective hydrogen oxidation material have yet to be successfully addressed based on the studies carried out to date (Grasselli et al. 1999a,b, 2000; Taikoyiannis et al. 1999).

4.3.2.2 Chemical Looping Oxidative Dehydrogenation of Propane

Similar to methane- and ethane-based olefin production technologies, propane can be oxidatively dehydrogenated into propylene via an O$_2$-cofeed scheme or a chemical looping scheme. Conventional propane ODH catalysts investigated to date are mostly vanadium oxide-based and molybdenum oxide-based (Watson and Ozkan 2000; Solsona et al. 2001, 2006b; Bañares and Khatib 2004; Klisińska et al. 2004; Demoulin et al. 2005). More recently, boron nitride (BN) has been identified as an efficient catalyst for propane ODH with high propylene selectivity (79%) with a propane conversion of 14% (Grant et al. 2016; Shi et al. 2017b). Despite the intensive investigations of propane ODH, propylene yield is still limited to 20% based on laboratory research in open literature. Further improvement on propylene yield is limited due to secondary oxidation of propylene to CO$_x$ and propane cracking to ethylene and methane (Chen et al. 1999). In addition to the extensive catalysis research, it is worth noting that platinum supported on basic oxides (e.g. ZnAl$_2$O$_4$) was used in a commercial STAR process for propane ODH. The likely reaction pathway is catalytic dehydrogenation with selective hydrogen combustion (Buyanov and Pakhomov 2001).

Chemical looping ODH of propane (CL–ODHP) has been proposed as a potential strategy to achieve higher propylene selectivity and to avoid cryogenic air separation. In a manner similar to CL–OCM and CL–ODH, propane is selectively oxidized to propylene with a redox catalyst in the absence of gaseous oxygen, and the redox catalyst is subsequently reoxidized with air in another reaction step. Ballarini et al. (2004) investigated Al_2O_3, TiO_2 and SiO_2-supported V_2O_5 for CL–ODHP. Up to 90% propylene yield at nearly 20% propane conversion was achieved, corresponding to an approximately 18% propylene yield. As a comparison, the same redox catalyst conducted in an O_2-cofeed scheme can only achieve up to a 10% propylene yield with low propylene selectivity (20–50%) (Ballarini et al. 2004). Other studies conducted on supported V_2O_5 catalysts also showed improved propylene selectivity in CL–ODH of propane (Creaser et al. 1999; Grabowski et al. 2002; Ballarini et al. 2003a, b; Kondratenko et al. 2005). These investigations indicate that CL–ODHP can potentially obtain higher propylene yield with an optimized redox catalyst. It is noted that existing studies on CL–ODHP are scarce and have focused exclusively on V and Mo oxide-based redox catalysts. The limitations in cost, toxicity, volatility, and oxygen carrying capacity of these oxides may limit the potential attractiveness of CL–ODHP. Moreover, the single-pass yield of propylene is still quite low. Therefore, further improvements in catalyst performance and exploration of other low-cost oxides as redox catalysts are highly desirable and represent an exciting research direction. In addition, CL–ODH of higher alkanes such as butane, pentane, and cycloalkanes can also be interesting and relevant research topics.

4.3.3 Naphtha-Based Ethylene Production

Naphtha can be produced from natural gas liquids and is an important feedstock for olefin production. Steam cracking of naphtha accounts for approximately 45% of ethylene produced worldwide but is a highly endothermic process, leading to high energy intensity and significant CO_2 emissions (Tao et al. 2006). Naphtha conversion from steam cracking is limited by thermodynamic equilibrium constraints, and coke formation is unavoidable. Oxidative conversion of naphtha, or oxy-cracking, combines oxidation reactions with cracking reactions and can lead to an autothermal process with higher olefin yields at lower reaction temperatures.

Conventional oxy-cracking schemes use a cofeed of gaseous oxygen with naphtha. These include noncatalytic gas-phase oxidative cracking processes and oxy-cracking with a heterogeneous catalyst. Liu et al reported noncatalytic oxy-cracking using cyclohexane as a model compound. At 750 °C, an 87.2% conversion of cyclohexane was obtained but CO_x yield was 15%. It was indicated that oxygen participates in the chain reactions of gaseous oxy-cracking, thus leading to an inevitable formation of CO_x (Liu et al. 2004; Zhu et al. 2006). On the other hand, oxy-cracking with a heterogeneous catalyst has the potential to achieve higher olefin selectivities at lower reaction temperatures. A number of representative heterogeneous catalysts have been demonstrated, including oxychlorides (e.g. BiOCl) (Kijima et al. 2001), gold containing catalysts (e.g. Au/La$_2$O$_3$) (Sá et al. 2011; Narasimharao and Ali 2013), alkali/alkali earth/rare earth oxides (e.g. Li doped MgO) (Leveles et al. 2002, 2003; Boyadjian et al. 2010; Gaab et al. 2004; Wu et al. 2016) (156, 158, 161, 162,164), boria catalysts (Buyevskaya et al. 1998; Buyevskaya and Baerns 1998), Mo and V containing catalysts (Cavani and Trifiro 1999), and Pt or other precious metal (Ag, Rh)-containing monolith (Subramanian et al. 2004; Basini et al. 2012). In spite of their potential advantages over conventional steam cracking, oxy-cracking processes still face challenges including safety concerns over mixing gaseous oxygen with fuels, and high CO$_x$ selectivity. Additionally, gaseous oxygen obtained from air separation is both capital and energy intensive.

Unlike conventional oxy-cracking, chemical looping oxy-cracking avoids the direct contact between gaseous oxygen and alkanes by using lattice oxygen from a redox catalyst. The advantages are: (i) less safety concerns due to concurrent operation of fuel and O$_2$; (ii) inhibited CO$_x$ formation and (iii) facilitated integrated air separation via the cyclic redox reactions. Although light olefin production via chemical looping ethane ODH has been widely investigated, studies on oxy-cracking of naphtha via chemical looping are relatively limited. To date, most research reported uses thermal cracking of naphtha and the H$_2$ co-product is selectively oxidized to water in situ by a redox catalyst. Dudek et al. (2019) and Tian et al. (2019) investigated chemical-looping oxy-cracking of *n*-hexane using perovskites (CaMnO$_3$, SrMnO$_3$) and hexa-aluminates (BaFe$_x$Al$_{12-x}$O$_{19}$, x = 1, 2, 3, 4, 6) as base materials and Na$_2$WO$_4$ as a promoter. Olefin yields up to 58% were reported at 725 °C and 4500 h^{-1} along with much lower CO$_x$ yields (~1.7 %) compared to oxy-cracking in the presence of gaseous oxygen. Na$_2$WO$_4$ acts as a shell material to improve selective hydrogen oxidation and to suppress CO$_x$ formation. Haribal et al. (2018) demonstrated that chemical-looping oxy-cracking of naphtha can achieve a 52 % energy reduction and a 50 % decrease in CO$_2$ emissions comparing to conventional steam cracking. Other studies focus on naphtha catalytic cracking and a following catalytic ODH using redox catalyst. Elbadawi and coworkers (2017) demonstrated VO$_x$/CeO$_x$–Al$_2$O$_3$ redox catalyst in a circulating fluidized bed simulator for chemical-looping oxy-cracking of naphtha and obtained 30% hexane conversion and 60% olefin selectivity at 525–600 °C.

4.4 CHEMICAL LOOPING LIGHT OLEFIN PRODUCTION – ADVANTAGES AND POTENTIAL OPPORTUNITIES

In spite of the exciting, recent advances in chemical looping–based OCM and ODH catalysts and processes, one must recognize that light olefin production processes, especially the production of ethylene, are some of the most established processes in the chemical industry. As such, they are extensively optimized and are difficult to be displaced by a newer technology unless the latter demonstrates significant improvements. Although the CL–ODH/OCM approaches are still at a relatively early stage of development, we believe that they possess a number of potential advantages that justify further research and development. The most significant advantage for the chemical looping–based approaches resides in the energy and CO$_2$ savings.

As discussed in Section 3.2.1, process analysis indicates that CL–ODH of ethane can lead to 82% energy and emission savings when compared to the conventional ethane cracking process (Haribal et al. 2017b). This can be somewhat counterintuitive since commercial cracking is almost perfectly optimized from a thermodynamic first law standpoint, with close to 95% efficiency (first law basis) (Zimmermann and Walzl, 2012). At a first glance, one would anticipate that any significant improvements in energy conversion efficiency would not have been possible given the highly optimized commercial plant. A recent second law analysis, also known as exergy analysis, presented by Neal et al. (2019b), provided an in-depth illustration of the fundamental process improvements and intensifications facilitated by CL–ODH. Figure 4.7 provides a breakdown of exergy loss for a conventional steam cracking process (Figure 4.7a) and a reference case for CL–ODH (Figure 4.7b). As is shown, the preheating and reactor sections (upstream) in CL–ODH account for a 3.4 GJ reduction in exergy loss (or lost work, LW) for each tonne of olefin produced. This has largely resulted from: (a) the lower exergy loss for in situ combustion of H_2 in CL–ODH; (b) the significantly smaller

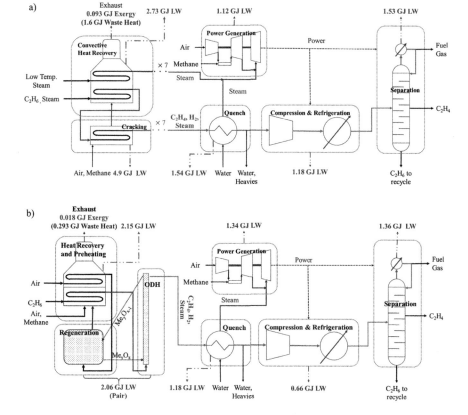

FIGURE 4.7 Process schematic and exergy/lost work (LW) schematic of a) Steam cracking of ethane and b) CL–ODH of Ethane [Results are in GJ/tonne of high value products (HVP), defined as C_{2+} olefins]. (Reproduced with permission from iScience, 19, Copyright 2019 Cell Press Neal et al. 2019b.)

temperature difference in CL–ODH reactors compared to cracker furnaces; and (c) the elimination of steam dilution.

CL–ODH also leads to notable exergy savings downstream of the reactors, which includes the compression and refrigeration block, the product separation block, and the power generation block (assuming on-site power generation for self-sustaining processes). Firstly, in situ combustion of hydrogen in CL–ODH and the improved single-pass product yield lead to significantly decreased compression loads since the volumetric flow rates of gaseous products from the CL–ODH reactor is much lower than that from the steam cracker (37% reduction). As such, 0.9 GJ/tonne of exergy savings can be anticipated within the compression and refrigeration block. In terms of the separation block, approximately 11% exergy savings (~0.17 GJ/tonne) was esti-mated. The limited exergy saving in this section is resulted from the need for CO and CO_2 removal in the CL–ODH process. Further improvements in redox catalyst performance can reduce the exergy loss in CL–ODH but still cannot eliminate the challenges for the demanding specifications of polymer grade ethylene. Nevertheless, CL–ODH provides up to 58% potential exergy savings compared to conventional steam cracking even though the latter exhibits near perfect first law efficiency. The significant exergy savings, coupled with zero emission from in situ hydrogen com-bustion in CL–ODH, leads to up to 85% emission reductions. This highlights the significant advantages of the chemical looping approach, especially under a carbon-constrained scenario. The above discussions also indicate that conventional first law analysis may provide somewhat incomplete information for the optimization of chem-ical production processes in the case that products with "low grade energy" such as co-generated steam are included in the efficiency calculation. In comparison, second law analysis provides additional insights on process optimization and intensification.

Besides the application of CL–ODH in centralized facilities for polymer grade olefin production, the increased availability of shale gas in North America, especially at remote locations, also offers significant opportunities for intensified, modular tech-nologies to convert C_1 and C_{2+} from distributed production sites. Such technologies, if successfully developed and deployed, would significantly reduce the emissions from flaring (> 6 billion m^3 each year in the United States) while also creating signifi-cant value (Wilkramanayake and Bahadur 2016). Compared to methane, which is the largest component in shale gas, challenges associated with over-supply of ethane are arguably more significant. Due to the limits in pipeline transportation capacity (up to 10 vol.%) and difficulty to liquefy ethane, it is often subjected to rejection via reinjec-tion to the well, flaring, or rejected at a fuel value. Each year, more than 210 million barrels of ethane (liquid equivalent) are rejected in the lower 48 states. Upgrading the low- to negative-value ethane to easily transportable liquid fuels is a promising solution to this supply glut. At a distributed scale, converting the light alkanes (e.g. C_1/C_2) into olefins as the final product would be less desirable since they are not eas-ily transportable. However, these light olefins can be oligomerized into gasoline and mid-distillates with high yields based on previous demonstrations by Mobil, ARCO, and others (Jones et al. 1985, 1986; Gaffney 1989; Neal et al. 2019a). Compared to the Fischer–Tropsch process, oligomerization can achieve much higher product selec-tivity and yields, leading to simpler product separation steps. The simpler overall process also enables a modular design for distributed scale applications. Figure 4.8

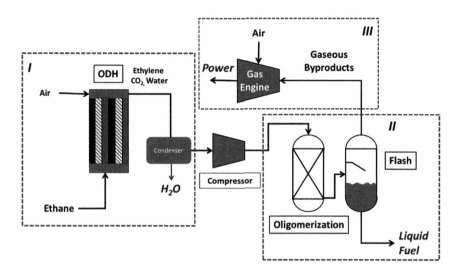

FIGURE 4.8 A simplified schematic of the CL–ODH based M-ETL scheme. (Reproduced with permission from *J. Adv. Manuf. Process, 1,* Copyright 2019, *Wiley* Neal et al. 2019a.)

exemplifies a modular ethane to liquids (M-ETL) process currently being investigated in our research group. Preliminary techno-economic analyses based on lab-scale experimental data indicate notable economic attractiveness (Neal et al. 2019a).

Section I: CL–ODH

$$C_2H_6 + MeO_x \rightarrow C_2H_4 + MeO_{x-1} + H_2O \qquad \left(\text{reaction 1: ODH}\right)$$

$$MeO_{x-1} + 1/2O_2 \rightarrow MeO_x + Heat \qquad \left(\text{reaction 2: Regeneration}\right)$$

Section II: Oligomerization, $4C_2H_4 = C_8H_{16}$ (gasoline and mid-distillates production)

Section III: Power Generation Gaseous fuel/heat $\rightarrow$ power

4.5 SUMMARY

The chemical industry is responsible for 2.5 gigatons of CO_2 emission each year and olefin production represents some of the most energy and carbon intensive technologies within the chemical sector. However, carbon management via process intensification for olefin production has yet to be thoroughly investigated, when compared to the power generation sector. From a heat integration standpoint, most state-of-the-art technologies for olefin production are near perfectly optimized with minimal waste heat generation. Conventional wisdom would have indicated that substantial reduction of energy consumption and CO_2 emission from these processes are unlikely, given their near 100% thermal (first law) efficiencies. The stringent requirements for the production of polymer grade olefins also adds the complexity for process optimization/intensification since any improved light olefin production processes should consider (a) alkane conversion reactions and the corresponding catalysts and/ or operating conditions; (b) feedstock preparation such as air separation for conventional ODH; and (c) complex product separation steps. These interrelated factors need to be accounted for at both the overall process and fundamental reaction/catalyst design levels. As illustrated using CL–ODH of ethane as an example in Section 4, the chemical approach offers excellent opportunities to reduce the exergy loss and CO_2 emissions for light olefin production. The large material design space for mixed oxide-based redox catalysts and the recent findings in promoting the oxide surface for selectivity enhancements offer exciting opportunities to develop improved redox catalysts for CL–ODH reactions. Meanwhile, practical redox catalysts should also be demonstrated to possess excellent redox stability, physical strength, and fluidizability (when operated under a circulating fluidized bed mode) as well as low-cost and minimal environmental impacts.

The significant increase in shale gas production within the United States has presented excellent opportunities for light olefin production due to the increased availability and lowered costs for light alkanes. Among the various alkane feedstocks, ethane is probably the most attractive under the current market conditions. The extensive recent studies on ethane CL–ODH have generated promising results. However, additional R&D efforts are still needed in order to advance these novel concepts to transformational technologies for more efficient ethane utilization, at both centralized and distributed scales. Meanwhile, methane coupling, as a direct methane to olefin technology, should not be overlooked considering the abundance of methane and its significance from the scientific standpoint. It is noted that a few simulation studies have pointed to a practical limit for C_{2+} yields (~30%) in conventional OCM with oxygen cofeed. One must keep this in mind when carrying out catalyst optimization efforts. While CL–OCM may also be subjected to yield limitations, its ability to eliminate gas phase nonselective reactions makes CL–OCM potentially more promising for significantly improved olefin yields. Achieving this would require in-depth studies on redox catalysts design and optimizations based on fundamental understanding of the reaction pathways at the gas-redox catalyst interface, in the gas phase, and within the redox catalyst particles. Compared to CL–OCM and CL–ODH of ethane, studies on chemical looping OCM of propane, butane, and higher alkanes are scarce. This represents an important area for future research given the significantly higher values of C_{3+} olefins and the scientific challenges associated with redox catalyst design for better selectivity at higher conversions. Given the opportunities provided by shale gas revolution and the excellent potential for process intensification offered by the chemical looping strategy, the authors are optimistic with respect to the prospects of potential breakthroughs in light olefin production via the aforementioned approaches.

REFERENCES

Adanez, J., Abad, A., Garcia-Labiano, F., Gayan, P., and F. Luis. 2012. Progress in Chemical-Looping Combustion and Reforming Technologies. *Prog. Energy Combust. Sci.* 38:215–282.

Al-Ghamdi, S., Volpe, M., Hossain, M. M., and H. de Lasa. 2013. VO x/c-Al$_2$O$_3$ Catalyst for Oxidative Dehydrogenation of Ethane to Ethylene: Desorption Kinetics and Catalytic Activity. *Appl. Catal. Gen.* 450:120–130.

Bakare, I. A., Mohamed, S. A., Al-Ghamdi, S., Razzak, S. A., Hossain, M. M., and H. I. de Lasa. 2015. Fluidized Bed ODH of Ethane to Ethylene over VO$_x$–MoO$_x$/γ-Al$_2$O$_3$ Catalyst: Desorption Kinetics and Catalytic Activity. *Chem. Eng. J.* 278:207–216.

Ballarini, N., Cavani, F., Cericola, A., Cortelli, C., Ferrari, M., Trifiro, F., Capannelli, G., Comite, A., Catani, R., and U. Cornaro. 2004. Supported Vanadium Oxide-Based Catalysts for the Oxidehydrogenation of Propane under Cyclic Conditions. *Catal. Today* 91:99–104.

Ballarini, N., Cavani, F., Cortelli, C., Giunchi, C., Nobili, P., Trifiro, F., Catani, R., and U. Cornaro. 2003b. Reactivity of V/Nb Mixed Oxides in the Oxidehydrogenation of Propane under Co-Feed and under Redox-Decoupling Conditions. *Catal. Today* 78:353–364.

Ballarini, N., Cavani, F., Ferrari, M., Catani, R., and U. Cornaro. 2003a. Oxydehydrogenation of Propane Catalyzed by VSiO Cogels: Enhancement of the Selectivity to Propylene by Operation under Cyclic Conditions. *J. Catal.* 213:95–102.

Bañares, M. A., and S. J. Khatib. 2004. Structure–Activity Relationships in Alumina-Supported Molybdena–Vanadia Catalysts for Propane Oxidative Dehydrogenation. *Catal. Today* 96:251–257.

Basini, L., Cimino, S., Guarinoni, A., Russo, G., and V. Arca. 2012. Olefins via Catalytic Partial Oxidation of Light Alkanes over Pt/LaMnO3 Monoliths. *Chem. Eng. J.* 207:473–480.

Bhasin, M. M., McCain, J. H., Vora, B. V., Imai, T., and P. R. Pujado. 2001. Dehydrogenation and Oxydehydrogenation of Paraffins to Olefins. *Appl. Catal. Gen.* 221:397–419.

Bhavsar, S., and G. Veser. 2014. Chemical Looping beyond Combustion: Production of Synthesis Gas via Chemical Looping Partial Oxidation of Methane. *RSC Adv.* 4:47254–47267.

Bhavsar, S., Najera, M., Solunke, R., and G. Veser. 2014. Chemical Looping: To Combustion and Beyond. *Catal. Today* 228:96–105.

Botella, P., García-González, E., Dejoz, A., López Nieto, J. M., Vázquez, M. I., and J. González-Calbet. 2004. Selective Oxidative Dehydrogenation of Ethane on MoVTeNbO Mixed Metal Oxide Catalysts. *J. Catal.* 225:428–438.

Boyadjian, C., van der Veer, B., Babich, I. V., Lefferts, L., and K. Seshan. 2010. Catalytic Oxidative Cracking as a Route to Olefins: Oxidative Conversion of Hexane over MoO$_3$-Li/MgO. *Catal. Today* 157:345–350.

Buyanov, R. A., and N. A. Pakhomov. 2001. Catalysts and Processes for Paraffin and Olefin Dehydrogenation1. *Kinet. Catal.* 42:64–75.

Buyevskaya, O. V., and M. Baerns. 1998. Catalytic Selective Oxidation of Propane. *Catal. Today* 42:315–323.

Buyevskaya, O. V., Müller, D., Pitsch, I., and M. Baerns. 1998. Selective Oxidative Conversion of Propane to Olefins and Oxygenates on Boria-Containing Catalysts. *Stud. Surf. Sci. Catal.* 119:671–676.

Cavani, F., and F. Trifiro. 1999. Selective Oxidation of Light Alkanes: Interaction between the Catalyst and the Gas Phase on Different Classes of Catalytic Materials. *Catal. Today* 51:561–580.

Cavani, F., Ballarini, N., and A. Cericola. 2007. Oxidative Dehydrogenation of Ethane and Propane: How Far from Commercial Implementation? *Catal. Today* 127:113–131.

Chan, M. S., Marek, E., Scott, S. A., and J. S. Dennis. 2018. Chemical Looping Epoxidation. *J. Catal.* 359:1–7.

Chao, Z.-S., and E. Ruckenstein. 2003. A Comparison of Catalytic Oxidative Dehydrogenation of Ethane Over Mixed V-Mg Oxides and Over Those Prepared via a Mesoporous V-Mg-O Pathway. *Catal. Lett.* 88:147–154.

Chen, K., Khodakov, A., Yang, J., Bell, A. T., and E. Iglesia. 1999. Isotopic Tracer and Kinetic Studies of Oxidative Dehydrogenation Pathways on Vanadium Oxide Catalysts. *J. Catal.* 186:325–333.

Chen, S., Zeng, L., Tian, H., Li, X., and J. Gong. 2017. Enhanced Lattice Oxygen Reactivity over Ni-Modified WO_3-Based Redox Catalysts for Chemical Looping Partial Oxidation of Methane. *ACS Catal.* 7:3548–3559.

Cheng, Z., Baser, D. S., Nadgouda, S. G., Qin, L., Fan, J. A., and L.-S. Fan. 2018. C2 Selectivity Enhancement in Chemical Looping Oxidative Coupling of Methane over a Mg–Mn Composite Oxygen Carrier by Li-Doping-Induced Oxygen Vacancies. *ACS Energy Lett.* 3:1730–1736.

Choudhary, V. R., Mulla, S. A., and V. H. Rane. 1998. Surface Basicity and Acidity of Alkaline Earth-Promoted La_2O_3 Catalysts and Their Performance in Oxidative Coupling of Methane. *J. Chem. Technol. Biotechnol.* 72:125–130.

Choudhary, V. R., Uphade, B. S., and S. A. Mulla. 1997. Oxidative Coupling of Methane over a Sr-Promoted La2O3 Catalyst Supported on a Low Surface Area Porous Catalyst Carrier. *Ind. Eng. Chem. Res.* 36:3594–3601.

Chung, C., Qin, L., Shah, V., and L.-S. Fan. 2017. Chemically and Physically Robust, Commercially-Viable Iron-Based Composite Oxygen Carriers Sustainable over 3000 Redox Cycles at High Temperatures for Chemical Looping Applications. *Energy Environ. Sci.* 10:2318–2323.

Chung, E. Y., Wang, W. K., Nadgouda, S. G., Baser, D. S., Sofranko, J. A., and L. S. Fan. 2016. Catalytic Oxygen Carriers and Process Systems for Oxidative Coupling of Methane Using the Chemical Looping Technology. *Ind. Eng. Chem. Res.* 55:12750–12764.

Creaser, D., Andersson, B., Hudgins, R. R., and P. L. Silveston. 1999. Transient Study of Oxidative Dehydrogenation of Propane. *Appl. Catal. Gen.* 187:147–160.

Dai, H. X., Ng, C. F., and C. T. Au. 2000. Perovskite-Type Halo-Oxide $La_{1-x}Sr_xFeO_{3-\delta}X_\sigma$ (X=F, Cl) Catalysts Selective for the Oxidation of Ethane to Ethene. *J. Catal.* 189:52–62.

Demoulin, O., Seunier, I., Dury, F., Navez, M., Rachwalik, R., Sulikowski, B., Gonzalez-Carrazan, S. R., Gaigneaux, E. M., and P. Ruiz. 2005. Modulation of Selective Sites by Introduction of N_2O, CO_2 and H_2 as Gaseous Promoters into the Feed during Oxidation Reactions. *Catal. Today* 99:217–226.

Donsì, F., Williams, K. A., and L. D. Schmidt. 2005. A Multistep Surface Mechanism for Ethane Oxidative Dehydrogenation on Pt- and Pt/Sn-Coated Monoliths. *Ind. Eng. Chem. Res.* 44:3453–3470.

Driscoll, D. J., Martir, W., Wang, J. X., and J. H. Lunsford. 1985. Formation of Gas-Phase Methyl Radicals over Magnesium Oxide. *J. Am. Chem. Soc.* 107:58–63.

Dry, M. E. 2002. High Quality Diesel via the Fischer–Tropsch Process – A Review. *J. Chem. Technol. Biotechnol.* 77:43–50.

Dudek, R. B., Gao, Y., Zhang, J., and F. Li. 2018. Manganese-Containing Redox Catalysts for Selective Hydrogen Combustion under a Cyclic Redox Scheme. *AIChE J.* 64:3141–3150.

Dudek, R. B., Tian, X., Blivin, M., Neal, L. M., Zhao, H., and F. Li. 2019. Perovskite Oxides for Redox Oxidative Cracking of N-Hexane under a Cyclic Redox Scheme. *Appl. Catal. B Environ.* 246:30–40.

Elbadawi, A. H., Ba-Shammakh, M. S., Al-Ghamdi, S., Razzak, S. A., and M. M. Hossain. 2016b. Reduction Kinetics and Catalytic Activity of VO_x/γ-Al_2O_3-ZrO_2 for Gas Phase Oxygen Free ODH of Ethane. *Chem. Eng. J.* 284:448–457.

Elbadawi, A. H., Ba-Shammakh, M. S., Al-Ghamdi, S., Razzak, S. A., Hossain, M. M., and H. I. de Lasa. 2016a. A Fluidizable VO_x/γ–Al_2O_3–ZrO_2 Catalyst for the ODH of Ethane to Ethylene Operating in a Gas Phase Oxygen Free Environment. *Chem. Eng. Sci.* 145:59–70.

Elbadawi, A. H., Khan, M. Y., Quddus, M. R., Razzak, S. A., and M. M. Hossain. 2017. Kinetics of Oxidative Cracking of N-Hexane to Olefins over $VO_x/Ce-Al_2O_3$ under Gas Phase Oxygen-Free Environment. *AIChE J.* 63:130–138.

Elkins, T. W., and H. E. Hagelin-Weaver. 2013. Oxidative Coupling of Methane over Unsupported and Alumina-Supported Samaria Catalysts. *Appl. Catal. Gen.* 454:100–114.

Elkins, T. W., Roberts, S. J., and H. E. Hagelin-Weaver. 2016. Effects of Alkali and Alkaline-Earth Metal Dopants on Magnesium Oxide Supported Rare-Earth Oxide Catalysts in the Oxidative Coupling of Methane. *Appl. Catal. Gen.* 528:175–190.

Fan, L.-S., and F. Li. 2010. Chemical Looping Technology and Its Fossil Energy Conversion Applications. *Ind. Eng. Chem. Res.* 49:10200–10211.

Fan, L.-S., Zeng, L., Wang, W., and S. Luo. 2012. Chemical Looping Processes for CO_2 Capture and Carbonaceous Fuel Conversion–Prospect and Opportunity. *Energy Environ. Sci.* 5:7254–7280.

Fang, X., Li, S., Lin, J., and Y. Chu. 1992. Oxidative Coupling of Methane on W–Mn Catalysts. *J. Mol. Catal. China* 6:427–433.

Farrell, B. L., Igenegbai, V. O., and S. Linic. 2016. A Viewpoint on Direct Methane Conversion to Ethane and Ethylene Using Oxidative Coupling on Solid Catalysts. *ACS Catal.* 6:4340–4346.

Fleischer, V., Simon, U., Parishan, S., Colmenares, M. G., Görke, O., Gurlo, A., Riedel, W., Thum, L., Schmidt, J., Risse, T., Dinse, K.-P., and S. Reinhard. 2018. Investigation of the Role of the $Na_2WO_4/Mn/SiO_2$ Catalyst Composition in the Oxidative Coupling of Methane by Chemical Looping Experiments. *J. Catal.* 360:102–117.

Fleischer, V., Steuer, R., Parishan, S., and R. Schomäcker. 2016. Investigation of the Surface Reaction Network of the Oxidative Coupling of Methane over $Na_2WO_4/Mn/SiO_2$ Catalyst by Temperature Programmed and Dynamic Experiments. *J. Catal.* 341:91–103.

Frank, B., Morassutto, M., Schomäcker, R., Schlögl, R., and D. S. Su. 2010. Oxidative Dehydrogenation of Ethane over Multiwalled Carbon Nanotubes. *ChemCatChem* 2:644–648.

Gaab, S., Find, J., Grasselli, R. K., and J. A. Lercher. 2004. Oxidative Ethane Activation over Oxide Supported Molten Alkali Metal Chloride Catalysts. *Stud. Surf. Sci. Catal.* 147:673–678.

Gaffney, A. M. 1989. *Hydrocarbon Production.* US4849571A. Published July 18, 1989.

Gaffney, A. M. 1985. *Methane Conversion.* US4499323A. Published February 12, 1985.

Gaffney, A. M. 1992. *Methane Conversion Process.* US5093542A. Published March 3, 1992.

Gaffney, A. M., Jones, C. A., and J. A. Sofranko. 1989. *Methane Conversion Process.* US4795849A. Published January 3, 1989.

Gaffney, A. M., Jones, C. A., Leonard, J. J., and J. A. Sofranko. 1988. Oxidative Coupling of Methane over Sodium Promoted Praseodymium Oxide. *J. Catal.* 114:422–432.

Gao, Y., Haeri, F., He, F., and F. Li. 2018. Alkali Metal-Promoted $La_xSr_{2-x}FeO_{4-\delta}$ Redox Catalysts for Chemical Looping Oxidative Dehydrogenation of Ethane. *ACS Catal.* 8:1757–1766.

Gao, Y., Neal, L., Ding, D., Wu, W., Baroi, C., Gaffney, A. M., and F. Li. 2019. Recent Advances in Intensified Ethylene Production—A Review. *ACS Catal.* 9:8592–8621.

Gao, Y., Neal, L. M., and F. Li. 2016. Li-Promoted $La_xSr_{2-x}FeO_{4-\delta}$ Core–Shell Redox Catalysts for Oxidative Dehydrogenation of Ethane under a Cyclic Redox Scheme. *ACS Catal.* 6:7293–7302.

Grabowski, R., Pietrzyk, S., Słoczyński, J., Genser, F., Wcisło, K., and B. Grzybowska-Świerkosz. 2002. Kinetics of the Propane Oxidative Dehydrogenation on Vanadia/Titania Catalysts from Steady-State and Transient Experiments. *Appl. Catal. Gen.* 232:277–288.

Grant, J. T., Carrero, C. A., Goeltl, F., Venegas, J., Mueller, P., Burt, S. P., Specht, S. E., McDermott, W. P., Chieregato, A., and I. Hermans. 2016. Selective Oxidative Dehydrogenation of Propane to Propene Using Boron Nitride Catalysts. *Science* 354:1570–1573.

Grasselli, R. K., Stern, D. L., and J. G. Tsikoyiannis. 1999a. Catalytic Dehydrogenation (DH) of Light Paraffins Combined with Selective Hydrogen Combustion (SHC): I. DH → SHC → DH Catalysts in Series (Co-Fed Process Mode). *Appl. Catal. Gen.* 189:1–8.

Grasselli, R. K., Stern, D. L., and J. G. Tsikoyiannis. 1999b. Catalytic Dehydrogenation (DH) of Light Paraffins Combined with Selective Hydrogen Combustion (SHC): II. DH+ SHC Catalysts Physically Mixed (Redox Process Mode). *Appl. Catal. Gen.* 189:9–14.

Grasselli, R. K., Stern, D. L., and J. G. Tsikoyiannis. 2000. Strategies for Combining Light Paraffin Dehydrogenation (DH) with Selective Hydrogen Combustion (SHC). *Stud. Surf. Sci. Catal.* 130:773–778.

Hardie, D. W. F., and J. D. Pratt. 1966. *A History of the Modern British Chemical Industry.* Oxford: Pergamon Press.

Haribal, V. P., Chen, Y., Neal, L., and F. Li. 2018. Intensification of Ethylene Production from Naphtha via a Redox Oxy-Cracking Scheme: Process Simulations and Analysis. *Engineering* 4:714–721.

Haribal, V. P., He, F., Mishra, A., and F. Li. 2017a. Iron-Doped $BaMnO_3$ for Hybrid Water Splitting and Syngas Generation. *ChemSusChem* 10:3402–3408.

Haribal, V. P., Neal, L. M., and F. Li. 2017b. Oxidative Dehydrogenation of Ethane under a Cyclic Redox Scheme–Process Simulations and Analysis. *Energy* 119:1024–1035.

Haribal, V. P., Wang, X., Dudek, R., Paulus, C., Turk, B., Gupta, R., and F. Li. 2019. Modified Ceria for "Low-Temperature" CO_2 Utilization: A Chemical Looping Route to Exploit Industrial Waste Heat. *Adv. Energy Mater.* 9:1901963.

He, F., and F. Li. 2015. Perovskite Promoted Iron Oxide for Hybrid Water-Splitting and Syngas Generation with Exceptional Conversion. *Energy Environ. Sci.* 8:535–539.

He, F., Galinsky, N., and F. Li. 2013.Chemical Looping Gasification of Solid Fuels Using Bimetallic Oxygen Carrier Particles–Feasibility Assessment and Process Simulations. *Int. J. Hydrog. Energy* 38:7839–7854.

He, F., Wang, H., and Y. Dai. 2007. Application of Fe_2O_3/Al_2O_3 Composite Particles as Oxygen Carrier of Chemical Looping Combustion. *J. Nat. Gas Chem.* 16:155–161.

Henning, D. A., and L. D. Schmidt. 2002. Oxidative Dehydrogenation of Ethane at Short Contact Times: Species and Temperature Profiles within and after the Catalyst. *Chem. Eng. Sci.* 57:2615–2625.

Hepworth, T. C. 1892. Oxygen for Limelight. *Nature* 47:176–177.

Hou, S., Cao, Y., Xiong, W., Liu, H., and Y. Kou. 2006. Site Requirements for the Oxidative Coupling of Methane on SiO_2-Supported Mn Catalysts. *Ind. Eng. Chem. Res.* 45:7077–7083.

Huang, P., Zhao, Y., Zhang, J., Zhu, Y., and Y. Sun. 2013. Exploiting Shape Effects of La_2O_3 Nanocatalysts for Oxidative Coupling of Methane Reaction. *Nanoscale* 5:10844–10848.

Huff, M., and L. D. Schmidt. 1993. Ethylene Formation by Oxidative Dehydrogenation of Ethane over Monoliths at Very Short Contact Times. *J. Phys. Chem.* 97:11815–11822.

Ishida, M., and H. Jin. 1994. A New Advanced Power-Generation System Using Chemical-Looping Combustion. *Energy* 19415–19422.

Ishida, M., and H. Jin. 1996. A Novel Chemical-Looping Combustor without NOx Formation. *Ind. Eng. Chem. Res.* 35:2469–2472.

Ishida, M., Zheng, D., and T. Akehata. 1987. Evaluation of a Chemical-Looping-Combustion Power-Generation System by Graphic Exergy Analysis. *Energy* 12:147–154.

Ito, T., and J. H. Lunsford. 1985. Synthesis of Ethylene and Ethane by Partial Oxidation of Methane over Lithium-Doped Magnesium Oxide. *Nature* 314:721–722.

Ito, T., Wang, J., Lin, C. H., and J. H. Lunsford. 1985. Oxidative Dimerization of Methane over a Lithium-Promoted Magnesium Oxide Catalyst. *J. Am. Chem. Soc.* 107:5062–5068.

Ji, S., Xiao, T., Li, S., Chou, L., Zhang, B., Xu, C., Hou, R., York, A. P., and M. L. Green. 2003. Surface WO_4 Tetrahedron: The Essence of the Oxidative Coupling of Methane over M–W–Mn/SiO_2 Catalysts. *J. Catal.* 220:47–56.

Jiao, F., Li, J., Pan, X., Xiao, J., Li, H., Ma, H., Wei, M., Pan, Y., Zhou, Z., and M. Li. 2016. Selective Conversion of Syngas to Light Olefins. *Science* 351:1065–1068.

Jin, H., and M. Ishida. 2001. Reactivity Study on a Novel Hydrogen Fueled Chemical-Looping Combustion. *Int. J. Hydrog. Energy* 26:889–894.

Jones, C. A., Leonard, J. J., and J. A. Sofranko. 1985. *Methane Conversion.* US4554395A. Published November 19, 1985.

Jones, C. A., Leonard, J. J., and J. A. Sofranko. 1986. *Two-Step Methane Conversion Process.* US4567307A. Published January 28, 1986.

Jones, C. A., Leonard, J. J., and J. A. Sofranko. 1987. The Oxidative Conversion of Methane to Higher Hydrocarbons over Alkali-Promoted $MnSiO_2$. *J. Catal.* 103:311–319.

Kang, J., Cheng, K., Zhang, L., Zhang, Q., Ding, J., Hua, W., Lou, Y., Zhai, Q., and Y. Wang. 2011. Mesoporous Zeolite-Supported Ruthenium Nanoparticles as Highly Selective Fischer–Tropsch Catalysts for the Production of C5–C11 Isoparaffins. *Angew. Chem. Int. Ed.* 50:5200–5203.

Keller, G. E., and M. M. Bhasin. 1982. Synthesis of Ethylene via Oxidative Coupling of Methane: I. Determination of Active Catalysts. *J. Catal.* 73:9–19.

Kijima, N., Matano, K., Saito, M., Oikawa, T., Konishi, T., Yasuda, H., Sato, T., and Y. Yoshimura. 2001. Oxidative Catalytic Cracking of N-Butane to Lower Alkenes over Layered BiOCl Catalyst. *Appl. Catal. Gen.* 206:237–244.

Klisińska, A., Haras, A., Samson, K., Witko, M., and B. Grzybowska. 2004. Effect of Additives on Properties of Vanadia-Based Catalysts for Oxidative Dehydrogenation of Propane: Experimental and Quantum Chemical Studies. *J. Mol. Catal. Chem.* 210:87–92.

Kondratenko, E. V., Cherian, M., and M. Baerns. 2005. Mechanistic Aspects of the Oxidative Dehydrogenation of Propane over an Alumina-Supported $VCrMnWO_x$ Mixed Oxide Catalyst. *Catal. Today* 99:59–67.

Korf, S. J., Roos, J. A., De Bruijn, N. A., Van Ommen, J. G., and J. R. H. Ross. 1987. Influence of CO_2 on the Oxidative Coupling of Methane over a Lithium Promoted Magnesium Oxide Catalyst. *J. Chem. Soc. Chem. Commun.* 19,. 1433–1434.

Korf, S. J., Roos, J. A., De Bruijn, N. A., Van Ommen, J. G., and J. R. H. Ross. 1990. Lithium Chemistry of Lithium Doped Magnesium Oxide Catalysts Used in the Oxidative Coupling of Methane. *Appl. Catal.* 58:131–146.

Kwapien, K., Paier, J., Sauer, J., Geske, M., Zavyalova, U., Horn, R., Schwach, P., Trunschke, A., and R. Schlögl. 2014. Sites for Methane Activation on Lithium-Doped Magnesium Oxide Surfaces. *Angew. Chem. Int. Ed.* 53:8774–8778.

Labinger, J. A. 1988. Oxidative Coupling of Methane: An Inherent Limit to Selectivity? *Catal. Lett.* 1:371–375.

Lee, M. R., Park, M.-J., Jeon, W., Choi, J.-W., Suh, Y.-W., and D. J. Suh. 2012. A Kinetic Model for the Oxidative Coupling of Methane over Na_2WO_4/Mn/SiO_2. *Fuel Process. Technol.* 96:175–182.

Leveles, L., Fuchs, S., Seshan, K., Lercher, J. A., and L. Lefferts. 2002. Oxidative Conversion of Light Alkanes to Olefins over Alkali Promoted Oxide Catalysts. *Appl. Catal. Gen.* 227:287–297.

Leveles, L., Seshan, K., Lercher, J. A., and L. Lefferts. 2003. Oxidative Conversion of Propane over Lithium-Promoted Magnesia Catalyst: I. Kinetics and Mechanism. *J. Catal.* 218:296–306.

Li, F., and L.-S. Fan. 2008. Clean Coal Conversion Processes–Progress and Challenges. *Energy Environ. Sci.* 1:248–267.

Liu, X., Li, W., Xu, H., and Y. Chen. 2004. A Comparative Study of Non-Oxidative Pyrolysis and Oxidative Cracking of Cyclohexane to Light Alkenes. *Fuel Process. Technol.* 86:151–167.

Luo, S., Zeng, L., Xu, D., Kathe, M., Chung, E., Deshpande, N., Qin, L., Majumder, A., Hsieh, T.-L., and A. Tong. 2014. Shale Gas-to-Syngas Chemical Looping Process for Stable Shale Gas Conversion to High Purity Syngas with a H_2: CO Ratio of 2:1. *Energy Environ. Sci.* 7:4104–4117.

Mihai, O., Chen, D., and A. Holmen. 2012. Chemical Looping Methane Partial Oxidation: The Effect of the Crystal Size and O Content of $LaFeO_3$. *J. Catal.* 293:175–185.

Mishra, A., and F. Li. 2018. Chemical Looping at the Nanoscale—Challenges and Opportunities. *Curr. Opin. Chem. Eng.* 20:143–150.

Mishra, A., Dudek, R., Gaffney, A., Ding, D., and F. Li. 2019a. Spinel Oxides as Coke-Resistant Supports for NiO-Based Oxygen Carriers in Chemical Looping Combustion of Methane. *Catal. Today* (In Press).

Mishra, A., Galinsky, N., He, F., Santiso, E. E., and F. Li. 2016. Perovskite-Structured $AMn_xB_{1-x}O_3$ (A = Ca or Ba, B = Fe or Ni) Redox Catalysts for Partial Oxidation of Methane. *Catal. Sci. Technol.* 6:4535–4544.

Mishra, A., Shafiefarhood, A., Dou, J., and F. Li. 2019b. Rh Promoted Perovskites for Exceptional "Low Temperature" Methane Conversion to Syngas. *Catal. Today* (In Press).

Mitchell III, H. L., and R. H. Waghorne. 1980. *Catalysts for the Conversion of Relatively Low Molecular Weight Hydrocarbons to Higher Molecular Weight Hydrocarbons and the Regeneration of the Catalysts.* US4239658A. Published December 16, 1980.

Mulla, S. A. R., Buyevskaya, O. V., and M. Baerns. 2001. Autothermal Oxidative Dehydrogenation of Ethane to Ethylene Using $Sr_xLa_{1.0} Nd_{1.0}O_y$ Catalysts as Ignitors. *J. Catal.* 197:43–48.

Nakamura, K.-I., Miyake, T., Konishi, T., and T. Suzuki. 2006. Oxidative Dehydrogenation of Ethane to Ethylene over NiO Loaded on High Surface Area MgO. *J. Mol. Catal. Chem.* 260:144–151.

Narasimharao, K., and T. T. Ali. 2013. Catalytic Oxidative Cracking of Propane over Nanosized Gold Supported $Ce_{0.5} Zr_{0.5} O_2$ Catalysts. *Catal. Lett.* 143:1074–1084.

Neal, L., Haribal, V., McCaig, J., Lamb, H. H., and F. Li. 2019a. Modular-Scale Ethane to Liquids via Chemical Looping Oxidative Dehydrogenation: Redox Catalyst Performance and Process Analysis. *J. Adv. Manuf. Process.* 1:e10015.

Neal, L. M., Haribal, V. P., and F. Li. 2019b. Intensified Ethylene Production via Chemical Looping through an Exergetically Efficient Redox Scheme. *iScience* 19:894–904.

Neal, L. M., Shafiefarhood, A., and F. Li. 2014. Dynamic Methane Partial Oxidation Using a Fe_2O_3@ $La_{0.8}Sr_{0.2}FeO_{3-\delta}$ Core–Shell Redox Catalyst in the Absence of Gaseous Oxygen. *ACS Catal.* 4:3560–3569.

Noon, D., Seubsai, A., and S. Senkan. 2013. Oxidative Coupling of Methane by Nanofiber Catalysts. *ChemCatChem* 5:146–149.

Novotný, P., Yusuf, S., Li, F., and H. H. Lamb. 2018. Oxidative Dehydrogenation of Ethane Using MoO_3/Fe_2O_3 Catalysts in a Cyclic Redox Mode. *Catal. Today* 317:50–55.

Pak, S., and J. H. Lunsford. 1998. Thermal Effects during the Oxidative Coupling of Methane over $Mn/Na_2WO_4/SiO_2$ and $Mn/Na_2WO_4/MgO$ Catalysts. *Appl. Catal. Gen.* 168:131–137.

Palermo, A., Vazquez, J. P. H., and R. M. Lambert. 2000. New Efficient Catalysts for the Oxidative Coupling of Methane. *Catal. Lett.* 68:191–196.

Palermo, A., Vazquez, J. P. H., Lee, A. F., Tikhov, M. S., and R. M. Lambert. 1998. Critical Influence of the Amorphous Silica-to-Cristobalite Phase Transition on the Performance of $Mn/Na_2WO_4/SiO_2$ Catalysts for the Oxidative Coupling of Methane. *J. Catal.* 177:259–266.

Peng, X., Cheng, K., Kang, J., Gu, B., Yu, X., Zhang, Q., and Y. Wang. 2015. Impact of Hydrogenolysis on the Selectivity of the Fischer–Tropsch Synthesis: Diesel Fuel Production over Mesoporous Zeolite-Y-Supported Cobalt Nanoparticles. *Angew. Chem. Int. Ed.* 54:4553–4556.

Sá, J., Ace, M., Delgado, J. J., Goguet, A., Hardacre, C., and K. Morgan. 2011. Activation of Alkanes by Gold-Modified Lanthanum Oxide. *ChemCatChem* 3:394–398.

Scher, E. C., Zurcher, F. R., Cizeron, J. M., Schammel, W. P., Tkachenko, A., Gamoras, J., Karshtedt, D., and G. Nyce. 2017. *Production of Ethylene with Nanowire Catalysts.* US9718054B2. Published August 1, 2017.

Shafiefarhood, A., Clay Hamill, J., Michael Neal, L., and F. Li. 2015. Methane Partial Oxidation Using $FeO_x@La_{0.8}Sr_{0.2}FeO_{3-\delta}$ Core–Shell Catalyst – Transient Pulse Studies. *Phys. Chem. Chem. Phys.* 17:31297–31307.

Shi, L., Wang, D., Song, W., Shao, D., Zhang, W.-P., and A.-H. Lu. 2017b. Edge-Hydroxylated Boron Nitride for Oxidative Dehydrogenation of Propane to Propylene. *ChemCatChem* 9:1788–1793.

Shi, L., Yan, B., Shao, D., Jiang, F., Wang, D., and A. H. Lu. 2017a. Selective Oxidative Dehydrogenation of Ethane to Ethylene over a Hydroxylated Boron Nitride Catalyst. *Chin. J. Catal.* 38:389–395.

Simon, U., Görke, O., Berthold, A., Arndt, S., Schomäcker, R., and H. Schubert. 2011. Fluidized Bed Processing of Sodium Tungsten Manganese Catalysts for the Oxidative Coupling of Methane. *Chem. Eng. J.* 168:1352–1359.

Solsona, B., Blasco, T., Nieto, J. L., Pena, M. L., Rey, F., and A. Vidal-Moya. 2001. Vanadium Oxide Supported on Mesoporous MCM-41 as Selective Catalysts in the Oxidative Dehydrogenation of Alkanes. *J. Catal.* 203:443–452.

Solsona, B., Dejoz, A., Garcia, T., Concepción, P., Nieto, J. M. L., Vázquez, M. I., and M. T. Navarro. 2006b. Molybdenum–Vanadium Supported on Mesoporous Alumina Catalysts for the Oxidative Dehydrogenation of Ethane. *Catal. Today* 117:228–233.

Solsona, B., Nieto, J. M. L., and U. Díaz. 2006a. Siliceous ITQ-6: A New Support for Vanadia in the Oxidative Dehydrogenation of Propane. *Microporous Mesoporous Mater.* 94:339–347.

Su, Y. S., Ying, J. Y., and W. H. Green. 2003. Upper Bound on the Yield for Oxidative Coupling of Methane. *J. Catal.* 218:321–333.

Subramanian, R., Panuccio, G. J., Krummenacher, J. J., Lee, I. C., and L. D. Schmidt. 2004. Catalytic Partial Oxidation of Higher Hydrocarbons: Reactivities and Selectivities of Mixtures. *Chem. Eng. Sci.* 59:5501–5507.

Takanabe, K., and E. Iglesia. 2009. Mechanistic Aspects and Reaction Pathways for Oxidative Coupling of Methane on $Mn/Na_2WO_4/SiO_2$ Catalysts. *J. Phys. Chem. C* 113:10131–10145.

Tang, M., Xu, L., and M. Fan. 2015. Progress in Oxygen Carrier Development of Methane-Based Chemical-Looping Reforming: A Review. *Appl. Energy* 151:143–156.

Tao, R., Patel, M., and K. Blok. 2006. Olefins from Conventional and Heavy Feedstocks: Energy Use in Steam Cracking and Alternative Processes. *Energy* 31:425–451.

Tian, X., Dudek, R. B., Gao, Y., Zhao, H., and F. Li. 2019. Redox Oxidative Cracking of N-Hexane with Fe-Substituted Barium Hexaaluminates as Redox Catalysts. *Catal. Sci. Technol.* 9:2211–2220.

Tsikoyiannis, J. G., Stern, D. L., and R. K. Grasselli. 1999. Metal Oxides as Selective Hydrogen Combustion (SHC) Catalysts and Their Potential in Light Paraffin Dehydrogenation. *J. Catal.* 184:77–86.

Van Der Laan, G. P., and A. Beenackers. 1999. Kinetics and Selectivity of the Fischer–Tropsch Synthesis: A Literature Review. *Catal. Rev.* 41:255–318.

Wang, D. J., Rosynek, M. P., and J. H. Lunsford. 1995. Oxidative Coupling of Methane over Oxide-Supported Sodium-Manganese Catalysts. *J. Catal.* 155:390–402.

Watson, R. B., and U. S. Ozkan. 2000. K/Mo Catalysts Supported over Sol–Gel Silica–Titania Mixed Oxides in the Oxidative Dehydrogenation of Propane. *J. Catal.* 191:12–29.

Wikramanayake, E. D., and V. Bahadur. 2016. Flared Natural Gas-Based Onsite Atmospheric Water Harvesting (AWH) for Oilfield Operations. *Environ. Res. Lett.* 11:034024.

Wu, J., Zhang, H., Qin, S., and C. Hu. 2007. La-Promoted $Na_2WO_4/Mn/SiO_2$ Catalysts for the Oxidative Conversion of Methane Simultaneously to Ethylene and Carbon Monoxide. *Appl. Catal. Gen.* 323:126–134.

Wu, Y., Shuo, L. I., and C. Li. 2016. Influence of Li Loading on the Catalytic Performance of Li/MgO in the Oxidative Dehydrogenation of Propane to Olefins. *J. Fuel Chem. Technol.* 44:1334–1340.

Xie, Q., Chen, L., Weng, W., and H. Wan. 2005. Preparation of MoVTe(Sb)Nb Mixed Oxide Catalysts Using a Slurry Method for Selective Oxidative Dehydrogenation of Ethane. *J. Mol. Catal. Chem.* 240:191–196.

Yildiz, M., Simon, U., Otremba, T., Aksu, Y., Kailasam, K., Thomas, A., Schomäcker, R., and S. Arndt. 2014. Support Material Variation for the Mn_xO_y-Na_2WO_4/SiO_2 Catalyst. *Catal. Today* 228:5–14.

Yokoyama, C., Bharadwaj, S. S., and L. D. Schmidt. 1996. Platinum-Tin and Platinum-Copper Catalysts for Autothermal Oxidative Dehydrogenation of Ethane to Ethylene. *Catal. Lett.* 38:181–188.

Yunarti, R. T., Lee, M., Hwang, Y. J., Choi, J.-W., Suh, D. J., Lee, J., Kim, I. W., and J.-M. Ha. 2014. Transition Metal-Doped TiO2 Nanowire Catalysts for the Oxidative Coupling of Methane. *Catal. Commun.* 50:54–58.

Yusuf, S., Neal, L., Haribal, V., Baldwin, M., Lamb, H. H., and F. Li. 2018. Manganese Silicate Based Redox Catalysts for Greener Ethylene Production via Chemical Looping–Oxidative Dehydrogenation of Ethane. *Appl. Catal. B Environ.* 232:77–85.

Yusuf, S., Neal, L. M., and F. Li. 2017. Effect of Promoters on Manganese-Containing Mixed Metal Oxides for Oxidative Dehydrogenation of Ethane via a Cyclic Redox Scheme. *ACS Catal.* 7:5163–5173.

Yusuf, S., Neal, L. M., Bao, Z., Wu, Z., and F. Li. 2019. Effects of Sodium and Tungsten Promoters on Mg_6MnO_8 Based Core-Shell Redox Catalysts for Chemical Looping–Oxidative Dehydrogenation of Ethane. *ACS Catal.* 9:3174–3186.

Zeng, L., Cheng, Z., Fan, J. A., Fan, L.-S., and J. Gong. 2018. Metal Oxide Redox Chemistry for Chemical Looping Processes. *Nat. Rev. Chem.* 2:349–364.

Zeng, Y., Akin, F. T., and Y. S. Lin. 2001. Oxidative Coupling of Methane on Fluorite-Structured Samarium–Yttrium–Bismuth Oxide. *Appl. Catal. Gen.* 213:33–45.

Zhang, J., Haribal, V., and F. Li. 2017. Perovskite Nanocomposites as Effective CO_2-Splitting Agents in a Cyclic Redox Scheme. *Sci. Adv.* 3:e1701184.

Zhang, X., Liu, J., Jing, Y., and Y. Xie. 2003. Support Effects on the Catalytic Behavior of NiO/Al_2O_3 for Oxidative Dehydrogenation of Ethane to Ethylene. *Appl. Catal. Gen.* 240:143–150.

Zhu, H., Liu, X., Ge, Q., Li, W., and H. Xu. 2006. Production of Lower Alkenes and Light Fuels by Gas Phase Oxidative Cracking of Heavy Hydrocarbons. *Fuel Process. Technol.* 87:649–657.

Zhu, X., Li, K., Neal, L., and F. Li. 2018. Perovskites as Geo-Inspired Oxygen Storage Materials for Chemical Looping and Three-Way Catalysis: A Perspective. *ACS Catal.* 8:8213–8236.

Zimmermann, H., and R. Walzl. 2009. Ethylene. In *Ullmann's Encyclopedia of Industrial Chemistry*, 465–526. Weinheim, Germany: Wiley-VCH.

Zimmermann, H., and R. Walzl. 2012. *Ullmann's Encyclopedia of Industrial Chemistry*, Weinheim, Germany: Wiley-VCH.

5 Oxidative Coupling of Methane

Hamid Reza Godini, Madan Mohan Bhasin, and Fausto Gallucci

CONTENTS

ABBREVIATIONS

Short Form	Long Form
BET	Measuring the specific surface based on Brunauer–Emmett–Teller theory
DME	DiMethyl Ether (Methoxymethane)
DRM	Dry Reforming of Methane
FT	Fischer Tropsch
GC	Gas Chromatography
IR	Infrared
MS	Mass Spectroscopy
OCM	Oxidative Coupling of Methane
ROI	Return on Investment

Definitions (Latin Letters)	Unit
Ethane + Ethylene (C_2)	-
Flux or membrane permeability (J)	mol/(m² s)
Molar flow (n)	mol/s
Pressure (P)	Pa, bar
Volume-based reaction rate (r)	mol/(m³ s)
Catalytic reaction rate ($\dot{r}$)	mol/(gr-cat s)
Moles/fraction of the consumed methane which appears in the desired products (*S: Selectivity*)	-
Temperature (T)	K, °C
Portion of inlet methane converted to desired and undesired products (*X: Methane conversion*)	-
Amount of methane appears in the (desired) product per whole inlet methane (*Y: Yield*)	-

Indices/Formulation/Short Form	Description
CO_3	Carbonate
Internal	Inside the membrane
l-Ax	Axial direction
OCM	OCM section in dual-membrane reactor
rad	Radial direction
sh, SH	Shell: Shell 1 (OCM), Shell 2 (DRM)
t, T	Tube
w	Wall
z	Axial coordinate
Inlet	Inlet stream
Outlet	Outlet stream

5.1 INTRODUCTION

Ethylene is the largest hydrocarbon chemicals produced in the world as it serves as a building block for a variety of chemical compounds. Current global ethylene production capacity is over 160 million tons per year (Mtpy) predicted to be growing 2–4% annually for the next decade. Two main feedstocks for industrial-scale ethylene production are ethane and naphtha, which are converted to ethylene using noncatalytic thermal-steam cracking processes.

Easy access to feedstock is one of the main factors affecting the decision of investors and strategic planners to choose the place, feedstock, and the technology used for ethylene production around the world. For instance, in the United States, the so-called wet shale gas contains a significant amount of ethane, which can be converted to ethylene via mature ethane cracking technology. On the other hand, oil-based components such as naphtha fraction from the oil refining process can be considered as an alternative feedstock, wherever it can be transported specially in the places in long distances from natural resources, for example, from those resources located in Europe and Asia-Pacific. Despite the energy intensity of these commercial technologies, their significant environmental impacts and the severe operating difficulties, such as coke formation, these have remained as the mainstream technologies for olefin production mainly due to their high production yield of desired olefin products and their flexible potential in producing and separating the commercially interesting side products such as propylene and Butadiene.

Having considered the efficiency and economy of the available technologies, local factors including the availability and the cost of feedstocks and products as well as the capacity of market and transportation should be also considered to choose an optimum technology and feedstock for ethylene production. This is in fact the main motivation for continuing to develope alternative technologies for olefin production and diversifying the technology choices compatible with different local conditions.

5.1.1 NATURAL GAS AND METHANE AS FEEDSTOCKS FOR ETHYLENE PRODUCTION

In an effort to increase the added-value, the pattern of natural gas utilization for energy generation has been shifted toward chemical production in general. In specific the direct chemical conversion processes are receiving special interests due to their energy-efficiency potentials and relatively simpler required process structure. Having considered the required operating conditions, value of the targeted products, observed selectivity and conversion as well as level of maturity of these direct processes, their potentials for being industrially-implemented can be relatively compared. However, the local potentials and the demands of the markets should also be taken into analysis for selecting an optimum direct methane conversion process for each locality. This is especially interesting to be analyzed for the fast-growing market for olefin, in more specific, ethylene production.

Unlike the process concepts for directly converting methane for ethylene production, the indirect ethylene production from methane is an established process technology. In fact, a general comparative assessment based on the stability and selectivity of the catalysts, robust operation of the reactors, value of the side products and difficulty of downstream separation/purification tasks quickly indicates that indirect methane–syngas–methanol–ethylene, for instance, should be considered being relatively more attractive and certainly competitive technology for industrial-scale operation in comparison to direct Oxidative Coupling of Methane (OCM) process. However, the capital cost of indirect methane conversion processes are very high in general. All in all, with the currently achievable OCM catalyst selectivity and yield and assuring a stable reactor performance, OCM process under some local conditions might be advantageous. For the natural gas rich locations, OCM technology becomes certainly superior to the existing alternative technologies if ethylene selectivity of higher than 80% can be secured for more than 20% methane conversion per pass. Here in this chapter, a multiperspective analysis of direct methane utilization in the context of Methane Oxidative Coupling process will be reported. Some of other processes utilizing natural gas as feedstock have been reviewed in other chapters of this book.

5.1.2 EVOLUTION OF UNDERSTANDING OF OXIDATIVE COUPLING OF METHANE

Before introducing the OCM concept and publishing the first results by Keller and Bhasin (1982), Union Carbide R&D worked for a long time and investigated different catalytic materials and operating conditions both steady state and unsteady state. Keller and Bhasin have evaluated vast majority of the inorganic oxides of the periodic table supported on alpha–alumina. An unsteady state catalytic process (also called redox mode) was preferred over a steady state or cofeed process. The metal oxides identified for the superior catalytic performance were Mn, Cd, Pb, Sn, Sb, Bi and Tl, while Li, Mg, Zn, Ti, In, Mo, Fe, Cr, W, Cu, Ag, Pt, Ce, V, B, and Al showed little or no activity. The active metals seem to exhibit common characteristics; meaning that they can cycle between at least two oxidation states. These observations were supported by thermodynamic calculations, for each metal oxide reducing to a lower oxidation state after coupling of methane and/or burning to make CO_x and water (Keller and Bhasin 1982). Since then, various catalysts (Zavyalova et al. 2011; Langfeld et al. 2012) and reactor concepts have been proposed for this application (Cruellas Labella et al. 2017), so that in some intense and other studies have been reported. Relative price of natural gas and crude oil has been an important factor in that trend. In fact, the original OCM activity was conducted in the early 1970s, when the price of crude oil was significantly increased. Next major efforts started again by rise of crude oil in the late 1990s and early 2000s.

The progress of technologies that have created new possibilities in the field of catalyst synthesis-characterization as well as the basic knowledge gathered in chemistry-material characteristics of the OCM catalysts have been the driving force to first better understand the mechanism of OCM catalytic reaction and then to synthesize gradually improved selective stable catalysts.

The efforts made in improving the efficiency of the downstream separation/purification units along with the new ideas and concepts for reactor development and restructuring and integrating the OCM process with other processes have also been the core for further analysis and improving the OCM process performance (García et al. 2019; Godini et al. 2019).

5.1.2.1 Reactions, Important Aspects, and Performance Indicators

Methane, as a very stable molecule, typically is activated with energy-duty (435 kJ mol^{-1}) under high temperature, while it can be activated in much lower temperature and energy in presence of oxygen. Low operating temperatures in general are preferred in terms of catalysts material stability, reactor design and operation as well as the process energy performance. After methane activation in the presence of active oxygen surface species and forming methyl radicals ($\cdot CH_3$), they undergo coupling reactions and form ethane, which is then dehydrogenated to form ethylene. The product species mainly ethylene/ethane and propylene/propane in the presence of excess oxygen are converted to carbon dioxide, carbon monoxide, hydrogen, and water as other side products of this reaction system represented by a combination of combustion, reforming and dehydrogenation reactions. These reactions represent a net exothermic reaction system in which the oxidative coupling reaction at 800 °C releases 139 kJ mol^{-1}. Several kinetic modeling and mechanism analyses have been reported where the weights of each of these involved reactions have been highlighted.

As the main OCM reactor performance indicators, selectivity (S) and yield (Y) of the desired C_2 (ethane and most preferably ethylene) products are calculated using Equations 5.1–5.2:

$$Y_{C_2} = -\frac{2 \times \left(n_{C_2H_6} + n_{C_2H_4} \right)}{n_{CH_4}^{inlet}} \times 100 \tag{5.1}$$

$$S_{C_2} = \frac{Y_{C_2}}{X_{CH_4}} = \frac{2 \times \left(n_{C_2H_6} + n_{C_2H_4} \right)}{n_{CH_4}^{inlet} - n_{CH_4}^{outlet}} \times 100 \tag{5.2}$$

where (n) stands for the molar flow rate of the gaseous species and (X) represents the methane conversion. In order to calculate the molar flow of the components in the reactor outlet stream based on the measured products mole fractions, usually a reference to an inert component such as nitrogen with a known flow rate is introduced into the reactor. The mole fraction of this component is measured, for instance, using a GC and the whole reactor outlet flow and thereby the molar flow of other components will be so calculated to ensure the same molar flow of this inert component in the inlet and outlet gas streams.

The ultimate selectivity of the OCM set of gas-surface parallel-series reactions is affected by the local partial pressure of the components and the reaction temperature. For instance, at the end of the reactor, a further undesired conversion of the desired products will be intensified. This is a typical behavior of such reaction systems dictating that the maximum C_2-yield is achieved in average values of per-pass methane conversion. Usually even in very high methane conversion (35–50%), which can only be achieved by supplying oxygen in an staged-wise or dosing mode, more than 40% of the converted methane appears in the form of carbon oxide molecules in conventional reactors. Some researchers, therefore, have suggested using an in-situ separation of hydrocarbons to improve the overall C_2-yield of the OCM reactor (Kruglov et al. 1996). However, such concepts should be further analyzed and developed with regard to their possible implementation in industrial-scale applications.

Since ethane dehydrogenation reaction is an endothermic one, the observed ratio of ethane to ethylene can be considered as an indicator for the thermal performance of the OCM reaction. Thermal management of the OCM reactor is very important, which not only affects the methane conversion and ethylene selectivity, but also is crucial for maintaining a safe operation of the reactor by preventing the runaway temperature and hot-spot formation.

5.1.2.2 Limitations of Catalysts, Reactors, and Process Performance

Mainly due to the required high temperature for this catalytic reaction system, selecting a proper material and maintaining the structure and stable selective performance of the OCM catalyst is difficult. Various catalysts distinguished by their support material and structures as well as the type and amount of their active components have been suggested and tested for OCM application (Kondratenko and Baerns 2008; Zavyalova et al. 2011; Langfeld et al. 2012). Being selective and, more importantly, being stable are the main desired characteristics of OCM catalysts. Most of the investigated OCM catalysts have shown either a poor stability or a low selectivity. Prior to any scale-up consideration, being able to produce the catalyst in large scale is also another important issue yet to be addressed for the OCM application.

Sharp variations of gas composition and temperature profile along the OCM catalytic bed are also affecting the ultimate OCM reactor performance. This is related to the activity and selectivity of the catalyst from one side and the configuration and design of the OCM reactor on the other side. In fact, some physical characteristics of the catalyst, such as its dimension, mechanical stability and so on, should also be tuned depending on the type of reactor being considered for a large scale.

In selecting and designing an OCM reactor, being able to control the safe-stable-selective operation is the key objective. In so doing, several scientific and engineering aspects, should be simultaneously and coherently optimized and controlled. As specific items, a proper selective catalyst, reactants feeding strategy, thermal management for controlling the operating temperature and temperature profile along the reactor, as well as the flow and dimensional parameters of the reactor should be properly designed and controlled.

Similar to many other oxidative processes, the selectivity of the OCM reaction is the most important issue to be secured. The issue of selectivity should be analyzed in the context of catalyst and reactor performance with the view on the OCM reaction mechanism. However, its impacts on the performance and requirements of the downstream units and their consequences on the whole OCM process structure and performance should also be addressed in any analysis.

For a quick conceptual design-analysis review of the OCM process structure, it can be divided into three main sections namely reaction section, carbon dioxide removal section, and ethylene separation and purification section. After the reactor section, carbon dioxide should be quickly separated via absorption, membranes or, less preferably, adsorption in order to avoid having problematic CO_2 component to pass further down in the process. On the other side, the unreacted methane should also be separated and treated further either in a separate reactor or via recycling it back to the OCM reactor. Unreacted methane and the rest of the light gases can be separated using the cryogenic distillation. Similarly the remaining ethane and ethylene can be separated from each other. Alternatively, ethylene can be separated using a selective adsorbent before or most preferably after the CO_2-removal section. Relatively cheaper source of utilities and energies, yet a larger capital cost and complicated control system are needed for this alternative process structure.

5.1.2.3 Highest Impact Parameters in a Practically Relevant Analysis Context

In OCM reports, usually the focus is either on the catalyst, reactor, operating conditions, or on the process. The results of such researches have certainly contributed to improving the understanding of the OCM system in each of those specific aspects. However, these have strong interactions and therefore, it should be highlighted that the impact of the investigated parameters should be analyzed in a bigger picture rather than on a certain part of the process. Following are the key aspects that should be addressed in any practically relevant OCM research, which aims to improve the chance of OCM process to be implemented on an industrial scale:

a) Catalyst stability: it is a key factor because without a stable catalyst, no practical application for OCM can be expected. Since the approaches usually implemented for improving the catalyst selectivity will also affect its stability and vice versa, these should be simultaneously addressed. Tuning the operating conditions in terms of temperature, operating pressure, and gas composition should also be done in this interactive analysis context. The impact of the operating pressure and gas composition on the downstream units and the whole OCM process is significant. Therefore, the catalytic performance of the OCM reactors under elevated pressure processing the feeds with proper oxygen (and dilution used) content are practically relevant to be considered in the context of OCM process analysis. Usually the resulting reactor performances in terms of ethylene selectivity and yield become poorer in this manner, but still is justified (Cantrell et al. 2002, 2003). Unfortunately vast majority of the OCM catalysts and process studies have been performed at atmospheric pressure.

b) Ethylene and propylene selectivity: after securing a safe and practically relevant range of pressure and temperature, obtaining highest ethylene selectivity while maintaining the highest possible methane conversion per pass should be targeted in the design of an OCM reactor. This should be also reflected in screening the catalysts and comparing their performances.

It needs to be reemphasized that the interactive impacts of parameters on the whole OCM process performance should be taken into analysis even when the focus is on specific aspect of the process. This requires an interdisciplinary multi-perspectives analysis based on the expertise of multi-disciplinary teams.

5.1.3 REMARKS AND CONCLUSIONS FOR STRUCTURING THIS CHAPTER

Having reviewed the main characteristics of the OCM process and the aspects to be covered, following structure was designed to systematically analyze this system here:

- Section 5.1 – Introduction: The motivations and characteristics of OCM process along with the important elements of an OCM research strategy were summarized and highlighted.
- Section 5.2 – Catalyst Research: Current understanding of the material, structural and chemical contributions of catalysts on the OCM reaction performance are summarized and discussed with the focus on the state of the art OCM catalysts.
- Section 5.3 – Reactor Research: Various reactor configurations and concepts suggested for OCM are analyzed and the important characteristics in designing an efficient OCM reactor in different scales are highlighted.
- Section 5.4 – Interaction of Catalyst and Reactor Research: The interactive impacts of catalyst and reactor characteristics on the ultimate observed performance of an OCM reactor under different range of operating conditions are discussed and some practically relevant conclusions and suggestions are provided.
- Section 5.5 – Process and Reactor Integration, Environmental, and Industrial Prospects: The main aspects of a scientific and engineering OCM process-scale analysis are reviewed and discussed. The available OCM results and knowledge will be revisited with the view on industrial applications of this process and some conceptual and practical conclusions are provided. The impacts of operating pressure and feed composition will be discussed in this context. The motivations, potentials, and possibilities as well as the reported strategies for integrating the OCM process with other processes are also reviewed and discussed. Specific attention will be devoted to the energy and environmental aspects associated with the reactor and process integrations in this system. Alternative sources of methane to be treated by

OCM process for ethylene production are reviewed and analyzed in terms of their techno-economic-environmental performances.

5.2 CATALYST RESEARCH

In the very active field of OCM, different researchers have synthesized, characterized and tested thousands of OCM catalysts with different focused aims since the first publication (Keller and Bhasin 1982). For instance, some researchers have tried to understand the reaction mechanism or the impacts of catalysts' structural characteristics on its reaction performance by analyzing relatively simple catalysts – model catalysts – such as Li/MgO. The performed analyses in this context cover single crystal study from one side up to in-situ characterization during macro-scale relatively long-term reactor tests. Some parts of comparative characterizing and testing of different OCM catalytic materials, or even same materials synthesized differently, have been also conducted in this context.

Some researchers have focused on a relatively complex but more practically relevant types of OCM catalysts such as La-based catalysts (Vallet-Regi et al. 1988; Bhasin and Campbell 1995; Cantrell et al. 2002, 2003) Mn-based (Bagherzadeh et al. 2012) and perovskite-based (Kenneth 1991) catalysts. For instance, Cantrell et al. reported an improvement in C_2-selectivity at 10 bar over that observed one at 1 bar using the defect/disordered La-oxy carbonate structure catalysts (Cantrell et al. 2002, 2003). However, significant drop in selectivity have been observed for the most of OCM catalysts tested at higher than atmospheric pressure. For instance, Pinabiau-Carlier et al. demonstrated that selectivity losses of 3–4% were observed for every one-bar increase in pressure (Pinabiau-Carlier et al. 1991).

It should be highlighted that at high pressure ($\geq$10 bar) operation, independent from the type of catalyst, securing the methane conversion of more than 20% in co-feeding mode is not possible as it requires either an oxygen concentration of above the safe operating limits or utilizing an inert dilution or staged oxygen feeding.

On the other side, the lowest gas-phase oxygen concentration and a relatively more selective performance under redox/cyclic (separate methane) feeding mode can be secured for instance using Mn-based catalysts. In Redox mode, selectivity to C_2+C_3 as high as 80–90% have been reported (Kenneth 1991; Bhasin and Slocum 1995; Jones et al. 1987).

Even the results of testing mixed La-based and Mn-Na$_2$WO$_4$ catalysts supported on alpha–alumina, have been reported which indicates a promising OCM performance even in low temperature range of up to 550 °C (Cantrell et al. 2002, 2003).

Perovskite-based catalysts have also shown a relatively selective performance, but their performance is affected in the presence of a significant amount of CO_2. Table 5.1 lists the most extensively researched family of catalysts, including the most practically relevant ones, along with their industrially-relevant performance indicators.

In Table 5.1, the performance of model catalyst (Li/MgO) as well as the catalysts recommended for low temperature OCM operations have been also listed. The latter ones have shown a selective performance, mainly because there will be less radicals converted in the gas-phase towards carbon oxides over these catalysts under a low temperature range. These catalysts, similar to ethane oxidehydrogenation catalysts,

TABLE 5.1

Main Family of Catalysts Studied for OCM and Their Practically Relevant Observations

Catalyst Family	La-based*	Mn-based	Perovskite	Model (Li/MgO)	Low-temperature
Selectivity under high pressure	Positive	Negative	negative	None-Unstable	Some tested***
Low temperature performance	Promising	Promising	Various	unstable/ inactive	Preferred
Cyclic performance	Promising	Promising	Few tested	Not tested	Promising
Reported C_2-selectivity**	< 80%	< 70%	< 70%	Unstable < 70%	< 70%

* Only for defect disordered catalyst (Cantrell et al. 2002, 2003), ** For methane conversion above 20% (all reported results have been tested in atmospheric pressure); *** Low-temperature high pressure performance is practically preferred (Cantrell et al. 2002, 2003)

Promising: Observed reaction performance and material characteristics are compatible with the operation modes

utilize surface oxygen to generate enough radicals, which can undergo the desired surface reactions to ultimately produce olefin. In the alternating redox mode, these catalysts can largely keep their selective performance even at higher operating temperatures by suppression of the product's oxidation due to the absence of oxygen in the gas-phase (Jones et al. 1987).

The impact of catalyst support structure on the large-scale operation of these catalysts have been also investigated. For instance, nano-wire catalysts structure (Schammel et al. 2013) as well as macro-size pellets have been tested, for which the impact of pressure on the long-term operation should be also addressed in an efficient catalyst design.

Mn-Na_2WO_4 and La_2O_3 catalysts have been extensively investigated in order to analyze their functionality and performance primarily to improve the understanding of their material-chemical-structural contributions on its reaction performance and ultimately to further improve their stability and selectivity. Depending on the context, sometimes even contradictory conclusions have been made in different reports over the same material. Therefore, in reviewing the reported results in literature, the analysis context and the testing conditions should be taken into consideration. This will be further discussed in Section 2.4.

Providing a high productivity is another important and yet rarely reported characteristic for an efficient OCM catalyst.

Despite all conducted researches, developing an efficient and stable OCM catalyst remains to be the most challenging and in fact the most crucial hindrance towards commercialization of the OCM process. Even still, there is no general agreement with regard to the contribution of the support material and active components as well as the role and nature of the active sites on the performance of OCM catalysts.

This is mainly because of the fact that under such high temperature (Myrach et al. 2010), the dynamic transformation of structure and active phases all across the catalyst body and the interaction of active components of the catalysts with oxygen and methyl radicals is very fast and volatile and therefore, experimentally following the activation-reaction chain is currently very difficult if not impossible.

There is, however, a general agreement with regard to the desired characteristics of an efficient OCM catalyst, namely being stable and selective, within an engineering context of designing an efficient catalytic structure, which is the focus of this chapter. The parameters and phenomena impacting the selectivity and stability of the catalysts, for instance, the migration of active components, irreversible structural and phase transformation, and so on should be addressed in this context. In order to do so, here in this chapter, the OCM catalysts reports are categorized and discussed based on their focus on investigating the contribution of metal active components, support, dopants and promotors, and their interactions on the selectivity and stability of OCM catalysts.

5.2.1 Selecting Active Components for OCM Catalysts

Wide range of metal oxides have been suggested and tested as active components of OCM catalysts (Keller and Bhasin 1982; Bhasin 1988; Bhasin and Campbell 1995; Cantrell et al. 2003; Langfeld et al. 2012). Depending on the targeted operating strategy, namely co-feeding, oxygen-dosing, and sequential alternating feeding of methane and oxygen, the activity and reducibility of the metal component becomes a decisive factor in prescreening the alternative catalytic materials. For instance, in a chemical looping cycle, the time needed for completion of the reduction step of some materials (e.g. Mn and Co) in comparison to their oxidation step is relatively very short. This adversely affects the effective operation and productivity of the catalyst, which are important engineering aspects in designing the OCM reactor.

It is believed that the catalyst characteristics in terms of its relative acidity-basicity strength can be so tuned that the right oxide surface reducibility can be secured. As a result, methane activation rate on the surface can be optimized to secure the highest C_2-yield (Gambo et al. 2018). This will be explained briefly in section 2.4.

Rare earth oxides such as Sm, Nd, Tb, and Ce, on the other hand, perform well in the co-feeding mode (Cantrell et al. 2003) because the reducibility is not a decisive factor. The same is true for the catalyst chosen to operate under oxygen-dosing strategy, for instance, in OCM membrane reactors. However, extra factors should be taken into account such as the potential of catalyst operating under high methane-to-oxygen ratio, its resistance to coke formation and its possible interactions with the membrane materials. For instance, migration of Na and W in the Mn-Na$_2$WO$_4$/SiO$_2$ catalyst in general and specifically in membrane reactor is a serious issue (Vamvakeros et al. 2015). With lesser consequences, such migration and accumulation of these components on the reactor wall or on the reactor outlet has been also often observed in the fixed-bed reactor operation.

Beside the impacts of feeding strategy, the selection of catalytic components and analyzing their performances with regard to the presence of oxygen, water and carbon dioxide as oxidative agents have been also discussed in literature (Wolf 1991; Cai and

Hang 2019). Specifically the Perovskite family of OCM catalysts are usually prone to be poisoned and reduce their activity showing instability due to the interaction of CO_2, which is the side product of the OCM reaction or used in higher concentration as diluent.

It has been generalized that the parameters which intensify generating carbonate will improve the OCM catalysts performance (Maitra et al. 1992; Schmack et al. 2019). However, it should be clarified which type of carbonate represents such a desired characteristic and performance (Matras et al. 2018), as shown in Section 2.5.

Carbonate materials in molten form have been recommended and implemented as catalyst for OCM reaction and their contributions with regard to improving the C_2-yield have been investigated in detail elsewhere (Tashjian et al. 1994). Generating carbide or other stable phases, which prevents stabilizing the ionic transfer cycle, will adversely affect the performance of molten carbonate catalytic materials. Enhancement of metal-support interactions partially by introducing the promotors have been also investigated in this context. General analysis of the contribution of promotors and dopants are provided in subsection 2.2.

5.2.2 Selecting Dopants and Promotors for OCM Catalysts

In this section, general understanding of contribution of promotors, co-promotors and additives on the OCM catalytic performance is reviewed. Properly selected promotors in interaction with the metal oxide support contribute in improving the selectivity and/or activity of OCM catalysts, which can be analyzed mainly via following the transformation of chemical and physical characteristics of the support. Therefore, discussing the contribution of dopants and promotors should also cover the contribution of the support simultaneously. The contribution of support and its physical characteristics, partially affected by the dopants and promotors will be discussed in Section 2.3. Here in this section, the focus will be on the material characteristics of the dopants and their role in directly affecting the OCM catalytic activity or in affecting the chemical and characteristics of the catalyst structure.

Promotors stabilize the position of active metal on the catalysts structure, prevents their migration and improve the efficiency of electron or proton circulation between the support body and the active promotors. For instance, promoting the Lanthanum oxide catalyst with strontium (Sr) to be incorporated in the catalyst bulk structure, leads to strengthen the role of highly active oxygen-radical sites and their basicity. In that manner, the number of weakly adsorbed oxygen species and thereby their undesired reactions with methyl radicals and hydrocarbons will be attenuated (Cong et al. 2017). Similarly the effect of other promotors on the performance Mn-Na_2WO_4/SiO_2 catalyst and its mechanism as another practically relevant OCM catalyst has been investigated (Arndt et al. 2012; Kiani et al. 2019). In fact, strontium promoted sodium tungstate Sr has a promising potential not only in terms of stability (Shurdumov et al. 2006), but also with regard to its electrical characteristics, which can positively affect the activation cycle and the system selectivity.

The majority of the elements tested under representative OCM conditions that provide more than a 10% C_2^+ yield have been reported in few scattered references and majority of them have been listed in earlier reports such as the ones reported by Bhasin and Campbell (1995). As another example, the promoting effect of chlorine on the selectivity to ethylene, not necessarily on the long-term stability of the catalyst, in a solid form or as a gaseous specie, has been reported (Otsuka et al. 1985; Abbas et al. 2005; Hiyoshi and Ikeda 2015). The observed improvements are widely believed to be due to the role of these promotors in initiating the desired gas-phase dehydrogenation/oxidehydrogenation reactions. Other dopants such as Ba, Na, Mn, W, etc. have been also tested. Even a combination of them can be used in a single catalysts for instance as is the case for well-studied $Mn-Na_2WO_4$ catalyst. This catalyst and impact of each of those dopants on its selectivity and activity will be discussed in detail in Section 2.4.

5.2.3 IMPACTS OF SUPPORT ON THE PERFORMANCE OF OCM CATALYSTS

The type and structural characteristics of support materials can have a significant impact on the selectivity and stability of catalyst. The interaction of various supports materials with promotors and dopants is the key aspect in analyzing their impacts on the chemical and physical-structural of the catalysts. In order to emphasize on the core role of support in such analysis, it should be highlighted that the supports with different structural characteristics, made from the same material and impregnated with same active components can show markedly different behavior (Yildiz et al. 2016). It is assumed that structural characteristics of the support affects the chance of distribution of active components. Better distribution of the active components will facilitates a better distribution of defects across the catalyst. Since these defects are believed to be associated with the distribution and local intensity of active oxygen specie and the resulted methane activation intensity, better distribution of defects can be correlated to the better distribution of methyl radicals generation which controls the rate of methyl coupling and thereby the catalyst selectivity. This was discussed in section 2.2, where the impact of promotors was discussed. On the other hand, better distribution of the active sites provides a relatively more homogenous distribution of the generated reaction heat and thereby the distribution of the adsorbed/desorbed species on the surface and ultimately the selectivity are improved.

It should be emphasized that this is a simplified picture of the functionality of catalytic structure for an engineering analysis and the actual meso-micro scale phenomona are much more complicated. However, the capability of this simplified analysis approach in explaining the observed behaviors of OCM catalytic systems have been demonstrated by several researchers. For instance, it has been concluded that better distribution of active components on the sol-gel made $Mn-Na_2WO_4/SiO_2$ catalyst in comparison to the amorphous silica wet-impregnated with the same composition (Godini 2014) is one of the reasons for its improved performance in terms of selectivity. Similarly, even for the catalysts synthesized with the same synthesis procedure, namely wet-impregnation method, SBA-15 silica supported catalysts have shown a better selectivity than amorphous support to distribute the active components and thereby better selectivity (Yildiz et al. 2016).

These can be generalized for other types of catalyst materials: for instance, the improving effect of BaF_2 was ascribed to the creation of structural defects, which were considered to be essential for generating active oxygen species (Au et al. 1998). In fact, not only the recipe and synthesis method, but also the thermal treatment conditions for instance the calcination temperature and duration can also impact the OCM catalyst performance.

Fundamental physical characteristics of the catalysts such as its conductivity, which can be affected as an interaction of the support, promotors and dopants, can have a significant impacts on its catalytic selectivity (Machida and Enyo 1987).

Designing an efficient OCM catalytic structure requires considering the interactive effects of concentration and distribution of the active components as well as the fundamental characteristics of the material and its pretreatment conditions. These along with the mass- and more importantly the heat transfer characteristics of the synthesized structure should be taken into consideration in any attempt for designing a large-scale OCM catalytic structure.

A major part of any analysis method is to utilize the already proven understanding of the correlations and impacts and to use that knowledge in screening-analysis. For instance, such basic knowledge for $Mn-Na_2WO_4/SiO_2$ catalyst will be provided in the following section as a case study to review the above discussed aspects for this benchmark catalyst.

5.2.4 INTERACTIVE EFFECTS ON THE PERFORMANCE OF OCM CATALYSTS

Having considered the individual impact of each of above mentioned aspects does not provide the whole perspective to analyze the performance of the catalytic structure and ultimately improving its design and accordingly its performance. Therefore, there is a need for a comprehensive systematic analysis to do so. Some recent researches have started to do so. For instance, a meta-analysis based on the reported data in literature has been developed to correlate the property-performance of the investigated catalysts with their OCM reaction performance (Schmack et al. 2019). This includes the composition of the catalysts in terms of type and amount of active components and support as well as tested operating conditions. By analyzing the data reported so far, relatively general conclusions could be made. For instance, it has been indicated that well-performing OCM catalysts provide thermodynamically stable carbonate and thermally stable oxide support (Schmack et al. 2019). This can be considered as starting point for systematic analyzing the OCM catalysts. In Section 4, these will be revisited considering the interactive impacts of reactions conditions and the catalysts characteristics and performance.

From an engineering viewpoint, coating the active components in interior of the support pellet with appropriate size, can address the concerns regarding reduction of its activity due to abrasion and the need for tailoring the local intensity of reaction heat, which both contribute to the stability of catalysts. Preventing the structural deformation and maintaining the surface (Cantrell et al. 2003) can also contribute on reducing the pressure drop along the bed initially and hopefully over 1–2 years of expected catalyst life.

5.2.5 EFFICIENT CHARACTERIZATIONS STRATEGY FOR SYSTEMATIC ANALYSIS OF THE OCM CATALYST PERFORMANCE

Wide range of methods have been applied to characterize the material, physical, structural and chemical specifications of OCM catalysts, among them some have been applied under or close to OCM reaction conditions. The main purpose is to understand what the features of the efficient OCM catalysts are and how they contribute in improving the catalysts' stability and selectivity to be ultimately considered in design of an efficient OCM catalyst. Depending on the research strategy, some researchers for instance have focused on analyzing the contribution of the surface chemical characteristics on the activity and selectivity of the OCM catalysts. In such studies, attempts were made to correlate the surface chemical characteristics of the OCM catalysts, especially the measured quality and quantity of basicity strengths, with the observed rate of methane conversion and C_2+ selectivity (in form of C_2+ products formation rate) (Sokolovskii and Mamedov 1992).

Others have focused on the impact of operating conditions including the temperature, methane-to-oxygen ratio etc. on the catalyst performance. Some others have tried to correlate these impacts for instance by tracking down the impacts of operating parameters on the surface physical-chemical characteristics. By applying in situ characterizations such as XRD-CT (Computed Tomography), the impacts of structural parameters on the catalyst activity can by dynamically tracked not only on the surface but also across the catalytic structure. The impacts of different reaction atmospheres, as a result of implementing different sets of operating parameters, on the transformation of the phases and thereby on the observed catalytic performance during the reaction can also be analyzed in this way. This has been done for La_2O_3/CaO catalyst by analysing dynamic transformation of its active components and phases while changing the operating temperature, methane-to-oxygen ratio, and so on, as shown in Figure 5.1 (Matras et al. 2018).

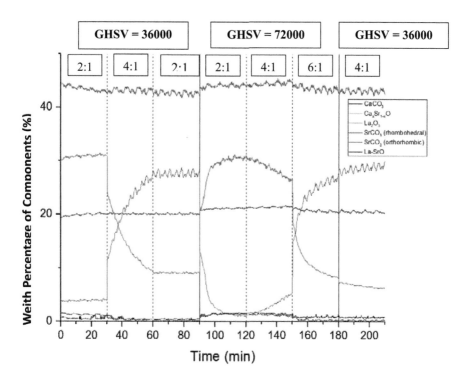

FIGURE 5.1   Tracking the dynamic behavior of La_2O_3/CaO catalyst under various reaction atmospheres.

Using this method or similar techniques, the impacts of single phase or co-presence of multiphases and dynamic transformation of them across the catalyst body on catalytic selectivity and activity can be studied. As seen in Figure 5.1, the intensity of carbonate formation ($SrCO_3$ rhombohedral) dynamically follows the thermally driven pattern of selective conversion of methane, which can be recognized by following the corresponding shifting of the methane-to-oxygen ratios. Such trend has been also predicted and confirmed even using conventional experimentations (Schmack et al. 2019).

In addition, such characterization methods can be combined with the more conventional characterization techniques to provide a comprehensive analysis. Even comparative analysis of the results of characterizations over the same material synthesized via different methods can also provide valuable information. For instance, distinguished features of Mn-Na_2WO_4/SiO_2 catalysts synthesized by sol-gel and wet-impregnation (Godini 2014) can be analyzed by considering the above mentioned approaches and prospects. The observed results should be interpreted by considering the intrinsic characterization of the catalysts, for instance the relative homogenous distribution of promotors and active components across the sol-gel catalyst in this case.

5.2.5.1 Comprehensive Analysis of an Mn-Na_2WO_4/SiO_2 Catalyst

In this subsection, the aspects discussed earlier will be revisited in analysing the characteristics and performance of Mn-Na_2WO_4/SiO_2 catalyst. The investigation of this catalyst and the efforts to improve its performance as one of the most investigated OCM catalyst have been continued since 1992 (Fang et al. 1992a, b;Wang et al. 1995; Ji et al. 2002, 2003; Dedov et al. 2011; Arndt et al. 2012; Lee et al. 2013; Sadjadi et al. 2015a).

The mechanistic aspects of the OCM reactions over the surface of this catalyst and the role of active sites (Ji et al. 1998, 1999a, 2000; Kwapień 2011), as well as the contributions of the gas-phase reactions and their interactions with the surface reactions in this catalyst (Korup et al. 2011; Takanabe and Iglesia 2009) have been discussed. The relative intensity of the gas-phase and surface reactions and their dependency on the reaction temperature and activity of the catalyst have been also investigated (to be discussed in detail in Section 4).

In the context of phase analysis, it has been explained that tungstate stabilizes sodium on the catalyst surface and sodium (Na^+) facilitates the transformation of amorphous silica to α-cristobalite, which is one of the few crystalline phases remaining on the surface under high temperature OCM reaction (Pak and Lunsford 1998; Pak et al. 1998; Palermo et al. 1998). Therefore, it is believed that α-cristobalite plays a central role facilitating forming and maintaining the active sites under OCM reaction. By following the dynamic transformation of α-cristobalite to quartz and Tridymite from one side and MnO_2, Na_2WO_4, WO_4, and Mn_2O_3 from the other side, the potential of selective conversion under the investigated range of operating conditions can be tracked qualitatively (Ji et al. 2003; Simon et al. 2011; Sadjadi et al. 2015a). It should be highlighted though that the targeted manganese oxide phases cannot be detected under the reduced state.

The role of tungsten oxides in its V and VI states and manganese oxides II, III and IV phases, Na–O–Mn and Na–O–W species, in connection with the quantity and quality of the oxygen lattice have been investigated under different oxidized and reduced atmospheres (Wang et al. 1995; Kou et al. 1998; Takanabe and Iglesia 2009; Ji et al. 2002, 2003; Simon et al. 2011; Dedov et al. 2011; Sadjadi et al. 2015a).

Different observed behavior of the sol-gel, amorphous silica supported and SBA-15 supported catalysts can be associated with distribution of the active components on these catalysts as well as their surface and structural (including the thermal and electron conductivity) characteristics (Pak and Lunsford 1998; Godini 2014; Yildiz et al. 2016). For instance, the intensity of the reaction across the catalyst can be tuned either by homogenous distribution of promotors or by changing the quantity of the active components in the catalyst recipe. Therefore, no specific catalyst composition, including the recommended catalyst composition 1.9–2% $Mn-5\%Na_2WO_4/SiO_2$ (Wang et al. 1995; Ji et al. 1999b; Wang et al. 2006), necessarily can represent the best catalyst composition synthesized by different methods in different scales ad under wide range of reaction conditions. Even the sequence of adding the promotors can impact its structural characteristics and its catalytic performance. For instance, it has been observed that micropores are formed when W and Na are implemented simultaneously, which can lead to different catalytic performances. The application of simple BET analysis is limited to prove the formation of these micropores. Nevertheless, consequence of adding the precursors is overwhelming and has been also observed when this catalyst is synthesized by different methods via incipient wetness impregnation (Wang et al. 1995) or sol-gel method (Godini 2014). It has been shown that not only better distribution of active components on the catalysts surface, but also across its bulk structure, can improve its selective performance. This can be measured and checked even using conventional Inductively Coupled Plasma (ICP) technique (Godini 2014) or operando XRD-CT.

5.2.5.2 General Conclusions Based on Catalyst Material-Chemical-Structural Characteristics

Distribution of the active components ultimately has its impact through creating defects which ultimately contribute in the dynamic regeneration cycle of active sites and thereby affects the selective activation of species. The interaction of OCM catalyst surface via its defects and active sites with the adsorbed species, is the function of reaction conditions and the chemical-structural characteristics of each catalyst. These are reflected in oxygen scrambling activity of the material which ultimately affects catalytic activity (Thum et al. 2019). Analysing the proton and electron conductivity of the catalyst structure, the activation mechanism and the rate of electron exchange between the catalyst surface and the adsorbed species under various reaction conditions can provide a valuable information for analysis and ultimately design of efficient OCM catalysts. It can be generally concluded that a well-tailored rate of methane activation (Gambo et al. 2018), which can be established by tuning the concentration of the oxygen species with the desired adsorption strength on the catalyst surface. This can be achieved by proper designing the reactor and catalyst properties leading to secure a selective performance. This is closely related to the location, quantity and the functioning roles of the involved oxide phases over the catalyst body as well as the macro dimensions of the catalytic structure. Being able to stabilize a proper temperature profile along the catalytic bed under the non-atmospheric pressure are also crucial.

Nevertheless, more investigations are needed to be able to link the observed characteristics of the catalysts, such as the distribution of surface metal components or the catalyst conductivity, to the observed performance of OCM catalysts in general.

5.2.6   OCM KINETIC ANALYSIS

Reaction mechanism study and kinetic analysis of the involved surface and gas-phase reactions in OCM system is an important part of the research and engineering activities ultimately targeting improving the OCM reactor performance. Specially, the contribution of the gas-phase reactions and its interaction with the surface reactions need to be well analyzed, understood and addressed in a proper reactor design. For instance, the results of such analysis can be utilized for (a) better designing of the gas-phase reaction chamber at the end of reactor where ethane can be efficiently converted to ethylene (Siluria reactor concept) and (b) proper utilizing a gas diluent either as excess methane or inert diluent such as carbon dioxide or steam. Similarly, the contribution of the design aspects associated to the external and internal mass and heat transfer characteristics of the catalytic bed and the reactor, such as the oxygen permeation and dimensions of the catalytic beds, should be addressed in a proper modeling (Jašo et al. 2011; Godini et al. 2013a, 2014a, b). In all of these analyses and designs, the thermal-reaction interactive aspects of this complicated gas-phase-catalytic system should be addressed.

The prediction potential of the available kinetic-models needs to be re-evaluated carefully because they either fail to (a) precisely predict the selectivity and yield of the products, (b) the trends with regard to the contribution of the gas-phase reactions and the impact of operating pressure, and c) to explain how reactor performance indicators are affected by varying operating parameters such as reaction temperature. Therefore, they should still be improved to reflect the actual OCM reaction mechanism (as discussed in previous sections) and the real weights and contributions of the gas-phase and surface reactions, especially under oxygen-dosing and feed alternating policy, where the average partial pressure of oxygen is varying.

In all cases, the interactive effects of operating parameters should also be analyzed, as will be discussed in Section 4.

5.3   REACTOR RESEARCH

Wide range of reactor concepts have been proposed and examined for Oxidative Coupling of Methane (OCM) process (Cruellas Labella et al. 2017). The OCM reactors can be categorized in three main groups as a) co-feeding of oxygen and methane (e.g. fixed-bed and fluidized-bed reactors), b) Dosing feeding of oxygen (e.g. membrane and stage-wise dosing reactors), c) alternating or periodic oxygen addition (e.g. chemical looping reactor), based on feeding policy that enhances selectivity and/or yield.

The status under which the gas and solid catalysts are coming into contact is also a decisive aspect not only affecting the reaction performance but also the thermal performance of the OCM reactor. For instance, fluidized-bed and conveying bed reactors, in which the catalyst particles are moving, provide close to isothermal thermal performance and controlling the temperature is relatively easier than for instance in a fixed-bed OCM reactor.

Followings are the main aspects in selection and design of an efficient OCM reactor and securing its stable, safe and robust operation, especially for industrial-scale operation:

1- Utilizing a stable and selective OCM catalytic structure
2- Heat management and proper thermal engineering
3- Securing a safe operation by avoiding an explosive atmosphere, particularly in the long inlet piping and any preheating zones and considering appropriate safeguards depending on the type of OCM reactor

Feed composition, levels of methane-to-oxygen ratio and dilution, temperature, pressure and gas hourly space velocity GHSV, which represents the relative dimensions of the catalytic bed and the feed flow, are the main operating and design parameters in an OCM reactor. For instance, targeting a higher methane conversion and ethylene yield using the feed with low methane-to-oxygen or more than 10% oxygen fraction, not only is not in line with a safe operation of OCM cofeed reactor as have been experimentally demonstrated (Khakpour et al. 2011), but also is not recommended due to poor expected selectivity.

Here in this chapter the impacts of these parameters along with other design and operating characteristics of OCM reactor, in a comparative analysis of fixed-bed, fluidized-bed and membrane reactor will be reviewed.

Moreover, the potentials of reactor integration in an integrated OCM process are analyzed to be further discussed in the context of integrated process in section 6.

5.3.1 FIXED-BED REACTOR

A fixed-bed reactor is a standard reactor concept for industrial applications in general. However, for an OCM application, special design features should be considered to address the high temperature and highly exothermic nature of this reaction, establishing an efficient preheating, proper temperature profile and efficient cooling at the end of reactor. The undesired contribution of the gas-phase reactions should be also suppressed as much as possible in a proper reactor design. Following is a set of practical measures to address these requirements in the design of a fixed-bed OCM reactor:

1- The activity of the catalytic bed and the intensity of reaction along the reactor should be tuned using inert component. This can be established by mixing the OCM catalyst pellets in proper size and shapes with inert pellets or by using an inert component as a binder for instance to construct a 3D catalytic structure or using a special type of support material with the desired thermal and reactive characteristics (Schammel et al. 2013). In this manner, even the issues of pressure drop and heat management along the catalytic bed can be better addressed. In fact, avoiding using a binder and instead directly coating the active components on the macro structure support pellet such as α-alumina have shown a promising performances demonstrated by Cantrell et al., for instance, in examples 47 and 48, columns 51–52 (Cantrell et al. 2003). This enables maintaining the activity of catalyst since the structure and surface area do not change. A pressure drop along the bed as well as the thermal management of the catalytic bed can be better handled.

2- The preheating section is designed to be right in the entrance of the catalytic bed where the methane-rich and oxygen-rich stream are mixed and heated up to the desired initiating OCM reactor temperature.

3- At the end of reactor, the contribution of gas-phase reaction that is primarily for further converting ethane to ethylene (Schammel et al. 2013) should be properly utilized right before the cooling section.

When a normal-sized catalyst pellet is used for OCM, a multistage multitubular structure seems to be a practical choice.

5.3.2 MEMBRANE REACTOR

In an OCM membrane reactor, usually an inorganic membrane is used to distribute oxygen along the catalytic bed or catalytic layer, where it reacts with methane to selectively produce ethylene mainly due to higher local methane-to-oxygen ratio and better distribution of reaction intensity and heat. Different membrane materials including ceramic, glass or metal membrane as dense or porous membranes can be used for such reactor design. Wide range of membranes representing different permeation mechanisms have been suggested and evaluated for such application including dense perm-selective (Czuprat et al. 2010; Wei et al. 2013) or proton exchange membranes (Langguth et al. 1997). In case of using dense oxygen-perm-selective membrane, air can be directly used as a source of oxygen and in this manner the need for using pure oxygen and an air separation unit to supply it can be avoided. As it will be discussed in details in Section 5, this cannot be the case for porous membrane which should use an oxygen-rich feed. However, in the OCM porous membrane reactors usually the catalytic bed can operate easily under practically relevant higher operating pressure of up to few bar (Godini et al. 2014c, d). However, the combustion of methane inside the porous membrane should be avoided, specially where it can meet significant amount of oxygen. Therefore, in most of the experimental research activities dealing with OCM porous membrane reactors, the characteristics of the initial commercial porous membrane have been adjusted to meet the requirements of this application, particularly being catalytically inactive and providing the desired level of permeation (Lafarga et al. 1994; Coronas et al. 1994a, b; Ramachandra et al. 1996; Lu et al. 2000; Godini et al. 2013a, 2014c). Usually these membrane reactors have been tested in dead-end structure feeding, to prevent back permeation of hydrocarbons into the shell side oxygen chamber, which is a severe safety issue for this type of reactors.

Metal membranes are to be preferred in commercial operation especially considering the targeted high pressure operation and enabling a relatively easier sealing technology.

Establishing a proper amount and profile of oxygen permeation along the catalytic bed is the main criteria for both the dense and porous membranes. There has been many efforts also for developing dense oxygen-perm-selective membrane for this application (Lu et al. 2000; Tan et al. 2007; Bhatia et al. 2009) especially with regard to presence of carbon dioxide which affect the performance of most of perovskite membrane materials. There are some other specific design aspects associated to the operation of such membrane including the compatibility of the catalyst with the membrane material, interaction of the sealing with the membrane, mechanical stability and proper thermal characteristics of the membrane for instance its thermal expansion, etc.

Similar general design concerns and aspects mentioned of the fixed-bed reactor are also relevant for the design of packed-bed membrane reactor. When a catalytic membrane reactor is synthesized by coating a catalytic layer over the membrane (Bhatia et al. 2009), the design concerns regarding its thermal performance is less, but securing the targeted amount of methane conversion and ethylene yield is the main design challenge.

For more detailed information about the operation and design of an OCM packed-bed membrane reactor, it is refereed to other references (Godini et al. 2018).

All in all, membrane reactor is a promising concept for OCM mainly because of two reasons. First, higher methane conversion can be theoretically secured without a need for surpassing a safety limitation regarding the local partial pressure of oxygen, which is severely limited in high pressure cofeeding reactors. Second, oxygen-dosing not only improves the C_2-selectivity because of increasing the local methane-to-oxygen ratio, but also by providing a more potential to control the thermal performance of the catalytic bed. Still some practically relevant concerns regarding the long-term operation of membrane reactor in general and oxygen selective dense membrane reactors in specific have to be addressed before being tested for 1–2 years period prior to a large industrial-scale operation for oxidative processes and in specific for OCM process. For instance, efficient sealing, stability and preventing local leakage of such membranes should be secured.

5.3.3 Fluidized-Bed Reactor

A fluidized-bed reactor has been suggested and investigated as promising OCM reactor primarily because of its thermal performance, which can be controlled to be close to isothermal performance trusting its beneficial impacts. This will certainly reduce the risks associated to the temperature control and runaway temperature, which is one of the main concerns in packed-bed reactors. However, the operation in fluidized-bed reactor causes special safety concerns mainly at the end of reactor, where ethylene, ethane and remaining oxygen due to slug flow regime can form an explosive atmosphere. Conceptually it should be highlighted that establishing an isothermal performance is not necessarily in favor of efficient OCM reactor operation targeting highest ethylene selectivity and yield. In fact, the desired temperature profile along an OCM reactor has shown to be not isothermal for instance as has been reported for OCM packed-bed membrane reactor (Godini et al. 2013a, 2014b, c). This in fact is one of the motivations to use a membrane for dosing oxygen along the fluidized-bed reactor to establish the desired temperature profile in a fluidized-bed membrane reactor. However, improving the selective performance of the OCM reactor due to known favorable effect of distributing oxygen along the bed, is the main motivation for such design (Jašo et al. 2011; Sadjadi et al. 2015b). The positive impact of non-isothermal performance on the OCM reaction performance can be seen even by following the experimentally observed performance of a fluidized-bed reactor while varying the feed flow rate. In this case, the best fluidized-bed reactor performance in terms of selectivity and ethylene yield have been observed for the low feed flow rate in the range close to establishing the minimum fluidization (Jašo et al. 2012; Sadjadi et al. 2015a, b). This is in fact a general phenomenon in other types of OCM reactors, which can be explained based on the interaction of the effects of operating and reactor parameters as will be discussed in detail in Section 4.

5.3.4 OTHER TYPES OF OCM REACTORS AND GENERAL DESIGN AND CONTROL ASPECTS

The performance of other types of OCM reactors have been also reviewed. Again, in addition to the general aspects valid in all types of OCM reactors, there are some specific aspects associated to each of those specific OCM reactor types. For instance, in chemical looping OCM reactor, as discussed in Section 2, the type of catalyst, its reducibility and oxidative potentials as well as the temperature control during the dynamic cycling are important aspects to be taken into analysis. More information about this reactor concept for OCM application can be found elsewhere (Parishan et al. 2018; Fleischer et al. 2016) to be reviewed in the context of discussed issues in this chapter book.

In general, other than fulfilling the kinetic requirements in term of establishing a right contact time and local gas composition, maintaining a suitable reaction intensity and temperature profile along the reactor are other key design factors for ensuring an efficient operation in an industrial-scale OCM reactor.

There are also some specific design characteristics to be considered for the lab-scale experimental reactors for OCM application. For instance, it is preferred to ensure:

1- Measuring temperature also inside the catalytic packed-bed
2- Establishing a thin catalytic bed, but still several times thicker than the diameter of single catalyst pellet for avoiding gas channeling and meeting the criteria to maintain adequate Aspect Ratio (Alonso et al. 2006).
3- Easy replacement of the catalysts and reactor parts using proper sealing and connections.

In order to develop and exploit a proper strategy for controlling the operating temperature inside the reactor, it should be highlighted that the reaction temperature inside the catalytic bed is affected by two factors, namely (1) the reaction intensity and generated heat of reaction, and (2) the value of the set temperature in the surrounding electrical furnace (TW) (Godini et al. 2014b, c; Salehi et al. 2016a, b) for the lab-scale reactors (the heating-duty of the furnace in the industrial-scale reactor). The second factor also affects the first factor and the reaction intensity inside the catalytic packed-bed. Therefore, the external set temperature in the surrounding environment or the heating duty of the furnace as well as the internally generated heat by reaction should be controlled interactively. Some of the reported reactor performances in literature should be reevaluated considering the possible impacts of the implemented temperature control strategy there. For instance, some reported catalytic performances are mainly affected by the thermal management (heating/cooling in an electrical heater) rather the intrinsic behavior of the catalyst.

Combination of employing an efficient preheating policy and distribution of reaction intensity along the reactor, which can be materialized by tailoring the activity of catalytic bed or distributing oxygen for instance via membrane along the reactor, can significantly enhance the reactor performance.

Infra-Red (IR) gas analyzer, Gas Chromatograph (GC) or Mass Spectroscope (MS) can measure the concentration of gaseous components in the reactor outlet gas stream. Measuring its flowrate and the concentration of inert component fed in a known molar flow enable precisely calculating the carbon balance, conversion, yield and selectivity.

5.4 INTERACTION OF THE CATALYST AND REACTOR RESEARCH

As discussed in the previous section, the interactive impacts of operating parameters and the reactor design characteristics affect the ultimate performance of OCM reactor. This may or may not be in line with the widely perceived known behavior of OCM catalysts. Choosing a proper catalyst, amount of catalysts, dimension and physical characteristics of the catalytic bed, reactor feeding policy, set of operating conditions and test running-procedure can significantly affect the OCM reactor performance. For instance, depending on the given set-values of operating parameters in OCM reactor, each parameter may affect the thermal performance of this system differently, which itself can result in a different reaction performance. Therefore, there are two main aspects associated to the ultimate performance of an OCM reactor, namely, thermal aspect which mainly reflects the reactor characteristics and the reaction aspects that reflects both catalyst and reactor characteristics. Accordingly, in the catalytic-characteristics dominated range of varying the operating parameters, the observed selectivity and yield are primarily concluded based on the intrinsic characters of the catalyst and its surface reaction mechanism. In other ranges, the dimension and characteristics of the reactor system such as its heat removal capacity also significantly affect the reaction performance. Therefore, in these ranges, the OCM performance of the reactors with different structural design and dimensional characteristics can be different even if their operating conditions are similar. These should be analyzed through a systematic thermal-reaction engineering (Godini et al. 2014b) in order to secure an efficient OCM reactor performance.

These aspects are reviewed in this section in order to provide a framework for analyzing the experimental data, framing the conclusions made accordingly, and using the resulted knowledge for an efficient reactor design.

5.4.1 OPERATING PARAMETERS

In the context of engineering analysis of effect of operating parameters, it should be highlighted that the range of these parameters should be chosen not only with regard to their direct impacts on the catalysts and reactor performance, but also on the whole OCM process performance. For instance, the operating pressure for industrial application of OCM cannot be atmospheric as reported in most of the lab-scale OCM reports so far. In fact, running the OCM reactor under low pressure imposes a significant cost of compression needed to pressurize the OCM reactor outlet gas, containing problematic components such as carbon dioxide, to make it ready to be treated in downstream units. On the other hand, the impact of elevated pressure on the performance of OCM reactor should be also taken into consideration (Albrecht et al. 2018) and be reevaluated in the general analysis concepts presented in this chapter.

5.4.1.1 Critical Importance of Pressure

Unfortunately, only few studies have investigated and considered the impact of operating pressure on the performance of the OCM catalysts or the whole OCM process. Most of the reported catalysts tests at elevated pressure has shown a reduction in C2-selectivity by increasing the pressure (Pinabiau-Charlier et al. 1991; Chou et al. 2004; Nipan et al. 2014). It was highlighted in section 2 that except for few reported catalysts, the so far reported OCM catalysts either have been tested only at atmospheric pressure or have shown a significant reduction while being tested at higher pressures. Operating pressure will impact the OCM reactor through the interaction of intensifying the gas-phase reactions and the activation rate of surface oxygen and thereby the radical generation.

From a practical view on the process scale operation, the OCM reactor should be operated at higher than atmospheric pressure so that the gas flow through the reactor and into the downstream units, is secured without a need for costly compression and handling the potentially corrosive gas species. For instance, 10 bar operating pressure in the OCM reactor can be synchronized with the required operation of CO_2 removal (Godini et al. 2019) or product adsorption units (García et al. 2019). Moreover, operating at such range of pressure helps to reduce the size of reactor and overall capital expenditures and thereby can improve the whole process economics.

5.4.1.2 Methane-to-Oxygen Ratio

The methane-to-oxygen ratio is a very important operating parameter with a significant impact on the OCM catalyst, reactor as well as the whole process performances. Higher methane-to-oxygen ratio usually will results in higher catalyst selectivity mainly because of the reduced partial pressure of oxygen in the reaction environment. In the reactor-scale however, a higher methane-to-oxygen ratio means less available stoichiometric oxygen for converting methane which consequently leads to lower methane conversion and ethylene yield. In fact, due to full conversion of oxygen, if the C_2-selectivity becomes less than 80% the obtaining methane conversion of more than 50% becomes impossible even for methane-to-oxygen ratio 2. These typical observations have been reported for most of the OCM reactor concepts.

On the other hand, working with higher methane-to-oxygen ratios decreases the probability of forming an explosive atmosphere and increases the chance for controlling the operating temperature. This however, can results in poor economic performance of the whole OCM process because separating the resulted huge amount of unreacted methane in that case significantly affects the performance and the operating cost of the downstream units. On the other hand, obtaining highest possible C_2-selectivity while securing enough methane conversion should be targeted in design of the OCM reactor. This is reflected in the whole economy of the OCM process. In fact, if the unreacted methane is recycled back to the OCM reactor, even using very high methane-to-oxygen ratios to secure the C_2-selectivity of more than 80% can be justified (Matherne and Culp 1992). Using higher methane-to-oxygen ratio, which stoichiometric means using excess methane, can impact the reactor performance also because of its diluting thermal effect.

5.4.1.3   Reactor Temperature

By conducting the exothermic OCM reactions along the reactor, the operating temperature continually increases until the temperature gradient between the reaction temperature and the temperature in the external surrounding environment becomes enough high to emit the excess generated heat. Approaching the reactor end, methane and oxygen (in reactors with co-feeding structure) are progressively consumed and the operating temperature decreases.

The exact reaction temperature usually cannot be measured simply using a thermocouple located inside the catalytic bed and cannot be directly set fixed as it is affected by the reactor set-temperature as well as the other parameters which affect the reaction intensity inside the reactor. For a given OCM reactor, there is a certain range of set-reactor temperature in which the C_2-selectivity and methane conversion both can be maintained high. On one side, lowering the temperature from that range may cause significantly reducing the methane conversion and thereby the C_2-yield. Since the reaction temperature is also affected in this case, the C_2-selectivity can be reduced too. On the other hand, increasing the reactor set-temperature outside that range will reduce the C_2-selectivity due to intensifying the undesired reactions, mainly the gas-phase combustion reactions.

It should be concluded that most of the experiments performed so far have been performed using electrical heaters in lab-scale using on/off mode operation. However, in the industrial-scale a close to adiabatic operation will be established using a furnace in which by burning the fuel, the required temperature for starting OCM reaction will be provided and yet the generated heat and the furnace duty are in balance to maintain the load of reaction and the required temperature profile along the catalytic bed.

5.4.1.4   Feed Flow or Gas Hourly Space Velocity (GHSV)

The feed flow to be processed with a given size of catalytic bed is usually represented as gas hourly space velocity (GHSV), which is determined based on various requirements. The first requirement is that the reaction contact time should be kept short. Nevertheless, the targeted practically relevant level of methane conversion should be also secured. The reaction atmosphere should be also kept far away from explosive zone, which normally will be more probable for a shorter contact time (Khakpour et al. 2011). The residence time and available surface area should be also enough for controlling the generated reaction heat, establishing the desired temperature profile and in the case of membrane reactor, the oxygen permeation profile.

5.4.1.5 Gas Dilution/Inert

The effects of the type and amount of dilution on the performance of an OCM reaction and, in general, on the performance of catalytic gas-phase exothermic and endothermic reactions have been usually the matter of debate not only on the reactor performance but, to a lesser extent, on the process-scale performance. There is a general understanding, however, that using higher dilution increases the C_2-selectivity mainly because of reducing the partial pressure of the reactants and reducing the undesired contribution of the gas-phase reactions. Not only because of this, but also because of the indirect impact of inert component gas on the thermal performance of the OCM reactor, mostly its impacts on improving the C_2-yield have been reported (Tye et al. 2002; Godini et al. 2013b, 2014c, d). It is therefore recommended that in explaining the observed experimental behaviors, the thermal effect of using inert diluent will be analyzed before jumping to a conclusion regarding its possible direct involvements in desired of undesired reactions. Such thermal impact in OCM reactor should be analyzed in connection to the intensity of generated reaction heat. This can be highlighted, for instance, under close to isothermal operation of the OCM fluidized-bed reactor, where it has been observed that introducing higher amounts of nitrogen dilution does not have a significant impact on the C_2-yield (Jašo et al. 2012). Interpreting the results of other types of OCM reactors in this regard is more complicated.

Choosing the diluent and its quantity depends also on the potentials and requirements of the process downstream units and how much of that diluent can be tolerated in the reactor feed composition with or without considering a recycle stream. It also depends on the source of inert component. For instance, in case of using air as a source of oxygen, setting the methane-to-oxygen ratio automatically sets the level of nitrogen that emerge. Using a dense membrane reactor is also beneficial in that perspective because it selectively separate oxygen from air and prevents the complication associated handling nitrogen in the reactor and in downstream units. Its impacts will be reflected in trade-off cost of separating-supplying oxygen and the operating cost of the downstream units.

Same analysis can be performed for using steam and carbon dioxide as diluent, which are preferred choices considering their relatively easier separation in downstream units or due to their potentials providing more potentials for an efficient reactor and process integration.

5.4.2 THERMAL-REACTION ANALYSIS

In the context of thermal-reaction analysis of the OCM reactor, it should be emphasized that each reactor setup has a specific heat transfer capacity with the surrounding environment. This will affect the thermal and reaction performance of the reactor interactively. For instance, the measured ethylene-to-ethane ratio can be considered not only as an indicator for the reaction performance but also the thermal performance of the reactor as ethane undergoes in an endothermic reaction to produce more ethylene.

Thermal effects are more pronounced in the practically relevant range of an OCM reactor performance, namely C_2-yield above 15% coming along with the C_2-selectivity higher than 50% and highlighted by ethylene concentration higher than ethane.

5.4.2.1 Stable, Transient, and Dynamic Behavior of OCM Reactors

Special attention should be devoted to analyzing the dynamic behavior of OCM reactor. As a part of implementing a proper control strategy, the procedure of heating should be also carefully designed to track the desired dynamic trajectory of this system. This has a crucial impact on the OCM reactor performance, as shown in Figure 5.2.

The results of the performed CFD simulation in Figure 5.2b shows the rates of desired oxidative reactions, namely Methane Coupling (MC) and Oxidative Dehydrogenation Ethane (ODE) as well as the ethylene and ethane fractions along the reactor. These have been reported in the Reference State (RS) and a Transient State (TS). Ethylene distribution along and across the catalytic bed in the membrane reactor, designed to achieve the highest ethylene, has been also presented there.

As is seen in these figures, the ethylene yield in an OCM reactor can be significantly improved by implementing a proper heating policy for instance a periodic temperature control mechanism.

C_2-yield [%]	C_2-selectivity [%]	C_2H_4-yield [%]	C_2H_4/ C_2H_6	ΔT (in reference to average T) [°C]	State
21	53	15.7	3	0	Reference State (RS)
29.5	66	23.6	4	+110	Transition State (TS)

(a)

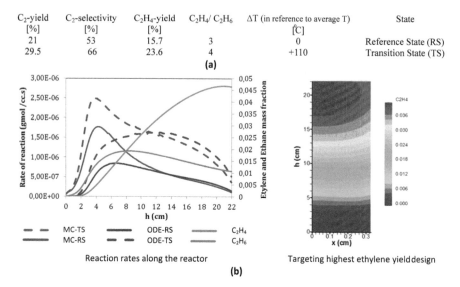

Reaction rates along the reactor | Targeting highest ethylene yield design

(b)

FIGURE 5.2 Tracking the performance of an OCM reactor (a) A dynamic trajectory of temperature and the reaction performance (Godini et al. 2018), (b) CFD simulated behavior in terms of distribution of the component concentrations and reaction rates along the reactor; Reactions: MC (Methane Coupling), ODE (Oxidative Dehydrogenation Ethane) (Salehi et al. 2016a).

5.5 PROCESS AND REACTOR INTEGRATION, ENVIRONMENTAL, AND INDUSTRIAL PROSPECTS

In the context of evaluating the performance of the OCM process, the possibility of treating the unreacted methane and further conversing the undesired CO_2 and less desired ethane, carbon oxide and hydrogen in other reactors such as methane reforming, CO_2 hydrogenation, ethane dehydrogenation reactors have been also taken into analysis (Godini et al. 2013b,c, 2019; Schammel et al. 2013). Some of the current challenges in the OCM process or in the processes integrated to OCM can be turned into potentials if they are properly integrated. The main and interactive impacts of integration will appear in the reactor-scale integration. However, the cost-energy performances of the downstream separation and purification units can be also improved in an efficient integrated OCM process design.

A quick review of the integration options and potentials in OCM process shows that this process shares either the products or the reactants with ethane dehydrogenation, methane reforming and some of the CO_2 hydrogenation reaction systems operating in similar range of operating pressure or temperature. Complementary exothermic or endothermic thermal natures of such integrations or the possibility for sharing the downstream units are other motivations for such integrations (Hoebnik 1995; Godini et al. 2013b, c, 2019; Schammel et al. 2013).

In order to analyze these, the techno-economic performance of the referenced stand-alone OCM process and distinguished integrated process structures in industrial-scale have been examined (Graf and Lefferts 2009; Godini et al. 2013c; Spallina et al. 2017). Selected lab-scale and miniplant-scale experimental data, which have been obtained under practically relevant conditions and for some downstream units even by using the exact industrial-scale data, have been utilized to develop and validate the representing models of the units at an industrial scale of some of these investigated process structures (Godini et al. 2019).

Conceptual analysis and selecting alternative unit operations, for instance using proper adsorption system (García et al. 2019) to effectively separate ethylene instead of demethanizer extends the possibilities and increases the potentials for an efficient process integration (Xuan Nghiem 2014; Godini et al. 2013b, c). It should be highlighted that the complication in the integrated process can also be increased by integrating the operating-challenging units as well as the number of integrated processes.

5.5.1 INTEGRATION POTENTIALS OF OCM REACTORS AND OTHER REACTION SYSTEMS

The aspects to be considered in further converting the remaining methane and generated carbon dioxide and ethane in integrated OCM process will be discussed here.

Further converting ethane in an individual or integrated reactor can be performed using noncatalytic ethane cracking (Vereshchagin et al. 1998; Schammel et al. 2013) or catalytic ethane dehydrogenation using oxygen or carbon dioxide as oxidative agents. Methane reforming can also consists of methane dry reforming, partial oxidation and methane steam reforming to be able to balance the heat management of the reactor as well as the product compositions (Song and Guo 2006).

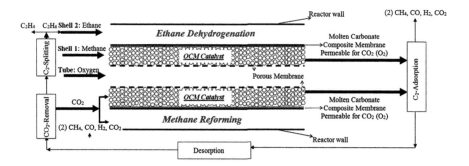

FIGURE 5.3 Schematic of integrated dual-membrae reactor representing its operating concept for OCM and ethane dehydrogenation integration or OCM and methane reforming integration (alternative (2)).

Promising aspects associated with using carbon dioxide as a diluent (Eng et al. 1995; Wang and Zhu 2004; Godini et al. 2014c, d) and the possibility of efficiently separating it under the reaction conditions (Chung et al. 2005; Rui et al. 2001) are the main motivations to integrate OCM using CO_2-diluted reactor with methane (dry) reforming (Godini et al. 2013b, c) or ethane CO_2 dehydrogenation, as shown in Figure 5.3 in the form of integrated dual-membrane reactors.

Such a structure enables efficient heat and mass integration and can result a higher ethylene production. The following criteria can be applied to the design of such an integrated reactor:

1- Assuring the compatibility of materials including the catalyst, membrane and gaseous species: Not only the nature and function of the materials, but also the targeted range of operating conditions should be carefully chosen within this perspective.

2- Assuring the compatibility of the reactor dimension with the feed flowrates in different section and the required contact times for the reaction and separation: It should be confirmed that the required heat transfer and mass transfer (membrane) area are in accordance with the required contact time for the reaction sections (determined by kinetic) and ensure a feasible reactor dimensions.

Plausibility of any reactor design should be carefully checked with the available literature data. For instance, for the reactors conceptually represented in Figure 5.3 (Tashjian et al. 1994; Anderson and Lin 2006, 2010; Rui et al. 2009, 2001; Gätner et al. 2012; Godini et al. 2013b) can be cited for detailed information and possible plausibility check.

5.5.2 Process Integration via Downstream Units

Considering the difficulties associated with ethylene separation and purification in downstream of the OCM process, several reports have suggested further converting ethylene to some complex products such as ethylene oxide and diethyl benzene,

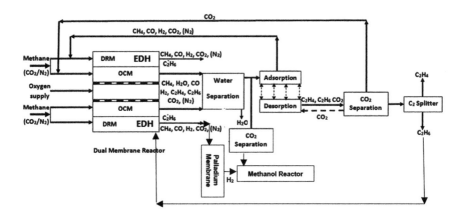

FIGURE 5.4 Integrated process: OCM, methane reforming, ethane dehydrogenation, carbon dioxide hydrogenation.

which in comparison to ethylene can be easier separated from the OCM reactor outlet gas stream (Edwards et al. 1991; Graf and Lefferts 2009). Other integration potentials are also available while targeting ethylene as the main ultimate product.

Using the shared downstream unit, for instance, demethanizer and C_2-splitter in the case of OCM-EDH, or alternative unit operations such as adsorption in case of OCM-DRM can improve the chance of process integration (Hoebink et al. 1995; Choudhary and Mulla 1997; Vereshchagin et al. 1998; Tiemersma 2010; Godini et al. 2013b, c, 2019).

Having considered the potentials in different sections of the process and specific reactor and process integration concepts discussed earlier, a flowsheet of an integrated OCM process partially addressing these potentials is shown in Figure 5.4.

The resulting H_2/CO ratio is an important factor in determining the subsequent application of the produced syngas ranging from Fischer Tropsch (FT) Gasoline and DiMethyl Ether (DME) (Dry and Steynberg 2004) all the way to targeting hydrogen production.

Proper palladium membrane, depending on the temperature and concentration of all available components, can be used for separating hydrogen, tuning H_2/CO ratio and supplying the hydrogen for the downstream CO_2 hydrogenation processes, for instance.

The amount of carbon dioxide entering the OCM and DRM sections is dictated by the performance of the reactor and downstream units. Therefore, the observed performance of the reactor sections also reflects the limitations imposed by the performance of the fully integrated process.

The balanced generation-recycling-consumption of carbon dioxide as the undesired product and diluting gas in OCM reactor, sweeping gas in adsorber as well as the feedstock for hydrogenation reactions, enable significantly reducing the net emission of carbon dioxide in such in integrated process.

Design and developing an efficient integrated OCM reactor is a multiperspective scientific and engineering analysis and design task. Not only for designing an efficient heat integration in the industrial-scale operation, especially on the preheating section, furnace and the quench section right after the reactor, but also on the separation and purification units, the practical limitations in terms of availability of the required utilities, compatibility of materials characteristics, etc. should be also taken into consideration.

5.5.3 TECHNO-ECONOMIC ANALYSIS OF STAND-ALONE AND INTEGRATED OCM PROCESSES

Considering the given selling price of ethylene, ethane, C_{3+}, and high-pressure (HP) steam as the main products of the OCM process, Return Of Investment (ROI) and the total energy consumption this process have been found to be below 15% (Eng et al. 1995; Godini et al. 2019) and more than 40 $GJ/t-C_2H_4$. Cryogenic distillation accounts for the most costly element of the OCM costly operation. The estimated relatively low capital and high operating costs are typical economic performance character of most of direct natural gas conversion processes.

Internal usage of available energy source for power and utility supply, more importantly further converting ethane to ethylene and carbon dioxide to value added oxygenates in relatively low pressure catalytic technologies, and in lesser extend converting unreacted methane to syngas can improve the economic balance of an integrated OCM process (Spallina et al. 2017; Godini et al. 2018, 2019). Techno-economic aspects of separation and recycling of methane to the OCM reactor has been also investigated elsewhere (Schammel et al. 2013).

In that perspective, the impact of price of methane and oxygen as the main raw materials on the economy of OCM process have been analyzed and demonstrated to highlight the importance of the local conditions in terms of price and quantity of the available methane resources as well as the source and technology used to supply oxygen. Therefore, primarily the availability of large quantity of local natural gas or other methane-rich resources such as biomethane and flared-gas in remote areas highlight the economic potentials of the OCM process for industrial-scale ethylene production with the production capacity of not less than few hundred tons of ethylene annually.

5.5.4 PRIORITIES FOR FUTURE RESEARCH

First, the material-structural-chemical characteristics of an efficient OCM catalyst should be better understood. Then, by properly addressing (possibly tailoring) these characteristics in selection, synthesis and tuning the support and active phases on it, a stable-selective catalyst structure (with proper shape, dimension etc.) should be developed. OCM reactor performance with high ethylene selectivity, stable catalyst and robust operation, preferably ≥ 10 bar, is the key on establishing an industrial-scale OCM process. The interactive effects of operating parameters and the reactor characteristics should be also taken into analysis to avoid misinterpreting the observed behaviors and correlate all effects to the catalyst specifications and develop an efficient operation strategy.

Still the SEC – Specific Energy Consumptions – (GJ/ton C_2H_4) of such process should be improved to be comparable with those for commercial ethane and naphtha crackers.

The perspective of using unit operation such as adsorption in the process downstream not only promises lowering the heavy cost of handling the unreacted methane, but also increases the potentials for an efficient process integration which is the key for industrial-scale implementation of OCM process. This also enables significantly improving the carbon footprint and environmental impacts of this process. All of these should be first further analyzed conceptually through a comprehensive con-current engineering along with techno-economic and life cycle analysis.

ACKNOWLEDGMENTS

This chapter is a review of the publications and reports we have accessed and evaluated, so it is also the contribution of all those authors and therefore we would like to thank them here. The deepest gratitude goes to Dr. George E.Keller, Prof. G. Wozny, and Prof. R. Schomaecker, with whom we have started and continued this OCM research/development journey. Other colleagues and coauthors, including Prof. J.-U. Repke, Dr. O. Görke, and coinventors/workers of Dr. Madan Mohan Bhasin in Union Carbide; Rick D. Cantrell, Anca Ghenciu, Kenneth D. Campbell, David M. A. Minahan, Alistair D. Westwood, Kenneth A. Nielsen, to name a few, as well as our PhD and Master's students, are specially acknowledged.

REFERENCES

Abbas, H., Azzis, H., and E. Bagherzadeh. 2005. *Patent WO 2005 005042*, assigned to HRD (USA) and National Petrochemical (Iran).

Albrecht, M., Rodemerck, U., Linke, D., and E. V. Kondratenko. 2018. Oxidative coupling of methane at elevated pressures: reactor concept and its validation, *Reaction Chemistry & Engineering* 3, 151–154.

Alonso, M., Patience, G. S., Fernandez, J. R., Lorences, M. J., Díez, F., Vega, A., and R. Cenni. 2006. Heat transfer studies in an inorganic membrane reactor at pilot plant scale, *Catalysis Today* 118, 32–38.

Anderson, M., and Y. S. Lin. 2006. Synthesis and characterization of carbonate-ceramic dual-phase membranes for carbon dioxide separation, *Proceeding of 9th International Conference on Inorganic Membranes*, 678–681.

Anderson, M., and Y. S. Lin. 2010. Carbonate-ceramic dual-phase membrane for carbon dioxide separation, *Journal of Membrane Science* 357, 122–129.

Arndt, S., Otremba, T., and U. Simon. 2012. $Mn-Na_2WO_4/SiO_2$ as catalyst for the oxidative coupling of methane. What is really known?, *Applied Catalysis A: General* 425–426, 53–61.

Au, C. T., Zhou, X. P., Liu, Y. W., Ji ,W. J., and C. F. Ng. 1998. The Characterization of BaF_2/Y_2O_3 catalysts for the OCM reaction, *Journal of Catalysis* 174, 153–163.

Bagherzadeh, E., Hassan, A., Anthony, R. G., Hassan, A., Bozkurt, B., and J. Zhang. 2012. Catalyst and method for converting natural gas to higher carbon compounds, *US Patent 8129305*.

Bhasin, M. M. 1988. Feasibility of etrhylene synthesis via oxidative coupling of methane, *Studies in Surface Science and Catalysis* 36, 343–357.

Bhasin, M. M., and K. D. Campbell. 1995. Oxidative coupling of methane – a progress report, in *Methane and AlkaneConversion Chemistry*, Edited by Bhasin M. M. and Slocum D. W. 1995. Springer Science Business Media, New York, NY, 3–17.

Bhatia, S., Thien, C. Y., and A. R. Mohamed. 2009. Oxidative coupling of methane (OCM) in a catalytic membrane reactor and comparison of its performance with other catalytic reactors, *Chemical Engineering Journal* 148, 525–532.

Cai, X., and H. Y. Hang. 2019. Advances in catalytic conversion of methane and carbon dioxide to highly valuable products, *Energy Science & Engineering* 7, 4–29.

Cantrell, R. D., Ghenciu, A., Campbell, K. D., Minahan, D. M. A., Bhasin, M. M., Westwood, A. D., and K. A. Nielsen. 2002. Catalyst for the oxidative dehydrogenation of hydrocarbons, *US Patent 6,403,523 B1*.

Cantrell, R. D., Ghenciu, A., Campbell, K. D., Minahan, D. M. A., Bhasin, M. M., Westwood, A. D., and K. A. Nielsen. 2003. Catalyst for the oxidative dehydrogenation of hydrocarbons, *US Patent 6,576,803 B2*.

Chou, L., Cai, Y., Zhang, B., Yang, J., Zhao, J., Niu, J., and S. Li. 2004. Role of SnO_2 over Sn-W-Mn/SiO_2 catalysts for oxidative coupling of methane reaction at elevated pressure, *Studies in Surface Science and Catalysis* 147, 619–624.

Choudhary, V. R., and S. A. R. Mulla. 1997. Coupling of exothermic and endothermic reactions in oxidative conversion of natural gas into ethylene/olefins over diluted SrO/La_2O_3/SA5205 catalyst, *Industrial Engineering Chemistry Research* 36, 3520–3527.

Chung, S. J., Park, J. H., Li, D., Ida, J.-I., Kumakiri, I., and J. Y. S. Lin. 2005. Dual-phase metal-carbonate membrane for high-temperature carbon dioxide separation, *Industrial & Engineering Chemistry Research* 44, 7999–8006.

Cong, L., Zhao, Y., Li, S., and Y. Sun. 2017. Sr-doping effects on La_2O3 catalyst for oxidative coupling of methane, *Chinese Journal of Catalysis* 38, 899–907. doi:10.1016/S1872-2067(17)62823-7.

Coronas, J., Menendez, M., and J. Santamaria. 1994a. Development of ceramic membrane reactors with a non-uniform permeation pattern: application to methane oxidative coupling, *Chemical Engineering Science* 49, 4749–4757.

Coronas, J., Menendez, M., and J. Santamaria. 1994b. Methane oxidative coupling using porous membrane reactors II – reaction studies, *Chemical Engineering Science* 49, 2015–2025.

Cruellas Labella, A., Melchiori, T., Gallucci, F., and M. van Sint Annaland. 2017. Advanced reactor concepts for oxidative coupling of methane, *Catalysis Reviews: Science and Engineering* 59, 234–294.

Czuprat, O., Schiestel, T., Voss, H., and J. Caro. 2010. Oxidative coupling of methane in a BCFZ perovskite hollow fiber membrane reactor, *Industrial Engineering Chemistry Research* 49, 10230–10236.

Dedov, A. G., Nipan, G. D., Loktev, A. S., Tyunyaev, A. A., Ketsko, V. A., Parkhomenko, K. V., and I. I. Moiseev. 2011. Oxidative coupling of methane: influence of the phase composition of silica-based catalysts, *Applied Catalysis A: General* 406, 1–12.

Dry, M. E., and A. P. Steynberg. 2004. Chapter book 5 – commercial FT process applications, *Studies in Surface Science and Catalysis* (Fischer-Tropsch Technology) 152, 406–481. ISBN: 978-0-444-51354-0.

Edwards, J. H., Do, K. T., and R. J. Tyler. 1991. The OXCO process. A new concept for the production of olefins from natural gas, *Fuel* 71, 325–334.

Eng, D., Chiang, P.-H., and M. Stoukides. 1995. Methane oxidative coupling: technical and economic evaluation of a chemical cogenerative fuel cell, *Energy Fuels* 9, 794–801.

Fang, X., Li, S., Lin, J., and Y. Chu. 1992a. Oxidative Coupling of Methane on W-Mn Catalysts, *Journal of Molecular Catalysis* (in Chinese) 6, 427–433.

Fang, X., Li, S., Lin, J., Gu, J., and J. D. Yan. 1992b. *Journal of Molecular Catalysis* (in Chinese) 8, 255–262.

Fleischer, V., Littlewood, P., Parishan, S., and R. Schomäcker. 2016. Chemical looping as reactor concept for the oxidative coupling of methane over a Na_2WO_4/Mn/SiO_2 catalyst, *Chemical Engineering Journal* 306, 646–654.

Gambo, Y., Jalil, A. A., Triwahyono, S., and A. A. Abdulrasheed. 2018. Recent advances and future prospect in catalysts for oxidative coupling of methane to ethylene: a review, *Journal of Industrial and Engineering Chemistry* 59, 218–229.

García, L., Poveda, Y. A., Rodríguez, G., Esche, E., Godini, H. R., Wozny, G., Repke, J.-U., and A. Orjuela. 2019. Adsorption separation of oxidative coupling of methane effluent gases. Mini-plant scale experiments and modeling, *Journal of Natural Gas Science and Engineering* 61, 106–118.

Gätner, C., Van veen, A. C., and J. A. Lercher. 2012. Mechanistic understanding and kinetic studies of highly selective oxidative dehydrogenation of ethane over novel supported molten chloride catalysts, *Proceeding of DMGK Conference* (Title: Reducing the carbon footprint of fuels and petrochemicals alternative feedstocks and innovative technologies).

Godini, H. R. 2014. *Analysis of Individual and Integrated Porous Packed-Bed Membrane Reactord for Oxidative Coupling of Methane*, TU Berlin PhD Dissertationn.

Godini, H. R., Azadi, M., Khadivi, M., Penteado, A., Wozny, G., and J.-U. Repke. 2019. A multi-perspectives analysis of methane oxidative coupling process based on miniplant-scale experimental data, *Journal of Chemical Engineering Research and Design* 151, 56–69.

Godini, H. R., Fleischer, V., Görke, O., Jašo, S., Schomäcker, R., and G. Wozny. 2014a. Thermal-reaction analysis of oxidative coupling of methane, *Chemie Ingenieur Technik* 86, 1906–1915.

Godini, H. R., Gili, A., Görke, O., Arndt, S., Simon, U., Thomas, A., Schomäcker, R., and G. Wozny. 2014b. Sol-gel method for synthesis of $Mn-Na_2WO_4/SiO_2$ catalyst for methane oxidative coupling, *Catalysis Today* 236, 12–22.

Godini, H. R., Gili, A., Görke, O., Simon, U., Hou, K., and G. Wozny. 2014c. Performance analysis of a porous packed-bed membrane reactor for oxidative coupling of methane: structural and operational characteristics, *Energy Fuels* 28, 877–890.

Godini, H. R., Kim, M., Fleischer, V., Görke, O., Khadivi, M., Schomäcker, R., and J.-U. Repke. 2018. *Chapter Book: Oxidative Coupling of Methane in Membrane Reactors*, RSC Publication. Print ISBN: 978–1–78262–875–0.

Godini, H. R., Trivedi, H., Gili de Villasante, A., Görke, O., Jašo, S., Simon, U., Berthold, A., Witt, W., and G. Wozny. 2013a. Design and demonstration of an experimental membrane reactor set-up for oxidative coupling of methane in UniCat mini-plant, *Chemical Engineering Research and Design* 91, 2671–2681.

Godini, H. R., Xiao, S., Jašo, S., Stünkel, S., Salerno, D., Xuan, Nghiem S., Song, S., and G. Wozny. 2013b. Techno-economic analysis of integrating the methane oxidative coupling and methane reforming processes, *Fuel Processing Technology* 106, 684–694.

Godini, H. R., Xiao, S., Kim, M., Görke, O., Song, S., and G. Wozny. 2013c. Dual-membrane reactor for methane oxidative coupling and dry methane reforming: reactor integration and process intensification, *Chemical Engineering and Processing* 74, 153–164.

Godini, H. R., Xiao, S., Kim, M., Holst, N., Jašo, S., Görke, O., Steinbach, J., and G. Wozny. 2014d. Experimental and model-based analysis of membrane reactor performance for methane oxidative coupling: effect of radial heat and mass transfer, *Journal of Industrial and Engineering Chemistry* 20, 1993–2002.

Graf, P. O., and L. Lefferts. 2009. Reactive separation of ethylene from the effluent gas of methane oxidative coupling via alkylation of benzene to ethylbenzene on ZSM-5, *Chemical Engineering Science* 64, 2773–2780.

Hiyoshi, N., and T. Ikeda. 2015. Oxidative coupling of methane over alkali chloride–Mn–Na_2WO_4/SiO_2 catalysts: promoting effect of molten alkali chloride, *Fuel Processing Technology* 133, 29–34.

Hoebink, J. H. B. J., Venderbosch, H. M., van Geem, P. C., van den Osterkamp, P. F., and G. B. Marin. 1995. Economics of the oxidative coupling of methane as an add-on unit for naphtha cracking, *Chemical Engineering Technology* 18, 12–16.

Jašo, S., Arellano-Garcia, H., and G. Wozny. 2011. Oxidative coupling of methane in flu-idized-bed reactor: infuence of feeding policy, hydrodynamics and reactor geometry, *Chemical Engineering Journal* 171, 255–271.

Jašo, S., Sadjadi, S., Godini, H. R., Simon, U., Arndt, S., Görke, O., Berthold, A., Arellano-Garcia, H., Schubert, H., Schomäcker, R., and G. Wozny. 2012. Experimental investiga-tion of fluidized-bed reactor performance for oxidative coupling of methane, *Journal of Natural Gas Chemistry* 21, 534–543.

Ji, S., Li, S., Gao, L., Liu, Y., Niu, J., and C. Xu. 1998. *Chinese Journal of Catalysis* (in Chinese) 11, 526–529.

Ji, S., Li, S., Gao, L., Xu, C., Zhang, B., and Y. Liu. 2000. *Chinese Journal of Catalysis* (in Chinese) 14, 1–5.

Ji, S., Li, S., Liu, Y., Gao, L., Niu, J., and C. Xu. 1999. Role of sodium in the oxidative cou-pling of methane over Na-W-Mn/SiO$_2$ catalysts, *Journal of Natural Gas Chemistry* 8, 1–8.

Ji, S., Xiao, T., Li, S., Chou, L., Zhang, B., Xu, C., Liu, Y., Hou, R., York, A. P. E., and M. L. H. Green. 2003. Surface WO$_4$ tetrahedron: the essence of the oxidative coupling of methane over M-W-Mn/SiO$_2$ catalyst, *Journal of Catalysis* 220, 47–56.

Ji, S., Xiao, T., Li, S., Xu, C., Hou, R., Coleman, K. S., and M. L. H. Green. 2002. The rela-tionship between the structure and the performance of Na-W-Mn/SiO$_2$ catalysts for the oxidative coupling of methane, *Applied Catalysis A: General* 225, 271–284.

Jones, C. A., Leonard, J. J., and J. A. Sofranko. 1987. The oxidative conversion of meth-ane to higher hydrocarbons over alkali-promoted Mn/SiO$_2$, *Journal of Catalysis* 103, 311–319.

Keller, G. E., and M. M. Bhasin. 1982. Synthesis of ethylene via oxidative coupling of meth-ane: I. Determination of active catalysts, *Journal of Catalysis* 73, 9–19.

Kenneth, D. Campbell. 1991. Double perovskite catalysts for oxidative coupling, *Patent No. 4,988,660.*

Khakpour, T., Holst, N., Holtappels, K., and J. Steinbach. 2011. Ignition behavior of methane-oxygen mixtures at elevated conditions, *AIChE Spring National Meeting*, March.

Kiani, D., Sourav, S., Baltrusaitis, J., and I. E. Wachs. 2019. Oxidative coupling of methane (OCM) by SiO$_2^-$ supported tungsten oxide catalysts promoted with Mn and Na, *ACS Catalysis* 9, 5912–5928.

Kondratenko, E., and M. Baerns. 2008. Oxidative coupling of methane – Chapter 13.17, in *Handbook of Heterogeneous Catalysis*, Wiley-VCH, 3010–3023.

Korup, O., Mavlyankariev, S., Geske, M., Goldsmith, C. F., and R. Horn. 2011. Measurement and analysis of spatial reactor profiles in high temperature catalysis research, *Chemical Engineering Processing: Process Intensification* 50, 998–1009.

Kou, Y., Zhang, B., Niu, J., Li, S., Wang, H., Tanaka, T., and S. Yoshida. 1998. Amorphous features of working catalysts: XAFS and XPS characterization of Mn/Na$_2$WO$_4$/SiO$_2$ as used for the oxidative coupling of methane, *Journal of Catalysis* 173, 399–408.

Kruglov, A. V., Bjorklund, M. C., and R. W. Carr. 1996. Optimization of the simulated coun-tercurrent moving-bed chromatographic reactor for the oxidative coupling of methane, *Chemical Engineering Science* 51, 2945–2950.

Kwapień, K. 2011. *Acitve Sites for Methane Activation in MgO and Li doped MgO*, PhD Dissertation in Humboldt-Universität zu Berlin.

Lafarga, D., Santamaria, J., and M. Menendez. 1994. Methane oxidative coupling using porous membrane reactors II – reactor development, *Chemical Engineering Science* 49, 2005–2015.

Langfeld, K., Frank, B., Strempel, V. E., Berger-Karin, C., Weinberg, G., Kondratenko, E. V., and R. Schomäcker. 2012. Comparison of oxidizing agents for the oxidative coupling of methane over state-of-the-art catalysts, *Applied Catalysis A: General* 417, 145–152.

Langguth, J., Dittmeyer, R., Hofmann, H., and G. Tomandl. 1997. Studies on oxidative coupling of methane using high-temperature proton-conducting membranes, *Applied Catalysis A: General* 158, 287–305.

Lee, J. Y., Jeon, W., Choi, J.-W., Suh, Y.-W., Ha, J.-M., Suh, D. J., and Y.-K. Park. 2013. Scaled-up production of C_2 hydrocarbons by the oxidative coupling of methane over pelletized Na_2WO_4/Mn/SiO_2 catalysts: observing hot spots for the selective process, *Fuel* 106, 851–857.

Lu, Y. P., Dixon, A. G., Moser, W. R., and Y. H. Ma. 2000. Oxidative coupling of methane in a modified γ-alumina membrane reactor, *Chemical Engineering Science* 55, 4901–4912.

Machida, K., and M. Enyo. 1987. Oxidative dimerization of methane over cerium mixed oxides and its relation with their ion-conducting characteristics, *Journal of the Chemical Society, Chemical Communications* 21,1639–1640.

Maitra, A. M., Campbell, I., and R. J. Tyler. 1992. Influence of basicity on the catalytic activity for oxidative coupling of methane, *Applied Catalysis A: General* 85, 27–46.

Matherne, J. L., and G. L. Culp. 1992. Direct conversion of methane to C_2's and liquid fuels: process economics, Chpater 14, in *Methane Conversion by Oxidative Processes*, Van Nostrand Reinhold.

Matras, D., Jacques, S. D. M., Godini, H. R., Khadivi, M., Drnec, J., Poulain, A., Cernik, R. J., and A. M. Beale. 2018. Real-time operando diffraction imaging of La–Sr/CaO during the oxidative coupling of methane, *Journal of Physical Chemistry C* 122, 2221–2230.

Myrach, P., Nilius, N., Levchenko, S. V., Gonchar, A., Risse, T., Dinse, K.-P., Boatner, L. A., Frandsen, W., Horn, R., Freund, H. J., Schlögl, R., and M. Scheffler. 2010. Temperature-dependent morphology: magnetic and optical properties of Li-doped MgO, *ChemCatChem* 2, 854–862.

Nipan, G. D., Artukh, V. A., Yusupov, V. S., Loktev, A. S., Spesivtsev, N. A., Dedov, A. G., and I. I. Moiseev. 2014. Pressure effect on the formation of active components of a catalyst for methane oxidative coupling, *Doklady Physical Chemistry* 455, 60–63.

Otsuka, K., Yokoyama, S., and A. Morikawa. 1985. Catalytic activity and selectivity-control for oxidative coupling of methane by oxygen-pumping through yttria-stabilized zirconia, *Chemistry Letters* 3, 319–322.

Pak, S., and J. H. Lunsford. 1998. Thermal effects during the oxidative coupling of methane over Mn/Na_2WO_4/SiO_2 and Mn/Na_2WO_4/MgO catalysts, *Applied Catalysis A: General* 168, 131–137.

Pak, S., Qiu, P., and J. H. Lunsford. 1998. Elementary reactions in the oxidative coupling of methane over Mn/Na_2WO_4/SiO_2 and Mn/Na_2WO_4/MgO catalysts, *Journal of Catalysis* 179, 222–230.

Palermo, A., Holgado, J., Lee, A., Tikhov, M., and R. Lambert. 1998. Critical influence of the amorphous silica-to-cristobalite phase transition on the performance of Mn/Na_2WO_4/SiO_2 catalysts for the oxidative coupling of methane, *Journal of Catalysis* 177, 259–266.

Parishan, S., Littlewood, P., Arinchtein, A., Fleischer, V., and R. Schomäcker. 2018. Chemical looping as a reactor concept for the oxidative coupling of methane over the MnxOy-Na_2WO_4/SiO_2 catalyst, benefits and limitation, *Catalysis Today* 311, 40–47.

Pinabiau-Carlier, M., Ben Hadid, A., and C. J. Cameron. 1991. The effect of total pressure on the oxidative coupling of methane reaction under cofeed conditions, *Studies in Surface Science and Catalysis* 61, 183–190.

Ramachandra, A. M., Lu, Y., Ma, Y. H., Moser, W. R., and A. G. Dixon. 1996. Oxidative coupling of methane in porous Vycor membrane reactors, *Journal of Membrane Science* 116, 253–264.

Rui, Z., Anderson, M., Lin, Y. S., and Y. Li. 2009. Modeling and analysis of carbon dioxide permeation through ceramic-carbonate dual-phase membranes, *Journal of Membrane Science* 345, 110–118.

Rui, Z., Ji, H., and Y. S. Lin. 2001. Modeling and analysis of ceramic carbonate dual-phase membrane reactor for carbon dioxide reforming with methane, *International Journal of Hydrogen Energy* 36, 8292–8300.

Sadjadi, S., Jašo, S., Godini, H. R., Arndt, S., Wollgarten, M., Blume, R., Görke, O., Schomäcker, R., Wozny, G., and U. Simon. 2015a. Feasibility study of the Mn–Na_2WO_4/SiO_2 catalytic system for the oxidative coupling of methane in a fluidized-bed reactor, *Catalysis Science & Technology* 5, 942–952.

Sadjadi, S., Simon, U., Godini, H. R., Görke, O., Schomäcker, R., and R. Wozny. 2015b. Reactor material and gas dilution effects on the performance of miniplant-scale fluidized-bed reactors for oxidative coupling of methane. *Chemical Engineering Journal* 12, 678–687.

Salehi, M.-S., Askarishahi, M., Godini, H. R., Görke, O., and G. Wozny. 2016a. Sustainable process design for oxidative coupling of methane (OCM): comprehensive reactor engineering via computational fluid dynamics (CFD) analysis of OCM packed-bed membrane reactors, *Industrial & Engineering Chemistry Research* 55, 3287–3299.

Salehi, M.-S., Askarishahi, M., Godini, H. R., Schomäcker, R., and G. Wozny. 2016b. CFD Simulation of oxidative coupling of methane in fluidized-bed reactors: a detailed analysis of flow-reaction characteristics and operating conditions, *Industrial & Engineering Chemistry Research* 55, 1149–1163.

Schammel, W. P., Wolfenbarger, J., Ajinkya, M., Ciczeron, J. M., Mccarty, J., Weinberger, S., Edwards, J. D., Sheridan, D., Scher, E. C., and J. McCormick. 2013. Oxidative coupling of methane systems and methods, *WO Patent 2013177433 A2*.

Schmack, R., Friedrich, A., Kondratenko, E. V., Polte, J., Werwatz, A., and R. Kraehnert. 2019. A meta-analysis of catalytic literature data reveals property-performance correlations for the OCM reaction, *Nature Communications* 10, 441–450.

Shurdumov, G. K., Shurdumova, Z. V., Cherkesov, Z. A., and A. M. Karmokov. 2006. Synthesis of alkaline-earth metal tungstates in melts of $(NaNO_3–M(NO_3)_2)_{eut}$–Na_2WO_4(M = Ca, Sr, Ba) systems, *Russian Journal of Inorganic Chemistry* 51, 531–532.

Simon, U., Görke, O., Berthold, A., Arndt, S., Schomäcker, R., and H. Schubert. 2011. Fluidized bed processing of sodium tungsten manganese catalysts for the oxidative coupling of methane, *Chemical Engineering Journal* 168, 1352–1359.

Sokolovskii, V. D., and E. A. Mamedov. 1992. Oxidative dehydrodimerization of methane, *Catalysis Today* 14, 415–419.

Song, X., and Z. Guo. 2006. Technologies for direct production of flexible H_2/CO ratio synthesis gas, *Energy Conversion and Management* 47, 560–569.

Spallina, V., Campos, Velarde I., Antonio Medrano Jimenez, J., Godini, H. R., Gallucci, F., and M. Van Sint Annaland. 2017. Techno-economic assessment of different routes for olefins production through the oxidative coupling of methane (OCM): advances in benchmark technologies, *Energy Conversion and Management* 154, 244–261.

Takanabe, K., and E. Iglesia. 2009. Mechanistic aspects and reaction pathways for oxidative coupling of methane on Mn/Na_2WO_4/SiO_2 catalysts, *Journal of Physical Chemistry C* 113, 10131–10145.

Tan, X., Pang, Z., Gu, Z., and S. Liu. 2007. Catalytic perovskite hollow fibre membrane reactors for methane oxidative coupling, *Journal of Membrane Science* 302(1–2), 109–114.

Tashjian, V., Cassir, M., Devynck, J., and W. Rummel. 1994. Oxidative dimerization of methane in molten Li_2CO_3-Na_2CO_3 eutectic supported by lithium aluminate at 700–850 °C, *Appllied Catalysis* A 108, 157–169.

Thum, L., Rudolph, M., Schomäcker, R., Wang, Y., Tarasov, A., Trunschke, A., and R. Schlögl. 2019. Oxygen activation in oxidative coupling of methane on calcium oxide, *Journal of Physical Chemistry C* 123, 8018–8026.

Tiemersma T. P. 2010. *Integrated Autothermal Reactor Concepts for Oxidative Coupling and Reforming of Methane*, PhD thesis in University of Twente.

Tye, C. T., Mohamed, A. R., and S. Bhatia. 2002. Modeling of catalytic reactor for oxidative coupling of methane using La$_2$O$_3$/CaO catalyst, *Chemical Engineering Journal* 87, 49–59.

Vallet-Regi, M., et al. 1988. Synthesis and characterization of a new double perovskite: LaCaMnCoO$_6$, *Journal of the Chemical Society, Dalton Transactions* 3, 775.

Vamvakeros, A., Jacques, S. D. M., Middelkoop, V., Di Michiel, M., Egan, C. K., Ismagilov, I. Z., Vaughan, G. B. M., Gallucci, F., van Sint Annaland, M., Shearing, P. R., Cernik, R. J., and A. M. Beale Real. 2015. Time chemical imaging of a working catalytic membrane reactor during oxidative coupling of methane, *Chemical Communications* 51, 12752–12755.

Vereshchagin, S. N., Gupalov, V. K., Ansimov, L. N., Terekhin, N. A., Kovrigin, L. A., Kirik, N. P., Kondratenko, E. V., and A. G. Anshits. 1998. Evaluation of the process of oxidative coupling of methane using liquefied natural gas from deposits of Krasnoyarsk region, *Catalysis Today* 42, 361–365.

Wang, D., Rosynek, M. P., and J. H. Lunsford. 1995. Oxidative coupling of methane over oxide-supported sodium-manganese catalysts, *Journal of Catalysis* 155, 390–402.

Wang, J., Li, S., Chou, L., Zhang, B., Song, H., Zhao, J., and Y. Jian. 2006. Comparative study on oxidation of methane to ethane and ethylene over Na$_2$WO$_4$–Mn/SiO$_2$ catalysts prepared by different methods, *Journal of Molecular Catalysis A: Chemical* 245, 272–277.

Wang, S., and Y. H. Zhu. 2004. Catalytic conversion of alkanes to olefins by carbon dioxide oxidative dehydrogenations, *Energy Fuels* 18, 1126–1139.

Wei, Y., Yang, W., Caro, J., and H. Wang. 2013. Dense ceramic oxygen permeable membranes and catalytic membrane reactors, *Chemical Engineering Journal* 220, 185–203.

Wolf, E. E. 1991. *Methane Conversion by Oxidative Processes: Fundamental and Engineering Aspects.* ISBN: 978–94–015–7451–8. doi:10.1007/978–94–015–7449–5.

Xuan Nghiem, S. 2014. *Ethylene Production by Oxidative Coupling of Methane: New Process Flow Diagram Based on Adsorptive Separation*, TU Berlin PhD Dissertation.

Yildiz, M., Aksu, Y., Simon, U., Otremba, T., Kailasam, K., Göbel, C., Girgsdies, F., Görke, O., Rosowski, F., Thomas, A., Schomäcker, R., and S. Arndt. 2016. Silica material variation for the MnxOy-Na$_2$WO$_4$/SiO$_2$, *Applied Catalysis A: General* 525, 168–179.

Zavyalova, U., Holena, M., Schlögl, R., and M. Baerns. 2011. Statistical analysis of past catalytic data on oxidative methane coupling for new insights into the composition of high-performance catalysts. *ChemCatChem* 3, 1935–1947.

6 Direct Natural Gas Conversion to Oxygenates

Consuelo Álvarez Galván

CONTENTS

6.1 INTRODUCTION

Natural gas (a fossil resource, based on a hydrocarbon gas mixture consisting primarily of methane) is found in many places around the world; however, its reserves are more geographically spread out across the world than oil reserves. These days, the main C1 oxygenates (methanol and formaldehyde) are industrially produced from methane by an expensive two-step process involving high-temperature endothermic reforming, followed by high-pressure methanol synthesis (and finally by the selective oxidation of methanol to formaldehyde). The direct oxidation of methane to oxygenates could provide economic profits for industries, offering an environmental-friendly alternative to methane flaring. The existing methane to methanol process is very efficient, with an overall carbon yield of ~70%. However, the most important disadvantage of this process is the high production cost. This fact hinders that natural gas displaces petroleum as the primary feedstock for the production of chemicals and fuels.

Thus, a large amount of natural gas reserves in the world are not being exploited, since the current technology is too expensive to overcome the cost of its transportation by pipeline from remote reservoirs, which can be up to five times as expensive as transporting liquid product. Therefore, there is a great interest on developing methods to boost the value of natural gas, either by synthesizing chemicals or fuels or more easily transportable products (Foulds and Gray 1995; Alvarez-Galván et al. 2011).

Among the different processes to valorize natural gas, so far, the most economically competitive route is through the production of syngas or hydrogen by steam reforming, These products can also be obtained by dry reforming (by reaction with carbon dioxide) and by partial oxidation with O_2 or air, but steam reforming is the industrially established one (Crabtree 1995). The reforming of methane over Ni catalysts with steam, carbon dioxide, or oxygen, separately, generates syngas at different stoichiometries. Table 6.1 displays these reactions and their enthalpies obtained at 1173 K (Choudhary and Choudhary 2008; da Silva 2016).

The value of natural gas can be enhanced by other processes; however, their industrial implementation is still far due to different limitations. For instance, in the oxidative coupling of methane, the yield to ethylene is limited to 30%, due to gasphase reactions. Another interesting process to produce methanol or formaldehyde by direct oxidation of methane (in continuous reactor) has produced yields of 8% as maximum. Some other reactions to valorize methane, such as the production of hydrogen cyanide by reaction of methane with ammonia (Degussa) or ammonia and oxygen (Andrussow), and production of acetylene by pyrolysis, are limited by the extremely high temperatures requirements (York, Xiao, and Green 2003).

6.2 CHALLENGES IN METHANE ACTIVATION AND FUNCTIONALIZATION

The main challenges in the direct partial oxidation of methane to C1 oxygenates are (Zhang, He, and Zhu 2003; Gunsalus et al. 2017; Horn and Schlogl 2015; Sun et al. 2014):

(i) Methane molecule has a low intrinsic reactivity. This molecule presents a symmetrical tetrahedron structure with four uniform C–H bonds, which makes it the most stable hydrocarbon molecule. The dissociation energy

---

TABLE 6.1

Processes for Syngas Production from Natural Gas (Choudhary and Choudhary 2008)

Process	Reaction	H_2:CO molar ratio	$\Delta H_{1173 K}$ (kJ mol^{-1})
Steam reforming	$CH_4 + H_2O \rightarrow 3H_2 + CO$	3:1	225.7
CO$_2$ reforming	$CH_4 + CO_2 \rightarrow 2H_2 + 2CO$	1:1	258.8
Partial oxidation reforming	$CH_4 + 0.5O_2 \rightarrow 2H_2 + CO$	2:1	– 23.1

of the first C–H bond is very high (440 kJ/mol), requiring the activation and reaction of methane harsh conditions. Methane is resistant to nucleophilic attacks since electron donation into the C–H σ* orbital is energetically challenging and sterically hindered. A little more favored, although still a challenge, is the removal of electrons from the C–H σ bond by strong electrophiles. The most facile way to activate methane molecule is by the homolytic C–H bond cleavage and hydrogen atom transfer to a radical. An example of this process is the oxidation of methane to methanol by heterogeneous catalysis, using Fe-HZSM-5 (Wood et al. 2004).

(ii)   Another drawback is the higher reactivity of desired products under the reaction conditions, which can lead to overoxidized products such as CO and CO_2. For instance, the dissociation energy of the $H–CH_2OH$ bond is less than the $H–CH_3$ bond.

(iii) Moreover, in contrast to the CH_4 molecule that has no polarity, C1-oxygenates are polar and are much easier to be adsorbed on the catalyst surface, being activated and oxidized (the stickiness of a small alkane is smaller than in a larger molecule).

(iv)   In addition, the comparison of Gibbs free energies of different methane oxidation reactions indicates that the formation of carbon oxides is more thermodynamically favorable than the oxidation to C1 oxygenates, being very difficult to control the deeply oxidation of these products of interest to CO and/or CO_2.

The abovementioned challenges are the consequence of the small number of industrial processes for the direct conversion of methane. So far, the most prominent industrial use of methane is for the production of synthesis gas, mainly by methane steam reforming. The obtained CO + H_2 mixture is mostly used for methanol and Fischer–Tropsch syntheses, ammonia synthesis (removing first CO), or hydrogenation reactions.

6.3 DIRECT METHANE OXIDATION TO C1-OXYGENATES

6.3.1 CRITERIA TO CLASSIFY THE DIRECT METHANE OXIDATION PROCESSES

There are different criteria to classify the different studied CH_4 direct oxidation reactions. These could be classified as (Gunsalus et al. 2017): (a) chain or nonchain reactions (b) catalytic and stoichiometric (c) O_2 or an oxidant than can be regenerated from O_2 (d) high and low temperature approaches (Scheme 6.1). Noncatalytic (stoichiometric) and catalytic direct methane oxidation processes to oxygenates will be discussed in Sections 6.4.1 and 6.4.2, respectively.

Mechanism-based classifications consider whether the elementary reactions with methane take place by chain or nonchain reactions and with stoichiometric reagents or with catalysts. The nonchain reactions are further classified as CH activation or CH oxidation. CH activation reactions consist of elementary reactions with methane that result in a methyl intermediate where the formal oxidation state on the carbon remains unchanged at −4. On the contrary, CH oxidation reactions are defined as

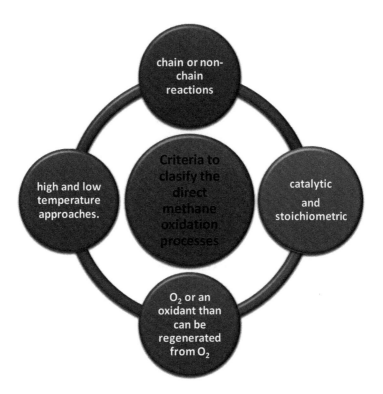

SCHEME 6.1 Criteria to classify the direct methane oxidation processes.

elementary reactions with methane where the carbon atom of the product is oxidized, being its oxidation state less negative than −4.

The approach that reaches highest yields are based on CH activation chemistry (Conley et al. 2006). CH activation consists of a two-step process in which (1) the CH bond coordinates to an open site at a transition metal center followed by (2) cleavage of the CH bond to form a metal-carbon bond. The advantage of the CH activation reaction is that can cleave the CH bond of alkanes using moderately energetic conditions. The high selectivity and mild reaction conditions of CH activation reaction is motivated by the atom transfer within the coordination sphere of carbon and the reactant.

The key steps in the most efficient activation-based catalysts for alkane oxidation are: (i) coordination of the CH bond to an electrophilic center, (ii) nucleophilic attack to generate organometallic intermediates (CH activation) and, finally, (iii) oxidative functionalization. Some of the CH activation reactions are: sigma-bond metathesis, oxidative addition, and electrophilic substitution (see Figure 6.1).

On the contrary, CH oxidation by chain reactions typically involves the generation of highly energetic and reactive intermediates such as free radicals, carbocations, carbanions, and carbenes, being formed under very high temperatures or using very energetic reagents (i.e. superacids or peroxides). In these reactions, some species (hydrogen peroxide, persulfate, etc.) are consumed to generate chain-propagating

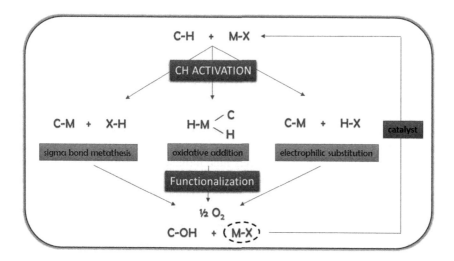

FIGURE 6.1 Catalyzed CH activation reactions and functionalization. (Reprinted from Conley et al. 2006, Copyright 2006, with permission from Elsevier.)

species, such as a methyl radical. These species react with oxygen or another reagent through a propagation sequence to generate the desired functionalized product and regenerate the chain-carrying species. Termination reactions stop the propagation sequence by quenching the radical (Hashiguchi et al. 2010).

Without taking into account the practical consideration of oxygen regenerability, a scheme of a possible methane functionalization processes classification is depicted in Figure 6.2. Nonchain reactions by CH activation or CH oxidation can be, in turn, catalytic or stoichiometric.

Another criterion to classify the direct partial oxidation of methane is the reaction temperature (high and low approaches). The huge amount of research done supports the fact that at high temperature (500–800 degrees C), and moderate pressures (up

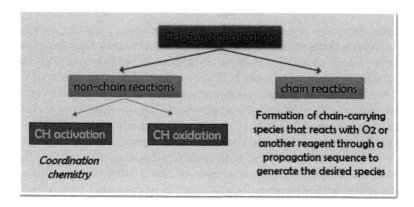

FIGURE 6.2 Classification of methane functionalization on the basis of reaction mechanism approaches.

to 10 bar, 1 MPa), the oxidation of methane occurs mainly by the homolytic activation of the C–H bond (Hammond et al. 2012), where the formation of products is controlled by a complex pathway of radical reactions. The oxygenated products (obtained from methyl radicals) are more reactive than methane, being mainly over-oxidized to CO and CO_2. As demonstrated by Hargreaves, Hutchings, and Joyner (1990) by decreasing the pressure (working at 1 atm), and thus the concentration of methyl radicals formed by the catalyst, the selectivity to C1 oxygenates could be increased. This conclusion has pushed the research working at low pressures and high temperatures (600–800°C), mainly using metal oxide catalysts. However, although under these reaction conditions relatively high formaldehyde selectivities have been found, methanol selectivities at these operational conditions are very low, being the poor stability of methanol at high temperatures one of the main causes of this fact.

Attempts to increase the yield toward methanol, decreasing the overoxidation selectivity to formaldehyde and carbon oxides (CO and CO_2) have done working at lower reaction temperatures. Low-temperature approaches use to involve the electrophilic activation of methane over high-valent electrophiles, such as those pioneered by Goldshleger et al. (1972) using homogeneous metal complexes (such as Pt^{II}, Pd^{II}, $Au^{I/III}$, Tl^{III}, and Hg^{II}) or elemental I_2, typically performed in highly acidic reaction media. Methanol is not directly obtained, being formed functionalized products such as methanosulfonic acid and methyl bisulphate (Hammond, Conrad, and Hermans 2012). In the framework of this approach, Conley et al. (2006) have reported the most efficient catalytic systems to convert methane to methyl bisulfate using a bipyrimidinylplatinum catalyst at 220°C in oleum. The maximum one-pass yield at 90% selectivity achieved by direct oxygenation of methane with CH activation catalysts using this Pt complex and Hg(II) complex are around 70% and 40% respectively, values much higher than those yields achieved with the most active catalysts by direct CH oxidation (around 5%) (Conley et al. 2006).

These two high and low temperature approaches will be more extensively addressed in Section 6.4.2.

6.3.2 Nature of Oxidants

The direct conversion of methane to oxygenates requires a co-reactant, typically an oxidant. Among different oxidants, air is the only economically feasible co-reactant for the large-scale chemical conversion of methane to products with high volume density, such as methanol or other bulk chemicals or fuels (Gunsalus et al. 2017). This would be a challenged process that would use methane to produce chemicals or fuels, instead of by petroleum-based processes. The use of pure oxygen can significantly increase the cost of the process, given the requirement of an O_2-separation plant. Therefore, the big challenge would be the development of a large-scale process that uses methane and (ideally) air as an oxidant co-reactant.

Among oxidants, hydrogen peroxide could be a good option because of simplicity of management, environmentally friendly nature of its co-product (water), high oxygen atom efficiency and versatility. Therefore, the oxidation of methane with hydrogen peroxide is interesting although both the conversion of methane and efficiency of the utilization of this oxidant are still low (Seki, Mizuno, and Misono 1997).

This approach will be commented in Section 6.4.2.2. In situ synthesis of H_2O_2 (through the reaction between H_2 and O_2) has been also addressed (Williams, Kolaczkowski, and Plucinski 2003; Park, Choi, and Lee 2000).

Nitrogen oxides can be also used as oxidant in catalytic and noncatalytic methane oxidation processes (Falconer, Hoare, and Overend 1973). The effects of nitrogen oxides as oxidants on the catalytic performance depend on the studied catalysts, being reported that the nature of reactive oxygen species is related to the kind of catalyst (Barbaux et al. 1988; Tabata et al. 2002). For instance, it has been reported that highly dispersed tetrahedral surface vanadia would be involved in primary oxidation reactions to form HCHO, while bulk V_2O_5 would participate in secondary oxidation reactions, producing CO and/or CO_2 (Tabata et al. 2002).

In order to classify the different possibilities for the direct oxidation of methane, a pertinent scheme could be that reported in the review by Gunsalus et al. (2017), in which the classification is based on whether the reactions use (i) O_2 or an oxidant that can be regenerated from O_2 as the only net co-reactant or (ii) O_2 or oxidants that cannot be generated from O_2 (based only on thermodynamic potentials). In both groups, not only catalytic systems but also stoichiometric systems can be included.

Ce^{4+}, Co^{3+}, Tl^{3+}, Cl_2, H_2O_2, IO_4^-, and $S_2O_8^{2-}$ can be used as oxidants in stoichiometric reaction systems for the functionalization of CH_4 to CH_3OH, but cannot be regenerated by O_2. The reactions with superacids such as SbF_5 are considered at non-O_2 regenerable, given the unsuitability of SbF_5 with H_2O that would be formed when using O_2 for the regeneration. Other non-O_2 regenerable systems include the use of diboranes and diazo compounds as reactants.

An interesting system that is generally considered as O_2-regenerable is that based on H_2SO_4 and SO_3, separately or as mixtures, constituting stoichiometric oxidants for methane functionalization. This pathway converts methane to methanol using O_2 as the overall co-reactant, although two main drawbacks have to be considered: (a) the irreversible and easy reaction of SO_3 and H_2O to form H_2SO_4, being both S compounds potentially sulfonating and dehydrating agents and (b) the formation of sulfuric acid/water azeotrope. These processes would lead to the undesirable conversion of methanol to methyl bisulfate and conversion of methanesulfonic acid to methanesulfonic anhydride and H_2SO_4. This process presents other disadvantages that makes it very expensive such as side oxidation reactions with SO_3, very expensive separation of methanol from concentrated sulfuric acid and high material investment for the plant (Gunsalus et al. 2017).

6.4 NONCATALYZED AND CATALYZED DIRECT PARTIAL OXIDATION OF METHANE TO OXYGENATES

6.4.1 Homogeneous (Noncatalyzed) Direct Partial Oxidation of Methane to Oxygenates

It is possible to obtain C1-oxygenates by direct partial oxidation of methane, under certain reaction pressures and temperatures. These processes have been widely reviewed by Foulds and Gray (1995) for the partial oxidation of methane to methanol.

The yield to C1-oxygenates depends on several factors such as: (i) type of reactor wall material, (ii) reaction conditions, and (iii) use of reaction initiators in the feed. The influence of these parameters on above reaction will be commented hereafter.

6.4.1.1 Reactor Wall Material

In the reactions that were carried out under high pressures, a stainless-steel reactor was used. However, the obtained C1-oxygenates yield were very low due to the deep oxidation of compounds of interest to carbon monoxide and carbon dioxide, reactions that are catalyzed by the metal surface of the reactor wall. Thus, it has been widely reported that the selectivity to methanol decreases with a reactor wall made of stainless steel (Zhang, He, and Zhu 2003). This drawback could be diminished if quartz or Pyrex linings are used in the reaction equipment (Foulds and Gray [1995] and references therein). Therefore, in order to ensure that the reaction is purely homogeneous, it would be imperative to use the glass-lined reactors (Burch, Squire, and Tsang 1989).

Hunter et al. (1990) have studied the influence of several coatings of reactor wall on the activity. They used materials comprised Pyrex glass, Teflon, stainless steel, silver, and copper. They concluded that the material of the reactor wall has little effect on the methanol selective in partial oxidation of methane, although it can influence the reaction temperature and reaction rate. A passive or inert wall is desirable in order to take advantage of the sensitizing action of higher hydrocarbons present in natural gas, which leads to a significant decrease of reaction temperature. The influence of the wall composition on the reaction is lesser at higher reactor pressures and lower surface/volume reactor ratios.

Burch, Squire, and Tsang (1989) have observed that no methanol is produced when using a steel reactor until the pressure is greater than 20 bar (2 MPa), while with a glass-lined reactor, methanol is formed at all pressures.

Zhang et al. (2002) designed a reactor in which the gap between the inner quartz line and the external stainless-steel tube was encapsulated using an O-ring. With this system, the methanol yield was increased up to 8% (with a conversion around 13% and a selectivity of 60%).

6.4.1.2 Reaction Conditions

The effects of different reaction parameters (pressure, temperature, CH_4/O_2 ratio, and residence time) over the yield to C1 oxygenates will be discussed in this subsection.

The reaction temperature is one of the parameters that should be considered to analyze its influence on the conversion and selectivity. Since this is a highly exothermic reaction, it has to be considered that there is going to be a large difference between the reactor wall temperature (which used to be between 300°C and 500°C) and the reaction temperature. Sometimes, this difference is not considered and some discrepancies in the results can be observed. It has been observed that there is a transition temperature whose value depends on the other reaction parameters, after which the oxygen conversion dramatically increases to 100% in a very narrow temperature range, and below which methane conversion was very low. If the reaction temperature is increased above this transition temperature, the selectivity to the desired products (mainly CH_3OH and $HCHO$) decreased since deep oxidation products were produced. This behavior is related to a reaction that evolves under a free radical mechanism (Zhang, He, and Zhu 2003). Moreover, an increase in the reaction temperature along with a total consumption of oxygen produces the oxidative coupling product, C_2H_6. The decomposition of HCHO, once is formed, to CO and H_2 has been also reported (Lodeng et al. 1995; Zhang et al. 2002). A hysteresis effect in methane conversion for increasing and decreasing reactor temperature has been observed (Yarlagadda et al. 1990).

Reaction pressure has also a key role in the conversion and selectivity to oxidized products. An increase in the reaction pressure shifts the transition temperature to lower values. For instance, an increase from 1 to 3 MPa, produces, a drop in the transition temperature in more than 30°C was observed. Nevertheless, if the pressure increases more than 5 MPa, the effect of pressure on the transition temperature would be less pronounced (Chun and Anthony 1993; Zhang et al. 2002). It has been observed an increase of methanol selectivity with increasing pressure. A decrease of pressure results in an increase of carbon monoxide and in a small increase in methane coupling products and formaldehyde (Foulds and Gray 1995).

It has been observed an increase of methanol selectivity with an increase of reaction pressure, being found higher production of carbon monoxide and carbon dioxide at lower pressures (Yarlagadda et al. 1988). On the contrary, Burch et al. (1989) reported that the selectivity to methanol is insensitive to pressure and other experimental parameters. Moreover, they obtained values around 40%, lower than other published results.

CH_4/O_2 ratio also affects the oxidation reaction. In order to increase the selectivity to oxygenates, this ratio is high. A low ratio favors deep oxidation and the risk of an explosion (Zhang, He, and Zhu 2008). At the temperatures and pressures normally used, (300–500°C; 0.1–3 MPa) the optimal oxygen limit is around 10%, above which, the explosion limit will be surpassed (Foulds and Gray 1995). Considering this fact, most studies are carried out using feed concentrations ranging from 1 to 10%. For the direct oxidation of methane to methanol, it has been widely reported that a decrease in this ratio (increasing the feed oxygen concentration), increases methane conversion but, in turn, decreases the selectivity to methanol (see Figure 6.3).

On the contrary, Burch, Squire, and Tsang (1989) reported that methane conversion did not change and methanol selectivity shows little dependency on feed oxygen concentration, although it has to be highlighted that the O_2 concentration used didn't significantly change (2.5, 4.9, and 6.6%). Another interesting result is related to the occurrence of hysteresis at O_2 feed concentrations > 7.5% (Foulds et al. 1993).

The effect of residence time on the homogeneous partial oxidation of methane to methanol and formaldehyde has been studied by Rytz and Baiker (1991). They found an increase in the conversion with increasing the residence time from 2 to 7 seconds, after which it remains constant. Foulds et al. (1993) reported that an increase in the gas flow rate leads to an increase in reaction temperature. The conversion was not affected by the feed rate, which is in accordance with the results obtained by Rytz and Baiker (1991), the residence time was greater than 7 seconds. Burch, Squire, and Tsang (1989) found an relationship between residence time and temperature, with an decrease of selectivity as the residence time increases, at low temperatures (350–400°C), while, at high temperature (close to 500°C), the selectivity increases as the residence time increases.

A way to decrease the residence time of the gases in the reactor is by introducing Pyrex or quartz packings inside it (Foulds and Gray 1995). However, it

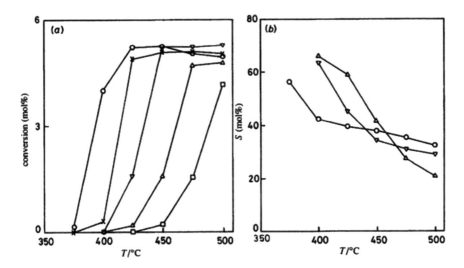

FIGURE 6.3 Effect of reaction temperature on (a) methane conversion (b) methanol selectivity in a Pyrex reactor at various pressures: □ 5, △ 10, ▽ 20, × 30, ○ 50 bar (1 bar = 0.1 MPa). (Reprinted from Burch, Squire, and Tsang 1989, Copyright 1989, with permission from the Royal Society of Chemistry.)

has been observed that the use of small beads results in some radical quenching (Baldwin et al. 1991). This fact would indicate that the homogeneous process depends on the surface reactions. Thomas, Willi, and Baiker (1992) conducted a quantitative analysis of the effect of the surface/volume (S/V) of the reactor, both experimentally and theoretically, by using a model developed by Vedeneev et al. (1988). The obtained results shown that the predicted methanol selectivity was higher at lower S/V ratios. Moreover, the Vedeneev model proposes that nearly all the overoxidized products are formed via formaldehyde, being methanol relatively stable and does not decompose significantly under reaction conditions. This stability of methanol was also experimentally observed by Burch, Squire, and Tsang (1989).

It is interesting to visualize the relationship between methane conversion and selectivity to methanol and formaldehyde represented in Figure 6.4. These data, which are obtained from several sources in the literature, show that even using some different reaction conditions, a trend between these two variables is found (Tabata et al. 2002).

In summary, it can be concluded that (i) high reaction temperatures decrease methanol selectivity; (ii) a relative high oxygen concentration lowers methanol selectivity and increases methane conversion, leading to a net increase in methanol yield; and (iii) high pressure favors both methane conversion and methanol selectivity.

6.4.1.3 Additives (Reaction Initiators)

In order to decrease the reaction temperature and, in this way, to increase the C1-oxygenates selectivity, some additives (H-donors and NO_x) can be added to the reaction (Zhang, He, and Zhu 2008).

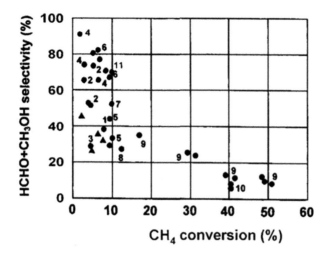

FIGURE 6.4 Selectivity to CH₃OH and HCHO vs. CH₄ conversion from several sources in the literature. (▲) homogeneous reaction; data from literature: (1) Lott and Sliepcevich; (2) Dowden and Walker; (3) Tripathy; (4) Brokhaus; (5) Brokhaus and Franke; (6) Hunter et al.; (7) Gesser et al.; (8) Rytz and Baiker; (9) Bañares et al.; (10) Takemoto et al.; (11) Tabata et al. Figure and references numbers from reference Tabata et al. (2002). (Reprinted from Tabata et al. 2002, Copyright 2002, with permission from Taylor & Francis.)

H-donors (such as hydrogen, and hydrocarbons (Omata, Fukuoka, and Fujimoto 1994)) will act providing hydrogen to the methoxy radical, thus stabilizing methanol. Higher molecular weight hydrocarbons sensitize the oxidation of methane because of their lower C–H bond strength, which provide free radicals easily. This fact decreases the oxidation reaction temperature, compared to the oxidation of pure methane, without feeding H-donors additives (Tabata et al. 2002). NO_x would produce a faster abstraction of hydrogen from methane, and, in this manner, would lower the temperature at which the initiation occurs (Foulds and Gray 1995). The role of nitrogen oxides in noncatalytic and catalytic methane activation was been widely reviewed by Tabata et al. (2002) reporting experimental and modelling studies.

Hunter et al. (1990) have reported a systematic study on the effects of different H-donors (hydrocarbons, oxygen-containing compounds, sulfur-containing compounds, and amines) on the partial oxidation of methane to oxygenates at 30 atm. The formation of formaldehyde is greater than with nonsensitized reaction, being, probably, at the expense of methanol. However, in most cases the relatively large number of sensitizers that had to be added would compromise any cost benefit. In conclusion, the addition of these sensitizers to the reaction feed decreases the reaction temperature and increases the selectivity to methanol and/or formaldehyde compared to the nonsensitized reaction.

The same group has reported the effect of using ozone to initiate the gas-phase reaction, founding that O_3 decomposes to form O_2 and oxygen radical, which abstract hydrogen from methane, thus forming a methyl radical (Gesser, Hunter, and Das 1992). In these studies, it must be ensured that the additive is, actually, sensitizing methane and not being converted to products by partial oxidation.

The use of nitrogen oxides as reaction initiators arises from the fact that these compounds and nitric acid have shown a promotion effect on the lowering of the reaction temperatures for light alkane partial oxidation (Smith and Milner 1931).

Bromly et al. (1996) reported a method for comparing the effect of both kinetic and thermodynamic parameters on the results of the experimental $CH_4 + NO + O_2$ reaction, at atmospheric pressure. The estimation of the kinetic model agrees with the experimental data. It is suggested that NO_2 and CO are the main products. The addition of NO provides a source of chain carriers at lower temperatures, being the reaction: $CH_3O_2 + NO \rightarrow CH_3O^\bullet + NO_2$, the most important step in the oxidation mechanism.

Teng et al. (2000) have reported an increase in methane conversion and in the selectivity to formaldehyde and/or methanol, using a small concentration of NO or NO_2, working at atmospheric pressure, assuring that NO_x promoted the formation of $CH_3^\bullet$ and $CH_3O^\bullet$ in the gas-phase reaction. Thus, the role of NO can be considered as follows:

$$NO + \tfrac{1}{2}O_2 \rightarrow NO_2$$

$CH_4 + NO_2 \rightarrow CH_3^\bullet + HNO_2$ For the formation of oxygenates, it is pointed out that after the abstraction of hydrogen from CH_4, $CH_3O_2O_2^\bullet$ is produced easily by oxygen insertion as

$$CH_3^\bullet + O_2 \rightarrow CH_3O_2^\bullet$$

$$CH_3O_2^\bullet + NO \rightarrow CH_3O^\bullet + NO_2$$

Regarding the formation of C1-oxygenates, several reaction routes concerning $CH_3O^\bullet$ were suggested as follows, which are focused on the formation of methanol (Teng et al. 2000):

$$CH_3O^\bullet + O_2 \rightarrow HCHO + HO_2^\bullet$$

$$CH_3O^\bullet + CH_3O^\bullet \rightarrow CH_3OH + HCHO$$

$$CH_3O^\bullet + HCHO \rightarrow CH_3OH + CHO^\bullet$$

$$CH_3O^\bullet + CH_4 \rightarrow CH_3OH + CH_3^\bullet$$

Otsuka et al. (1998) and Otsuka, Takahashi, and Yamanaka (1999) studied the promotion of the partial oxidation of CH_4, C_2H_6, C_3H_8, and iso-C_4H_{10} with O_2 by NO, reporting a remarkable increase in the yield to oxygenates with the addition of NO to the feed. They suggested that NO_2 generated from NO and O_2 initiated the oxidation of alkanes, accelerating the C–C bond scission and increasing the formation of C1-oxygenates. It is also suggested a possible reaction mechanism where NO_2 (formed by reaction between $NO + O_2$) is an initiator for the formation of alkyl radicals. The alkyl radicals ($R^\bullet$) reacts with NO_2 producing RONO as a reaction intermediate for the formation of oxygenates. Then, RONO decomposes to RO$^\bullet$ regenerating NO. RO$^\bullet$ decomposes to aldehydes or acetone and CH$_3$$^\bullet$ which is further oxidized into HCHO, CO, and CO$_2$. The reactivity of methane in the reactions of methane + NO was very enhanced by the presence of O$_2$, indicating that NO promotes the oxidation of CH$_4$ only in the presence of oxygen, which indicates that NO$_2$ acts as an initiator.

Irusta, Lombardo, and Miro (1994) also found an increase in formaldehyde yield using NO as an additive in the feed for the partial oxidation of methane (at 1 bar, 0.1 MPa), explaining the result by a radical reaction mechanism initiated by NO, which was first suggested by McConkey and Wilkinson (1967).

The lowering effect of the ignition temperature of methane conversion with the addition of NO$_2$ in the mixture of methane and oxygen have been also studied by Tabata et al. (2002, 2000). The possible mechanisms and kinetic modelling for the reactions CH$_4$ + O$_2$+NO$_x$ are widely compiled in a review by Tabata et al. (2002).

The influence of NO$_x$ concentration has been examined by Teng et al. (2000) and Tabata et al. (2000). Reactivity increased with the concentration of NO$_2$ up to 0.75% and decreased a little at 1%. Concerning NO addition, it was found that the selectivity to formaldehyde had a maximum value at 1% NO and for methanol, at 0.5% NO.

6.4.2 Catalyzed Direct Oxidation of Methane to C1-Oxygenates

There are two approaches to oxidize methane in one step to C1 oxygenates: by heterogeneous and homogeneous catalytic oxidation. The first one is based on the use of solid catalysts operating at high temperatures (400–800°C) and at atmospheric pressure, being the first stage, the C–H cleavage and the formation of the CH$_3$* radical. The second approach consist of the activation of C–H bond under mild temperature conditions but at high pressure, without radical reaction, by reaction of metalorganic complexes and metal salts with the sigma-bond electron pairs of methane molecule under a homogeneous process.

6.4.2.1 Homogeneous (Nonsolid) Catalyzed Direct Oxidation

The homogeneous catalytic systems for the selective oxidation of methane have been deeply reviewed by Balcer (2015) and by Conley et al. (2006). Homogeneous C–H activation by molecular complexes (organometallic) can involve four different mechanisms: (a) oxidative addition (b) σ-bond metathesis and in few cases (c) 1,2-addition or (d) electrophilic activation.

The homogeneous selective catalytic oxidation of methane can be achieved at mild temperatures (lower than 200°C) using transition metal-mediates procedures, which is a challenge, considering the high stability of methane molecule. These catalysts can activate and functionalize C–H bonds under low temperatures (avoiding gas-phase radical reactions), producing the oxidation with high energy and atom efficiency.

6.4.2.1.1 Methane Oxidation in Different Solvents

In 1972, Goldshleger et al. (1972) (Shilov's Group) developed a method to oxidize methane (at 120°C and around 4 MPa) producing methanol and halogenalkanes, catalyzed by PtCl$_2$ in an aqueous solution with [PtCl$_6$]$^{2-}$ acting as the ultimate oxidant.

Lin, Hogan, and Sen (1996) used metallic palladium supported on an active carbon or RhCl$_3$ to oxidize CH$_4$ with O$_2$ and CO to acids in an aqueous solvent (0.1 N

HCl) at 70–200°C and high partial methane pressure (~5 MPa). A catalyst based on $RuCl_3$ produced mainly acetic acid, but also formaldehyde and methanol.

Methane oxidation has also been carried out using organic solvent systems. Methane and other alkanes have been oxidized by contact with an oxygen-containing gas in the presence of catalyst based on a metalloporphyrin in which hydrogen atoms in the porphyrin ring have been substituted with one or more cyano groups (Ellis and Lyons 1992). Metals in the porphyrin are transition metals: Fe, Cr, Mn, Ru, Cu, or Co. Hydrogen atoms in the porphyrin ring may also be substituted with halogen atoms. The reaction was performed in liquid phase using organic solvents such as: C_6H_6, CH_3COOH, acetonitrile, methylacetate, and others.

Methane has been oxidized by air using acetonitrile as solvent, with a catalyst based on $[nBu_4]VO_3$-pyrazine-2-carboxylic acid and H_2O_2 as promoter (Nizova, SussFink, and Shulpin 1997a, b). The main product, under 35°C, was methyl hydroperoxide, which can be easily reduced to ethanol. At higher reaction temperatures, the main products were formaldehyde and formic acid.

The production of methanol by direct oxidation of methane using water as solvent has been patented by Sherman (1999). The patent involves both the method of and the equipment, in which a catalytic surface is formed on the exterior of a stainless-steel tube. Methane is kept inside the tube at determined pressure, and water is continuously flowing across the exterior surface forming small sized methane bubbles. UV-light energy is focused onto the catalytic surface to form hydroxyl radicals from the flowing water. The hydroxyl radicals cleave the carbon-hydrogen bonds of the methane to form methyl ions which react with the hydroxyl ions to form methanol (Sherman 1999).

Organic solvents may take part in the reaction or be converted under the conditions used for oxidation or oxidative functionalization of methane. Therefore, the use of aqueous solvents is desirable. Several transition metal chlorides ($FeCl_3$, $CoCl_2$, $RuCl_3$, $RhCl_3$, $PdCl_2$, $OsCl_3$, $IrCl_3$, H_2PtCl_6, $CuCl_2$, and $HAuCl_4$) were studied for the selective oxidation of methane using hydrogen peroxide in aqueous medium. Among the metal chlorides explored, $OsCl_3$ showed the highest turnover frequency for the formation of organic oxygenates (mostly alcohols and aldehydes). UV-Vis spectroscopic measurements suggested that $OsCl_3$ was oxidized into Os(IV) species by hydrogen peroxide in aqueous medium, and these species could be involved in the oxygenation of CH_4. It is suggested that the reactions proceeded via a radical pathway, because adding a radical scavenger produced a decrease in conversion (Yuan et al. 2007).

Another chloride, CuCl, was used as a catalyst by Stauffer (1993), who develops a method for the oxidation of methane with oxygen, using organic chlorine compounds as solvent. Methanol is produced from methane using two reaction steps operated in tandem.

Seki, Mizuno, and Misono (1997) studied methane oxidation at low temperature (50–80°C) and high pressure (5 MPa) with H_2O_2 catalyzed by 12-molybdovanadophosphoric acid catalyst precursor (heteropoly acid), which were dissolved in the anhydride of trifluoroacetic acid, methyl cyanide, and water. The reaction products were methanol, formic acid, methyl acetate, ethane, methyl trifluoroacetate, and carbon dioxide.

Periana et al. (1993a) has developed a method to convert hydrocarbons, such as methane, into alcohols, thiols, esters, and halogenoalkanes by a two-step reaction process at low temperature. In the first step, methane is put in contact with an organic

$$CH_4 + 2H_2SO_4 \xrightarrow[\text{220°C, oleum}]{\substack{\text{catalyst:} \\ \text{Pt complex}}} CH_3OSO_3H + SO_2 + 2H_2O$$

SCHEME 6.2 High yield production of methyl bisulphate by the selective catalytic oxidation of methane using a platinum(II) complex catalyst in oleum. (Adapted from Periana et al. 1998).

or inorganic acid (being preferred the strong inorganic acids) and an oxidizing agent (preferred O_2, H_2SO_4, SO_3, and HNO_3) in the presence of a catalyst, based on a transition metal. This reaction converts methane in the methyl oxyester of the acid and in the reduced oxidant. The second step involves the conversion of methyl ester to methanol and the regeneration of the acid. Platinum complexes are reported to be active for the selective conversion of methane to methyl esters, with a yield greater than 70% per pass, using sulfuric acid as oxidant, in which the catalyst was completely dissolved (Periana et al. 1998), which represents a breakthrough in this process (see Scheme 6.2). It has been demonstrated that this process occurs by an electrophilic attack on CH_4 by solvated $[Pt(NH_3)_2^-(OSO_3H)^+]+$ complex (Mylvaganam, Bacskay, and Hush 2000).

H_2SO_4/SO_3 (oleum) acts as both the solvent and the oxidative species. This allows a catalytic cycle based on Pt^0/Pt^{2+} to proceed and also increases the stability of the reaction product, protecting the product from overoxidation by the electrophilic transition metal (Hammond, Conrad, and Hermans 2012).

Molten salts (sodium and potassium nitrate) have been also used as solvents in the direct oxidation of methane with oxygen mainly producing methanol and carbon oxides (Lee et al. 1998).

6.4.2.1.2 Catalysts Nature for Methane Oxidation

Different compounds have been reported as active catalysts in the homogeneous catalytic oxidation of methane to C1 oxygenates, such as mercury, palladium, platinum, iodine compounds, etc., which are well summarized by Balcer (2015).

This section is focused on methane partial oxidation using homogeneous organometallic and/or inorganic complexes as catalysts, in systems in which the reaction with methane occurs in the gaseous or liquid phase.

The reaction comprises the formation of the metal-alkyl intermediate via electrophilic attack of the transition metal, with a successive series of oxidative addition/reductive eliminations that produce the oxidized alkane species. A limitation of this process is that methanol is not directly obtained (Hammond, Conrad, and Hermans 2012).

Different groups (Table 6.2) have used salts as dry oxidants catalysts and fuming sulfuric acid (a mixture of H_2SO_4 and SO_3) as oxidant for the selective oxidation of methane to methanol, methane sulfonic acid and methyl bisulfate:

$$CH_4 + 2H_2SO_4 \rightarrow CH_3OSO_3H + 2H_2O + SO_2$$

TABLE 6.2

Homogeneous Catalytic Systems for Selective Oxidation of Methane (Adapted from Balcer 2015)

Compounds group	Catalyst	Reactants/medium	Products	Reaction conditions	Ref.
Hg compounds	Hg(II) salt*	CH_4 + fuming H_2SO_4	$CH_3OSO_3H \rightarrow$ CH3OH	180°C, 3.45 MPa	(Periana et al. 1993b)
	HgSO$_4$*	CH_4 + fuming H_2SO_4 (very conc.)	$CH_3OSO_3H \rightarrow CH_3OH$	150–200°C, 4–10 MPa	(Gang et al. 2000)
	HgSO$_4$	CH_4 + fuming H_2SO_4	CH_3SO_3H (at 90°C) CH_3OSO_3H (at 160°C)	90–160°C 6.9 MPa	(Basickes, Hogan, and Sen 1996)
	HgSO$_4$	CH_4 + fuming H_2SO_4	CH_3OSO_3H	180°C 6.9 MPa	(Sen et al. 1994)
	Hg(CF$_3$SO$_3$)$_2$	CH_4 + fuming H_2SO_4 + O_2	CH_3SO_3H CH_3OSO_3H (higher proportion of SO3 and O$_2$)	148°C 2.5 MPa	(Mukhopadhyay and Bell 2004b)
Pd compounds	Pd(II) salt	CH_4 + H_2O_2 (initiator) CF$_3$COOH	CH_3OH	90°C 5.5 MPa	(Kao, Hutson, and Sen 1991)
	Pd (II) acetate	CH_4 + CF$_3$COOH	CH_3OH/CF$_3$COOH	80°C 5.5 MPa	(Taylor, Anderson, and Noceti 1997)
	Pd^0	CH_4 + fuming H2SO4	CH_3OSO_3H + $(CH_3)_2SO_4 \rightarrow CH_3OH$	160–180°C 1.5–4.5 MPa	(Michalkiewicz, Kalucki, and Sosnicki 2003; Michalkiewcz and Kalucki 2003; Michalkiewicz 2003)
	PdSO$_4$	CH_4 + liquid H_2SO_4	CH_3COOH	180°C	(Periana et al. 2003)

(Continued)

TABLE 6.2 (CONTINUED)
Homogeneous Catalytic Systems for Selective Oxidation of Methane (Adapted from Balcer 2015)

Compounds group	Catalyst	Reactants/medium	Products	Reaction conditions	Ref.
Pt compounds	Pt-bipyrimidine	$CH_4 + H_2SO_4$	Methanol derivative	100°C	(Periana et al. 1998)
	$PtCl_4$	CH_4 + fuming H2SO4 (25% SO3)	CH_3OSO_3H	0.1 MPa, 130–220	(Michalkiewicz and Kosowski 2007; Michalkiewicz 2006)
	$PtCl_2$	$CH_4 + CO + H_2SO_4$	CH_3COOH	180°C 2.75 MPa (for CH4 and CO)	(Zerella and Bell 2006)
I compounds	I, KI, NaI, CH_3I, I_2O_5, KIO_3; KIO_4	CH_4 + fuming H_2SO_4 (65% SO3)	CH_3OSO_3H	4 MPa, 170–190°C	(Gang et al. 2004)
	I	CH_4 + fuming H_2SO_4 (2% SO_3)	CH_3OSO_3H	3.4 MPa, 195°C	(Periana et al. 2002)
	I	CH_4 + fuming H_2SO_4 (16–25% SO_3)	CH_3OSO_3H	130°C	(Michalkiewicz 2011)
Other compounds	$Co(OCOCF_3)_3$	$CH_4 + O_2$ In CF_3CO_2H solution	CF_3COOCH_3	150–180°C, 10–40 atm	(Vargaftik, Stolarov, and Moiseev 1990)
	$EuCl_3$ dissolved in CF_3CO_2H	$CH_4 + O_2 + Zn$	CH_3OH	40°C, CH_4 0.1–3 MPa, O2 (0.4 MPa)	(Yamanaka, Soma, and Otsuka 1995)
	$EuCl_3$ dissolved in CF_3CO_2H+ Bis(2,4,-pentanedionate)TiO and TiO_2	$CH_4 + O_2 + Zn$	CH_3OH	40°C, CH_4 0.1–3 MPa, O2 (0.4 MPa)	(Yamanaka, Soma, and Otsuka 1996)
	$K_2S_2O_8$	CH_4 + fuming H_2SO_4	CH_3OSO_3H	180°C 6.9 MPa	(Sen et al. 1994)

(Continued)

TABLE 6.2 (CONTINUED)
Homogeneous Catalytic Systems for Selective Oxidation of Methane (Adapted from Balcer 2015)

Compounds group	Catalyst	Reactants/medium	Products	Reaction conditions	Ref.
	Metal peroxide (M: Ca, Mg, Ba, ···), CaO_2, as the best	CH_4 + H2SO$_4$ (30% SO$_3$)	CH_3SO_3H	70°C, 4.5 MPa	(Mukhopadhyay and Bell 2003a)
	$Ce(SO_4)_2$	CH_4 + O_2 in fuming H_2SO_4 (27–33% SO3)	CH_3SO_3H	130°C, 2.5 MPa CH_4, 0.3 MPa O_2	(Mukhopadhyay and Bell 2004a)
	$K_4P_2O_8$	CH_4 in fuming H_2SO_4 (27–33% SO3)	CH_3SO_3H	95°C, 1.4 MPa CH_4	(Mukhopadhyay and Bell 2003b)
	Au^{3+} cations (from Au_2O_3 dissolution)	CH_4 + H_2SeO_4 + H_2SO_4	CH_3OH	180°C, 2.7 MPa CH_4	(Jones et al. 2004)
	thallium(III) and lead(IV) cations	CH_4 + trifluoroacetic acid solvent	Ester of CH_3OH	180°C	(Hashiguchi et al. 2014)
	2.5% Ba 2.5% La 0.5% Ni 0.1% Ru/SiO$_2$	CH_4 + O_2 + HBr	CH3COBr, CH$_3$COOH	600°C	(Wang et al. 2006)
	HgO/Zn-MCM-41	CH_4 + O_2 + Ar + Br$_2$	HCOOH, CH$_3$OH, (CH$_3$)$_2$–O	220°C	(Li et al. 2008)
	Cu^{I}Cu^{I}Cu^I complex	CH_4 + O_2 (or +H$_2$O$_2$)	CH_3OH	room temperature	(Nagababu et al. 2014)
	I$_2$, KI, KBr, KCl, NaCl, HCl	CH_4 + H$_2$SO$_4$ (25% SO$_3$)	CH_3OSO_3H → CH_3OH	160°C, 4.3 MPa CH_4	(Jarosinska et al. 2008)

* free radical initiators

$$CH_3OSO_3H + H_2O \rightarrow CH_3OH + H_2SO_4$$

$$SO_2 + 0.5O_2 + H_2O \rightarrow H_2SO_4$$

Noble metals salts and in their metallic state have been used as catalysts for homogeneous selective methane oxidation. This method could be based on methane esterification and further hydrolysis of methyl ester to methanol. Michalkiewicks, Ziebro, and Tomaszewska (2006) investigated the impact of catalyst nature on the esterification process. It has been also studied the separation of the ester-fuming sulfuric acid mixture by a low-pressure membrane distillation. There is a handicap in case of the conversion of methane to methanol in oleum (fuming sulfuric acid) via methyl bisulphate (CH_3OSO_3H), due to the formation of sulfuric acid and a huge amount of water consumption, being recommended the extraction of the ester before hydrolysis.

Table 6.2 compiles the main contributions in catalytic homogeneous methane oxidation.

6.4.2.2 Solid-Catalyzed Direct Oxidation

The direct partial oxidation of methane to C1-oxygenates, such as methanol and formaldehyde has been studied for decades, but, so far, the low yields hinders an industrial application:

$$CH_4(g) + \tfrac{1}{2}\,O_2(g) \leftrightarrow CH_3OH\,(g) \quad \Delta H_r^0(298\,K) = -126\,kJ \cdot mol^{-1}$$

$$CH_4(g) + O_2(g) \leftrightarrow HCHO(g) + H_2O \quad \Delta H_r^0(298\,K) = -368\,kJ \cdot mol^{-1}$$

The low yields achieved by direct partial oxidation under high temperature heterogeneous catalytic reaction is motivated by the fact that this process consist of a consecutive reaction proceeding through the homolytic C–H bond cleavage, but the C–H bond strength in H–CH$_2$OH (397 kJ/mol) and in H–CHO (364 kJ/mol) are lower than in H–CH$_3$ (439 kJ/mol) (Labinger 2004).

Silica is one of the most used substrates for active phases typically based on oxides from group V and VI. MoO$_x$/SiO$_2$ and VO$_x$/SiO$_2$ (mainly MoO$_3$ and V$_2$O$_5$ phases) are among the most studied catalytic compositions for gas-phase methane oxidation to C1-oxygenates (formaldehyde and methanol). However, the highest yields are around 4% at the most (de Vekki and Marakaev 2009). It is not easy to stablish structure-activity correlations, since silica has also a role in the reactivity and gas-phase occur in parallel to catalytic surface reactions (Horn and Schlogl 2015). In the formation of formaldehyde, for the catalysts based on Mo and V oxide supported on silica, there is clear evidence that the lattice oxygen is involved in the process, being formed on the surface, rather than in the gas phase; although it would by possible that the methyl radical may escape into the gas phase. As widely referenced in de Vekki and Marakaev review (2009), numerous authors adhere to the purely radical reaction or the ion-radical pathways.

Banares et al. (1997) applied high surface area silica-supported redox oxide catalysts (MO(x)/SiO$_2$; M = V, Mo, W, and Re) for the partial oxidation of methane, being found as primary products: formaldehyde, C_2H_n, and CO, while CO is formed from further oxidation of HCHO and hydrocarbons. Supported vanadium oxide was found to be the most active and selective due to its higher reducibility, providing more sites for oxygen activation. RhO_x-catalyst exhibited relative high yield to HCHO, however it deactivated from sublimation of metal oxide under reaction at high temperature.

6.4.2.2.1 *Molybdena*

Most of the catalysts based on molybdenum oxide used for the direct partial oxidation of methane into C1-oxygenates are supported on different substrates such as ZnO, Fe_2O_3, UO_2, VO_2, SiO_2, zeolites, and molybdenum oxometalates (Tabata et al. 2002).

MoO_3/SiO_2 has been used as catalyst for the partial oxidation of CH_4, being the major products: HCHO, CH_3OH, CO, CO_2, and H_2O. Kinetic analysis indicates that CH_4 is directly oxidized to HCHO and CO_2. HCHO is oxidized to CO, which is itself further oxidized, providing an alternative route to CO_2 (Spencer, Pereira, and Grasselli 1990).

The role of the different Mo species in MoO_3/SiO_2 catalysts was studied by Bañares and Fierro (1993), varying the surface concentration of Mo atoms. The results show that molybdenum oxide is highly dispersed on the silica at low molybdenum concentrations forming two-dimensional polymolybdates. It was found a correlation between relatively high methane conversion and formaldehyde selectivity and the presence of highly dispersed polymolybdate structures on the silica surface, with optimum around 1–2 Mo atom/nm^2 (Banares and Fierro 1993; Banares, Fierro, and Moffat 1993; Tabata et al. 2002) (Figure 6.5). Similar results were reported by Suzuki et al. (1995).

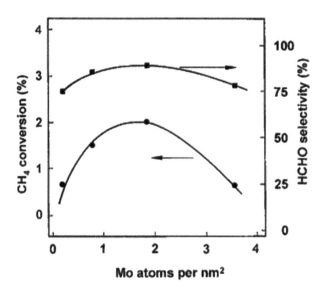

FIGURE 6.5   Influence of molybdenum content in MoO_3/SiO_2 catalysts on CH_4 conversion and HCHO selectivity at 873 K, a contact time of 3 sec, and a CH_4/O_2 molar ratio of 11. (Reprinted from Tabata et al. 2002, Copyright 2002, with permission from Taylor & Francis.)

It was shown that oxygen was incorporated into formaldehyde from the lattice of MoO_3, whereas carbon dioxide is produced from methane. The role of O_2 from the gas phase is restoring the oxidation state of Mo, being the process under a Mars–van Krevelen mechanism (Banares et al. 1993).

Isolated molybdate species supported on SiO_2 are reported to have the highest specific activity and selectivity for the direct oxidation of CH_4 to HCHO. X-ray absorption fine structure (EXAFS), isotopic labeling experiments and in situ Raman spectroscopy were used to study the Mo oxide structure and the reducibility of the dispersed molybdate species and the exchange of oxygen atoms between the gas phase and the catalyst using SiO_2-supported Mo oxide catalysts for the direct oxidation of CH_4 to HCHO. It is suggested that peroxides are formed by the reaction of O2 with a small concentration of reduced molybdate species and that the reaction of CH_4 with these peroxide species leads to the formation of HCHO. It was found oxygen exchange between the gas phase and the catalyst and a positive effect of low concentrations of H_2O on the HCHO formation rate (Ohler and Bell 2006).

Theoretical studies based on DFT calculations suggest that the active phase for the selective oxidation of CH4 to HCHO consists of di-oxo molybdate structures (Chempath and Bell 2007). Both mono-oxo and di-oxo molybdate structures were used to represent the active centers. The energetics for each elementary reaction and the entropy changes were determined. The results show that the mechanism based on di-oxo molybdate species is in accordance with observed rates of methane oxidation. It was also observed that the formation of HCHO occurs via the reaction of methane with peroxide species formed by the adsorption of O_2 on reduced Mo-IV sites.

The contribution of CH_4 and HCHO reactions occurring in the gas phase and on the surfaces of silica and dispersed MO_x during the selective oxidation of CH_4 to HCHO over MoO_x/SiO_2 were investigated (Ohler and Bell 2005). It was found that homogeneous oxidation of CH_4 was insignificant, and the homogeneous oxidation of HCHO contributed scarcely. The formation of HCHO was found to occur largely over MoO_x and only to a limited extension over silica. For SiO_2, the only process occurring was HCHO decomposition, which was independent of the CH_4/O_2 ratio, while in the case of MoO_x, HCHO decomposition was accompanied by direct HCHO oxidation, principally to CO.

Methanol has been also produced by direct partial oxidation of methane using catalysts based on Mo oxide, being found that $Fe_2O_3/(MoO_3)_3$ was the most active (working at 403–573 K, 97% methane-3% O_2 and 5 MPa). The yield to methanol may be enhanced if the temperature is cooled down lower than 473 K in less than 0.03 s (Dowden and Walker 1971).

The influence of MoOx aggregation (with catalysts prepared by different methods and different silicas) on the oxidation of methanol has been studied by XPS and Raman spectroscopy, being found that surface molybdenum oxide coverage on SiO_2 is the only important factor that determine the catalytic properties during methanol oxidation. It is found that an increase in the surface Mo coverage produces a decrease in the methanol oxidation catalytic activity (Banares, Hu, and Wachs 1994).

CuO/MoO_3 has been used as catalyst 2 MPa and 758 K), obtaining a yield of 490 g/kg cat-1·h-1 to oxygenates (methanol and formaldehyde). And some amount of ethanol and acetaldehyde because the feed stream contained also ethane (Stroud 1975).

The N_2O was assessed as oxidant in methane oxidation using MoO3/SiO2 catalyst, being found the highest yield to methanol and formaldehyde (84.6%, total selectivity) with 1.7% of MoO_3 (weight) working under 1 bar (0.1 MPa) and 833 K (Liu, Iwamoto, and Lunsford 1982). The use of 1.1 wt% MoO_3/SiO_2 (Banares, Fierro, and Moffat 1993) achieved higher CH4 conversion and HCHO selectivity with O_2 as oxidant in comparison to N_2O at 863 K and atmospheric pressure. The catalytic differences were explained by the different oxidizing capability of N_2O and O_2. Considering the Mars–van Krevelen mechanism, the re-oxidation of the catalyst was less effective with N_2O than with O_2.

The influence of other metal oxides as dopants (Cu, Co, Fe, Ga and Zn) of MoO_3 was studied (Taylor et al. 1998), being found that the net yield to methanol was higher with the catalyst doped with Ga_2O_3, which was explained by a synergism between MoO_3 and Ga_2O_3. Ga_2O_3 could activate methane to produce surface methyl activated species. These species could migrate to MoO_3 sites, which favor oxygen insertion, thus leading to the formation of methanol. The intimate contact between these two phases provides boundaries for the cooperation of both processes.

Considering the higher specific activity of the monomeric Mo species, attempts to deposit very dispersed Mo species have been done over different supports, such as HY and HZSM-5. Finely dispersed MoO_3 species can be obtained using different methods: (i) solid state reaction between $MoCl_5$ and hydroxyl groups of the zeolite, (ii) impregnation with ammonium heptamolybdate, followed by thermal decomposition under very low pressure for long periods of time, and (iii) adsorption-decomposition of $Mo(CO)_6$ complex (Banares, Pawelec, and Fierro 1992). Isolated oxo-species of Mo(VI) have been also obtained by grafting of MoO_2(acetylacetonate)$_2$ on the zeolite surface (Antinolo et al. 2000).

Ordered mesoporous SiO_2 has been also used as substrate of MoO_x in catalysts for selective oxidation of methane to formaldehyde. The activity of polymeric MoO_x species was significantly higher than that of isolated MoO_x, with a maximum HCHO yield achieved over 20 wt% MoO_x/SBA-15 catalyst at 873 K under atmospheric pressure (Lou et al. 2008). Using also SBA-15 support, small P-modified molybdenum oxide clusters encapsulated inside the mesoporous channels of SBA-15 were found to show superb catalytic performance for the partial oxidation of CH_4 to HCHO with O_2 (selectivity of 90% at a single-pass CH_4 conversion of $\approx$ 6%) (Yang et al. 2003).

Zirconia has been also used as support of MoO_x in catalysts for the selective oxidation of methane to formaldehyde with O2. The maximum yield to HCHO (ca. 4%) was obtained at 673 K, with 12 wt% Mo, and under the following reaction conditions: 5.0 MPa, $CH_4/O_2/N_2$ = 10/1/3, and 12000 ml·g$^{-1}_{catalyst}$ ·h^{-1}. The formation is related to the formation of $Zr(MoO_4)_2$ that depends on the Mo loading. The specific activity was greater with an increase of molybdenum oxide density. The Mo = O species of $Zr(MoO_4)_2$ enable selective oxidation of CH_4 to HCHO, while the more lattice oxygen species and bulk MoO_3 accelerated the deep oxidation of HCHO to CO and/ or CO_2 (Zhang et al. 2003).

The reactivity of MoO_x supported on La–Co–O and ZrO_2 was studied under high pressure (4.2–5 MPa) and temperatures of 653–693 K. The catalyst supported on zirconia gave only trace amounts of methanol. On the contrary, the catalyst supported on La–Co–O (7% of MoO_3) yielded around 6% of methanol (Zhang et al.

2005a). The authors explained the results attending to the reducibility, related to the $(O^-)/(O^{2-})$ ratio. This ratio is linked to the deficiency of oxygen lattice sites, being concluded that a certain deficiency contributes to the migration and transformation of O_2 from the gas phase to give O_2^- and O^- species on the catalyst surface.

6.4.2.2.2   Vanadia

Vanadia has been widely studied in catalysts for the selective oxidation of methane to methanol and formaldehyde, especially in the 1990s (Spencer and Pereira 1989; Faraldos et al. 1996; Kartheuser et al. 1993; Koranne, Goodwin, and Marcelin 1994; Parmaliana et al. 1995; Faraldos et al. 1995; Banares et al. 2000).

The partial oxidation of CH_4 to HCHO by molecular oxygen has been studied using vanadium oxide-silica catalysts in the temperature range 773–873 K at atmospheric pressure. CO was the major reaction product at the highest achieved conversion. CH_4 oxidation was found to follow a sequential reaction path, where HCHO is formed from CH_4, CO from HCHO, and CO_2 from CO. This reaction pathway is different than the observed for molybdenum oxide-silica-catalyzed reaction, where CO_2 is formed directly from CH_4 in a parallel pathway (Spencer and Pereira 1989).

As it happened for silica-supported Mo oxide catalysts, the reactivity for methane conversion using silica-supported vanadium oxide catalysts, seems to be essentially related to dispersed isolated surface metal oxide species. The reactivity of both active phases has been compared, being found that silica-supported vanadium oxide catalysts are more active (to convert methane) but less selective to HCHO than SiO_2-supported Mo oxide catalysts (Faraldos et al. 1996).

In order to study the role of lattice oxygen, getting insight into the reaction mechanism, $^{18}O_2$ isotopic technique was used, being confirmed that the reaction proceeds via a Mars–van Krevelen mechanism, by which the lattice oxygen forms HCHO, while O_2 from the gas phase restored the oxygen vacancy (Koranne, Goodwin, and Marcelin 1994), similarly to the results obtained for MoO_3/SiO_2 catalysts. It was also suggested that highly dispersed tetrahedral vanadia species would be involved in the primary oxidation reaction that produces C1 oxygenates, while secondary oxidation reactions, produced with oxygen associated with bulk V_2O_5 particles, would lead to deeper oxidized products.

Bañares et al. (1998) reported a significant increase in the yield to formaldehyde and methanol adding a small concentration of NO to a feed of CH_4+O_2 at 900 K. The yield to CH_3OH + HCHO was around 7% at atmospheric pressure, 883 K, CH_4/O_2 = 2, 0.39% NO molar, and a flow of 89.6 mL·min^{-1}. Moreover, any increase in the NO concentration leaded to increases in CH_3OH production, which suggests that this compound can be also formed through gas-phase reaction, while HCHO requires the heterogeneous catalytic activation. On the other hand, it has to be highlighted that, with and without NO in the feed, the selectivity-conversion data on V_2O_5/SiO_2 catalyst show a clear dependence on reaction temperature (Figure 6.6).

As widely reported, high space velocity and high CH_4/O_2 ratio result in a lower CH_4 conversion, while the HCHO space–time yield follows the opposite trend. The effect of space velocity, $CH_4{:}O_2$ ratio and the absence of addition of 1% NO to the feed was studied using a series of VO_x/SiO_2 catalysts prepared by a single step sol-gel method (with different proportions of V). The highest yield to formaldehyde (1600 g

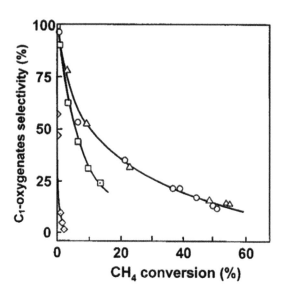

FIGURE 6.6 Selectivity to C1 oxygenates vs CH_4 conversion at different NO concentration levels in the feed. Reaction conditions: Flow: 89.6 mL·min^{-1}; catalyst weight: 0:20 g; atmospheric pressure; and 883 K. ($\Diamond$) 0.0% NO; ($\square$), 0.03% NO; ($\triangle$), 0.14% NO; ($\bigcirc$), 0.39% NO. (Reprinted from Tabata et al. 2002, Copyright 2002, with permission from Taylor & Francis.)

HCHO·h^{-1}·kg$^{-1}_{cat}$) was achieved for the catalyst with 1.5 wt % V, at high space velocity (2830 mLN/min·g$_{cat}$), and high CH_4/O_2 molar ratio (9). The most active-selective catalyst showed highly dispersed isolated/slightly oligomerized surface vanadium species, with a higher proportion of V–O–Si species present in isolated VO_4 tetrahedra (Loricera et al. 2017). The addition of nitric oxide increases the conversion of methane and has a strong effect on the product distribution being methanol the major C1-oxygenate compound observed. When the concentration of NO exceeds a limit, carbon oxides formation dominates (Banares et al. 1998).

V$_2$O$_5$–SiO$_2$ xerogels with wide-ranging V loadings and high surface areas were found to be selective for methane oxidation (Wang et al. 2003a). Among the three types of V species found: (i) accessible surface tetrahedral V^{5+}, (ii) inaccessible V^{5+} species, and (iii) crystalline V_2O_5. However, the accessible surface tetrahedral V^{5+} species were the active sites for CH_4 conversion to oxygenates. The calculation of turnover frequencies allowed the optimization of vanadium amount, being 1 wt.% V_2O_5–SiO$_2$ the catalyst with the highest yield to oxygenates.

Mesoporous SiO_2 has been uses as substrate for VO_x species, being found that D2 silica defects could be the preferential anchoring sites of the hydroxylated monomeric vanadium species, which would be the most efficient CH_4 selective oxidation sites. The positive effect of a water co-feed on the catalytic performance was explained by an increase in the number of those hydroxylated monomeric species (Launay et al. 2007a). In summary, the high performance has been attributed to the higher dispersion of vanadium-isolated species, which are very active and selective, and to a greater silanol surface content. The novel preparation method used by the authors favors the formation of isolated species (Nguyen et al. 2006). The molecular structure of these more active and selective catalytic sites was investigated by infrared and Raman spectroscopies. In situ and operando experiments have been conducted in order to understand these hydroxylation/dehydroxylation and redox reactions of the vanadium species (Launay et al. 2007b).

Ordered mesoporous silica, such as SBA-15 and MCM-41 have been compared as supports of vanadium oxide in the catalytic partial oxidation of methane to formaldehyde. These supports favor the formation of highly dispersed (monomeric and oligomeric) V species. For VOx/SBA-15, the maximum methane conversion is achieved for a V loading of $\approx$ 4 wt% (corresponding to the theoretical monolayer capacity for SBA-15), while the maximum activity was for 1–2% V, for amorphous silica-supported V catalysts, which is influenced by the higher surface area of this mesoporous support (Fornes et al. 2003). The optimized reaction conditions were: 891 K, GHSV = 417,000 L(N) kg^{-1} h^{-1}, and CH_4:O_2 molar ratio of 8:1. The obtained HCHO yield is significantly higher than the maximum reported for amorphous VO_x/SiO_2, and comparable to that recently reported for VO_x/MCM-41.

This excellent behavior of VO_x supported on SBA-15 for the selective oxidation of methane to formaldehyde was confirmed in other studies (Lin et al. 2003). Highly dispersed V species have been achieved using MCM-41 as support, being found an improved catalytic performance compared to an amorphous silica support. The high concentration of isolated V oxide sites is crucial to minimize the consecutive oxidation of formaldehyde to carbon oxides. The highest HCHO yield (2103 $g{\cdot}kg^{-1}$ h^{-1}) was obtained with the V-MCM-41 catalyst containing 1.86 wt.% vanadium loading at 933 K, CH_4:O_2 ratio of 13:1, GHSV of 1.3×10^6 $L{\cdot}kg^{-1}$ h^{-1} (Du et al. 2006).

The coordination structures of V in VO_x/MCM-41 catalysts for the catalytic partial oxidation of methane were studied by X-ray absorption spectroscopy (XANES and EXAFS). Vanadium is tetrahedrally coordinated with oxygen in V-MCM-41 prepared by direct hydrothermal synthesis and by template-ion exchange (TIE) methods, but the location of vanadium is different. It seems that vanadium is mainly incorporated inside the framework of MCM-41 by the DHT method if the V loading is lower than 1 wt%, while the vanadyl tetrahedra are probably dispersed on the wall surface of MCM-41 for the samples prepared by the TIE method. The V-MCM-41 prepared by the TIE method showed superior catalytic performance (Zhang et al. 2005b).

De Vekki and Marakaev (2009) addressed mechanistic insights into selective oxidation of methane to oxygenates on MoO_3/SiO_2 and VO_x/SiO_2 attending to homogeneous and catalyzed reactions. The oxidation kinetics of methane, methanol, and formaldehyde at 643–923 K were studied on MoO_3/SiO_2 and V_2O_5/SiO_2, a detailed parallel-consecutive reaction scheme is suggested (Lund 1992) (Scheme 6.3).

The limiting yields of formaldehyde are influenced by the fact that it is an intermediate in oxidation to CO. CO_2 is formed by two pathways: directly from methane and by oxidation of CO in the gas phase. In the first case, the yield of CO_2 is independent of the degree of methane conversion (MoO_3 catalyst), and in the second, it raises with the conversion (V_2O_5 catalyst). The source of oxygen in the oxygenate can be oxygen from the gas phase or from lattice oxygen. In some catalysts, where oxygen is firmly bound, it is not exchanged. However, for MoO_3/SiO_2 and V_2O_5/SiO_2, it has

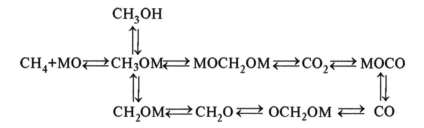

SCHEME 6.3 Possible reaction mechanism of methane oxidation to methanol and formaldehyde at 370–650°C (643–923 K) on MoO_3/SiO_2 and V_2O_5/SiO_2 (Reprinted from de Vekki and Marakaev (2009), Copyright (2009), with permission from Springer Nature.)

been demonstrated that the lattice oxygen is involved in the formation of partial and full oxidized products (Banares et al. 1994; Smith et al. 1993).

6.4.2.2.3 Zeolites

Zeolite-based catalysts (containing transition metals, such as Cu, Fe, and Co) offers a selective route to oxidize methane to methanol at relative low temperature (Smeets et al. 2010; Horn and Schlogl 2015) with O_2; H_2O_2, and N_2O as oxidants. The first works, reported by Groothaert et al. (2005) and Smeets, Groothaert, and Schoonheydt (2005), indicate that Cu/ZSM-5, activated with O_2 or N_2O, was able to convert methane into methanol. Some of the main contributions in this topic are summarized hereafter.

Fe/HZSM-5 was used in a flow reactor with O_2 or N_2O as oxidants. When the oxidant was O_2, only CO_2 was obtained. However, reaction with nitrous oxide resulted in the formation of carbon oxides, methanol, and formaldehyde (Anderson and Tsai 1985). The conversion decreased quickly due to the formation of coke.

Raman spectroscopy, O_2 labeling, X-ray absorption spectroscopy and DFT studies have been used to define the active center. For Cu-Mordenite, it was found that the structure of the active Cu site depends on the reaction conditions. In a dry environment, $[Cu_2O]^{2+}$ was suggested to be site that activates methane molecule; and in a moist oxidation environment, the activation of CH_4 is favored by a water stable Cu(II) oxide species (Woertink et al. 2009; Alayon et al. 2014).

The influence of the reactor system has been studied, being found that with Fe-zeolite and using H_2O_2 as oxidant in aqueous medium and at low temperature ($\approx$ 323 K), a maximum conversion of 10% and a methanol selectivity of 93% was achieved, while in a continuous system, a maximum conversion of 0.5% and a methanol selectivity of 92.2% was reported (Dinh et al. 2018; Hammond et al. 2012).

Hammond et al. (2012) studied the direct catalytic conversion of methane to methanol in an aqueous medium by using copper-promoted Fe-ZSM-5. The reaction was carried out with aqueous hydrogen peroxide in a batch reactor. Iron is involved in the activation of the C–H bond, while copper favor the formation of methanol. The results are promising since the catalyst is stable and reusable, giving 10% of conversion and a selectivity to methanol higher than 90%.

The stability and reactivity of copper oxo-clusters in ZSM-5 zeolite for methane oxidation to methanol has been approached using a periodic density functional theory study complemented by ab initio thermodynamic analysis in order to identify the active sites and reaction mechanism in Cu/ZSM-5 catalysts (Li et al. 2016). It was demonstrated that depending on the conditions of catalyst activation, binuclear [Cu(mu-O) Cu]$^{2+}$ species and trinuclear oxygenated [Cu$_3$(mu-O)$_3$]$^{2+}$ clusters can preferentially be stabilized in ZSM-5. The trinuclear Cu sites are the most stable extra-framework Cu species in Cu/ZSM-5 activated by calcination, whereas the formation of the binuclear complexes is favored under an oxygen-poor atmosphere. Binuclear Cu sites react with CH$_4$ stoichiometrically to form methoxy groups strongly bound in the support micropores. In contrast, trinuclear clusters favors the direct conversion of CH$_4$ to CH$_3$OH, pointing out that these clusters in ZSM-5 are potential candidates to promote the low temperature direct conversion of CH$_4$ to CH$_3$OH (Li et al. 2016).

The influence of Si/Al ratio in ZSM-5 has been studied for the selective oxidation of methane in aqueous hydrogen peroxide. Methyl hydroperoxide was confirmed as the initial product of methane oxidation. Formaldehyde was identified as an intermediate in the oxidation reaction pathway from methanol to formic acid and finally, carbon dioxide. Oxidation of the HCHO intermediate to formic acid was demonstrated to be the source of evolved hydrogen and HCHO appeared to be also the source of low levels of polyoxomethylene (Al-Shihri, Richard, and Chadwick 2017).

The influence of Si/Al ratio has been also studied in a series of Cu-, Fe- and Fe/ Cu-containing zeolite (ZSM-5, beta, Y) catalysts for the selective oxidation of methane to methanol using hydrogen peroxide as oxidizing agent. The acid sites strength and concentration were proved to have a strong influence on the yield to oxygenates. In particular, a significant increase in the methanol production was observed when lowering Si/Al ratio in the ZSM-5 supported catalysts. This can be explained by the increased amount of Brønsted acid sites. The Fe-only ZSM-5 catalysts exhibited the highest catalytic yield to oxygenates with HCOOH being the major product (Kalamaras et al. 2016). Combining catalytic and spectroscopic studies, Hammond et al. (2013) demonstrated that extra-framework Fe species are the active constituents of the catalyst for the selective methane oxidation to methanol in Cu–Fe–ZSM-5.

Reaction kinetics studies on the direct catalytic conversion of CH$_4$ into CH$_3$OH with O$_2$ at low temperature over copper-exchanged zeolites exhibited continuous catalytic activity and high selectivity for some commercial zeolite topologies under mild conditions (e.g., 483 K and atmospheric pressure). The catalytic turnover frequencies were confirmed by steady state and transient tests with isotopically labeled molecules. The catalytic rates and apparent activation energies are influenced by the zeolite topology, with caged-based zeolites showing the highest rates. Although the reaction rates are still low, this research opens the pathway for new developments on the catalytic oxidation of methane to methanol under mild reaction conditions (Narsimhan et al. 2016).

Recently, a broad overview of the direct conversion of methane to methanol using catalysts based on cation-exchanged zeolites has been published (Kulkarni et al. 2018), highlighting the role of theory in developing a molecular-level understanding of the reaction, analyzing a large database of DFT calculations for many transition metal cations, active site types and zeolite topologies. The authors present a unifying approach of CH$_4$ activation processes in terms of C–H bond activation, active site stability and CH$_3$OH extraction and define the key challenges and future strategies of this topic.

The recent advances in revealing the methane oxidation mechanism with inorganic biomimic Cu–ZSM-5 systems have also been reviewed. This study highlights the reactive intermediates that are involved, which is relevant in order to understand the working mechanism of Cu-oxidation enzymes, thus facilitating the development of more efficient catalysts for this process (Vanelderen et al. 2011).

Recently, some other metals, such as Rh in the form of isolated sites (in ZSM-5) that activate C–H bond in methane molecule, have been explored by Shan et al. (2017), opening new pathways of improving the efficiency.

6.4.2.2.4   Other Solid Catalysts

Heteropolycompounds (or Polyoxometalates)

Sun et al. (2014) have published a review about the application of heteropoly compounds (HPC) in catalytic oxidation of light alkanes, including methane, emphasizing the possible reaction mechanism and the effect of several parameters. These compounds are an important type of nanosized clusters containing transition metals, main group elements and a certain number of protons and/or water molecules. They constitute like "oxide molecules", which can favor the understanding of the molecular reaction mechanism. In this review, the studies on unsupported and supported HPC applied to the catalytic oxidation of CH_4 to CH_3OH, HCHO and HCOOH have been reported.

There are several structures that are active: (i) the primary one constituted by the heteropolyanion itself, (ii) the secondary structure, that is the 3D arrangement of polyanions, and (iii) the tertiary structure, represented by the way by which the secondary structures assembles into solid particles (Sun et al. 2014). The conversion seems to be related to the reaction temperature. The selectivity is controlled by the nature of the HPC and by the oxidant employed (determined by the type of ions, acidity and oxidizing properties). It has been reported that both conversion and selectivity to C1 oxygenates is improved by including vanadium or iron into the structure of heteropolyanion.

Some of these HPC catalysts: $(NH_4)_6HSiMo_{11}FeO_{40}$, $(NH_4)_4PMo_{11}FeO_{39}$, and $H_4PMo_{11}VO_4$ have been used in the partial oxidation of methane to formaldehyde and methanol at 700–750°C at atmospheric pressure using O_2 or N_2O as oxidant (Benlounes et al. 2008). The maximum yield (6.4 and 2.6% for formaldehyde and methanol, respectively) was obtained using $(NH_4)_6HSiMo_{11}FeO_{40}$ as catalyst, and O_2 as oxidant, at 700°C.

Polyoxometalates have been also used as selective catalyst in the oxidation of methane with H_2O_2 in aqueous phase, which constitutes a good example of the direct oxidation of methane under environmental-friendly conditions. Thus, for instance, Mizuno et al. (2000) studied the oxidation of CH_4 with H_2O_2 in water using a potassium salt of $SiW_{10}[Fe(OH)_2]_2O_{38}^{6-}$ polyoxometalate at 353 K, with the main products being methylformate and carbon dioxide.

Low-temperature oxygenation of methane into formic acid with molecular oxygen was achieved in the presence of H_2 catalyzed by $Pd_{0.08}Cs_{2.5}H_{1.34}PVMo_{11}O_{40}$ (Mizuno et al. 1997). No selective oxidation proceeded in the absence of Pd or when carbon monoxide was used instead of H_2. The selectivity to formic acid was higher than 67% and the highest yield was achieved at 300°C. The addition of 1.9 vol% hydrogen peroxide instead of a gaseous mixture of H_2 and O_2 led to similar conversion and selectivity to oxygenates.

Silica-supported heteropolyacids have been investigated in methane oxidation by oxygen or nitrous oxide (Kasztelan and Moffat 1987), although they reached a very low methanol selectivity (fixed-bed continuous flow reactor, 843 K).

Mansouri et al. (2013) have applied catalysts based on heteropolyoxometalates (anion: $[PW_{11}MO_{39}]$, M: Co(II), Ni(II), Fe(III)) for methane oxidation to methanol using N_2O or O_2. When the oxidant was nitrous oxide, the highest methanol yield was achieved by the catalyst $CsPW_{11}Fe$. When the temperature was raised to 923 K, the selectivity decreased. When the reactions were carried out using oxygen as oxidant, the conversion was lower, and the methanol selectivity was similar. In this case, $CsPW_{11}Co$-catalyzed reaction achieved the highest CH_4 conversion. The authors explained these results in basis to the different reaction pathways for methane oxidation by nitrous oxide or oxygen. In the first case, N_2O activation involves MO_2-active sites which provide methoxy intermediates for methanol and formaldehyde formation. On the contrary, the activation using oxygen could involve a possible hydrogen abstraction step by the lattice oxygen atoms of MO groups, producing organometallic intermediates (such as methyl-metal compounds).

Mizuno et al. (2000) have investigated the performance of di-iron silicontungstate complexes (water-soluble potassium salt of $SiW_{10}[Fe(OH)_2]_2O_{38}^{6-}$) for the oxidation of methane with hydrogen peroxide in water. Seki, Mizuno, and Misono (2000) studied the catalytic performance of various keggin-type heteropolyacids and V-containing complexes in trifluoroacetic acid anhydride for methane oxidation with hydrogen peroxide. The most efficient catalyst precursor was 11-molybdo-1-vanadophosphoric acid. Bar-Nahum, Khenkin, and Neumann (2004) obtained interesting results studying the aerobic partial oxidation of CH_4 in water catalyzed by $Pt(Mebipym)Cl_2]^+[H_4 PV_2Mo_{10}O_{40}]^-$ supported on SiO_2. The reactions were carried out at 50°C under a pressure of 3000 kPa methane and 200 kPa of O_2, being the TON reached 33 mol product/mol of catalyst. The presence of the polyoxometalate in the $[Pt(Mebipym)-Cl_2]^+[H_4PV_2Mo_{10}O_{40}]$-hybrid catalyst plays a key role, since enables mild aerobic oxidation of CH_4 and facilitates both (a) oxidation of Pt(II) to Pt(IV) intermediates and (b) the addition of methane and methanol to a Pt(II) center, improving oxidation of intermediate hydride species. Acetaldehyde is likely formed by the oxidation of methane to formaldehyde via methanol, followed by its coupling with methane, possibly occurring in the coordination sphere of the catalyst.

Metals

The selective oxidation of methane to methanol under mild conditions using H_2O_2 as oxidant has been also addressed using metals, although the low activity and selectivity of these phases are the reason of the scarce studies found in literature.

Williams et al. (2018) studied this process using bimetallic $AuPd/TiO_2$, prepared by stabilizer-free sol-immobilization. The catalyst showed low activity and selectivity, as well as deleterious hydrogen peroxide decomposition, which was explained by the small particle size of the metal nanoparticles. The key to get highly active catalysts seem to be related to the control of nanoparticle size, metal loading and metal oxidation state. These factors influence the harmful H_2O_2 decomposition. For instance, $AuPd/TiO_2$ catalysts were found to be more active after heat treatments, which facilitate particle size growth that, in turn, yields an improvement in the turn-over frequency and in a decrease in the H_2O_2 decomposition rate.

Metal Organic Frameworks

Recently, metal organic frameworks (MOF)–based catalyst precursors have been applied to the oxidation of methane with H_2O_2. The thermal and chemical stabilities of some of these types of materials, together with the scalability of their syntheses, make them attractive catalysts for the selective low-temperature conversion of light alkanes to higher-value oxygenates.

The reaction paths associated to Fe containing MIL-53(Al) MOF (MIL is Materials of Institute Lavoisier) used for the catalytic oxidation of methane with hydrogen peroxide were studied by DFT calculations (Szecsenyi et al. 2018). These results show that the activation barrier for the activation of the Fe sites is comparable to that of the subsequent C–H activation. The pronounced selectivity of the oxidation reaction over MIL-53(Al,Fe) towards the target CH_3OH and CH_3OOH products is attributed to the limited coordination freedom of the Fe species, which efficiently prevents the direct oxidation paths. These species are encapsulated in the extended octahedral $[AlO_6]$ structure-forming chains.

Copper-exchanged MOFs have been also reported to be active for selectively converting methane to methanol at 150–200°C. An increase in the reaction pressure from 0.1 to 4 MPa bar resulted in an increase of methanol yield. Combining studies of the stoichiometric activity with characterization by in situ X-ray absorption spectroscopy and density functional theory. It was concluded that dehydrated dinuclear copper oxyl sites formed after activation at 200°C are responsible for the activity (Zheng et al. 2019).

In this line, Fe(II) ions present as structural moieties in several metalorganic frameworks (e.g., MIL-100, MIL-101, and MIL-808) were identified by density functional calculations as promising catalysts for C–H bond activation for ethane and methane following the oxidative activation of iron. The authors have identified key changes in the chemical composition of the node (tri-iron oxide) that would modulate catalytic activity (Vitillo et al. 2019). Thus, considering the C–H bond activation enthalpies, Fe-(IV) = O units generated from the Fe(II) centers are predicted to oxidize methane to methanol and ethane to ethanol Moreover, it should be possible to desorb product alcohols from the catalysts by thermal activation without compromising the structure of the MOF.

Copper oxide clusters on the nodes of a MOF (NU-1000) are also active for the oxidation of methane to methanol under mild reaction conditions (Ikuno et al. 2017).

Transition Metal Complexes

Alkane oxidation (including methane) with hydrogen peroxide in acetonitrile has been addressed using catalysts based on simple iron(III) complexes, such as perchlorates, acetates and a catalyst based on a binuclear iron(III) complex with 1,4,7-triazacyclo-nonane. This can be considered as a structural model of some iron-containing biologically active compounds, and especially enzymes (Shul'pin et al. 2004). The same group also studied the oxidation of methane (also ethane, propane, n-butane, and isobutene) in acetonitrile solution by air and H_2O_2 at 20–75°C using a catalyst based on vanadium complex-pyrazine-2-carboxylic acid (Nizova, SussFink, and Shulpin 1997a, b). In addition to alkyl hydroperoxides, which are the primary oxidation products, alcohols, aldehydes or ketones, and carboxylic acids were also obtained.

Other Metal Oxides

A great part of the research on methane partial oxidation has been addressed using metal oxides. Considering that the most common catalysts for oxidizing methanol into formaldehyde are based on silver metal or on a mixture of an iron and molybdenum or vanadium oxides, several studies were focused in these oxides and also on Cr and W oxides (Tabata et al. 2002).

Otsuka and Hatano (1987) have tested numerous metal oxides, making an attempt on correlating activity with electronegativity of the respective cation. Ga_2O_3 and Bi_2O_3 reached the maximum methane conversion, whose electronegativity is in the middle of the electronegativity scale.

In order to maximize the yield to C1 oxygenates, step 3 has to be inhibited. These authors proposed a binary oxide mixture of Be and B supported on silica. They achieved a yield of 1% HCHO, working at 873 K and using a CH_4/O_2 ratio of 3 in the feed (0.43 $g \cdot h \cdot L^{-1}$).

Certain polymorphic forms of Nb_2O_5 (H and M phases, with a block type structure) resulted more selective to obtain formaldehyde, although the highest yield was obtained over $T\text{-}Nb_2O_5$ supported on silica (Michalkiewicz et al. 2008).

Phosphates

Iron phosphates are active for direct partial oxidation of methane. Wang and Otsuka (1995) have reported high selectivities using $FePO_4$ in the oxidation of methane with O_2, being observed an increase to C1 oxygenates yield in the presence of N_2O. These authors found that the ferrous pyrophosphate $(Fe_4(P_2O_5)_2)$ was less active than $FePO_4$ (bulk) for methane oxidation, since the coordination of Fe^{2+} in this compound does not activate CH_4. They propose that the active site for the formation of methanol by methane oxidation with a mixture of H_2 and O_2 is the iron site isolated by acid phosphate groups. The study of silica-supported $FePO_4$ catalysts by diffraction and spectroscopic techniques confirmed that iron forms two phases: nearly half is forming a crystalline $FePO_4$ structure, while the other half is in a ferric orthophosphate, with Fe^{3+} occupying the center of a trigonal bipyramid of oxygen atoms (Deguire et al. 1987; Tabata et al. 2002).

Cu–Fe pyrophosphates have been used as catalysts for methane oxidation to methanol by N_2O at high temperatures (903 K) and atmospheric pressure (Polniser et al. 2011). The maximum methane conversion was 5% and the maximum yield to oxygenates was 1.6%, being much higher than the selectivity to formaldehyde compared to methanol (in general, one order of magnitude higher). The yield to oxygenates depended on the molar ratio Fe/(Fe+Cu), being suggested a possible synergism between copper and iron cations. On the other hand, an increase of the reaction temperature, resulted in an increase of methane conversion, but with a concomitant decrease of methanol selectivity. The replacement of nitrous oxide by oxygen starts with methanol selectivity of around 10%, but drops to almost zero when the conversion reached 30%.

The possible synergism between Cu and Fe was investigated using similar active phases, but supported on ZnO (Sojka, Herman, and Klier 1991). The authors achieved a formaldehyde yield of 76 g/kg$_{cat}$·h using a feed mixture methane: air equal to 1 and working between 773 and 1123 K. They suggested that the oxidation of methyl intermediates to methoxide could be favored by promotion of both Lewis acidity and the Fe(II)/(III) and Cu(I)/(II) redox coupling.

The activity of FePO$_4$ catalysts supported on Al$_2$O$_3$, TiO$_2$, and SiO$_2$ has been tested for the selective oxidation of methane with oxygen as oxidant. The catalytic behavior depended on the type of support, being silica the support that gave the highest yield to oxygenates, while alumina-supported catalyst exhibited the lowest selectivity to HCHO. The best performance of the silica-supported catalyst is attributed to the presence of easily reducible Fe, in high coordination, and isolated by phosphate groups (McCormick and Alptekin 2000).

Iron phosphate/MCM-41 has been studied for partial oxidation of methane with both oxygen and nitrous oxide. It is suggested that the supported-iron phosphate species that are dispersed in the mesopores of MCM-41 can be more easily reduced than unsupported iron phosphate. Supported-iron phosphate onto MCM-41 increases both methane conversion and selectivity to oxygenates with both oxygen and nitrous oxide. The supported catalyst increases the activation of nitrous oxide, thus inhibiting the formation of carbon deposits (Wang et al. 2003b).

6.5   DIRECT METHANE OXIDATION TO HIGHER OXYGENATES

Homogeneous selective oxidation of methane to methyl formate at low temperatures has been widely studied over heteropolycompound catalysts in highly acidic media by Seki, Mizuno, and Misono (1997), Seki et al. (2000) and reviewed by Sun et al. (2014). Seki, Mizuno, and Misono (1997) first reported the oxygenation of CH$_4$ with H$_2$O$_2$ using H$_4$PVMo$_{11}$O$_{40}$ in (CF$_3$CO)$_2$O solvent at 80°C. and 50 atm, with methyl formate as the main product.

Kitamura et al. (1998) have studied the partial oxidation of methane to methyl acetate and methyl triofluoroacetate at low temperatures using heteropolyacids as solvents (typical reaction conditions: 80°C, 20 atm, trifluoroacetic acid as solvent) and K$_2$S$_2$O$_8$ as the most efficient solvent. H$_5$PV$_2$W$_{10}$O$_{40}$ was one of the most active catalysts. Vanadium appear to promote the reaction, that it is believed to go through a radical mechanism (Kitamura et al. 1998; Sun et al. 2014).

Hereafter, some heterogeneous catalytic approaches are mentioned. Methane has been oxidized to trifluoromethyl acetate (obtaining formic acid as by-product) in an aqueous solution of trifluoroacetic, using metallic palladium on an active carbon support and Cu(CH$_3$COO)$_2$ catalysts (Park, Choi, and Lee 2000).

Peng et al. used molten salts (silver and sodium nitrate) as solvent for the direct oxidation of methane, producing methyl trifluoroacetate and propanone, using the following catalysts: Cu(CF$_3$CO$_2$)$_2$, Co(CF$_3$CO$_2$)$_2$, Ce(SO$_4$)$_2$, MnSO$_4$ and CuSO$_4$ (Peng and Deng 2000).

Another approach to produce higher oxygenates by reaction methane with oxygen is by combining different catalytic processes. Thus, high-yield of propanal has been obtained from methane and air by combining three catalytic routes (Green et al.

1992): (i) ethene is produced from the oxidative coupling reaction, (ii) carbon monoxide and hydrogen are produced from partial oxidation of methane, and (iii) these gases are converted into propanal using a hydroformylation catalyst.

6.6 CONCLUSIONS AND OUTLOOK

Today, greener technologies that minimize pollutants emissions while maximizing energy, fuel and, chemicals output are desired. This trend is supported by the need of reducing the dependence on petroleum and by increasing the use of natural gas, whose major component is methane.

Different noncatalytic and catalytic (homogeneous and heterogeneous) routes have been studied during the last decades in order to develop an industrial process to produce oxygenates (being alcohols the highest volume commodity chemicals) from methane by direct oxidation, without going through the production of syngas. This established technology is based on reactions at high temperature, with multiple steps that lead to an inefficient and expensive process.

Thus, the development of direct alkane oxidation processes at low temperature and highly selective could promote the transition from petroleum dependence towards new technologies based on the valorization of natural gas reserves as primary feedstocks for the production of chemicals and fuels. One of the main natural gas resources that could be used to valorize them are those associated to petroleum drilling operations, since, nowadays, these reserves are just flared (or combusted). This change of concept not only would have economic incentives but also would have positive environmental consequences, due to a concomitant decrease in carbon dioxide emissions, which are directly related with global climate change.

In spite of the enormous research work on this field, so far, it does not exist a catalytic system that can convert methane to C1 oxygenates with high yield using oxygen as the only oxidant in a single step. The main problem is associated to overoxidation of methane to carbon dioxide, due to the high C–H bond energy of CH_4 molecule compared to the lower C–H bond energy of the partially oxidized products. This would be one of the main reasons of the low yield to oxygenates by these direct processes: the high reactivity of the desired oxygenate product under the reaction conditions. An industrial application of the direct conversion of methane to oxygenates would require an increase in selectivity (> 90%), at a minimum conversion of around 20% for methane and oxidant per pass. On the other hand, another key parameter, the volumetric productivity ($mol \cdot cm^{-3} \cdot s^{-1}$), would determine the size and cost of the reactor. In order to reduce the expenses, it will be desired to decrease the reaction temperature. Moreover, the use of air as oxidant, instead of pure O_2, will reduce the cost of investing in an oxygen separation plant. Also, in order to reduce the cost, the process should have few steps and use inexpensive reactors; the product should be isolated easily, and the reaction pressure should be lower than 3.5 MPa. The loss of selectivity typically is due to the formation of CO_2, which could be alleviated investing in products separation and heat management technologies.

In comparison to methane oxidation to formaldehyde, the progress in the selective oxidation of methane to methanol is significant. Concerning the approach by homogeneous catalytic oxidation, Goldshleger et al. (1972) (Shilov's group) and Periana et al. (1998) have applied different complexes that cleavage C–H bond of methane at low temperatures with high selectivity. Periana's group (1998) reported a process that represents a breakthrough, being achieved more than 70% one-pass yield of methanol by direct, low-temperature oxidative conversion of methane, using platinum complexes derived from the bidiazine ligand family in concentrated sulfuric acid. This process produced methyl bisulfate, which can be hydrolyzed to methanol and sulfuric acid. However, it would be desirable that the process could be carried out using less corrosive and harmless solvents and with heterogeneous catalysts. In this approach, there are several key issues that have to be further investigated such as the influence over CH activation reaction by water and reaction products, design of catalysts that are stable under the conditions required for methane functionalization and the development of new functionalization reactions that can be used with facile activation of C–H bond. Using this approach, Palkovitz et al. (2009) developed heterogeneous catalysts, which were able to catalyze the selective methane oxidation (in oleum) to methyl sulfate and after, to methanol, at low temperature using a catalyst based on Pt-covalent triazine framework formed by the trimerization of aromatic nitriles in molten $ZnCl_2$. High activity at high selectivity and stability over several recycling steps was achieved, contributing to a breakthrough for this reaction.

Concerning the results obtained by gas-phase oxidation of methane to methanol or formaldehyde, using classical heterogeneous catalysts based on oxides (such as MoOx or VO_x, mainly supported on silica), the highest yields per pass are around 4–5%, values that do not favor an industrial implementation. It has been widely reported that the presence of a small amount of nitrogen oxides in the feed promotes the selectivity to C1 oxygenates. On the other hand, progress in the development of most efficient catalyst formulations for this type of processes, derived by stablishing structure-activity correlations, is hampered by the support reactivity and by the occurrence of gas-phase reactions in parallel to the catalytic surface reactions.

Promising heterogeneous catalytic processes are based on the use of zeolite catalysts containing transition metal ions. It has been reported that methanol can been produced by direct partial oxidation of methane over Fe-zeolite using H_2O_2 as an oxidant at low temperature, in an aqueous medium (with yield ~ 10%). However, the industrial application of this system is, to date, unfeasible due to the high cost of the oxidant. Moreover, recovering dilute methanol from aqueous solution constitutes a challenge. In a gas-phase system, using O_2 as oxidant, a looping process with temperature and feed changes would be required, in order to minimize deeper oxidation.

In general, a proper strategy for developing efficient catalysts for the direct oxidation of methane to oxygenates should consider all structure-function correlations in order to increase stability, rate and selectivity.

REFERENCES

Al-Shihri, S., C. J. Richard, and D. Chadwick. 2017. Selective oxidation of methane to methanol over ZSM-5 catalysts in aqueous hydrogen peroxide: role of formaldehyde. *Chemcatchem* 9 (7):1276–1283.

Alayon, E. M. C., M. Nachtegaal, A. Bodi, and J. A. van Bokhoven. 2014. Reaction conditions of methane-to-methanol conversion affect the structure of active copper sites. *ACS Catal.* 4 (1):16–22.

Alvarez-Galvan, M. C., N. Mota, M. Ojeda, S. Rojas, R. M. Navarro, and J. L. G. Fierro. 2011. Direct methane conversion routes to chemicals and fuels. *Catal. Today* 171 (1):15–23.

Anderson, J. R. and P. Tsai. 1985. Oxidation of methane over H-ZSM5 and other catalysts. *Appl. Catal.* 19 (1):141–152.

Antinolo, A., P. Canizares, F. Carrillo-Hermosilla, et al. 2000. A grafted methane partial oxidation catalyst from $MoO_2(acac)_2$ and HZSM-5 zeolite. *Appl. Catal. A Gen.* 193 (1–2):139–146.

Balcer, S. 2015. Homogeneous catalytic systems for selective oxidation of methane: state of the art. *Polish J. Chem. Technol.* 17 (3):52–61.

Baldwin, T. R., R. Burch, G. D. Squire, and S. C. Tsang. 1991. Influence of homogeneous gas-phase reactions in the partial oxidation of methane to methanol and formaldehyde in the presence of oxide catalysts. *Appl. Catal.* 74 (1):137–152.

Banares, M. A. and J. L. G. Fierro. 1993. Selective oxidation of methane to formaldehyde on supported molybdate catalysts. *Catal. Lett.* 17 (3–4):205–211.

Banares, M. A., B. Pawelec, and J. L. G. Fierro. 1992. direct conversion of methane to C-1 oxygenates over MoO_3-USY zeolites. *Zeolites* 12 (8):882–888.

Banares, M. A., H. C. Hu, and I. E. Wachs. 1994. Molybdena on silica catalysts – role of preparation methods on the structure selectivity properties for the oxidation of methanol. *J. Catal.* 150 (2):407–420.

Banares, M. A., I. Rodriguezramos, A. Guerreroruiz, et al. 1993. Mechanistic aspects of the selective oxidation of methane to C1-oxygenates over MoO_3/SiO_2 catalysts in a single catalytic step. *Stud. Surf. Sci. Catal.* 75:1131–1144.

Banares, M. A., J. H. Cardoso, F. Agullo-Rueda, J. M. Correa-Bueno, and J. L. G. Fierro. 2000. Dynamic states of V-oxide species: reducibility and performance for methane oxidation on V_2O_5/SiO_2 catalysts as a function of coverage. *Catal. Lett.* 64 (2–4):191–196.

Banares, M. A., J. H. Cardoso, G. J. Hutchings, J. M. C. Bueno, and J. L. G. Fierro. 1998. Selective oxidation of methane to methanol and formaldehyde over V_2O_5/SiO_2 catalysts. Role of NO in the gas phase. *Catal. Lett.* 56 (2–3):149–153.

Banares, M. A., J. L. G. Fierro, and J. B. Moffat. 1993. The partial oxidation of methane on MoO_3/SiO_2 catalysts – influence of the molybdenum content and type of oxidant. *J. Catal.* 142 (2):406–417.

Banares, M. A., L. J. Alemany, M. L. Granados, M. Faraldos, and J. L. G. Fierro. 1997. Partial oxidation of methane to formaldehyde on silica-supported transition metal oxide catalysts. *Catal. Today* 33 (1–3):73–83.

Banares, M. A., N. D. Spencer, M. D. Jones, and I. E. Wachs. 1994. Effect of alkali-metal cations on the structure of Mo(VI) SiO_2 catalysts and its relevance to the selective oxidation of methane and methanol. *J. Catal.* 146 (1):204–210.

Barbaux, Y., A. R. Elamrani, E. Payen, L. Gengembre, J. P. Bonnelle, and B. Grzybowska. 1988. Silica supported molybdena catalysts – characterization and methane oxidation. *Appl. Catal.* 44 (1–2):117–132.

Bar-Nahum, I., A. M. Khenkin, and R. Neumann. 2004. Mild, aqueous, aerobic, catalytic oxidation of methane to methanol and acetaldehyde catalyzed by a supported bipyrimidinylplatinum-polyoxometalate hybrid compound. *J. Am. Chem. Soc.* 126 (33):10236–10237.

Basickes, N., T. E. Hogan, and A. Sen. 1996. Radical-initiated functionalization of methane and ethane in fuming sulfuric acid. *J. Am. Chem. Soc.* 118 (51):13111–13112.

Benlounes, O., S. Mansouri, C. Rabia, and S. Hocine. 2008. Direct oxidation of methane to oxygenates over heteropolyanions. *J. Nat. Gas Chem.* 17 (3):309–312.

Bromly, J. H., F. J. Barnes, S. Muris, X. You, and B. S. Haynes. 1996. Kinetic and thermo-dynamic sensitivity analysis of the NO-sensitised oxidation of methane. *Comb. Sci. Technol.* 115 (4–6):259–296.

Burch, R., G. D. Squire, and S. C. Tsang. 1989. Direct conversion of methane into methanol. *J. Chem. Soc. Faraday Trans. I* 85:3561–3568.

Chempath, S. and A. T. Bell. 2007. A DFT study of the mechanism and kinetics of meth-ane oxidation to formaldehyde occurring on silica-supported molybdena. *J. Catal.* 247 (1):119–126.

Choudhary, T. V. and V. R. Choudhary. 2008. Energy-efficient syngas production through, catalytic oxy-methane reforming reactions. *Angew. Chem. Int. Ed.* 47 (10):1828–1847.

Chun, J. W. and R. G. Anthony. 1993. Catalytic oxidations of methane to methanol. *Ind. Eng. Chem. Res.* 32 (2):259–263.

Conley, B. L., W. J. Tenn, K. J. H. Young, et al. 2006. Design and study of homogeneous catalysts for the selective, low temperature oxidation of hydrocarbons. *J. Mol. Catal. A Chem.* 251 (1–2):8–23.

Crabtree, R. H. 1995. Aspects of methane chemistry. *Chem. Rev.* 95 (4):987–1007.

da Silva, M. J. 2016. Synthesis of methanol from methane: challenges and advances on the multi-step (syngas) and one-step routes (DMTM). *Fuel Proc. Technol.* 145:42–61.

de Vekki, A. V. and S. Marakaev. 2009. Catalytic partial oxidation of methane to formalde-hyde. *Russian J. Appl. Chem.* 82 (4):521–536.

Deguire, M. R., T. R. S. Prasanna, G. Kalonji, and R. C. Ohandley. 1987. Phase-equilibria in the iron-oxide cobalt oxide phosphorus oxide system. *J. Am. Ceram. Soc.* 70 (11):831–837.

Dinh, K. T., M. M. Sullivan, P. Serna, R. J. Meyer, M. Dinca, and Y. Roman-Leshkov. 2018. Viewpoint on the partial oxidation of methane to methanol using Cu- and Fe-exchanged zeolites. *ACS Catal.* 8 (9):8306–8313.

Dowden, D. A. and G. T. Walker. 1971. Oxygenated hydrocarbons production, *Brit. Pat. GB1,244,001.*

Du, G., S. Lim, Y. Yang, C. Wang, L. Pfefferle, and G. L. Haller. 2006. Catalytic performance of vanadium incorporated MCM-41 catalysts for the partial oxidation of methane to formaldehyde. *Appl. Catal. A Gen.* 302 (1):48–61.

Ellis, Jr. P. E. and J. E. Lyons. 1992. Cyano- and polycyanometalloporphyrins as catalysts for alkane oxidation, *United States Pat. 5,118,886.*

Falconer, J. W., D. E. Hoare, and R. Overend. 1973. Oxidation of methane by nitrous-oxide. *Combust. Flame* 21 (3):339–344.

Faraldos, M., M. A. Banares, J. A. Anderson, and J. L. G. Fierro. 1995. Structural features of silica-supported vanadia catalysts and their relevance in the selective oxidation of methane to formaldehyde. In *Methane and Alkane Conversion Chemistry*, ed. M. M. Bhasin and D. W. Slocum, 241–247. New York, NY: Plenum Press Div Plenum Publishing Corp.

Faraldos, M., M. A. Banares, J. A. Anderson, H. C. Hu, I. E. Wachs, and J. L. G. Fierro. 1996. Comparison of silica-supported MoO_3 and V_2O_5 catalysts in the selective partial oxida-tion of methane. *J. Catal.* 160 (2):214–221.

Fornes, V., C. Lopez, H. H. Lopez, and A. Martinez. 2003. Catalytic performance of mesopo-rous VOx/SBA-15 catalysts for the partial oxidation of methane to formaldehyde. *Appl. Catal. A Gen.* 249 (2):345–354.

Foulds, G. A. and B. F. Gray. 1995. Homogeneous gas-phase partial oxidation of methane to methanol and formaldehyde. *Fuel Proc. Technol.* 42 (2–3):129–150.

Foulds, G. A., B. F. Gray, S. A. Miller, and G. S. Walker. 1993. Homogeneous gas-phase oxidation of methane using oxygen as oxidant in an annular reactor. *Ind. Eng. Chem. Res.* 32 (5):780–787.

Gang, X., H. Birch, Y. Zhu, H. A. Hjuler, and N. J. Bjerrum. 2000. Direct oxidation of methane to methanol by mercuric sulfate catalyst. *J. Catal.* 196 (2):287–292.

Gang, X., Y. Zhu, H. Birch, H. A. Hjuler, and N. J. Bjerrum. 2004. Iodine as catalyst for the direct oxidation of methane to methyl sulfates in oleum. *Appl. Catal. A Gen.* 261 (1):91–98.

Gesser, H. D., N. R. Hunter, and P. A. Das. 1992. The ozone sensitized oxidative conversion of methane to methanol and ethane to ethanol. *Catal. Lett.* 16 (1–2):217–221.

Goldshleger, N. F., A. A. Shteinma, A. E. Shilov, and V. V. Eskova. 1972. Reactions of alkanes in solutions of chloride complexes of platinum. *Russian J. Phys. Chem. USSR* 46 (5):785–786.

Green, M. L. H., S. C. Tsang, P. D. F. Vernon, and A. P. E. York. 1992. High-yield synthesis of propanal from methane and air. *Catal. Lett.* 13 (4):341–347.

Groothaert, M. H., P. J. Smeets, B. F. Sels, P. A. Jacobs, and R. A. Schoonheydt. 2005. Selective oxidation of methane by the bis(mu-oxo)dicopper core stabilized on ZSM-5 and mordenite zeolites. *J. Am. Chem. Soc.* 127 (5):1394–1395.

Gunsalus, N. J., A. Koppaka, S. H. Park, S. M. Bischof, B. G. Hashiguchi, and R. A. Periana. 2017. Homogeneous functionalization of methane. *Chem. Rev.* 117 (13):8521–8573.

Hammond, C., M. M. Forde, M. H. Ab Rahim, et al. 2012. Direct catalytic conversion of methane to methanol in an aqueous medium by using copper-promoted Fe-ZSM-5. *Angew. Chem. Int. Ed.* 51 (21):5129–5133.

Hammond, C., N. Dimitratos, R. L. Jenkins, et al. 2013. Elucidation and evolution of the active component within Cu/Fe/ZSM-5 for catalytic methane oxidation: from synthesis to catalysis. *ACS Catal.* 3 (4):689–699.

Hammond, C., S. Conrad, and I. Hermans. 2012. Oxidative methane upgrading. *Chemsuschem* 5 (9):1668–1686.

Hargreaves, J. S. J., G. J. Hutchings, and R. W. Joyner. 1990. Control of product selectivity in the partial oxidation of methane. *Nature* 348 (6300):428–429.

Hashiguchi, B. G., C. H. Hövelmann, S. M. Bischof, K. S. Lokare, C. H. Leung, and R. A. Periana. 2010. Methane to methanol conversion. In *Energy Production and Storage: Inorganic Chemical Strategies for a Warming World*, ed. R. H. Crabtree, 101–142. West Sussex: Wiley & Sons Ltd.

Hashiguchi, B. G., M. M. Konnick, S. M. Bischof, et al. 2014. Main-group compounds selectively oxidize mixtures of methane, ethane, and propane to alcohol esters. *Science* 343 (6176):1232–1237.

Horn, R. and R. Schlogl. 2015. Methane activation by heterogeneous catalysis. *Catal. Lett.* 145 (1):23–39.

Hunter, N. R., H. D. Gesser, L. A. Morton, P. S. Yarlagadda, and D. P. C. Fung. 1990. Methanol formation at high-pressure by the catalyzed oxidation of natural-gas and by the sensitized oxidation of methane. *Appl. Catal.* 57 (1):45–54.

Ikuno, T., J. Zheng, A. Vjunov, et al. 2017. Methane oxidation to methanol catalyzed by Cu-Oxo clusters stabilized in NU-1000 metal-organic framework. *J. Am. Chem. Soc.* 139 (30):10294–10301.

Irusta, S., E. A. Lombardo, and E. E. Miro. 1994. Effects of no and solids on the oxidation of methane to formaldehyde. *Catal. Lett.* 29 (3–4):339–348.

Jarosinska, M., K. Lubkowski, J. G. Sosnicki, and B. Michalkiewicz. 2008. Application of halogens as catalysts of CH_4 esterification. *Catal. Lett.* 126 (3–4):407–412.

Jones, C. J., D. Taube, V. R. Ziatdinov, et al. 2004. Selective oxidation of methane to methanol catalyzed, with C-H activation, by homogeneous, cationic gold. *Angew. Chem. Int. Ed.* 43 (35):4626–4629.

Kalamaras, C., D. Palomas, R. Bos, A. Horton, M. Crimmin, and K. Hellgardt. 2016. Selective oxidation of methane to methanol over Cu- and Fe-exchanged zeolites: the effect of Si/Al molar ratio. *Catal. Lett.* 146 (2):483–492.

Kao, L. C., A. C. Hutson, and A. Sen. 1991. Low-temperature, palladium(II)-catalyzed, solution-phase oxidation of methane to a methanol derivative. *J. Am. Chem. Soc.* 113 (2):700–701.

Kartheuser, B., B. K. Hodnett, H. Zanthoff, and M. Baerns. 1993. Transient experiments on the selective oxidation of methane to formaldehyde over V_2O_5/SiO_2 studied in the temporal-analysis-of-products reactor. *Catal. Lett.* 21 (3–4):209–214.

Kasztelan, S. and J. B. Moffat. 1987. The oxidation of methane on heteropolyoxometalates 1. Catalytic properties of silica-supported heteropolyacids. *J. Catal.* 106 (2):512–524.

Kitamura, T., D. G. Piao, Y. Taniguchi, and Y. Fujiwara. 1998. Heteropolyacid-catalyzed partial oxidation of methane in trifluoroacetic acid. In *Natural Gas Conversion V*, ed. A. Parmaliana, D. Sanfilippo, F. Frusteri, A. Vaccari, and F. Arena, vol. 119, 301–305. Amsterdam: Elsevier Science Publ BV.

Koranne, M. M., J. G. Goodwin, and G. Marcelin. 1994. Partial oxidation of methane over silica-supported and alumina-supported vanadia catalysts. *J. Catal.* 148 (1):388–391.

Kulkarni, A. R., Z. J. Zhao, S. Siahrostami, J. K. Norskov, and F. Studt. 2018. Cation-exchanged zeolites for the selective oxidation of methane to methanol. *Catal. Sci. Technol.* 8 (1):114–123.

Labinger, J. A. 2004. Selective alkane oxidation: hot and cold approaches to a hot problem. *J. Mol. Catal. A Chem.* 220 (1):27–35.

Launay, H., S. Loridant, A. Pigamo, J. L. Dubois, and J. M. M. Millet. 2007a. Vanadium species in new catalysts for the selective oxidation of methane to formaldehyde: specificity and molecular structure dynamics with water. *J. Catal.* 246 (2):390–398.

Launay, H., S. Loridant, D. L. Nguyen, A. M. Volodin, J. L. Dubois, and J. M. M. Millet. 2007b. Vanadium species in new catalysts for the selective oxidation of methane to formaldehyde: activation of the catalytic sites. *Catal. Today* 128 (3–4):176–182.

Lee, B. J., S. Kitsukawa, H. Nakagawa, S. Asakura, and K. Fukuda. 1998. The partial oxidation of methane to methanol with nitrite and nitrate melts. *Z. Naturforsch. B J. Chem. Sci.* 53 (7):679–682.

Li, F., G. Yuan, F. Yan, and F. Yan. 2008. Bromine-mediated conversion of methane to C1 oxygenates over Zn-MCM-41 supported mercuric oxide. *Appl. Catal. A Gen.* 335 (1):82–87.

Li, G., P. Vassilev, M. Sanchez-Sanchez, J. A. Lercher, E. J. M. Hensen, and E. A. Pidko. 2016. Stability and reactivity of copper oxo-clusters in ZSM-5 zeolite for selective methane oxidation to methanol. *J. Catal.* 338:305–312.

Lin, B., X. Wang, Q. Guo, W. Yang, Q. Zhang, and Y. Wang. 2003. Excellent catalytic performances of SBA-15-supported vanadium oxide for partial oxidation of methane to formaldehyde. *Chem. Lett.* 32 (9):860–861.

Lin, M., T. E. Hogan, and A. Sen. 1996. Catalytic carbon-carbon and carbon-hydrogen bond cleavage in lower alkanes. Low-temperature hydroxylations and hydroxycarbonylations with dioxygen as the oxidant. *J. Am. Chem. Soc.* 118 (19):4574–4580.

Liu, R. S., M. Iwamoto, and J. H. Lunsford. 1982. Partial oxidation of methane by nitrous-oxide over molybdenum oxide supported on silica. *J. Chem. Soc. Chem. Commun.* 11 (1):78–79.

Lodeng, R., O. A. Lindvag, P. Soraker, P. T. Roterud, and O. T. Onsager. 1995. Experimental and modeling study of the selective homogeneous gas-phase oxidation of methane to methanol. *Ind. Eng. Chem. Res.* 34 (4):1044–1059.

Loricera, C. V., M. C. Alvarez-Galvan, R. Guil-Lopez, A. A. Ismail, S. A. Al-Sayari, and J. L. G. Fierro. 2017. Structure and reactivity of sol-gel V/SiO$_2$ catalysts for the direct conversion of methane to formaldehyde. *Topics Catal.* 60 (15–16):1129–1139.

Lou, Y. C., Q. H. Tang, H. R. Wang, B. T. Chia, Y. Wang, and Y. H. Yang. 2008. Selective oxidation of methane to formaldehyde by oxygen over SBA-15-supported molybdenum oxides. *Appl. Catal. A Gen.* 350 (1):118–125.

Lund, C. R. F. 1992. Improving selectivity during methane partial oxidation by use of a membrane reactor. *Catal. Lett.* 12 (4):395–404.

Mansouri, S., O. Benlounes, C. Rabia, R. Thouvenot, M. M. Bettahar, and S. Hocine. 2013. Partial oxidation of methane over modified Keggin-type polyoxotungstates. *J. Mol. Catal. A Chem.* 379:255–262.

McConkey, B. H. and P. R. Wilkinson. 1967. Oxidation of methane to formaldehyde in a fluidized bed reactor. *Ind. Eng. Chem. Proc. Design Dev.* 6 (4):436–440.

McCormick, R. L. and G. O. Alptekin. 2000. Comparison of alumina-, silica-, titania-, and zirconia-supported $FePO_4$ catalysts for selective methane oxidation. *Catal. Today* 55 (3):269–280.

Michalkiewicz, B. 2003. Methane conversion to methanol in condensed phase. *Kin. Catal.* 44 (6):801–805.

Michalkiewicz, B. 2006. The kinetics of homogeneous catalytic methane oxidation. *Appl. Catal. A Gen.* 307 (2):270–274.

Michalkiewicz, B. 2011. Methane oxidation to methyl bisulfate in oleum at ambient pressure in the presence of iodine as a catalyst. *Appl. Catal. A Gen.* 394 (1–2):266–268.

Michalkiewcz, B. and K. Kalucki. 2003. The role of pressure in the partial methane oxidation process in the Pd-oleum environment. *Chem. Papers* 57 (6):393–396.

Michalkiewicz, B. and P. Kosowski. 2007. The selective catalytic oxidation of methane to methyl bisulfate at ambient pressure. *Catal. Commun.* 8 (12):1939–1942.

Michalkiewicz, B., J. Srenscek-Nazzal, P. Tabero, B. Grzmil, and U. Narkiewicz. 2008. Selective methane oxidation to formaldehyde using polymorphic T-, M-, and H-forms of niobium(V) oxide as catalysts. *Chem. Papers* 62 (1):106–113.

Michalkiewicz, B., J. Ziebro, and M. Tomaszewska. 2006. Preliminary investigation of low pressure membrane distillation of methyl bisulphate from its solutions in fuming sulphuric acid combined with hydrolysis to methanol. *J. Membrane Sci.* 286 (1–2):223–227.

Michalkiewicz, B., K. Kalucki, and J. G. Sosnicki. 2003. Catalytic system containing metallic palladium in the process of methane partial oxidation. *J. Catal.* 215 (1):14–19.

Mizuno, N., H. Ishige, Y. Seki, et al. 1997. Low-temperature oxygenation of methane into formic acid with molecular oxygen in the presence of hydrogen catalysed by $Pd_{0.08}Cs_{2.5}H_{1.34}PVMo_{11}O_{40}$. *Chem. Commun.* (14):1295–1296.

Mizuno, N., M. Misono, Y. Nishiyama, Y. Seki, I. Kiyoto, and C. Nozaki. 2000. Comparison of catalytic activity and efficiency of hydrogen peroxide utilization for di-iron-containing silicotungstate with those for iron containing complexes and the oxidation of methane and ethane. *Res. Chem. Intermed.* 26 (2):193–199.

Mukhopadhyay, S. and A. T. Bell. 2003a. Direct sulfonation of methane at low pressure to methanesulfonic acid in the presence of potassium peroxydiphosphate as the inhibitor. *Org. Proc. Res. Dev.* 7 (2):161–163.

Mukhopadhyay, S. and A. T. Bell. 2003b. Direct liquid-phase sulfonation of methane to methanesulfonic acid by SO_3 in the presence of a metal peroxide. *Angew. Chem. Int. Ed.* 42 (9):1019–1021.

Mukhopadhyay, S. and A. T. Bell. 2004a. Catalyzed sulfonation of methane to methanesulfonic acid. *J. Mol. Catal. A Chem.* 211 (1–2):59–65.

Mukhopadhyay, S. and A. T. Bell. 2004b. Direct sulfonation of methane to methanesulfonic acid by sulfur trioxide catalyzed by cerium(IV) sulfate in the presence of molecular oxygen. *Adv. Syn. Catal.* 346 (8):913–916.

Mylvaganam, K., G. B. Bacskay, and N. S. Hush. 2000. Homogeneous conversion of methane to methanol. 2. Catalytic activation of methane by cis- and trans-platin: a density functional study of the Shilov type reaction. *J. Am. Chem. Soc.* 122 (9):2041–2052.

Nagababu, P., S. S. F. Yu, S. Maji, R. Ramu, and S. I. Chan. 2014. Developing an efficient catalyst for controlled oxidation of small alkanes under ambient conditions. *Catal. Sci. Technol.* 4 (4):930–935.

Narsimhan, K., K. Iyoki, K. Dinh, and Y. Roman-Leshkov. 2016. Catalytic oxidation of methane into methanol over copper-exchanged zeolites with oxygen at low temperature. *ACS Central Sci.* 2 (6):424–429.

Nguyen, L. D., S. Loridant, H. Launay, A. Pigamo, J. L. Dubois, and J. M. M. Millet. 2006. Study of new catalysts based on vanadium oxide supported on mesoporous silica for the partial oxidation of methane to formaldehyde: catalytic properties and reaction mechanism. *J. Catal.* 237 (1):38–48.

Nizova, G. V., G. SussFink, and G. B. Shulpin. 1997a. Oxidations by the reagent O_2-H_2O_2 vanadium complex pyrazine-2-carboxylic acid 8. Efficient oxygenation of methane and other lower alkanes in acetonitrile. *Tetrahedron* 53 (10):3603–3614.

Nizova, G. V., G. SussFink, and G. B. Shulpin. 1997b. Catalytic oxidation of methane to methyl hydroperoxide and other oxygenates under mild conditions. *Chem. Commun.* 4 (4):397–398.

Ohler, N. and A. T. Bell. 2005. Selective oxidation of methane over MoO_x/SiO_2: isolation of the kinetics of reactions occurring in the gas phase and on the surfaces of SiO_2 and MoO_x. *J. Catal.* 231 (1):115–130.

Ohler, N. and A. T. Bell. 2006. Study of the elementary processes involved in the selective oxidation of methane over MoO_x/SiO_2. *J. Phys. Chem. B* 110 (6):2700–2709.

Omata, K., N. Fukuoka, and K. Fujimoto. 1994. Methane partial oxidation to methanol 1. Effects of reaction conditions and additives. *Ind. Eng. Chem. Res.* 33 (4):784–789.

Otsuka, K. and M. Hatano. 1987. The catalysts for the synthesis of formaldehyde by partial oxidation of methane. *J. Catal.* 108 (1):252–255.

Otsuka, K., R. Takahashi, and I. Yamanaka. 1999. Oxygenates from light alkanes catalyzed by NOx in the gas phase. *J. Catal.* 185 (1):182–191.

Otsuka, K., R. Takahashi, K. Amakawa, and I. Yamanaka. 1998. Partial oxidation of light alkanes by NOx in the gas phase. *Catal. Today* 45 (1–4):23–28.

Palkovits, R., M. Antonietti, P. Kuhn, A. Thomas, and F. Schuth. 2009. Solid catalysts for the selective low-temperature oxidation of methane to methanol. *Angew. Chem. Int. Ed.* 48 (37):6909–6912.

Park, E. D., S. H. Choi, and J. S. Lee. 2000. Characterization of Pd/C and Cu catalysts for the oxidation of methane to a methanol derivative. *J. Catal.* 194 (1):33–44.

Parmaliana, A., F. Arena, F. Frusteri, D. Miceli, and V. Sokolovskii. 1995. On the nature of active-sites of silica-based oxide catalysts in the partial oxidation of methane to formaldehyde. *Catal. Today* 24 (3):231–236.

Peng, J. J. and Y. Q. Deng. 2000. Direct catalytic conversion of methane in molten salt medium system under mild conditions. *Appl. Catal. A Gen.* 201 (2):L55–L57.

Periana, R. A., D. J. Taube, E. R. Evitt, et al. 1993b. A mercury-catalyzed, high-yield system for the oxidation of methane to methanol. *Science* 259 (5093):340–343.

Periana, R. A., D. J. Taube, H. Taube, and E. R. Evitt. 1993a. Conversion of lower alkane(s) into higher molecular wt. cpds. – by oxidative conversion to oxy-ester(s) using metal catalyst, conversion to alkyl intermediates and catalytic conversion to hydrocarbon(s), *Patent US2003120125-A1.*

Periana, R. A., D. J. Taube, S. Gamble, H. Taube, T. Satoh, and H. Fujii. 1998. Platinum catalysts for the high-yield oxidation of methane to a methanol derivative. *Science* 280 (5363):560–564.

Periana, R. A., O. Mironov, D. Taube, G. Bhalla, and C. J. Jones. 2003. Catalytic, oxidative condensation of CH_4 to CH_3COOH in one step via CH activation. *Science* 301 (5634):814–818.

Periana, R. A., O. Mirinov, D. J. Taube, and S. Gamble. 2002. High yield conversion of methane to methyl bisulfate catalyzed by iodine cations. *Chem. Commun.* (20):2376–2377.

Polniser, R., M. Stolcova, M. Hronec, and M. Mikula. 2011. Structure and reactivity of copper iron pyrophosphate catalysts for selective oxidation of methane to formaldehyde and methanol. *Appl. Catal. A Gen.* 400 (1–2):122–130.

Rytz, D. W. and A. Baiker. 1991. Partial oxidation of methane to methanol in a flow reactor at elevated pressure. *Ind. Eng. Chem. Res.* 30 (10):2287–2292.

Seki, Y., J. S. Min, M. Misono, and N. Mizuno. 2000. Reaction mechanism of oxidation of methane with hydrogen peroxide catalyzed by 11-molybdo-1-vanadophosphoric acid catalyst precursor. *J. Phys. Chem. B* 104 (25):5940–5944.

Seki, Y., N. Mizuno, and M. Misono. 1997. High-yield liquid-phase oxygenation of methane with hydrogen peroxide catalyzed by 12-molybdovanadophosphoric acid catalyst precursor. *Appl. Catal. A Gen.* 158 (1–2):L47–L51.

Seki, Y., N. Mizuno, and M. Misono. 2000. Catalytic performance of 11-molybdo-1-vanadophosphoric acid as a catalyst precursor and the optimization of reaction conditions for the oxidation of methane with hydrogen peroxide. *Appl. Catal. A Gen.* 194:13–20.

Sen, A., M. A. Benvenuto, M. Lin, A. C. Hutson, and N. Basickes. 1994. Activation of methane and ethane and their selective oxidation to the alcohols in protic media. *J. Am. Chem. Soc.* 116 (3):998–1003.

Shan, J., M. Li, L. F. Allard, S. Lee, and M. Flytzani-Stephanopoulos. 2017. Mild oxidation of methane to methanol or acetic acid on supported isolated rhodium catalysts. *Nature* 551 (7682):605–608.

Sherman, J. H. 1999. Method and apparatus for manufacturing methanol, *United States Pat. US5,954,925.*

Shul'pin, G. B., G. V. Nizova, Y. N. Kozlov, L. G. Cuervo, and G. Suss-Fink. 2004. Hydrogen peroxide oxygenation of alkanes including methane and ethane catalyzed by iron complexes in acetonitrile. *Ad. Syn. Catal.* 346 (2–3):317–332.

Smeets, P. J., J. S. Woertink, B. F. Sels, E. I. Solomon, and R. A. Schoonheydt. 2010. Transition-metal ions in zeolites: coordination and activation of oxygen. *Inorg. Chem.* 49 (8):3573–3583.

Smeets, P. J., M. H. Groothaert, and R. A. Schoonheydt. 2005. Cu based zeolites: a UV-vis study of the active site in the selective methane oxidation at low temperatures. *Catal. Today* 110 (3–4):303–309.

Smith, D. F. and R. T. Milner. 1931. Partial oxidation of methane in the presence of oxides of nitrogen. *Ind. Eng. Chem.* 23:357–360.

Smith, M. R., L. Zhang, S. A. Driscoll, and U. S. Ozkan. 1993. Effect of surface species on activity and selectivity of MoO_3/SiO_2 catalysts in partial oxidation of methane to formaldehyde. *Catal. Lett.* 19 (1):1–15.

Sojka, Z., R. G. Herman, and K. Klier. 1991. Selective oxidation of methane to formaldehyde over doubly copper-iron doped zinc-oxide catalysts via a selectivity shift mechanism. *J. Chem. Soc. Chem. Commun.* (3):185–186.

Spencer, N. D. and C. J. Pereira. 1989. V_2O_5-SiO_2-catalyzed methane partial oxidation with molecular-oxygen. *J. Catal.* 116 (2):399–406.

Spencer, N. D., C. J. Pereira, and R. K. Grasselli. 1990. The effect of sodium on the MoO_3-SiO_2-catalyzed partial oxidation of methane. *J. Catal.* 126 (2):546–554.

Stauffer, J. E. 1993. Process for methyl alcohol, *United States Pat. 5,185,479.*

Stroud, H. J. F. 1975. Improvements in or relating to the oxidation of gases which consist principally of hydrocarbons, *Brit. Pat. 1,398,385.*

Sun, M., J. Z. Zhang, P. Putaj, et al. 2014. Catalytic oxidation of light alkanes (C1-C4) by heteropoly compounds. *Chem. Rev.* 114 (2):981–1019.

Suzuki, K., T. Hayakawa, M. Shimizu, and K. Takehira. 1995. Partial oxidation of methane over silica-supported molybdenum oxide catalysts. *Catal. Lett.* 30 (1–4):159–169.

Szecsenyi, A., G. N. Li, J. Gascon, and E. A. Pidko. 2018. Unraveling reaction networks behind the catalytic oxidation of methane with H_2O_2 over a mixed-metal MIL-53(Al,Fe) MOF catalyst. *Chem. Sci.* 9 (33):6765–6776.

Tabata, K., Y. Teng, T. Takemoto, et al. 2002. Activation of methane by oxygen and nitrogen oxides. *Catal. Rev. Sci. Eng.* 44 (1):1–58.

Tabata, K., Y. Teng, Y. Yamaguchi, H. Sakurai, and E. Suzuki. 2000. Experimental verification of theoretically calculated transition barriers of the reactions in a gaseous selective oxidation of CH_4-O_2-NO_2. *J. Phys. Chem. A* 104 (12):2648–2654.

Taylor, C. E., R. R. Anderson, and R. P. Noceti. 1997. Activation of methane with organopalladium complexes. *Catal. Today* 35 (4):407–413.

Taylor, S. H., J. S. J. Hargreaves, G. J. Hutchings, R. W. Joyner, and C. W. Lembacher. 1998. The partial oxidation of methane to methanol: an approach to catalyst design. *Catal. Today* 42 (3):217–224.

Teng, Y., H. Sakurai, K. Tabata, and E. Suzuki. 2000. Methanol formation from methane partial oxidation in CH_4-O_2-NO gaseous phase at atmospheric pressure. *Appl. Catal. A Gen.* 190 (1–2):283–289.

Thomas, D. J., R. Willi, and A. Baiker. 1992. Partial oxidation of methane – the role of surface-reactions. *Ind. Eng. Chem. Res.* 31 (10):2272–2278.

Vanelderen, P., R. G. Hadt, P. J. Smeets, E. I. Solomon, R. A. Schoonheydt, and B. F. Sels. 2011. Cu-ZSM-5: a biomimetic inorganic model for methane oxidation. *J. Catal.* 284 (2):157–164.

Vargaftik, M. N., I. P. Stolarov, and I. I. Moiseev. 1990. Highly selective partial oxidation of methane to methyl trifluoroacetate. *J. Chem. Soc. Chem. Commun.* (15):1049–1050.

Vedeneev, V. I., M. Y. Goldenberg, N. I. Gorban, and M. A. Teitelboim. 1988. Quantitative model of the oxidation of methane at high-pressures 3. Mechanism of formation of reaction-products. *Kin. Catal.* 29 (6):1121–1126.

Vitillo, J. G., A. Bhan, C. J. Cramer, C. C. Lu, and L. Gagliardi. 2019. Quantum chemical characterization of structural single Fe(II) sites in MIL-type metal-organic frameworks for the oxidation of methane to methanol and ethane to ethanol. *ACS Catal.* 9 (4):2870–2879.

Wang, C. B., R. G. Herman, C. Shi, Q. Sun, and J. E. Roberts. 2003a. V_2O_5-SiO_2 xerogels for methane oxidation to oxygenates: preparation, characterization, and catalytic properties. *Appl. Catal. A Gen.* 247 (2):321–333.

Wang, K. X., H. F. Xu, W. S. Li, C. T. Au, and X. P. Zhou. 2006. The synthesis of acetic acid from methane via oxidative bromination, carbonylation, and hydrolysis. *Appl. Catal. A Gen.* 304 (1):168–177.

Wang, X., Y. Wang, Q. Tang, Q. Guo, Q. Zhang, and H. Wan. 2003b. MCM-41-supported iron phosphate catalyst for partial oxidation of methane to oxygenates with oxygen and nitrous oxide. *J. Catal.* 217 (2):457–467.

Wang, Y. and K. Otsuka. 1995. Catalytic-oxidation of methane to methanol with H_2-O_2 gas-mixture at atmospheric-pressure. *J. Catal.* 155 (2):256–267.

Williams, C., J. H. Carter, N. F. Dummer, et al. 2018. Selective oxidation of methane to methanol using supported AuPd catalysts prepared by stabilizer-free sol-immobilization. *ACS Catal.* 8 (3):2567–2576.

Williams, G. R., S. T. Kolaczkowski, and P. Plucinski. 2003. Catalyst instabilities during the liquid phase partial oxidation of methane. *Catal. Today* 81 (4):631–640.

Woertink, J. S., P. J. Smeets, M. H. Groothaert, et al. 2009. A Cu_2O^{2+} core in Cu-ZSM-5, the active site in the oxidation of methane to methanol. *Proc. Natl. Acad. Sci. U. S. A.* 106 (45):18908–18913.

Wood, B. R., J. A. Reimer, A. T. Bell, M. T. Janicke, and K. C. Ott. 2004. Methanol formation on Fe/Al-MFI via the oxidation of methane by nitrous oxide. *J. Catal.* 225 (2):300–306.

Yamanaka, I., M. Soma, and K. Otsuka. 1995. Oxidation of methane to methanol with oxygen catalyzed by europium trichloride at room-temperature. *J. Chem. Soc. Chem. Commun.* (21):2235–2236.

Yamanaka, I., M. Soma, and K. Otsuka. 1996. Enhancing effect of titanium(II) for the oxidation of methane with O_2 by an $EuCl_3$-Zn-CF_3CO_2H-catalytic system at 40 degrees C. *Chem. Lett.* (7):565–566.

Yang, W., X. Wang, Q. Guo, Q. Zhang, and Y. Wang. 2003. Superior catalytic performance of phosphorus-modified molybdenum oxide clusters encapsulated inside SBA-15 in the partial oxidation of methane. *New J. Chem.* 27 (9):1301–1303.

Yarlagadda, P. S., L. A. Morton, N. R. Hunter, and H. D. Gesser. 1988. Direct conversion of methane to methanol in a flow reactor. *Ind. Eng. Chem. Res.* 27 (2):252–256.

Yarlagadda, P. S., L. A. Morton, N. R. Hunter, and H. D. Gesser. 1990. Temperature oscillations during the high-pressure partial oxidation of methane in a tubular flow reactor. *Combust. Flame* 79 (2):216–218.

York, A. P. E., T. Xiao, and M. L. H. Green. 2003. Brief overview of the partial oxidation of methane to synthesis gas. *Topics Catal.* 22 (3–4):345–358.

Yuan, Q., W. Deng, Q. Zhang, and Y. Wang. 2007. Osmium-catalyzed selective oxidations of methane and ethane with hydrogen peroxide in aqueous medium. *Adv. Syn. Catal.* 349 (7):1199–1209.

Zerella, M. and A. T. Bell. 2006. Pt-catalyzed oxidative carbonylation of methane to acetic acid in sulfuric acid. *J. Mol. Catal. A Chem.* 259 (1–2):296–301.

Zhang, Q., D. He, and Q. Zhu. 2003. Recent progress in direct partial oxidation of methane to methanol. *J. Nat. Gas Chem.* 12:81–89.

Zhang, Q., D. He, and Q. Zhu. 2008. Direct partial oxidation of methane to methanol: reaction zones and role of catalyst location. *J. Nat. Gas Chem.* 17 (1):24–28.

Zhang, Q., D. He, J. Li, B. Xu, Y. Liang, and Q. Zhu. 2002. Comparatively high yield methanol production from gas phase partial oxidation of methane. *Appl. Catal. A Gen.* 224 (1–2):201–207.

Zhang, Q., W. Yang, X. Wang, Y. Wang, T. Shishido, and K. Takehira. 2005b. Coordination structures of vanadium and iron in MCM-41 and the catalytic properties in partial oxidation of methane. *Micro. Meso. Mater.* 77 (2–3):223–234.

Zhang, X., D. He, Q. Zhang, B. Xu, and Q. Zhu. 2005a. Comparative studies on direct conversion of methane to methanol/formaldehyde over La-Co-O and ZrO_2 supported molybdenum oxide catalysts. *Topics Catal.* 32 (3–4):215–223.

Zhang, X., D. He, Q. Zhang, Q. Ye, B. Xu, and Q. Zhu. 2003. Selective oxidation of methane to formaldehyde over Mo/ZrO_2 catalysts. *Appl. Catal. A Gen.* 249 (1):107–117.

Zheng, J., J. Ye, M. A. Ortuno, et al. 2019. Selective methane oxidation to methanol on Cu-Oxo dimers stabilized by zirconia nodes of an NU-1000 metal-organic framework. *J. Am. Chem. Soc.* 141 (23):9292–9304.

7 Hydrogen and Solid Carbon Products from Natural Gas

A Review of Process Requirements, Current Technologies, Market Analysis, and Preliminary Techno Economic Assessment

Robert Dagle, Vannesa Dagle, Mark Bearden,
J Holladay, Theodore Krause, and Shabbir Ahmed

CONTENTS

7.1   INTRODUCTION

The use of abundant natural gas and its efficient conversion to fuels and chemicals has been an important area of research. In addition to energy independence, the development of new technologies that produce no emissions and carbon dioxide (CO_2)-free energy and fuels has also received significant interest in recent decades (Abbas and Wan Daud 2010a). Hydrogen (H_2) in particular has received substantial attention in part because it can be produced from diverse domestic resources, used in multiple applications across sectors, and its use in fuel cell applications is completely pollution-free. Furthermore, because fuel cells convert H_2 and oxygen directly to electricity more efficiently than internal combustion engines, they can reduce systemic energy losses. The ability to produce H_2 cost effectively will be a major determining factor for future implementation of this energy resource (Xu et al. 2003). Natural gas is an abundant resource in the United States, and coupled with its available and growing infrastructure, it offers a pathway to building H_2 infrastructure. Today, 95% of the H_2 produced in the United States is made by natural gas reforming in large central plants. This is an important technology pathway for near-term H_2 production (Rostrup-Nielsen et al. 2002). This paper first covers the conventional process of SMR, and then provides an assessment of natural gas to solid carbon and H_2 processes.

7.1.1   STEAM-METHANE REFORMING

SMR is a mature production process in which high-temperature steam (700°C–1000°C) is used to produce H_2 from a methane (CH_4) source such as natural gas. In SMR, CH_4 reacts with steam under 3–25 bar pressure in the presence of a catalyst to produce H_2, carbon monoxide (CO), and a relatively small amount of CO_2. Steam reforming is endothermic – that is, heat must be supplied to the process for the reaction to proceed (Equation (7.1)). Subsequently, in what is called the water-gas shift (WGS) reaction, the CO and steam are reacted using a catalyst to produce CO_2 and more H_2 (Equation (7.2)). Combining the SMR and WGS reactions results in primary products H_2 and CO_2 (Equation (7.3)) (Rostrup-Nielsen et al. 2002). In a final process step called pressure-swing adsorption (PSA), CO_2 and other impurities are removed from the gas stream, leaving essentially pure H_2. Steam reforming can also be used to produce H_2 from other fuels such as ethanol, propane, or even gasoline.

$$CH_4 + H_2O = CO + 3H_2 \quad \Delta H^\circ = 206 \text{ kJ/mol} \qquad (7.1)$$

$$CO + H_2O = CO_2 + H_2 \quad \Delta H^\circ = -41 \text{ kJ/mol} \qquad (7.2)$$

Combining Equations (7.1) and (7.2), the net reaction is:

$$CH_4 + 2H_2O = CO_2 + 4H_2 \quad \Delta H^\circ = 165 \text{ kJ/mol} \qquad (7.3)$$

Currently, SMR is the technology with the greatest advantage in terms of the lowest cost and highest energy efficiency and is the preferred choice of industry today. SMR is used industrially to produce ~95% of the H_2 consumed in the United States (Abbas and Wan Daud 2010a). However, the process generates significant quantities of CO_2. The reaction stoichiometry (Equation (7.3)) suggests 5.5 kg_{CO2}/kg_{H2} is produced, but the quantity is much higher in practice. Being a strongly endothermic process, it requires additional energy that results in the release of additional CO_2. Depending on the energy sources used and efficiency of the process, SMR generates 9–14 kg_{CO2}/kg_{H2}. Based on low-cost shale gas, the cost of H_2 production from SMR can be < \$2/kg. If CO_2 capture is considered part of a SMR technology option, this presents additional cost.

7.1.2 STEAM-METHANE REFORMING WITH CARBON CAPTURE

To understand the cost of deploying a CO_2 capture system in a H_2 production plant, the International Energy Agency Greenhouse Gas Research and Development (R&D) Program commissioned Amec Foster Wheeler to undertake a study, which resulted in a report entitled, *Techno-Economic Evaluation of Hydrogen Production with CO_2 Capture* (International Energy Agency 2017). The primary purpose of the study was to evaluate the performance and cost of a greenfield modern SMR plant producing 100,000 Nm³/h of H_2 from natural gas as feedstock/fuel operating in merchant plant mode. (Note that 100,000 Nm³/h of H_2 corresponds to 216 tons per day production; for comparison, the Air Products' Port Arthur II plant has a capacity of 265 tons per day.) The International Energy Agency study focused on the economic evaluation of five different options to capture CO_2 from SMR, and these include the following cases:

- Base Case: Modern SMR Plant with feedstock pretreatment, prereforming, high-temperature WGS, and PSA

- Case 1A: SMR with capture of CO_2 from the shifted syngas using methyl diethanolamine
- Case 1B: SMR with burners firing H_2-rich fuel and capture of CO_2 from the shifted syngas using methyl diethanolamine
- Case 2A: SMR with capture of CO_2 from the PSA tail gas using methyl diethanolamine
- Case 2B: SMR with capture of CO_2 from the PSA tail gas using cryogenic and membrane separation
- Case 3: SMR with capture of CO_2 from the flue gas using monoethanolamine.

For this study, the price of natural gas was assumed to be 6€/GJ (LHV) (~$7/MMBtu) and the price of electricity 80€/MWh (~$0.09/kWh; 1€ = 1.1$ was assumed for cost conversion). The levelized cost of H_2 production (LCOH) was calculated for these six cases, and results are shown in Figure 7.1. Some export of electricity takes place in the base case (from excess steam fed to a steam turbine), thus lowering the LCOH, but this accounts for only a small fraction of the overall H_2 cost. The cost of natural gas is the major fraction in the LCOH.

Costs of CO_2 avoidance (CAC) were calculated by comparing the CO_2 emissions per Nm^3 H_2 and the LCOH of plants with capture and a reference plant without capture:

$$CAC = \left(LCOH_{CCS} - LCOH_{Ref}\right)/\left(CO_2 Emissions_{Ref} - CO_2 Emissions_{CCS}\right) \quad (7.4)$$

where CAC is expressed in € per tonne (metric ton) of CO_2 (or in $/tonne), LCOH is expressed in € per Nm^3/h H_2 (or in $/kg), and CO_2 emission is expressed in tonnes of CO_2 per Nm^3/h H_2 (or in tonnes CO_2 per kg H_2). The notation CCS stands for carbon capture and storage.

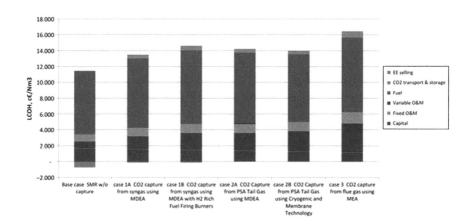

FIGURE 7.1   LCOHs for the six cases taken from Figure 8 in International Energy Agency (2017).

TABLE 7.1

Summary of Data for LCOH, CO$_2$ Emissions, and CO$_2$ Avoidance Cost (International Energy Agency 2017)

Case	CO$_2$ emissions	CO$_2$ avoided	LCOH		Cost of avoided CO$_2$	
	kg CO$_2$/ kg H$_2$	%	€c/Nm^3	$/kg	€/tonne	$/tonne
Base Case	9.0		11.4	1.39		
Case 1A	4.1	54.2	13.5	1.65	47.1	51.8
Case 1B	3.2	63.9	14.6	1.78	62.0	68.2
Case 2A	4.3	52.2	14.2	1.74	66.3	72.9
Case 2B	4.2	53.4	14	1.71	59.5	65.5
Case 3	1.0	89.0	16.5	2.02	69.8	76.8

Table 7.1 summarizes data for the LCOH, CO$_2$ emissions, and CAC. It can be seen that it would require a cost of €76.8/tonne of CO$_2$ emissions to make the higher capture rate option (Case 3) more attractive than the base case.

H$_2$ production cost when coupling SMR with CO$_2$ capture can be compared to renewable or pyrolytic H$_2$ production processes that do not produce CO$_2$. Equation (7.4), which defines the CAC for the cases with CCS relative to the base case SMR, can be used to compare the cost of H$_2$ produced from renewable sources or through NG decomposition (with no CO$_2$ emissions) to the cost of H$_2$ from SMR. The CAC is assumed to be $80/tonne (as for Case 3, where 90% of CO$_2$ emissions are avoided), and CO$_2$ emissions from the average SMR plant are assumed to be 10 kgCO$_2$/kgH$_2$. Note that the base case in this study uses 9 kgCO$_2$/kgH$_2$, but this is apparently a more efficient plant model than most commercial SMR plants. By comparison, SMR GREET Model suggests that greenhouse gas emissions are 11 kgCO$_2$/kgH$_2$. Substituting these values in Equation (7.5) results in the following price difference (International Energy Agency 2017):

$$H_{2\ renew} - H_{2\ SMR} = 0.8\,\$/kg \tag{7.5}$$

A $1.4/kg hydrogen cost in the base case SMR suggests that the cost of pyrolytic H$_2$ should be below $2.2/kg to compete with the SMR-CCS strategy. Sales of carbon co-produced with pyrolytic H$_2$ can help offset the prosecution price and reach the CO$_2$-free hydrogen price point that is competitive with the SMR-CCS.

7.1.3 METHANE PYROLYSIS

The prevalence of fracking has revolutionized the U.S. natural gas industry, making it economically competitive to recover large reserves of natural gas that were previously considered uneconomical to recover and leading to an unprecedented increase in natural gas production coupled with a drop in the price of natural gas to historical lows. As such, there is considerable interest in developing new process technologies that use natural gas for producing fuels and chemicals, including H_2. One potential technology is the thermal decomposition of natural gas. Methane can be decomposed into carbon (C) and H_2 according to the following endothermic reaction:

$$CH_4 \rightarrow C(s) + 2H_2 \quad \Delta H° = 74.8 \, kJ/mol \tag{7.6}$$

The reaction stoichiometry suggests that the molar ratio of H_2 and carbon to the feedstock CH_4 as 2 and 1, respectively. Actual processes will yield less because of the production of byproducts (e.g., olefins, aromatics) and the need to oxidize part of the feedstock to generate the heat for the process. Muradov and Veziroğlu (2005) estimate the H_2/CH_4 molar ratio from a process to be 1.7. Assuming the carbon-containing products will be sequestered or sold as carbon black, the efficiency of producing H_2 via decomposition using a molar yield of 1.7 translates to (1.7 × LHV of H_2/LHV of CH_4) = 51%.

For the pyrolysis reaction above, each mol of CH_4 can yield a mass-balanced maximum of 12 g of carbon and 4 g of H_2. This 3:1 mass ratio (2:1 molar ratio) is skewed in practice by the formation of byproducts and the process energy required. The heat of formation of CH_4 is 74.8 kJ, i.e., the energy required to decompose a mol of CH_4. If the energy for the decomposition is derived by the combustion of CH_4, then that requires another 0.093 mol of CH_4. Thus, the mass and energy-balanced maximum yields for H_2 and carbon are reduced to 10.98 g of carbon and 3.66 g of H_2 per mol of CH_4. Stated in terms of energy, the minimum energy that will be consumed for the production of the H_2 and carbon via CH_4 decomposition is 240 kJ/gH_2 (66.7 kWh/kgH_2, 5.96 kWh/m$^3_{H2}$) and 73 kJ/g$_{carbon}$ (20.3 kWh/kg$_{carbon}$), respectively. The yield of practical systems will necessarily require more energy as a result of inefficiencies that include incomplete CH_4 conversion, loss of products (H_2, C), formation of byproducts, heat transfer limitations, energy recovery, heat loss, etc.

The pyrolysis process does not produce CO or CO_2 as byproducts, and thus the need for WGS or CO_2 removal is eliminated (Abbas and Wan Daud 2010a). Without inclusion of the WGS and preferential oxidation of CO reactions, processing for CH_4 decomposition is greatly simplified relative to steam reforming (Amin et al. 2011). Thus, the energy requirement for CH_4 catalytic cracking is nearly half of that required for steam reforming on a per-mole CH_4 decomposed basis (Amin et al. 2011). One study reports that per mol of H_2 produced, the energy requirement is 37.4 kJ/mol for CH_4 decomposition, compared to 63.3 kJ/mol for steam reforming (Amin et al. 2011). However, about half of the amount of H_2 per mol CH_4 is produced.

CH_4 decomposition is an endothermic reaction, and because of the strong carbon-to-hydrogen (C–H) bonds, noncatalytic thermal cracking of CH_4 requires a temperature >1200°C to obtain a reasonable yield. By using a catalyst, the temperature can be significantly reduced (Abbas and Wan Daud 2010a). Thermodynamically, breaking all four C–H bonds of CH_4, such as with steam reforming CH_4, into synthesis gas or CH_4 decomposition into C and H, is much easier than breaking only one or two of the C–H bonds, under either oxidative or nonoxidative conditions (Xu et al. 2003). Direct conversion of CH_4 with the assistance of oxidants is thermodynamically more favorable than under non-oxidation conditions, and thus has received much more attention (Xu et al. 2003). However, there are potential benefits for nonoxidative conversion.

At high temperatures between 1300 and 1600°C, CH_4 and other hydrocarbons present in natural gas decompose to yield H_2 and solid carbon in an inert atmosphere. Compared to SMR, the reaction is less endothermic and produces no CO_2 emissions if the produced carbon is not burned in subsequent processes. Furthermore, there is the potential for a considerable reduction in capital and operating costs compared to SMR. This is because the gas stream exiting the reactor in the thermal decomposition process has a considerably higher H_2 concentration and does not contain CO_2 or CO, compared to the composition of the gas stream existing the steam reformer. Therefore, the gas stream should require considerably less downstream processing to produce commercial-grade H_2 compared to SMR. In addition, the sale of the carbon product (i.e., carbon black, carbon fibers, etc.) can be used to offset the cost of the H_2.

Commercially, the thermal decomposition of natural gas has been employed in the carbon black industry since the 1920s to produce selected grades of carbon for use in manufacturing tires and electrical components (Wang et al. 2003). The H_2 could, in theory, be recovered and sold as merchant H_2, but in practice, the carbon black industry uses a portion of the H_2 to provide the required process heat, and the remainder of the H_2 is used to generate process heat for the facility or nearby facilities adjacent to the fence line. The ability to offset the cost of producing H_2 by the sale of carbon is a promising scenario. To the best of our knowledge, there is no commercial process based on producing both carbon and H_2 for the carbon black and H_2 markets, respectively.

A number of excellent recent articles provide reviews of CH_4 decomposition. These include discussion of catalysts and reaction mechanisms and kinetics (Abbas and Wan Daud 2010a; Abánades et al. 2011; Alvarez-Galvan et al. 2011), technologies that use renewable energy options for producing H_2 by thermal decomposition (Muradov and Veziroğlu 2008), and engineering design considerations (Alvarez-Galvan et al. 2011). For example, an in-depth review by Amin et al. (2011) discusses the catalytic cracking of CH_4 for H_2 production and reports on the thermodynamics, catalysts and supports used, catalyst deactivation and regeneration, the growth of carbon product particles, and the cracking reaction mechanism with inclusion of kinetic models. Additionally, the cost of H_2 production is reported for a number of different pathways. Abbas and Wan Daud (2010a) also reviewed decomposition processes and discuss insights about heating sources, reactor types, catalysts employed, catalyst regeneration, and the quality of the carbon produced. The effects of temperature and flow rate on the product distribution, the energy sources, the reactors and their operating condition, and catalysts were also reported.

7.2 ENGINEERING REVIEW

Several approaches to supplying heat for the highly endothermic CH_4 pyrolysis reaction have been proposed. These include direct heating of the reaction zone, externally heated catalyst particles used as the heat carrier, similar to fluidized bed reactors (FBRs), and the addition of a small amount of oxygen to generate the necessary heat, which is an autothermal pyrolysis type process (Muradov 2005). The endothermic heat of reaction and sensible heat requirements also have led to some processes in which air or oxygen are co-injected to generate heat and thus lead to other byproducts that require separations downstream. Depending on the source of heat for the reaction, the process may be termed *autothermal*, in which the heat of reaction is generated by oxidation of the CH_4 or other hydrocarbons present in the natural gas, or *allothermal*, in which the energy is supplied from an external source and transferred via a heated surface or transmitted with microwaves or electric power. These reactions may be conducted with or without catalysts; the catalysts offer the benefits of faster kinetics at lower temperature and may improve the selectivity of the product streams.

The main products of CH_4 decomposition are solid carbon and H_2; therefore, the problem with FBRs is carbon deposition on the catalyst particles (Abbas and Wan Daud 2010a). As the reaction proceeds, carbon will deposit on the catalyst surface, blocking the pores and interparticle spaces. Because of this pore-blocking process, the pressure drop through the bed will rise as gas flow is impeded (Abbas and Wan Daud 2010a). Thus, deposited carbon must be removed periodically. Muradov studied different types of reactors and concluded that FBRs were the most promising type of reactor for large-scale operations because it provides a constant flow of solids through the reaction zone, making it suitable for continuous addition and withdrawal of catalyst from the reactor and catalyst regeneration (Muradov 2002). There are many reports about the use of electric furnaces to supply the energy needed to decompose CH_4. Other reports focus on the use of plasma, concentrated solar energy, and a molten-metal bath as alternative means of supplying the required energy. The following subsections provide brief descriptions of various means of providing the reactor energetics and the various reactor technologies as reported in the open and closed literature. In this section, we report in broad terms the thermodynamic and processing requirements for each of the classes of process technologies (e.g., thermal, catalytic, plasma).

7.2.1 Thermal (Noncatalytic) Reactions

Thermal cracking of CH_4 removes and separates carbon in a single step (Ahmed et al. 2009). A major drawback of this technique is that noncatalytic cracking is very slow for practical applications below 1000°C, while catalytic cracking of CH_4 can be conducted at temperatures as low as 500°C (Amin et al. 2011). The function of the catalyst is to reduce the activation energy required for CH_4 decomposition, thereby leading to lower operating temperatures. In this section, we focus our attention on thermal approaches.

Thermal decomposition of CH_4 at temperatures above 1000°C will yield H_2 gas and soot or coke (carbon-rich species), with higher temperatures favoring faster and higher conversion. The combination of higher temperature and longer residence times favors the production of H_2 and carbon (Wheeler and Wood 1928; Fischer and Pichler 1932; Holman et al. 1995). The product distribution favors more gaseous products with higher yields of olefins and aromatics at lower temperatures in short contact time reactors. Low pressure favors higher conversion of CH_4 but also tends to produce more olefins. The olefins decompose to carbon and H_2 as residence time increases. The morphology of the solid carbon is affected by the prevailing decomposition mechanism, which in turn is determined by temperature, pressure, catalyst,

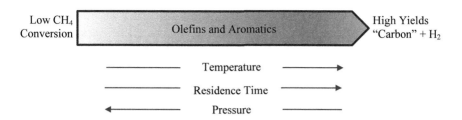

FIGURE 7.2 Conceptual representation of conversion and yields as a function of reactor conditions. "Carbon" represents particles of soot and graphite.

and residence time. Figure 7.2 shows a conceptual representation of CH_4 conversion during the decomposition process as a function of the operating conditions. The intermediates, olefins and aromatics, are initially formed and then decompose further to "carbon" and H_2. The soot part of "carbon" is essentially a C–H molecule where the C/H ratio is near 1 at low temperatures and increases with residence time at high temperatures (Blanquart and Pitsch 2009). The soot particles are considered to be aggregates formed from smaller spherical primary particles.

Holmen et al. (1976, 1995) studied the conversion of CH_4 in high-temperature processes and reported on the yields of products that include carbon, H_2, and olefins, with mechanisms and reaction kinetics.

Equilibrium product distributions of CH_4, H_2, and carbon (typically graphite) can be derived from calculations using correlations for the equilibrium constants or Gibbs free energy minimization (Amin et al. 2011). Snoeck et al. (1997) provides experimental results and correlations for catalytic cracking on nickel (Ni).

Figure 7.3 shows the equilibrium yields from the decomposition of CH_4 in the temperature range of 25–1000°C at a pressure of 1 atm (Outotec Pori Research Center 2002). The dominant products are CH_4, carbon, and H_2. A small amount ($<2 \times 10^{-5}$ mol/mol-CH_4) of ethane is formed and peaks at 500°C. Small amounts of ethylene and acetylene also are predicted at temperatures above 500 and 1300°C, respectively. Figure 7.4 shows the effect of pressure where the higher pressure shifts conversion of CH_4 and yields of carbon and H_2 toward a higher temperature.

7.2.2 CATALYTIC REACTIONS

7.2.2.1 Catalytic Reactors Types

The choice of reactor type is an important consideration in the design of an economically-viable process. Packed-bed reactors (PBRs) and FBRs are most commonly used for CH_4 decomposition. Abbas and Wan Daud (2010a) discuss different reactor technologies evaluated and reported upon in the literature. The in-depth review by Amin et al. (2011) specifically describes reaction kinetics and models available for predicting reaction rates. Here we discuss the highlights for the various reactor approaches reported in the literature, as well as the means for separation of the carbon product from the process.

FBRs offer advantages that include the ability to continuously add catalyst particles and withdraw carbonaceous solid products, avoidance of reactor clogging due to

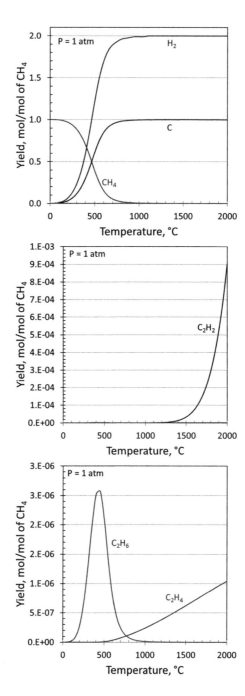

FIGURE 7.3 Equilibrium yields of methane and products as a function of temperature at p = 1 atm during methane pyrolysis.

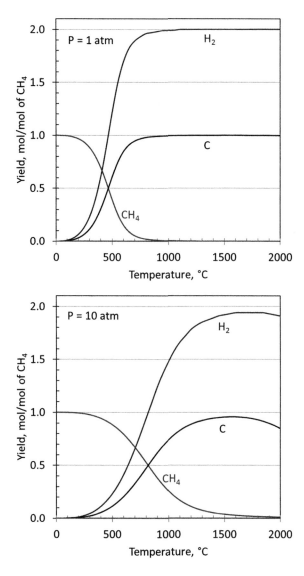

FIGURE 7.4 Effect of pressure and temperature on the equilibrium yields of methane and products during methane pyrolysis.

carbon deposits, uniform temperature distribution, and good mass and heat transfer. Challenges include the need to capture the fine particles and large gas flows required to maintain fluidization and the particle size reduction by attrition.

PBRs are simple in design; they require only the CH_4 reactant to flow through the reactor. The reactor may or may not contain catalysts, but the design demands a temperature profile conducive to fast conversion and effective removal of the carbon products.

In a rotary bed reactor, CH_4 flows through an inclined tube that moves the catalyst layer forward as the bed rotates. Carbon forms on the catalysts, and the carbon and

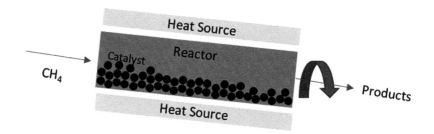

FIGURE 7.5   Simplified rotary bed reactor process flow diagram.

catalysts are withdrawn at the bottom of the tube. Energy to drive the reaction can be provided by heating sources either outside or inside the tube wall. Figure 7.5 shows a schematic of a rotary bed. This design allows a continuous withdrawal of solid products and possible regeneration of the catalyst (Chesnokov and Chichkan 2009; Pinilla et al. 2009).

Plasma reactors, described in more detail below, use gas and electrical energy to create a strong electromagnetic field to generate a gaseous mixture of electrons and highly charged positive ions. A plasma torch increases the temperature of the CH_4 to very high temperatures (~2000°C), causing its decomposition (Holmen et al. 1995; Fulcheri and Schwob 1995; Lynum et al. 1996; Fincke et al. 2002; Jabs et al. 2007).

7.2.2.2   Catalysts

Muradov and Veziroğlu (2005) summarize published data available on the conversion of CH_4 to carbon products as shown in Figure 7.6. Ni- and iron (Fe)-based catalysts operate in the 500–700°C and 700–950°C ranges, respectively, producing carbon filaments (including carbon nanotubes [CNTs]). Carbon-based catalysts operated in the 850–950°C range produce both carbon filaments and turbostratic carbon. Graphitic and turbostratic carbon are formed in the 700–1000°C range using a variety of catalysts. Amorphous carbon is produced non-catalytically at operating temperatures >1150°C (Muradov and Veziroğlu 2005). Amorphous carbon includes carbon black and thermal black. The key attributes reported for both metallic and carbon types of catalysts and the nature of carbon produced from each are discussed below.

Deactivation of the catalyst by coke formation is seen as the most important problem with CH_4 pyrolysis. The capacity of the catalyst to accumulate a significant amount of carbon filaments limits its time on stream or the residence time required. Two main types of catalysts – metallic and carbonaceous – have been the focus of most investigations (Abánades et al. 2012). In their comprehensive review, Abbas and Wan Daud (2010a) provide more detail on the various catalysts reported and discuss catalyst deactivation and regeneration.

7.2.2.2.1   Metal Catalysts

Numerous metal-based catalysts have been studied for CH_4 decomposition. Their practical application requires development to increase their specific activities and lifetimes. It has been reported that the rate of CH_4 decomposition activity of the

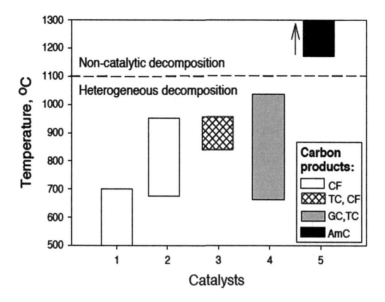

FIGURE 7.6 Summary of published data on catalysts, temperature, and carbon products from methane decomposition. Catalysts: 1-Ni-based, 2-Fe-based, 3-carbon-based, 4-Co, Ni, Pd, Pt, Cr, Ru, Mo, and W catalysts, 5-noncatalytic decomposition. Carbon products: CF-carbon filaments, TC-turbostratic carbon, GC-graphitic carbon, AmC-amorphous carbon (Muradov and Veziroğlu 2015). (Reproduced with permission from Elsevier.)

metals follows the order: Co, Ru, Ni, Rh > Pt, Re, Ir > Pd, Cu, W, Fe, Mo. However, much attention has been given to Ni- and Fe-based catalysts (Abbas and Wan Daud 2010a).

The effectiveness of a catalyst should be based on its activity and operating lifetime, given the large amount of carbon produced during conversion. A key finding regarding catalyst deactivation pertains to the mechanism by which carbon dissolution occurs on the metal surface and diffuses through the particle. The carbon then precipitates at the metal-support interface, detaching the metal particle from the support and forming a carbon filament with an exposed metal particle at its tip. This mode of carbon accumulation allows the catalyst to maintain its activity for an extended period of time without deactivation. It has been shown that thousands of carbon atoms can be deposited on the catalyst by surface nickel atoms. However, eventually the catalyst is deactivated as access to active site becomes limited (Abbas and Wan Daud 2010a). Regardless of the catalyst activity and lifetime, two overarching problems with the use of a metal catalyst are regeneration and practical separation of the produced carbon.

The operating conditions and kinetic results for metal catalysts, as reported from reference (Abbas and Wan Daud 2010a), are as follows:

- Reaction Order: 1 (Snoeck et al. 1997)
- Activation Energy (kJ/mol): 29.5–46 (Fukuda et al. 2004)
- Temperature (°C): 500–1000

- Space Velocity (per hr): 50–50,000
- Sustainability Factor (ratio of the reaction rate after 1 hour on stream divided by the initial reaction rate): 0.14–1.

Reaction Mechanisms. Various reaction mechanisms have been developed by different groups and can be summarized as follows (Amin et al. 2011):

- Nondissociative adsorption of CH_4
  - CH_4 + I (vacant site) = CH_4 (ad)
  - CH_4 (ad) = CH_3 (ad) + H (ad)
  - CH_3 (ad) = CH_2 (ad) + H (ad)
  - CH_2 (ad) = CH (ad) + H (ad)
  - CH (ad) = C (ad) + H (ad)
  - C (ad) = C (dissolved)
  - 2H (ad) = H_2 + 2I.
- Dissociative adsorption of CH_4
  - CH_4 + I (vacant site) = CH_3 (ad) + H (ad)
  - CH_3 (ad) = CH_2 (ad) + H (ad)
  - CH_2 (ad) = CH (ad) + H (ad)
  - CH (ad) = C (ad) + H (ad)
  - C (ad) = C (dissolved)
  - 2H (ad) = H_2 + 2I.

Holmen et al. (1995) summarizes the conversion of CH_4 to carbon, H_2, olefins, and aromatics, and presents values for the reaction rate parameters. Considerable mechanistic and kinetic information is available from work on CH_4 coupling reactions (Olsvik and Billaud 1994).

7.2.2.2.2 Carbon Catalysts

Industrial catalysts typically consist of metals supported on appropriate catalyst supports. Alumina, silica, and carbon are the most widely used industrial supports (Abbas and Wan Daud 2010a). One strategy for CH_4 decomposition is to use carbon as the support and also as the catalyst (Abbas and Wan Daud 2010a; Rodríguez-Reinoso 1998). The separation of carbon product from the carbon catalyst itself thus is not necessary. Also, the process can be autocatalytic with the carbon produced serving as the catalyst for further reaction (Abbas and Wan Daud 2010a; Muradov and Veziroğlu 2005). Additional benefits reported for using carbon include its relatively low-cost, high-temperature resistance, and tolerance to sulfur and other contaminants found in the feedstock (Abbas and Wan Daud 2010a).

Several types of carbon materials have been evaluated for catalyzing CH_4 decomposition: activated carbon, carbon black, glassy carbon, acetylene black, graphite, diamond powder, CNTs, and fullerenes (Muradov et al. 2005a). Among these materials, most studies have focused on activated carbon because of its activity, good stability, and availability. The catalytic activity of carbon forms for CH_4 decomposition has been reported to vary according to the ordered nature of carbon as follows: amorphous > turbostratic > graphite (Abbas and Wan Daud 2010a). With all carbon types, deactivation occurs through deposition of carbon on the catalyst surface that blocks active sites and reduces the catalyst surface area. The deposit has lower surface area and activity compared to the original carbon catalyst (Abbas and Wan Daud 2010a). Physical attributes of carbon, such as surface area, pore volume, and particle size, are important characteristics that affect deactivation. At higher temperatures (e.g., > 850°C), carbon catalysts tend to deactivate at a slower rate. The reason is that at higher temperatures, the rate of diffusion and deposition increases inside the pores where the majority of the surface area is located (Abbas and Wan Daud 2010a).

Operating conditions and results have been summarized for carbon catalysts as follows (Abbas and Wan Daud 2010a):

- Reaction order: 0.4 (activated carbon); 2 (carbon black) (Abbas and Wan Daud 2010b; Charlier et al. 2007)
- Activation energy (kJ/mol): 138 (activated carbon); 236 (carbon black) (Abbas and Wan Daud 2010a; Xu et al. 2003; Muradov et al. 2005b; Pinilla et al. 2008)
- Temperature (°C): 750–1500
- Space velocity (per hr): 360–36,000.

7.2.2.3 Catalyst Regeneration

Regeneration of deactivated catalysts is usually done by burning off the carbon with air or steam, both leading to CO_2 production comparable to the amount of CO_2 produced in the SMR process (Abbas and Wan Daud 2010a). Regeneration by oxidation generates high temperatures on the catalyst surface, leading to sintering and loss of active area. Regenerating with CO_2 also has been reported, but doing so produces CO (Fulcheri and Schwob 1995; Rostrup-Nielsen et al. 2002). Thus, carbon oxides are produced regardless which process is used. Other researchers have evaluated catalyst regeneration via physical removal of carbon deposits on the catalyst surface by attrition (Abbas and Wan Daud 2010a). However, results have been mixed in that only the carbon deposited on the external surface of the catalyst particle can be removed, and this carbon represents only a fraction of the total carbon produced during decomposition (Abbas and Wan Daud 2010a). Therefore, not all solid carbon product is separated, and this carbon will need to be removed from the catalyst surface by the methods described previously, thus leading to production of carbon oxides.

7.2.3 Plasma Reactions

Plasma processes can be classified as either thermal plasmas (hot plasma, usually greater than ~700°C) or nonequilibrium plasmas (cold plasma, less than ~700°C). Thermal plasmas include direct current (DC) arch torch, alternating current (AC), radio frequency (RF) inductively coupled torch, and high-frequency capacitive torch. Nonequilibrium plasma (cold plasma) includes microwave, corona discharge plasma, dielectric barrier discharge plasma, atmospheric pressure glow discharge plasma, and gliding arc discharge. Generally, thermal plasmas achieve higher CH_4 conversions than nonthermal plasmas. Operating under plasma conditions requires the use of electric power, temperatures in the range of 700–5500°C for thermal plasma (Fulcheri and Schwob 1995; Schwob et al. 2000; Charlier et al. 2007; Fabry et al. 2009), pressures close to atmospheric (Fulcheri et al. 2002; Charlier et al. 2007; Fabry et al. 2009; Dors et al. 2014), and the use of a plasma gas (e.g., nitrogen, argon, helium, H_2, CO) (International Energy Agency 2017). Plasma-assisted CH_4 conversion can be achieved with or without catalysts. Two types of catalysts evaluated for CH_4 dissociation with plasma include metal-based catalysts (e.g., Ni, Co, La, Fe/Si, Pd, and Pt) and carbon-based catalysts (Wang and Xu 2003; Cho et al. 2004; Indarto et al. 2006; Pinilla et al. 2008; Konno et al. 2013). Catalysts have been reported to not enhance CH_4 conversion significantly, although in one study of a microwave plasma-catalytic system, CH_4 conversion was reported to be improved by about one order of magnitude with the use of Pd and Pt catalysts (Wang and Xu 2003; Cho et al. 2004; Indarto et al. 2006). Cold plasmas are inhomogeneous in discharge space, resulting in a limited reaction region and restricted conversion (Tao et al. 2011). However, Longmier et al. (2012) developed an RF nonthermal plasma reaction chamber that was efficient for fully decomposing CH_4 into H_2 and carbon graphite. Thermal plasma processes enable high CH_4 conversion (> 80%) with H_2 as a major product (Fincke et al. 2002; Fulcheri et al. 2002; Fabry et al. 2009; Tsai and Chen 2009). Other products may include carbon, acetylene, benzene, and ethane (Tsai and Chen 2009).

Plasma technologies for converting CH_4 into H_2 and carbon present several advantages. High yields of H_2 and carbon (100% carbon yield) can be obtained. The size of plasma reactors is an order of magnitude smaller than thermal reactors (Vinokurov et al. 2005; Longmier et al. 2012). A large range of feedstocks can be used in the plasma process, thereby making the process flexible. Also, the use of nonthermal plasmas has been reported to work at temperatures in the 850–900°C range, which is ~500°C lower than temperatures used for noncatalytic CH_4 decomposition (Muradov et al. 2009). Thus, using a plasma to provide the high energetics required for the process is beneficial in many ways.

Overall, the following advantages of microwave energy over conventional thermochemical processes were reported by Abbas and Wan Daud (2010a):

- Noncontact heating
- Energy transfer without heat transfer
- Rapid heating
- Quick startup and shutdown
- Volumetric heating, which can be more uniform throughout the material body
- Higher level of safety and automation
- Transport of energy from the source through a waveguide (a hollow, nonmagnetic metal tube) to the applicator (where the electromagnetic energy interacts with the material)
- Different grades of product carbon can be made at higher temperatures (Tsai and Chen 2009)

Disadvantages of plasma processes relate to the energy requirements and the fact that energy is provided by externally supplied electricity. Longmier et al. (2012) indicate that the energy required per kilogram of H_2 produced is 37 times higher than that required for steam reforming of CH_4, but the energy comparison does not include the energy required to sequester the emitted CO_2 during SMR. Regardless, the high energy requirements for plasma processing have been a distinct drawback of this approach. Another potential issue with plasma processes is low solid carbon product yield. In the decomposition process, a relatively high fraction of methyl radical enables formation of stable hydrocarbons and polymeric species, and a significant amount of energy is needed to crack CH_4 all the way to solid carbon (Muradov 2002). However, as discussed above, some plasma processes provide sufficient energy to achieve complete decomposition.

7.2.4 SEPARATIONS REQUIRED

A major challenge for the continuous catalytic CH_4 pyrolysis process is separation of the solid carbon product and regeneration of the spent catalyst, which are critical in the overall economics of the process (Amin et al. 2011). Some have described the aggregation of carbon deposits in the reactor as a main obstacle; one report describes it as the main technological "show stopper" (Abánades et al. 2011). In a PBR, carbon buildup will eventually cause pressure increases and ultimately block process flow through the reactor. To maintain catalytic activity and to avoid plugging of the reactor, the deposited carbon reactant must be periodically removed. Muradov (2002) evaluated various reactor types (e.g., PBR, FBR, free-volume reactor, spouted-bed reactor, and tubular reactor) and concluded that the FBR is the most promising reactor for large-scale application because it provides a constant flow of solids through the reaction zone, making it suitable for continuous addition and withdrawal of catalyst particles from the reactor. Fluidization of catalyst particles also increases heat and mass transfer rates (Abbas and Wan Daud 2010a).

Different methods have been used to separate the carbon byproduct from the H_2 stream. Bag filters have been used to remove carbon elutriates and also have been employed in plasma processes (Abbas and Wan Daud 2010a). Economic analysis for CH_4 pyrolysis technology employing this separation method needs to be evaluated. Because of density differences, use of a molten-metal reactor makes it possible to skim carbon from the top of the metal surface. While this may offer one solution to the issue of solid carbon separation, relatively high temperatures are still required (as with noncatalytic thermal pyrolysis). Geibler et al. (2016) report that liquid metal temperatures in the range of 930–1175°C are required.

Conversion of CH_4 to carbon and H_2 ideally would not include any other chemical species in the reactor stream. However, natural gas contains impurities (moisture, air, sulfur species, etc.) so separations are required in the process. If the pyrolysis process is conducted at elevated pressures, then PSA is an effective solution for extracting the product H_2. For low pressure pyrolysis, alternative gas separation techniques must be adopted.

7.3 PROCESS TECHNOLOGIES

In this section, we discuss specific technologies reported in the literature at various stages of commercialization that use different approaches to tackle the issues described above. Methane decomposition processes considered in this study are summarized in Table 7.2. More detail is provided for each process, as classified within thermal, plasma, microwave, and hybrid technology classes.

TABLE 7.2

Summary of Methane Decomposition Processes

Wulff (1936)	• Thermal cracking
	• Intermittent: Natural gas combustion followed by cracking
	• Packed media: carborundum
	• 1200–1650°C
	• Desired product: olefins
	• Research and development phase
HYPRO	• Solid recirculating catalyst
	• Methane thermal cracking
	• Carbon deposited on catalyst burned to regenerate the catalyst
	• Pilot scale
Fluidized Bed	• Activated carbon catalyst
(Muradov 1998,	• Electrical heating or by alternately switching between combustion and
2004)	catalytic cracking
	• Research and development phase
EGT Enterprises	• Methane pyrolysis reactor
(Ennis 2009)	• 2000°C
	• Energy demand 14 kJ/g_C and 43 kJ/g_{H2}
	• Research and development phase
Institut Francais du	• Combustion followed by CH_4 pyrolysis on ceramic honeycomb
Petrole (Weill et	• 1200°C
al. 1990)	• Pilot-scale demonstration
	• Desired product: olefins
	• Research and development phase
Steinberg and	• Methane decomposition in molten metal (tin, 232°C or copper, 1083°C)
Arenius Systems	• Methane bubbled through bath, carbon floats to the surface, gas is mainly H_2
(Steinberg 1999)	• H_2 purified by PSA
	• Pilot scale
BASF (Bode et al.	• Methane pyrolysis, 1200–1400°C, produces carbon and H_2
2014; Kern et al.	• H_2 reacted with CO_2 from external process to produce syngas
2015; Maass et	• Fluidized bed with support materials and carbonaceous granules
al. 2016)	• Research and development phase
Monolith Materials	• Plasma reactor
(2017a,b)	• Produces carbon black and H_2
	• H_2 combustion to produce steam (replacing coal-fired boiler)
	• Commercial (early stage)
Kvaerner (Lynum	• Plasma torch decomposes CH_4 to form carbon
et al. 1996, 2000)	• H_2 is the plasma gas
	• 1000–2000°C
	• Research and development phase
GasPlas	• Co-axial vortex flow reactor
	• Microwave induced plasma
	• Research and development phase
Remarks on	• Use of inert gases such as argon
Plasma Processes	• Electrical power is a premium, available from the grid at efficiencies of
	< 40%

7.3.1 THERMAL CATALYTIC PROCESSES

One of the earliest technologies for thermally cracking CH_4 to produce useful products is the Wulff process (Wulff 1936). It uses high-temperature steam to crack natural gas to acetylene, producing significant byproducts of H_2 and carbon. The process begins with natural gas combustion to heat the chamber and the packed, carborundum-crystal media. In subsequent steps, combustion air is stopped, and CH_4 cracks in contact with the heated surfaces of the chamber and the media which can be at temperatures as high as 1650°C. This process continues until the temperature cools 1200°C. The products are primarily H_2, acetylene, ethylene, carbon oxides in the gas stream, and solid carbon deposited on the media. Valued gas products subsequently need to be separated from small amounts of nitrogen, oxygen, and unreacted CH_4.

Another pioneering process for producing H_2 by thermal decomposition of CH_4 is the HYPRO process, which was demonstrated by UOP. A solid recirculating catalyst was used. The process employed a Ni-based catalyst to lower the decomposition temperature and two FBRs with a solids-circulation system. The carbon deposited on the catalyst was burned to regenerate the catalyst and provide process heat. This process was not CO_2-free.

Muradov (1993) compared the thermodynamically predicted yield of H_2 as a function of temperature to experimental yields observed using different catalysts. Iron oxide was shown to achieve equilibrium-predicted H_2 yields at 900°C. Based on X-ray diffraction analysis, carbon in the form of graphite crystallites was observed on the catalyst. A progress report and journal article proposed the use of activated carbon or carbon as the catalyst in an FBR (Muradov 1998, 2004). The fluidized bed could be heated using either electric heaters or a dual bed reactor system where one of the reactors is decomposing CH_4, while the other is being regenerated and heated by combustion (Muradov and Veziroğlu 2005).

EGT Enterprises developed a pyrolysis reactor using a series of heating zones with an electric resistance heating screen capable of operating up to 2000°C (Ennis 2009). One configuration showed the process operating at a feed rate of 4600 ft³/hr at a temperature of 1370°C and producing 9180 ft³/hr of product gas containing 97% H_2 with the balance being mainly unreacted CH_4. The 97% H_2 in the product corresponds to 94% CH_4 conversion. The energy demand for the process was 260 kW, which is 128% more than the required heat of reaction (i.e., the process has an efficiency of 44%) equivalent to 14 kJ/g of carbon and 43 kJ/g of H_2.

Weill et al. (1990) used a ceramic honeycomb pyrolysis reactor in which combustion gas was used to heat the CH_4 to be pyrolyzed at 1200°C. Based on pilot-scale tests, the process yielded ethylene, acetylene, benzene, coke, and tars.

BASF, Linde Group, and ThyssenKrupp developed a pyrolysis-based process for producing syngas and solid carbon (Bode et al. 2014). The H_2 produced by this pyrolysis process is reacted with CO_2 to produce syngas via the reverse WGS reaction. The pyrolysis reaction occurs in a FBR containing support materials and carbonaceous granules and operating at 1200–1400°C.

Steinberg (1999) proposed a molten metal-based CH_4 decomposition process that bubbles CH_4 through a molten tin (melting point of 232°C) or copper (1083°C) bath.

Because of its lower density, the carbon floats to the surface of the bath which facilitates its recovery. The gas stream contains primarily H_2 with lesser amounts of unreacted CH_4 and other byproducts. Arenius Systems is developing such a process at a pilot scale to provide H_2 for fuel cells with carbon black as a byproduct.

7.3.2 CONCENTRATED SOLAR POWER

Concentrated solar power is a renewable source of high-temperature process heat, and direct solar irradiation of the reactants provides for a very efficient heat transfer directly to the reaction site (Steinfeld et al. 1997). In a review article, Steinfeld (2005) discusses several solar natural gas conversion approaches and provides a comparative analysis for the different approaches in terms of exergy efficiency which was reported to be ~30% for solar cracking of CH_4. Hirsch and Steinfeld (2004) also report the design of a solar chemical reactor that uses a cavity-receiver that directly irradiated a vortex flow of natural gas premixed with carbon particles which serves as both a radiant absorber and a catalyst. After exiting the reactor, the products are cooled and the solid products are collected on filters. Continuous operation was demonstrated at the 5 kW scale with 67% CH_4 conversion at 1600 K and 1 bar. The carbon formed was filamentary in nature (Hirsch and Steinfeld 2004).

In a different study, Steinfeld et al. (1997) used a sun-tracking heliostat and stationary parabolic concentrators to supply energy to a small-scale FBR. Direct irradiation of a Ni/Al_2O_3 catalyst provided effective heat transfer to the reaction site. Methane underwent pyrolysis at 850 K with the conversion decreasing with time on stream as the carbon product accumulated, which caused the catalyst to deactivate. Yields of 200 g carbon/g catalyst were reported.

Generally, catalytic-based solar thermochemical applications have issues similar to other catalytic-based thermochemical conversion processes, such as recovery of the carbon from the catalyst and regenerating the catalyst. One disadvantage of using solar power is difficulty in maintaining stable heat flux. However, there is great potential benefit in using solar for supplying the energy for highly endothermic reactions.

7.3.3 PLASMA PROCESSES

Fulcheri et al. (2002) extensively studied CH_4 conversion to H_2 and carbon. They developed a plasma reactor with three electrodes, creating a compound arc by applying an AC current to the electrodes (Schwob et al. 2000; Fulcheri et al. 2002; Charlier et al. 2007; Fabry et al. 2009). The major advantage of this process is the total conversion of the hydrocarbon into carbon (100% carbon yield) and pure H_2. This process allows production of different carbon materials with defined nanostructures (e.g., carbon black, fullerenes, and nanotubes) by controlling the operating conditions (Charlier et al. 2007; Fabry et al. 2009). In addition, the feedstock may contain or consist of CH_4, acetylene, propylene, C_4 hydrocarbons, light or heavy oil, waste oil, and pyrolysis fuel oil (Fabry et al. 2009). The feedstock can comprise solid carbon material (e.g., carbon black, acetylene) for production of CNTs (Charlier et al. 2007).

TABLE 7.3

Hydrogen from Thermal Plasma Decomposition of Methane (100% Conversion Assumed) (Kim et al. 2005)

Argon/ CH₄ ratio	Ar flow	CH₄ flow	Plasma power	Total input power	H₂ output
2	167 SLPM	83.5 SLPM	24 kW	40 kW	10 Nm³/hr

Recently, Monolith Materials partnered with Fulcheri and Schwob's group and Aker Solutions to build a pilot plant in Redwood City, California. The plant uses a plasma-based process that converts natural gas into H_2 and carbon black (Monolith 2017a). Monolith (2017b) is building a production-scale plant in Hallam, Nebraska, that will sell the H_2 to an adjacent power plant to produce CO_2-free electricity.

Kim et al. (2005) studied the thermal decomposition of CH_4 by using a DC-RF hybrid thermal plasma to demonstrate the production of H_2, a small amount of olefins, and 20–50 nm carbon particles. Their design simulation estimates a total power requirement of 40 kW to produce 10 Nm³/hr (0.89 kg/hr) of H_2 and 2.5 kg/hr of carbon black, assuming 60% thermal efficiency of the DC-RF torch (see Table 7.3). Their simulated process uses 83.5 SLPM of CH_4 with a LHV of 49.8 kW. Considering only the 40 kW of plasma power for the process, the specific energy demands are 161 kJ/g$_{H2}$ and 57.6 kJ/g$_{Carbon}$, or cumulatively 42 kJ/g$_{Carbon+H2}$ products. Comparing this to an SMR process operating with 70% efficiency (LHV of H_2 as a percentage of LHV of the CH_4 feed), the specific energy demand is 172 kJ/g$_{H2}$, which compares favorably with the plasma process. However, if we include the LHV of the CH_4 in the energy input to the plasma process, then the specific energy demands for that process increase by a factor of 2+ to 362 kJ/g$_{H2}$, 129 kJ/g$_{Carbon}$, and 95 kJ/g$_{Carbon+H2}$.

Moshrefi et al. (2012) reported on CH_4 conversion using a DC spark discharge plasma experiment designed to produce olefins, where the gaseous products included acetylene, ethylene, and ethane accompanied by the formation of solid carbon. With a conversion percentage of 45% and assuming all H_2 in the converted CH_4 is available, the process efficiency based on the LHV of H_2 only is 16% (Figure 7.7). Extrapolating to 100% conversion, the efficiency would increase to 35%. From a H_2-only perspective, the process energy efficiency is low; however, if the carbon can be sold as a

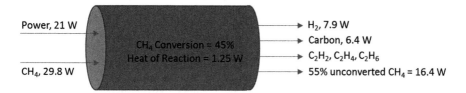

Efficiency = (7.9x100)/(21+29.8) = 15.6%

FIGURE 7.7 Direct current spark discharge plasma experiments produced hydrogen and carbon (Moshrefi et al. 2012).

commodity, the process may be economically attractive. The plasma was generated using argon gas, which is likely to pose an economical challenge to scale up.

Kvaerner developed the Carbon Black and Hydrogen Process using a plasma torch for CH_4 pyrolysis to produce H_2 and carbon black (Muradov and Veziroğlu 2005; Amin et al. 2011). H_2 is used as the plasma gas. A commercial plant was built and operated in Canada between 1999 and 2004. The reactor temperatures varied between 700°C at the primary gas inlet and reached as high as 2000°C. A secondary stream of CH_4 was then decomposed by contacting the hot gases.

GasPlas uses a microwave, nonthermal plasma to produce H_2 and carbon. Its noted features include an ambient pressure process, use of a highly nonequilibrium plasma, scalability, and the ability to control carbon formation and collection. One set of experimental results showed that a stream of CH_4 (3 L/min) mixed with Ni (17 L/min) when subjected to 1900 W of microwave power converted 53.9% of the CH_4 (Risby and Pennington 2012).

7.3.4 CHEMICAL PROCESSES FOR DECOMPOSING METHANE

ETCH, LLC is developing a low-temperature, quasi-catalytic process for producing carbon and H_2 from natural gas based on a process invented by Professor Jonah Erlebacher at Johns Hopkins University under a U.S. Department of Energy (DOE) Advanced Research Projects Agency-Energy (ARPA-E) grant (Erlebacher and Gaskey 2017). The process is based on the reaction of hydrocarbons, such as CH_4, with anhydrous nickel chloride to form nickel metal, carbon, and HCl at ~800°C. Lowering the temperature results in a back reaction between the nickel metal and HCl to regenerate nickel chloride and release H_2 gas. Carbon is separated from the nickel chloride by sublimation. The reaction is tolerant to impurities found in natural gas. The process requires no water and uses substantially less energy than SMR making it attractive for deployment in regions where process water is not readily available. Initial estimates indicate that the process could produce 4.7 kg of carbon for every kilogram of H_2, with a projected H_2 production cost competitive with SMR (Erlebacher and Krause 2017).

7.4 MARKET – CURRENT AND NEW OPPORTUNITIES

The United States produces ~10 million metric tons of H_2 annually, and SMR accounts for more than 95% of this production. New markets for H_2 such as fuel cell-powered vehicles are projected to significantly increase the demand for H_2 in the coming decades. A key challenge for fuel cell cars is the ability to produce H_2 at a cost that is competitive with the cost of producing gasoline on an equivalent energy basis. The ability to monetize the carbon produced during the thermal decomposition of natural gas offers the potential to reduce production costs of H_2 compared to those of SMR. In a 1987 study, Steinberg and Cheng (1989) compared the production cost of H_2 by several pathways, including CH_4 cracking, and estimated a cost of $5.10 per kilogram after taking credit for carbon black, exceeding SMR. To the best of our knowledge, there are still no viable commercial processes based on producing both carbon and H_2 for the carbon black and H_2 markets, respectively. This highlights the need for continued cost reductions through R&D, and through creative market strategies.

Carbon is marketed in a number of different forms including carbon black (the oldest and most mature market for carbon), CNTs, carbon nanofibers, carbon fibers, graphene, and needle coke. Table 7.4 summarizes the market potential for the different types of carbons. More details for each of the markets follow.

7.4.1 CURRENT MARKETS

7.4.1.1 Natural Gas

Development of hydraulic fracking and horizontal drilling has enabled the ability to recover natural gas from shale and tight formations at cost-competitive prices, resulting in significant growth in the natural gas supply while leading to lower prices. For example, U.S. natural gas production rose from 23.7 trillion cubic feet (tcf) in 2005 to 32.6 tcf in 2016; consumption rose from 21.7 tcf in 2005 to 27.5 tcf in 2016; and prices dropped from $8.12 per million cubic feet (mcf) to $2.35/mcf. U.S. natural gas reserves have increased from 213.3 tcf in 2005 to 307.7 tcf in 2015 (U. S. Energy Information Agency 2017a). Figure 7.8 shows the location and reserve potential for the major shale gas basins in the United States.

According to the U. S. Energy Information Administration (2017b) *Annual Energy Outlook (AEO) 2017*, natural gas prices are projected to increase in its Reference Case, rising modestly through 2030 to $5.00 to $6.00/mmBTU* as the use of natural gas for electric power generation increases. Prices are projected to remain relatively flat after 2030 as technology improvements keep pace with the rising demand (Figure 7.9). Under its High Oil and Gas Resource and Technology scenario, prices are projected to remain relatively low at $3.00 to $4.00/mmBTU due to lower production costs and higher resource availability. Under its Low Oil and Gas Resource and Technology case, prices are projected to approach $10.00/mmBTU, driving down domestic production and consumption. U.S. natural gas production is projected to grow at an annual growth rate of around 4% between 2016 and 2020 spurred by large, capital-intensive projects such as new liquefaction export terminals and petrochemical plants built in response to the low natural gas prices (Figure 7.10). Beyond 2020, natural gas consumption is expected to grow at a rate of ~1% annually as export growth moderates and domestic use becomes more efficient, causing prices to rise more slowly. Natural gas from shale gas and tight oil is projected to account for nearly two-thirds of U.S. production by 2040 (Figure 7.11).

7.4.1.2 Hydrogen

The United States produces ~10 million metric tons of H_2 annually and SMR accounts for 95% of the H_2 produced. The major consumers of H_2 are petroleum refining (47%) and fertilizer production (45%). Methanol production consumes ~4% and the electronics, metals, and food industries combined consume the remaining 3% (Brown 2016a).The U.S. demand for H_2 by industry was 13.8 million metric tons in 2015, and the demand in oil refining accounted for 10.5 million metric tons, which

---

* The unit mmBTU is a thousand-thousand, or million BTUs.

TABLE 7.4
Market Analysis for Potential Carbon Products (K = thousand, M = million, MT = metric ton)

Type of carbon	Types of applications	Expected price for carbon	Size of the market (current/ projected)	Corresponding hydrogen production[a]
Carbon black (Grand View Research 2017a,b)	Tires, printing inks, high-performance coatings and plastics	$0.4–2+/kg depending on product requirements	U.S. market • ~2M MT (2017) Global market • 12M MT (2014) • 16.4M MT (2022)	U.S. market • 0.67M MT Global market • 4M MT (2014) • 5.4M MT (2022)
Graphite (Benchmark Mineral Intelligence 2016)	Lithium-ion batteries	$10+/kg	Global market • 80K MT (2015) • 250K MT (2020)	Global market • 27K MT (2015) • 83K MT (2020)
Carbon fiber (Das et al. 2016; Witten et al. 2016; Mazumdar et al. 2017)	Aerospace, automobiles, sports and leisure, construction, wind turbines, carbon-reinforced composite materials, and textiles	$25–113/kg depending on product requirements	Global market • 70K MT (2016) • 100K MT (2020)	Global market • 23.3K MT (2016) • 33.3K MT (2020)
Carbon nanotubes (Sherman 2007; Grand View Research 2015)	Polymers, plastics, electronics, lithium-ion batteries	$0.10–600.00 per gram depending on application requirements	Global market • 5K MT (2014) • 20K MT (2022)	Global market • 1.7K MT (2014) • 6.7K MT (2022)
Needle coke (Halim et al. 2013)	Graphite electrodes for electric arc steel furnaces	~$1.5/kg	Global market • ~1.5M MT (2014)	Global market • ~0.50M MT (2014)

[a] Based on stoichiometric ratio of carbon to hydrogen present in methane. Does not take into account process efficiency or use of hydrogen to provide process heat or loss of hydrogen during hydrogen recovery.

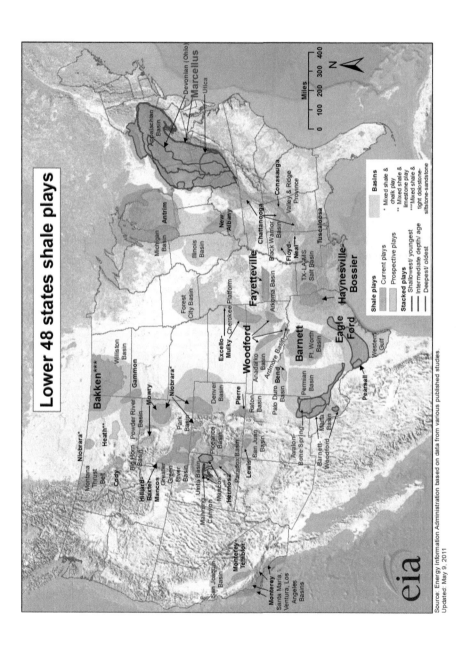

FIGURE 7.8 Location and projected reserves for major U.S. shale gas basins in the United States (Fincke et al. 2002).

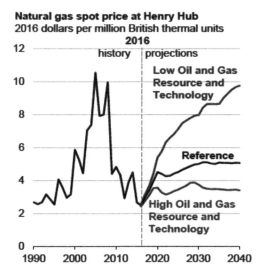

FIGURE 7.9 Historical and projected price of natural gas from 1990 through 2040. (U. S. Energy Information Administration 2017b.)

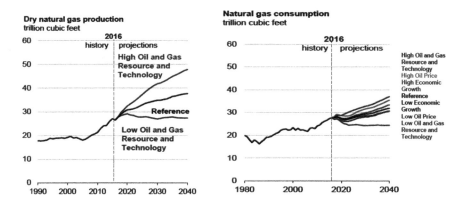

FIGURE 7.10 Total U.S. natural gas production and consumption from 1990 through 2040. (U. S. Energy Information Administration 2017b.)

is equivalent to 76% of the total demand (Brown 2016b). The industrial demand is met by either on-purpose production or the recovery of byproduct H_2. On-purpose H_2 is produced by both the consuming industry, referred to as captive production, and by merchant gas companies. In 2015, ~10 million tons of H_2 was produced; 6.4 million tons were produced by captive production, and 3.6 million tons were produced by merchant gas companies. Approximately 5.8 million metric tons of byproduct H_2 was recovered by the oil refining industry and from multiple other processes. Of this amount, ~3.8 million metric tons of H_2 were recovered for self-use or for sale to merchant gas companies and the remaining 2.0 million metric tons were combusted for process heat. Merchant gas companies recovered 0.29 million metric tons of

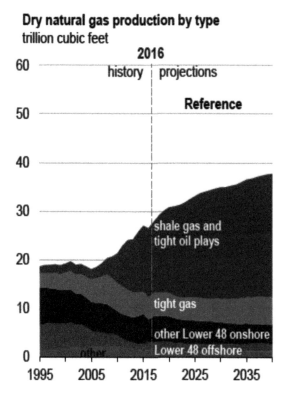

Dry natural gas production by type
trillion cubic feet

FIGURE 7.11   .Historic and projected growth in the production of natural gas from shale basins and tight oil between 1995 and 2040. (U. S. Energy Information Administration 2017b.)

H_2 (including ethylene, chlor-alkali, styrene, acetylene, and propylene) from a number of industries, and another 0.10 million metric tons were recovered from other unidentified sources. In terms of production cost, the price of H_2 is tied to the price of natural gas given that SMR accounts for > 90% of the H_2 produced in the United States. As a "rule of thumb," the cost of H_2 produced by SMR is approximately three times the cost of natural gas per unit of energy produced (T-Raissi and Block 2004). Given the current cost of natural gas of ~$3.00/mmBTU, the production cost of H_2 equates to $9.00/mmBTU or ~$1.00/kg based on the LHV of H_2.

Demand for H_2 worldwide is dominated by the oil refining, ammonia, and methanol industries (Brown 2016c). In the United States, captive H_2 production accounts for nearly all the H_2 used in ammonia and methanol production. Growth in these markets appears promising with the price of natural gas in the United States likely to be competitive with world prices over the next decade. In oil refining, slightly more than half of the H_2 demand is met by recovering byproduct H_2 from catalytic reforming units with the remaining H_2 demand met by captive production and merchant supply. While captive production has increased substantially over the past 25 years, merchant supply has increased from nearly nothing to 15% over the same period. Growth in demand in the methanol industry is due to a fourfold increase in methanol production since 1990. In oil refining, the growth in demand is due to stricter limits on the sulfur content in gasoline and diesel fuel resulting in an increase H_2 demand for hydrodesulfurization and the increasing need for H_2 for hydroprocessing heavier crudes to favor distillates over residuals. Growth in the merchant supply market has been driven by the increasing demand of oil refineries with > 90% of U.S. merchant H_2 production being delivered to oil refineries (Brown 2015). Long-term growth in the demand for H_2 for oil refining is uncertain given the projected flat or decreasing demand for petroleum products in the transportation sector in the coming decades due to increasing penetration of electric vehicles into the light-duty vehicle market coupled with projections that U.S. petroleum consumption peaked in 2006 and that annual miles driven is expected to peak in 2018 (U. S. Energy Information Administration 2017b).

Globally, the demand and supply of H_2 was estimated to be 87 million metric tons in 2015 with captive production accounting for > 60 million metric tons. Ammonia production accounted for ~30 million metric tons or nearly 50% of the captive H_2 production with methanol production and oil refining each accounting for ~25% of the total production. Merchant supply of H_2 accounted for < 10% of the total demand; however, it is the fast growing market, growing at an annual rate of 15.3% since 1990 (Cockerill 2016).The global market for H_2 is projected to grow from an estimated \$115.25B in 2017 to \$154.74B by 2022, at a compound annual growth rate of 6.07% (Markets and Markets 2017b). The growing demand for H_2 is due to 1) an increasing demand in oil refining to further reduce the sulfur content of gasoline and diesel and an increasing demand for a lighter mix of products (i.e., a higher percentage of distillates and a lower percentage of residual products) and 2) an increasing demand for methanol for use as a transportation fuel and as a feedstock for producing olefins (Brown 2016c). Fuel cell-powered vehicles represent a potential disruptive technology that could have a significant impact on the world demand for H_2. With about one billion light-duty vehicles on the road globally, the potential demand for H_2, assuming all vehicles are fuel cell-powered, is estimated to be ~150 million metric tons based on 10,000 miles driven annually and a fuel economy of 67 miles per kilogram.

7.4.1.3   Carbon

Carbon black is essentially pure carbon, typically 97% or greater by weight, in the form of colloidal particles formed by the thermal decomposition of gaseous or liquid hydrocarbons under controlled conditions (Park and Heo 2015). Because the thermal decomposition of CH_4 is endothermic, some processes add oxygen to the feed resulting in the partial oxidation of CH_4 to accelerate the reaction. Carbon black is used as filler and a strengthening/reinforcing agent in the manufacture of tires and other rubber and plastic products. It is used in printing inks to enhance formulations and in high-performance coatings to provide pigmentation, conductivity, and protection from ultraviolet light. Certain grades of carbon black are approved by the U.S. Food and Drug Administration for use in items such as coffee mugs, food trays, and cutlery (Monolith 2016). Approximately 90% of the carbon black produced is used in rubber applications, primarily tires, 9% as a pigment, and the remaining 1% in a number of diverse applications (Park and Heo 2015).

Carbon black can be produced by a number of processes including the furnace black or oil-furnace process, the thermal black process, and the acetylene black process.

7.4.1.3.1   Furnace Black Process

Heavy aromatic oils are atomized and injected into a natural gas-heated furnace under controlled conditions (U. S. Environmental Protection Agency 1983). The oils are partially combusted or undergo thermal decomposed at temperatures ranging from 1600 to 1800°C to form microscopic carbon particles. The exhaust gas containing the carbon particles is cooled using water quenches and heat exchangers. The carbon is separated from the cooled gas stream using water sprays and a fabric filter. The recovered carbon is densified and processed into pellets of various grades and sizes. Yields range from 35 to 65% depending on the composition of the grade of carbon black produced. The residual gas contains a variety of compounds, including H_2 and CO, and is used without further purification to produce heat, steam, or electric power.

7.4.1.3.2   Thermal Black Process

Natural gas is thermally cracked to produce carbon black (U. S. Environmental Protection Agency 1983). Two furnaces are used in the process and natural gas undergoes thermal cracking in the first furnace. Similar to the furnace black process, the exhaust from the first furnace is cooled and the carbon is filtered out. The exhaust gas, which contained upward of 90% H_2 and a balance of CH_4 and other light hydrocarbons, is burned to heat the second furnace. When the temperature of the first furnace becomes too low to promote cracking, the order of the furnaces is reversed. Typically, ~15 to 20% of the exhaust gas is used to heat the furnace, and the remaining gas is used without further processing to produce heat, steam, or electric power.

7.4.1.3.3   Acetylene Black Process

Acetylene is used instead of natural gas (Kühner and Voll 1993). Whereas the thermal black process is endothermic, the acetylene process is highly exothermic. Because of the higher purity of the feedstock and exothermic nature of the process compared to the thermal black process, the acetylene process produces a very pure carbon black that has a greater degree of crystallinity than that produced by the other methods.

Each process produces different grades of carbon black that have unique physical properties (e.g., particle size, agglomerate size, surface area, structure/void volume, color, etc.). Rubber-grade carbon black is classified as either ASTM or Specialty Grade, and each classification contains a large number of grades based on physical properties such as particle size and structure. Today, > 95% of the carbon black produced in the United States is produced using the furnace black process because it can produce nearly all of the different grades by carefully controlling the process conditions.

The price of carbon black depends on the product specifications (i.e., the grade and volume purchased). Prices can range from $400 to over $1000 per ton for ASTM-grade carbon black for use in tires to over $2000 per ton for specialty grade carbon black (Alibaba.com 2017). Given that the vast majority of carbon black produced worldwide is produced from oil, the price of crude oil significantly influences the carbon black prices.

The United States currently produces ~2 million metric tons of carbon black annually with a projected 4.2% annual growth rate through 2022 (Figure 7.12) (Grand View Research 2017a). Producers headquartered in the United States include

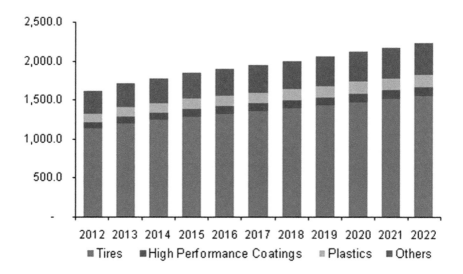

FIGURE 7.12   U.S. carbon black market volume by application from 2012 projected through 2022. Volume is in kilotons (Grand View Research 2017a). (Reproduced with permission from Grand View Research.)

Cabot Corporation (Boston, MA), Continental Carbon Company (Houston, TX), and Sid Richardson Carbon & Energy Co (Fort Worth, TX). Two other major industrial manufacturers are Birla Carbon, headquartered in India, and Orion Engineered Carbon, headquartered in Luxembourg. Given favorable market factors, new companies such as Monolith are entering the market. Monolith's plant located in Redwood City, California, is the first carbon black manufacturing facility built and licensed in the United States in the last 30 years.

The global demand for carbon black was nearly 12 million metric tons in 2014. Global demand is projected to increase to 16.4 million metric tons by 2022, representing a market value of $28B dollars with most of the growth in demand occurring in China and India (Grand View Research 2017b). Global demand is being spurred by the increasing demand for automobiles in the Asian market that will drive tire sales. Analysts predict that because of the projected global growth in demand for tires, U.S. tire manufacturers will face a shortfall in the availability of carbon black manufactured in the United States beginning in 2020 due. Specialty carbon black for use in plastics and lightweight automotive components is expected to be the fastest growing market segment.

7.4.2 New Market Opportunities for Methane Conversion to Carbon and Hydrogen

Other forms of carbon besides carbon black may be considered as products. Examples include graphite/graphene, carbon fiber precursors, CNTs, and needle coke. Hydrocarbons in the form of aromatics that can be produced from CH_4 pyrolysis also are discussed in this section.

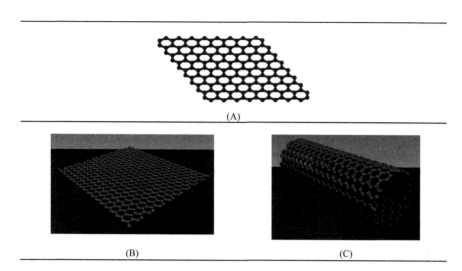

(A)

(B)   (C)

FIGURE 7.13 Comparison of the structure of (A) graphite, (B) graphene, and (C) a single-walled carbon nanotube (Tsai and Chen 2009).

7.4.2.1 Graphite/Graphene

Graphite/graphene materials are nearly 100% carbon and differ primarily in topology (Figure 7.13). Graphite is a naturally occurring material. It has a layered, planar structure. The individual layered sheets are called graphene. In each layer, the carbon atoms are arranged in a honeycomb lattice with a separation of 0.142 nm, and the distance between planes is 0.335 nm (Grand View Research 2015). Atoms in the plane are bonded covalently, with only three of the four potential bonding sites being satisfied. The fourth electron is free to migrate in the plane, making graphite electrically conductive. However, it does not conduct in a direction at right angles to the plane. Weak van der Waals forces bond the layers, allowing layers to be easily separated or to slide past each other. A rolled layer is a CNT.

Demand for natural graphite is 1–1.2 million tons per year and consists of several different forms of graphite – flake, amorphous, and lump. Historical applications primarily use amorphous and lump graphite, while most newly emerging technologies and applications use flake graphite. Of the up to 1.2 million tons of graphite that are processed each year, just 40% is the flake form. Historically, China was responsible for the large decline in graphite prices in the 1990s as product was dumped on the market to earn foreign exchange. Much like the market for rare elements, this essentially killed the industry in the West, making the United States highly dependent on supply from China. It is unlikely China can repeat that practice. The majority of Chinese graphite mines are small, many are seasonal, and labor and environmental standards are poor. Easily mined surface oxide deposits are being depleted, and mining is now moving into deeper and higher-cost deposits.

Graphite is used in electronic applications such as for Li-ion batteries where the market in 2015 was 80,000 tons/yr (Benchmark Mineral Intelligence 2016). The U.S. Geological Survey states that large-scale fuel cell applications being developed today could consume as much graphite as all other uses combined.

But even if only half of the U.S. Geological Survey demand is realized, graphite use is going to explode just because of fuel cells, let alone other known demand drivers and new applications. Thus, the natural graphite market is currently small, but it possibly will be very important for new fuel cell applications and/or batteries.

Graphene is a special form of graphite that is 100% carbon and only one layer or at most a few layers thick. It potentially has all of the application space that graphite has but currently has little production. The global market for graphene reached $9 million by 2012, with most sales in the semiconductor, electronics, battery energy, and composites industries (Wikipedia 2017). The market is projected to grow from 80,000 metric tons produced in 2015 to 250,000 metric tons by 2020 (Benchmark Mineral Intelligence 2016). Generation as a co-product with H_2 could be a potential benefit, but the current market would soon be saturated unless demand increases.

There is little reason to believe *a priori* that carbon produced from a pyrolysis process would be well-ordered. However, one study reported the formation of well-ordered pyrolytic graphite structures as well as fibrous carbon on transition metals. The pyrolytic carbon was reported to be of equal quality to recrystallized graphite normally produced at much higher temperatures. Results appeared to be repeatable (Robertson 1970). However, the value of the carbon will depend heavily on the treatment of the feed gas. Carbon black furnace processes operate by injecting a heated aromatic liquid hydrocarbon into the combustion zone of furnace fired by natural gas, where the hydrocarbon is decomposed to form carbon black at temperatures on the order of 1320–1540°C. Depending on the feed composition and the grade of black produced, process carbon black yields have ranged from 35–65% (U. S. Environmental Protection Agency 1983).

7.4.2.2 Carbon Fibers

Carbon fibers are polycrystalline, two-dimensional planar hexagonal networks of carbon containing between 92 and 100% carbon by weight formed by heating carbon-containing precursors at temperatures ranging from 1000 to 1500°C (Park and Lee 2015). If the fibers are heated above 2000°C, the hexagonal carbon network undergoes conversion to graphene with yields in excess of 99%. These fibers are referred to as "graphite fibers."

Carbon fibers have a number of favorable mechanical and chemical properties, such as high tensile strength and stiffness, low density, dimensional stability, low coefficient of thermal expansion, fatigue resistance, and chemical inertness and biological compatibility (Park and Lee 2015). Carbon fibers are finding increasing use in a variety of applications such as aerospace, automobiles, sports equipment, the chemical industry, wind turbines, carbon-reinforced composite materials, textiles, etc. (Holmes 2014; Park and Lee 2015; Mazumdar 2016; Witten et al. 2016). The physical properties (primarily tensile strength and modulus, etc.) determine the proper use of carbon fibers (Milbrandt and Booth 2016).

Carbon fibers are manufactured from precursor fibers using a combination of heat and stretching treatments. The most common precursors are polyacrylonitrile (PAN) and pitch, which is a complex blend of polyaromatic and heterocylcic compounds. Other linear and cyclic precursors include phenolic polymers, polyacenephthalene, polyamide, polyphenylene, poly-p-phenylene, benzobisthiazole, polybenzoxazole, polybenzimidazole, polyvinyl alcohol, and polyvinylidene chloride, and polystyrene (Park and Heo 2015). Currently, PAN is used to produce 95% of the carbon fibers worldwide, and pitch is used for the remaining 5%. Precursors, namely PAN, account for ~51% of the manufacturing cost of carbon fibers, and their high price is one of the barriers to their widespread use (Milbrandt and Booth 2016). Additionally, the current methods for manufacturing carbon fibers are slow and energy-intensive. Thus, both alternative methods of manufacturing and use of cheaper precursors are under exploration.

Alternative precursors to PAN under investigation include biomass precursors such as lignin, glycerol, and lignocellulosic sugars (Milbrandt and Booth 2016). There has been a particular interest in using lignin as a precursor because of its availability, low cost relative to other precursors, and enhanced structural properties. However, no biomass-based carbon fiber has been developed with the necessary structural properties required for use in the major carbon fiber applications (e.g., aerospace, wind, and automotive). The report authored by Milbrandt and Samuel Broth provides both technical and market information about each bio-based carbon fiber precursor (Milbrandt and Booth 2016). The report authored by Baker and Rials (2013) provides a comprehensive review of carbon fiber manufacture specifically from lignin, which includes a cost comparison of potentially low-cost carbon fibers.

The process for producing carbon fibers depends on the precursor and the desired physical properties. For PAN, the process starts by polymerizing the PAN-based precursor, which is then spun into fibers. The fibers are treated using an air-based oxidation process at temperatures between 200 and 400°C to stabilize the fiber. The stabilized fibers are heat treated in the absence of oxygen at temperatures ranging from 800 to 1600°C to remove noncarbon impurities such as H_2, oxygen, and nitrogen and to induce carbonization. Next, a surface treatment is used to improve the mechanical properties of the carbon fiber. Finally, the fiber is washed, dried, and sized (Park and Lee 2015; Park and Heo 2015). For pitch, the process starts by melting pitch so that it can be extruded and drawn into fibers. Similar to the PAN-based process, fibers are air-treated to stabilize the fiber; they then undergo a higher temperature heat treatment to induce carbonization (Park and Heo 2015). Pitch has the advantage of lower cost and producing a higher char yield than PAN, but the processing costs are higher to achieve carbon fibers of similar performance to PAN (Park and Heo 2015).

An alternative to the PAN- and pitch-based processes is the "vapor-grown" production process. The process involves exposing light-hydrocarbon gases, such as CH_4, acetylene, or ethylene or coal-gas, to a solid catalyst, such as Co, Fe, or Ni, to form carbon filaments with diameters as small as 0.1 μm as precursors for carbon fiber growth (Tibbets 1983; Alig et al. 2000; Endo and Dresselhaus 2003; Park and Lee 2015). The filaments consist of graphitizable carbon that is transformed into larger diameter graphite fibers by treatment at temperatures above 250°C. Exposing these carbon filaments to subsequent chemical vapor deposition using the same carbonaceous gases causes the filaments to grow in diameter ranging from 60 to 200 nm and ~100 μm in length, yielding vapor-grown carbon fibers or gas-phase-grown carbon fibers (Park and Lee 2015). Compared to the PAN-based process, the manufacturing process for vapor-grown carbon fibers is simpler, faster, and cheaper and could provide an innovative approach for fabricating high-performance fibers at lower costs.

The global market demand for carbon fibers in 2016 was 70,000 metric tons (Mazumdar et al. 2017). The market is projected to grow at an annual growth rate of 10 to 13% through 2020 with the market demand expected to exceed 100,000 metric tons in 2020 (Witten et al. 2016). The global market value was $2.15B in 2015, which is projected to grow to $4.2B by 2022 (Witten et al. 2016). Combined, the aerospace industry including defense, the automotive industry, and wind turbine manufacturing in the energy sector accounted for between 35,000 to 45,000 metric tons, and continuing growth in these industries is expected to support the high projected annual growth rate. Although worldwide demand for carbon fibers has increased significantly over the past decade, high production costs have limited wider spread use of carbon fibers (Park and Lee 2015). Because of the complex and multistage manufacturing processes required to produce carbon fibers, only a limited number of companies are engaged in their mass production. These companies include Toray (including its purchase of Zoltek), Toho Tenax, Mitsubishi Chemical, Formosa Plastics, SGL, Hexcel, DOW, and Kemrock (Park and Lee 2015; Das et al. 2016).

Analysts believe there is an excellent potential market opportunity for the use of carbon fibers in construction (Holmes 2014; Mazumdar 2016; Mazumdar et al. 2017). An increasing demand for fiber-reinforced plastic bathtubs, doors, windows, and panels is being spurred on by the continual growth in the U.S. housing market (Mazumdar 2016). Another application is carbon fiber reinforced concrete (i.e., "carbon concrete"), which is increasingly being used to repair bridges and other aging structures (Holmes 2014). Using carbon concrete to repair bridges that are in unsatisfactory condition is considered a major market opportunity. For example, more than 150,000 of the 600,000 bridges in the United States are considered unsuitable for the current or projected traffic demands, primarily because of corrosion of the steel reinforcement (Kawahara et al. 2012). Germany is projected to invest up to €16B–€17B to repair or replace bridges by 2030 (Holmes 2014). Although the cost of carbon-reinforced concrete is higher than steel-reinforced concrete, the higher cost is counterbalanced by its high specific properties such as lower weight relative to a steel-reinforced concrete deck (which reduces the load demand on the supporting structure), better corrosion resistance, better seismic protection, and lower erection and maintenance costs (Kawahara et al. 2012). Worldwide demand for carbon concrete in 2013 was 2300 metric tons, representing revenues totaling $590M. An annual growth rate of 6–9% is projected through 2020, and revenues are projected to exceed $1B by 2022. Lowering the cost of carbon concrete is considered key to more rapidly increasing its use in the construction industry.

The Institute for Advanced Composites Manufacturing Innovation (2017) is working to advance the composite industry, developing new manufacturing techniques and identifying potential new markets. Advancements will increase scrap and material costs and applications within the automotive industry has been identified.

7.4.2.3 Carbon Nanotubes

CNTs, including single-walled carbon nanotubes (SWCNTs) and multi-walled nanotubes (MWCNTs), are used in polymers, electronics, plastics, and energy storage.

The major application for CNTs is in composite fibers in polymers to improve thermal, electrical, and mechanical properties. This application accounted for over 60% of the market share in 2014. The high molecular complexity of graphene in MWCNTs increases their tensile strength. MWCNTs are increasingly being used in engineered polymers including polyetherimide, polycarbonate, and polyetheretherketone. The growing demand for polymers for use in the automotive and construction industries, particularly in China, India, Brazil, and the Middle East, is expected to spur market growth (Grand View Research 2015). CNTs are increasingly being used in the production of lithium-ion batteries. The application growth in lithium-ion batteries for use in grid and renewable energy storage is expected to increase the demand for CNTs for these applications.

The global CNT market demand for CNTs was slightly over ~5000 tons in 2014 and is projected to grow to over 20,000 tons by 2022 (Figure 7.14) (Grand View Research 2015). This is orders of magnitude lower than for other carbon products such as carbon black (12 million tons per year). Estimates of market value range from $3.4B (2020) to $5.6B in (2022) (Grand View Research 2015; Markets and Markets 2017a). MWCNTs are the largest production segment given that their manufacturing cost is significantly lower than that of SWCNTs. Selling prices range from $50/lb (MWCNT at Hyperion Catalysis) to $600 per gram (SWCNT in defense and niche markets) (Sherman 2007).

The Asia Pacific region is the fastest growing market due to increasing domestic demand coupled with lower manufacturing costs compared to the United States and Europe. Among the major manufacturers of CNTs are Arkema S.A. (France), Arry International Group LTD. (China), Carbon Solutions Inc. (United States), Cheap Tubes

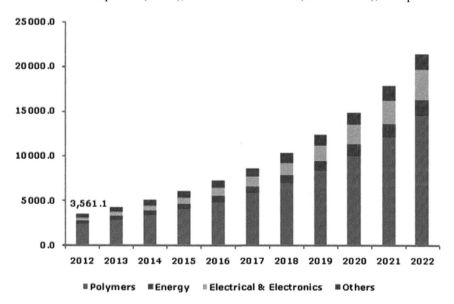

FIGURE 7.14 Global CNT market estimates and forecast by application from 2012 through 2022. Volume is in tons (Grand View Research 2015). (Reproduced with permission from Grand View Research.)

Inc. (United States), CNano Technology LTD. (United States), CNT Company Ltd. (Korea), Continental Carbon Company (United States), Hanwha Chemical Co. Ltd. (Korea), Hyperion Catalysis international Inc. (United States), KLEAN CARBON Inc. (Canada), Kumho Petrochemical Company LTD. (South Korea), Nano-C Inc. (United States), Nanocyl S.A. (Belgium), NanoIntegris Inc. (United States), NanoLab, Inc. (United States), Nanoshel LLC (United States), Nanothinx S.A. (Greece), Showa Denko K.K. (Japan), SouthWest NanoTechnologies Inc. (United States), Thomas Swan and Co. Ltd. (United Kingdom), and Toray Industries, Inc. (Japan).

7.4.2.4   Needle Coke

Needle coke is a premium-grade, high-value petroleum coke used in the manufacturing of graphite electrodes for electric arc furnaces in the steel industry. The main differences between needle coke and ordinary coke are their structural characteristics, coefficients of thermal expansion, electrical conductivity, and oxidizability. Needle coke has a high level of graphite resulting from its microcrystalline structure. A high level of anisotropy, large crystalline size, and large crystal areas must be achieved to obtain good quality needle coke (Halim et al. 2013). The term "needle" is used to describe the acicular morphology of the coke; it tends to form oriented needle-like structures that are visible to the naked eye. The coefficient of thermal expansion is one of the most important characteristics of petroleum coke in evaluating the feasibility of using a particular coke in the production of graphitized items that have a high resistance to shock. Carbon with its low coefficient of thermal expansion can dissipate thermal energy without cracking (Halim et al. 2013).

Needle coke is typically produced by delayed coking of the heavies remaining after catalytic cracking in a refinery. Delayed coking is a process for producing coke by transforming a complex mixture of aromatics to solid carbon (Halim et al. 2013). It provides thermal energy to form the mesophase of the precursor during carbonization. To achieve excellent quality needle coke, two major steps are needed: first, coalescence of the mesophase to its formation and second, rearrangement of the mesophase in the solidification stage. Different starting materials will have different chemical makeups, thus requiring different operating conditions. These conditions, particularly temperature and pressure, need to be optimized to achieve quality needle coke. Calcined needle coke typically is higher in carbon and lower in ash constituents, such as sulfur and metals, than standard calcined petroleum coke. A calcined form of needle coke is the raw material to produce graphite electrodes used in electric arc steel furnaces. The global market demand for needle coke is currently ~1.5M tons/yr. It has been reported that demand has increased in recent years and that this trend will continue.

7.5   TECHNO-ECONOMIC ASSESSMENT

7.5.1   BREAK-EVEN PRICE OF CARBON PRODUCTS VERSUS THE COST OF METHANE

Commercial production of H_2 and a carbon product via decomposition of natural ultimately gas will depend on process economics. Capital investment and operating costs will vary with the technology choice and scale of operations, but it is possible to set some target prices based on current information. The DOE Hydrogen and Fuel Cell Technologies Office (HFTO) has set the cost of H_2 dispensed and delivered at \$4/gge (~\$4/kg). The economic analysis presented below assumes the following simplifying assumptions:

- The selling price of H_2 is \$4/kg.
- The cost of CH_4 (as the raw material) represents a certain percentage, x, of the total sales revenue. The value of x will increase as the production process technology matures. That is, higher yield to desired products will lead to improvements in energy efficiency and lower capital investments. Larger plants also will improve the economics through economies of scale. The product yields of H_2 and carbon are the same and proportional to their ratio in the CH_4 (i.e., 2 moles of H_2 per g-atom of carbon, or 4 g of H_2 per 12 g of carbon).

These assumptions allow one to set the lowest possible price for the carbon product at which the process can be economically viable. Figure 7.15 shows the relationship between the break-even price of the carbon product and the cost of CH_4 in the plant as a percentage of the revenue from selling the H_2 and carbon products for three different prices of natural gas. For example, with $x = 50\%$ and a natural gas price of \$2.95 per million BTU (MBtu), the carbon product price needs to be greater than \$6.53/kg.

It was found with this analysis that at < \$1/kg, carbon black is not a viable final product. However, production of higher value products such as graphite, graphene, nanocarbon, and/or needle coke would make the process economically viable.

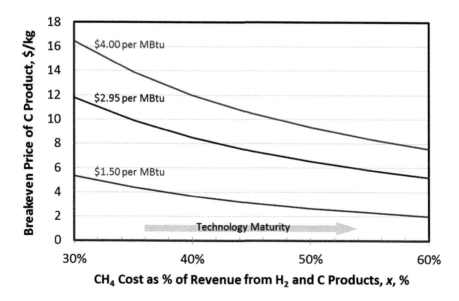

FIGURE 7.15   Break-even price of carbon products vs. the cost of methane as a percentage of the revenue from the sale of hydrogen and carbon products, as functions of the price of methane. Price of H_2 = \$4/kg.

7.5.2 ASPEN PROCESS MODELING AND ECONOMIC ANALYSIS

With this high-level pricing analysis in place, a separate set of process economics were evaluated in more detail for four direct CH_4 conversion processing options using ASPEN modeling. Processes for CH_4 plasma conversion to carbon black and SMR and electrolysis as benchmark processes were evaluated. Both distributed processing and large-scale processing for SMR were modeled for comparison. The following four cases were evaluated:

- Small-scale SMR (H_2 and CO_2 products)
- Small-scale electrolysis (H2 product)
- Small-scale low-temperature plasma (H_2 and C products)
- Large-scale SMR (H_2 and CO_2 products)

For the electrolysis and SMR options, information from the Process Economics Program (PEP) yearbook (2014) was used for capital and operating costs. PEP reports provide a consistent and credible methodology for evaluating process technologies, particularly for incumbent chemical and petrochemical processes as explored here. Large-scale SMR currently dominates the commercial H_2 production market. However, electrolysis systems are proposed in scenarios of anticipated surpluses of renewable electricity from sources such as wind and solar, so small-scale electrolysis information also was examined. Information from the PEP report for each of these H_2 production technologies was converted to an H2A (hydrogen analysis) spreadsheet; this spreadsheet is utilized by the DOE-HFTO for assessing economics on a standardized basis. Low-temperature plasma conversion was directly compared to conventional SMR at the small-scale of 1900 kg H_2/day. (Results from unpublished work performed previously at Pacific Northwest National Laboratory indicate that power supplies were limited to a size compatible with 1900 kg/day.) Conversion at 700°C with a 100°C approach to equilibrium was used to determine the reaction products. An ASPEN Plus® simulation was used to determine the material and energy balance and operating costs. The ASPEN Process Economic Analyzer® was used to assess capital cost. The capital cost and unit ratio information was then transferred to an H2A spreadsheet for this process for economic evaluation.

The bottom-line energy requirements, CAPEX, byproduct production, and resulting H_2 cost summary is presented in 2007 U.S. dollars in Table 7.5. It was found that the production and sale of carbon black in the plasma process reduces the net cost of H_2. This is due in part to the credit assigned to the sale of valuable aromatics. H_2 cost of $2.5/kg was assigned to the plasma pyrolysis (at a "small-scale" at the 1914 H_2 kg/day scale). By comparison the benchmark SMR produced H_2 at $7.4/kg at this same scale. However, it should be noted that the price assigned to the carbon black byproduct greatly affects the resulting H_2 cost, as illustrated in the sensitivity analysis shown in Figure 7.16 for the small-scale plasma case. In Table 7.5 we assume a carbon black selling price of $0.61/lb. We also demonstrate the impact of process scale on economics. Similarly, the cost of H_2 was reduced from $7.4/kg to $1.6/kg for the benchmark SMR at small- and large-scale processes, respectively. Nonetheless, these process economic results demonstrate how the cost of H_2 can be substantially

TABLE 7.5

Hydrogen Cost, Requirements, and Byproduct Production Summary for the Techno-Economic Cases Analyzed in This Study (Assumptions: Hydrogen Cost Before Compression and Delivery; Electricity Price = $0.056/kWh; Natural Gas Price = $5/MMBTU; Carbon Black Selling Price = $0.61/lb; 2007 $U.S.)

Technology	H_2 production scale (kg/day)	Electric power requirement (kWh/kgH_2)	Natural gas requirement (MMBtu/ kgH_2)	MeOH requirement (kg/kg H_2)	CAPEX (MM$)	Byproducts (kg/kgH_2)				H_2 cost ($/kg)
						Carbon black	Aromatics (BTX)	Fuel gas	Steam	
Plasma (Small Scale)	1,914	16.1	0.26		11.4	3.1				2.5
Electrolysis (Small Scale)	1,914	131.24	NA		19.5					15.5
SMR (Small Scale)	1,914	2.05	0.15		14.4					7.4
SMR (Large Scale)	112,222	0.32	0.17		163.4				9.67	1.6

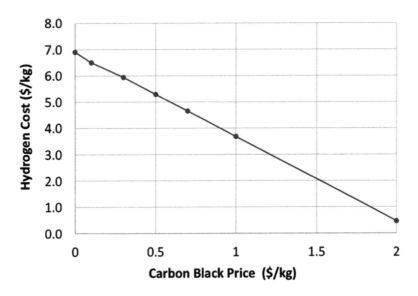

FIGURE 7.16 Sensitivity of carbon black selling price on net hydrogen cost for the small-scale plasma case study.

driven down by the sale of valuable carbon byproduct – even at smaller processing scales – and the carbon selling price is a critical factor.

7.6 TECHNOLOGY BARRIERS TO COMMERCIAL IMPLEMENTATION AND R&D OPPORTUNITIES

Thermal decomposition of natural gas is currently used in the carbon black industry to produce carbon black for use in tires and electrical equipment, but natural gas as the feedstock has been largely replaced by heavy oil fractions from crude oil processing. The high reaction temperature (> 1000°C) required for methane conversion contributes greatly to process inefficiencies and limits the choice of materials of construction, adversely impacts catalyst life, and exacerbates heat losses. Catalytic thermal decomposition has been extensively researched at the laboratory scale with the primary purpose of decreasing the temperature required for conversion. However, steam pressure buildup and loss of catalytic active sites due to carbon fouling are problematic. Nonthermal plasma processes for producing carbon and hydrogen have been reported as alternatives but, require a significant amount of electric power. Molten-metal technology has been reported with a major benefit being the relative ease of solid carbon separation from the molten metal due to density differences. However, a high conversion temperature is still required. Solar thermochemical processes leverage the use of inexpensive solar heat. However, noncatalytic processes require high temperature (e.g., 1600°C), and the high conversion temperature requires the use of expensive construction materials. Catalytic processes drive down operating temperature requirements, but solid carbon handling is still an issue.

In Table 7.6, we summarize potential solid carbon products, beyond just carbon black, and provide an assessment of the status of natural gas conversion technology,

TABLE 7.6

Summary of Carbon Products and Natural Gas Conversion Technology Status, Major Barriers to Commercial Implementation, and R&D Needs to Overcome Barriers

Type of carbon	Technology status	Major barriers in technology applications and scaling up	R&D needed to overcome barriers
Carbon black, thermal	Commercialized	• Ability to produce multiple grades of carbon black using natural gas as a feedstock to provide greater flexibility to meet to market demand • Lack of downstream processes to recover and produce industrial-grade high-purity H_2 • Impact of producing industrial-grade high-purity H_2 on the overall process efficiency and production costs of carbon black • Production cost of H_2 needs to be competitive with H_2 produced by SMR	• Reaction studies to define operating parameters and reactor designs for producing different grades of carbon black • Separations and recovery process technology development to produce industrial-grade high-purity H_2 at a cost-competitive price • Demonstrate an integrated process for producing carbon black and recovering industrial-grade high-purity H_2 that maximizes overall product yields of carbon black and H_2 and produces both products at market-competitive prices • Techno-economic analysis to determine the cost of recovering H_2 and its impact on producing H_2 and carbon black at market-competitive prices at various production capacities
Carbon black, plasma	Pilot-scale tests demonstrated, first commercial plant expected to be operational in 2018	• Ability to produce multiple grades of carbon black using natural gas as a feedstock to provide greater flexibility to meet to market demand • Lack of downstream processes to recover and produce industrial-grade high-purity H_2 • Impact of producing industrial-grade high-purity H_2 on the overall process efficiency and production costs of carbon black • Production cost of H_2 needs to be competitive with H_2 produced by SMR	• Reaction studies to define operating parameters and reactor designs for producing different grades of carbon black • Separations and recovery process technology development to produce industrial-grade high-purity H_2 at a cost-competitive price • Demonstrate an integrated process to produce carbon black and recover industrial-grade high-purity H_2 that maximizes overall product yields of carbon black and H_2 and produces both products at market-competitive prices • Techno-economic analysis to determine the cost of recovering H_2 and its impact on producing H_2 and carbon black at market-competitive prices at various production capacities

(Continued)

TABLE 7.6 (CONTINUED)

Summary of Carbon Products and Natural Gas Conversion Technology Status, Major Barriers to Commercial Implementation, and R&D Needs to Overcome Barriers

Type of carbon	Technology status	Major barriers in technology applications and scaling up	R&D needed to overcome barriers
Carbon fiber	Commercialized	• Lower production capacity of the vapor-grown carbon nanofiber process compared to PAN- and pitch-based processes • Downstream processing to recover and produce industrial-grade high-purity H_2 • Impact of producing industrial-grade high-purity H_2 on the overall process efficiency and production costs of carbon fibers • Production cost of H_2 needs to be competitive with H_2 produced by SMR	• Process development to increase production capacity to allow vapor-grown carbon fibers process to compete with PAN- and pitch-based processes • Separations and recovery process development to recover industrial-grade high-purity H_2 • Demonstrate an integrated process to produce carbon fibers and recover industrial-grade high-purity H_2 that maximizes overall product yields of carbon fibers and H_2 and produces both products at market-competitive prices • Techno-economic analysis to determine the cost of recovering H_2 and its impact on producing H_2 and carbon fibers at market-competitive prices
Carbon nanotubes	Commercialized	• Low production capacity of vapor-grown carbon nanotubes process • Downstream processing to recover and produce industrial-grade high-purity H_2 • Impact of producing industrial-grade high-purity H_2 on the overall process efficiency and production costs of carbon nanotubes • Production cost of H_2 needs to be competitive with H_2 produced by SMR	• Process development to increase production capacity of carbon nanotubes • Separations and recovery process development to recover industrial-grade high-purity H_2 • Demonstrate an integrated process to produce carbon nanotubes and recover industrial-grade high-purity H_2 that maximizes overall product yields of carbon nanotubes and H_2 and produces both products at market-competitive prices • Techno-economic analysis to determine the cost of recovering H_2 and its impact on producing H_2 and carbon nanotubes at market-competitive prices
Needle Coke	Commercialized	• Ability to produce highly crystalline needle coke from natural gas is not well understood • Downstream processing to recover and produce industrial-grade high-purity H_2 • Production cost of H_2 needs to be competitive with H_2 produced by SMR	• Process development aimed at producing needle coke needed • Understand quality of crystalline carbon formation and evaluate suitability as a needle coke precursor • Separations and recovery process development to recover industrial-grade high-purity H_2 • Techno-economic analysis to determine cost of recovering H_2 and its impact on producing H_2 and needle coke at market-competitive prices

major barriers to commercial implementation, and R&D needed to overcome these barriers.

7.7 CONCLUSIONS

The major findings and conclusions from this study are outlined below:

- Processes for decomposing natural gas to generate carbon and H_2 are attractive because of the current low cost of natural gas and the available infrastructure for delivering natural gas to the conversion plant. Production costs and demand for carbon and H_2 will be key factors in determining the optimal production capacity and location of plants.
- Noncatalytic thermal decomposition of natural gas currently is used in the carbon black industry to produce carbon black for use in tires and electrical equipment, but natural gas as the feedstock has been largely replaced by heavy oil fractions from crude oil processing. H_2 is burned to provide process heat. Because these processes operate at high temperatures, off-gas treatment that significantly increases the capital cost is required to reduce nitrogen oxides and other emissions. Additionally, the high reaction temperatures required for CH_4 conversion contributes greatly to process inefficiencies. The high reaction temperatures also limit the choice of reactor materials, adversely impact catalyst life, and exacerbate heat losses.
    - Despite historically low natural gas prices, it is not clear whether existing oil-furnace process facilities can be retrofitted to burn natural gas (as is done in the power industry).
    - Emissions control, particularly of nitrogen oxides, is a major issue with many of the major carbon black producers that have been fined recently by the U.S. Environmental Protection Agency and have had to install expensive gas cleanup systems. This could be a major technical issue for a small-scale distributed facility.
    - The purity of the H_2 is not known because it is burned to provide process heat. Cleanup will be necessary to produce fuel-cell-quality H_2.
- Catalytic thermal decomposition has been extensively researched in laboratories, with a primary goal of reducing the required conversion temperature. A wide range of metal and carbon-based catalysts have been reported in the literature. Nickel, cobalt, and iron have been studied the most, and they produce both amorphous and structured carbons. Amorphous carbons such as activated carbon and carbon black are more active than structured carbons such as graphite and diamond. Published data do not delineate the effect of catalysts and process conditions on the quality/morphology of the carbon product. Reactors have been demonstrated or proposed with intermittent fixed beds or fluidized beds.
    - At least one potentially commercial process, UOP's HYPRO™ process, was developed in the late 1960s, but it was not cost competitive with SMR. Carbon was burned during the catalyst regeneration stage to generate process heat.

- The National Aeronautics and Space Administration, the U.S. Department of Defense, and the DOE have funded a number of pilot-scale demonstrations for producing fuel-cell-quality H_2 in which the carbon is burned to regenerate the catalyst.
- Although several highly-active catalyst material formulations have been employed, further catalyst material development is needed because any process will require extensive catalyst recycling (i.e., carbon deposition/carbon removal). Catalyst mechanical stability will be an issue because carbon deposition on the catalyst surface leads to catalyst detachment from the support as the carbon is deposited. Separation of the catalyst and the carbon byproduct remains a challenge.
- Cost-competitive and process-effective technologies for recovering the carbon (separating the carbon from the catalyst) will require further R&D. It is not clear whether carbon can be recovered without having to burn a portion of it to fully regenerate the catalyst. Retention of a portion of the catalyst, particularly Ni-based catalysts, in the carbon product could be a regulatory problem.
- It is not clear what effect catalysts have on the grade of carbon produced and how much this can be controlled by the catalyst. Optimizing the catalyst/process to produce the desired grade of carbon represents an R&D opportunity. Preventing gas-deposition of carbon during a catalytic process could be a major issue depending on the grade of carbon being produced and product specifications.
- Optimizing the process to produce both the desired grade of carbon and fuel cell-grade H_2, particularly if the carbon being produced has a very tight product specification, could be a major challenge. It may require less than ~100% single-pass CH_4 conversion, which increases the complexity with added separations and recycle burdens. Current processes for producing carbon are not overly concerned with the H_2 purity because the hydrogen is combusted to provide heat. It will be a major challenge for processes targeting high-value carbons, such as nanofibers, to achieve both high-quality carbon and H_2 products.
- Plasma processes for producing carbon and H_2 require electric power (expensive and produced with low efficiencies) and plasma gas.
  - Nonthermal equilibrium plasmas are not considered energetic enough to produce carbon and H_2 at required production scales.
  - Studies have suggested that DC plasma processes could be economically competitive for producing H_2 with alternative processes. DC plasma processes operate at extremely high temperatures and producing large reaction volumes is an issue.
  - The AC three-phase plasma process is being commercialized by Monolith Materials. The technology has been demonstrated at various pilot scales in Europe.
- Molten-metal technology has been reported, and a major benefit is its relative ease of solid carbon separation from the molten metal. However, high temperature still is required for the conversion.

- Solar thermochemical processes rely on solar augmentation for heating. However, noncatalytic processes require high temperature (e.g., 1600°C) so the cost of reactor construction materials is high. Catalytic processes would drive down operating temperature requirements, but solid carbon handling remains an issue. Finally, the cost of a "solar" plant is an issue.
- Our techno-economic analysis suggests that when the carbon is sold as byproduct, the net cost of H_2 can be drastically reduced to < \$2/kg, which is the DOE H_2 cost target (before compression and delivery). However, the economics are driven by the value and quality of the carbon byproduct.
- No known commercial process produces both carbon and fuel-cell-quality H_2 as commercial products. There are commercial processes for producing carbon that burn some of the H_2 to provide heat for the reaction and burn the rest to provide process heat to the facility or other nearby facilities. There are also commercial processes for producing fuel-cell-quality H_2, but carbon is not recovered. Carbon is burned to regenerate the catalyst and to provide process heat.

Areas of opportunity for R&D in the CH_4 pyrolysis technical space include the following:

- The natural gas infrastructure would allow development of a distributed H_2-refueling network and distributed fuel cell power networks. Producing H_2 is also achievable using SMR or water electrolysis using wind (or other renewable) energy. These processes also produce pressurized H_2. The natural gas decomposition process will have to compete with the above alternatives to produce a higher value product – derivable from the carbon combined with any environmental credit for not producing CO_2.
- High-pressure, low-temperature decomposition would reduce the product cost. Technology breakthroughs in will be required to overcome thermodynamic limitations that restrict the high yields of H_2 and carbon at these conditions.
- Solid carbon separation, classification by value, and handling are key challenges. H_2 production is relatively easy because all of these processes use high temperatures. Even cleanup/purification is manageable with PSA. The next steps of compression and delivery drive up the cost of H_2. Therefore, a high-value byproduct is needed to offset this cost.
- High-value carbon markets do exist. Graphite is a high-value product used in lithium-ion batteries. Nanotube carbons also are high-value products (20 kilotons/yr by 2022). However, suitable technologies optimized for producing both H_2 and valuable carbon byproduct must be developed. Furthermore, solid carbon as a byproduct could reduce costs only if sufficiently large markets for the carbon products are found. Challenges include low conversion levels and catalyst stability, which are key technological barriers to commercial implementation.
- The overarching challenges for producing these high-value carbon products are (1) identification of the conditions at which the different grades of

carbon will be formed during the decomposition process, (2) separation of the different grades of carbon, and (3) competition from alternative feedstock sources. For example, petcoke from refineries is cheap and can be processed to produce graphite.

ACKNOWLEDGMENTS

This book chapter is based on a formal technical report entitled "R&D Opportunities for Development of Natural Gas Conversion Technologies for Co-Production of Hydrogen and Value-Added Solid Carbon Products," which was published in November of 2017. Although we note that in this latest contribution new techno economic analysis using ASPEN process modeling is provided. The authors thank the U.S. Department of Energy Fuel Cell Technologies Office within the Office of Energy Efficiency and Renewable Energy for funding the report that is the basis for this chapter. We also thank the staff from that office for technical assistance, including Maxim Lyubovsky, Eric Miller, Richard Farmer, and Sunita Satyapal. The report was reviewed by Professor John Hu from West Virginia University, Drs. Christopher Matranga and Peter Balash from the U.S. Department of Energy's National Energy Technology Laboratory, Dr. Marc Von Keitz from the U.S. Department of Energy's Advanced Research Projects Agency-Energy, as well as Joe Cresko and Rudy Kashar from the U.S. Department of Energy's Advanced Manufacturing Office.

REFERENCES

Abánades, A., Rubbia, C., and D. Salmieri. 2012. Technological challenges for industrial development of hydrogen production based on methane cracking. *Energy* 46:359–363.

Abánades, A., Ruiz, E., Ferruelo, E.M., Hernández, F., Cabanillas, A., Martínez-Val, J.M., Rubio, J.A., López, C., Gavela, R., Barrera, G., Rubbia, C., Salmieri, D., Rodilla, E., and D. Gutiérrez. 2011. Experimental analysis of direct thermal methane cracking. *Int. J. Hydrogen Energ.* 36:12877–12886.

Abbas, H.F., and W.M.A. Wan Daud. 2010a. Hydrogen production by methane decomposition: a review. *Int. J. Hydrogen Energ.* 35:1160–1190.

Abbas, H.F., and W.M.A. Wan Daud. 2010b. Hydrogen production by thermocatalytic decomposition of methane using a fixed bed activated carbon in a pilot scale unit: apparent kinetic, deactivation and diffusional limitation studies. *Int. J. Hydrogen Energ.* 35:12268–12276.

Ahmed, S., Aitani, A., Rahman, F., Al-Dawood, A., and F. Al-Muhaish. 2009. Decomposition of hydrocarbons to hydrogen and carbon. *Appl. Catal. A Gen.* 359:1–24.

Alibaba.com. 2017. *Market Price for Carbon Black*. https://www.alibaba.com/showroom/market-price-for-carbon-black.html (Accessed August 14, 2017).

Alig, R., Burton, D., Kennel, E., and M. Lake. 2000. *Development of Pilot Plant for the Production of Vapor Grown Carbon Fiber from Ohio Coal*. Final report, July 1997 to July 2000. Technical report, U.S. Department of Energy: Office of Scientific and Technical Information. https://www.osti.gov/scitech/servlets/purl/1185202/ (Accessed August 14, 2017).

Alvarez-Galvan, M.C., Mota, N., Ojeda, M., Rojas, S., Navarro, R.M., and J.L.G. Fierro. 2011. Direct methane conversion routes to chemicals and fuels. *Catal. Today* 171:15–23.

Amin, A.M., Croiset, E., and W. Epling. 2011. Review of methane catalytic cracking for hydrogen production. *Int. J. Hydrogen Energ.* 36:2904–2935.

Baker, D.A., and T.G. Rials. 2013. Recent advances in low-cost carbon fiber manufacture from lignin. *J. Appl. Polym. Sci.* 130:713–728.

Benchmark Mineral Intelligence. 2016. *Graphite Demand from Lithium Ion Batteries to More than Treble in 4 Years*. http://benchmarkminerals.com/graphite-demand-from-lithium-ion-batteries-to-more-than-treble-in-4-years/ (Accessed on August 14, 2017).

Blanquart, H., and H. Pitsch. 2009. Analyzing the effects of temperature on soot formation with a joint volume-surface-hydrogen model. *Combust. Flame* 156:1614–1626.

Bode, A., Agar, D.W., Buker, K., Goke, V., Hensmann, M., Janhsen, U., Klingler, D., Schlichting, J., and S.A. Schunk. 2014. Research cooperation develops innovative technology for environmentally sustainable syngas production from carbon dioxide and hydrogen. *Paper Presented at the 20th World Hydrogen Energy Conference*, Gwangju Metropolitan City, Korea.

Brown, D. 2015. Hydrogen production and consumption in the U.S. - the last 25 years. *CryoGas International*, September 2015, 38–39.

Brown, D. 2016a. U.S. and world hydrogen production - 2014. *CryoGas International*, March 2016, 32–33.

Brown, D. 2016b. U.S. hydrogen production - 2015. *CryoGas International*, July 2016, 22–23.

Brown, D. 2016c. Hydrogen supply and demand - past, present, and future. *Gasworld*, April 2016, 37–40.

Carbon Black Sales. *Carbon Black - World Consumption*. http://carbonblacksales.com/carbon-black-reinforcing-agent/ (Accessed August 14, 2017).

Charlier, J.-C., Fabry, F., Flamant, G., Fulcheri, L., Gonzalez, J., Grivei, E., Gruenberger, T.-M., Okuno, H., and N. Probst. 2007. Carbon nanostructures and process for the production of carbon-based nanotubes, nanofibres and nanostructures. *US20070183959A1*, Published August 9, 2007.

Chesnokov, V.V., and A.S. Chichkan. 2009. Production of hydrogen by methane catalytic decomposition over Ni–Cu–Fe/Al$_2$O$_3$ catalyst. *Int. J. Hydrogen Energ.* 34:2979–2985.

Cho, W., Lee, S.-H., Ju, W.-S., Baek, Y., and J.K. Lee. 2004. Conversion of natural gas to hydrogen and carbon black by plasma and application of plasma carbon black. *Catal. Today* 98:633–638.

Cockerill, R. 2016. Supply and demand - start-ups signal a shift in future hydrogen supply. *Gasworld*, October 2016.

Das, S., Warren, J., West, D., and S.M. Schexnayder. 2016. *Global Carbon Fiber Composites Supply Chain Competitiveness Analysis*. Technical report, Oak Ridge National Laboratory. https://www.nrel.gov/docs/fy16osti/66071.pdf (Accessed August 14, 2017).

Dors, M., Nowakowska, H., Jasinski, M., and J. Mizeraczyk. 2014. Chemical kinetics of methane pyrolysis in microwave plasma at atmospheric pressure. *Plasma Chem. Plasma Process.* 34:313–326.

Endo, M., and M. Dresselhaus. 2003. *Carbon Fibers and Carbon Nanotubes*. http://web.mit.edu/tinytech/Nanostructures/Spring2003/MDresselhaus/i789.pdf (Accessed August 14, 2017).

Ennis, B.P. 2009. Electric reaction technology for fuels processing. *US7563525B2*. Published July 21, 2009.

Erlebacher, J., and B. Gaskey. 2017. Method of carbon dioxide-free hydrogen production from hydrocarbon decomposition over metal salts. *US9776860B2*. Published Octorber 3, 2017.

Erlebacher, J., and T. Krause. 2017. *Private Communication*, June 15, 2017.

Fabry, F., Grivei, E., Fulcheri, L., Leroux, P., Probst, N., Smet, N., Peroy, J.-Y., Flamant G., and F. Fischer. 2009. Device and method for converting carbon containing feedstock into carbon containing materials having a defined structure. *US20090142250 A1*. Published June 4, 2009.

Fincke, J.R., Anderson, R.P., Hyde, T.A., and B.A. Detering. 2002. Plasma pyrolysis of methane to hydrogen and carbon black. *Ind. Eng. Chem. Res.* 41:1425–1435.

Fischer, F., and H. Pichler. 1932. Uber die Thermische Zersetzung von Methan. *Brennst. Chem.* 13:381–383.

Fukuda, S., Nakamura, N., Monden, J., and M. Nishikawa. 2004. Experimental study of cracking methane by Ni/SiO_2 catalyst. *J. Nucl. Mater.* 329–333:1365–1369.

Fulcheri, L., and Y. Schwob. 1995. From methane to hydrogen, carbon black and water. *Int. J. Hydrogen Energ.* 20:197–202.

Fulcheri, L., Probst, N., Flamant, G., Fabry, F., Grivei, E., and X. Bourrat. 2002. Plasma processing: a step towards the production of new grades of carbon black. *Carbon* 40:169–176.

Geibler, T., Abánades, A., Heinzel, A., Mehravaran, K., Müller, G., Rathnam, R.K., Rubbia, C., Salmieri, D., Stoppel, L., Stückrad, S., Weisenburger, A., Wenninger, H., and T. Wetzel. 2016. Hydrogen production via methane pyrolysis in a liquid metal bubble column reactor with a packed bed. *Chem. Eng. J.* 299:192–200.

Grand View Research. 2015. *Carbon Nanotubes (CNT) Market Analysis by Product (Single Walled Carbon Nanotubes (SWCNT), Multi Walled Carbon Nanotubes (MWCNT)), by Application (Polymers, Energy, Electrical & Electronics) and Segment Forecasts to 2022*. Market research report. http://www.grandviewresearch.com/industry-analysis /carbon-nanotubes-cnt-market (Accessed August 14, 2017).

Grand View Research. 2017a. *Carbon Black Market Analysis by Application (Tires, High Performance Coatings, Plastics) and Segment Forecasts to 2022*. http://www.gran dviewresearch.com/industry-analysis/carbon-black-market (Accessed August 14, 2017).

Grand View Research. 2017b. *Carbon Black Market Size to Reach USD 28.05 Billion by 2022*. http://www.grandviewresearch.com/press-release/global-carbon-black-market (Accessed August 14, 2017).

Halim, H.P., Im, J.S., and C.W. Lee. 2013. Preparation of needle coke from petroleum by-products. *Carbon Lett.* 14:152–161.

Hirsch, D., and A. Steinfeld. 2004. Solar hydrogen production by thermal decomposition of natural gas using a vortex-flow reactor. *Int. J. Hydrogen Energ.* 29:47–55.

Holmen, A., Olsvik, O., and O.A. Rokstad. 1995. Pyrolysis of natural gas: chemistry and process concepts. *Fuel Process. Technol.* 42:249–267.

Holmen, A., Rokstad, O.A., and A. Solbakken. 1976. High-temperature pyrolysis of hydrocarbons. I. Methane to acetylene. *Ind. Eng. Chem. Proc. Des. Dev.* 15:439–444.

Holmes, M. 2014. Global carbon fibre market remains on upward trend. *Reinf. Plast.* 58:38–45.

Indarto, A., Choi, J.-W., Lee, H., and H.K. Song. 2006. Methane conversion using dielectric barrier discharge: comparison with thermal process and catalyst effects. *J. Nat. Gas Chem.* 15:87–92.

Institute for Advanced Composites Manufacturing Innovation. 2017. https://iacmi.org/2017 /06/21/iacmi-the-composites-institute-and-material-innovation-technologies-partner -on-carbon-fiber-project/ (Accessed August 14, 2017).

International Energy Agency. 2017. *Techno-Economic Evaluation of SME Based Standalone (Merchant) Hydrogen Plants with CCS*. IEAGHG Technical Report. https://ieaghg.org /exco_docs/2017-02.pdf.

Jabs, H., Westerheim, D., Hennings, B., Soekamto, D., Shandy, S., Minevski, Z., and A.J. Cisar. 2007. Hydrogen production using plasma-based reformation. *US20070267289*. Published November 22, 2007.

Kawahara, B., Estrada, H., and L.S. Lee. 2012. Life-cycle cost comparison for steel reinforced concrete and fiber reinforced polymer bridge decks. In *Fiber Reinforced Polymer (FRP) Composites for Infrastructure Applications*, ed. Jain R., and L. Lee, 237–273. Dordrecht: Springer.

Kern, M., Glenk, F., Klingler, D., Bode, A., Kolios, G., Schunk, S., Wasserschaff, G., Bernnat, J., Zoels, B., Schmidt, S., and R. Koenig. 2015. Parallel preparation of hydrogen, carbon monoxide, and a carbon-comprising product. *US20150336795*. Published November 26, 2015.

Kim, S.K., Seo, J.H., Nam, J.S., Ju, W.T., and S.H. Hong. 2005. Production of hydrogen and carbon black by methane decomposition using DC-RF hybrid thermal plasmas. *IEEE Trans. Plasma Sci.* 33:813–823.

Konno, K., Onoe, K., Takiguchi, Y., and T. Yamaguchi. 2013. Direct preparation of hydrogen and carbon nanotubes by microwave plasma decomposition of methane over Fe/Si activated by biased hydrogen plasma. *Green Sust. Chem.* 3:19–25.

Kühner, G., and M. Voll. 1993. *Carbon Black: Science and Technology*. New York, NY: Marcel Dekker.

Longmier, B., Gallimore, A., and N. Hershkowitz. 2012. Hydrogen production from methane using an RF plasma source in total nonambipolar flow. *Plasma Sour. Sci. Technol.* 21(01):5007–5015.

Lynum, S., Haugsten, K., Hox, K., Hugdahl, J., and N. Myklebust. 1996. A method for decomposition of hydrocarbons. *EP0616599*. Published February 28, 1996.

Lynum, S., Myklebust, N., and K. Hox. 2000. Decomposition of hydrocarbon to carbon black. *US6068827*. Published May 30, 2000.

Maass, H.-J., Goeke, V., Machhammer, O., Guzmann, M., Schneider, C., Hormuth, W.A., Bode, A., Klingler, D., Kern, M., and G. Kolios. 2016. Method for the parallel production of hydrogen and carbon-containing products. *US9359200*. Published June 7, 2016.

Markets and Markets. 2017a. *Carbon Nanotubes (CNT) Market Worth $9.84 Billion by 2023*. http://www.marketsandmarkets.com/PressReleases/carbon-nanotubes.asp.

Markets and Markets. 2017b. *Hydrogen Generation Market Worth 154.74 Billion USD by 2022*. http://www.marketsandmarkets.com/PressReleases/hydrogen.asp (Accessed August 14, 2017).

Mazumdar, S. 2016. Annual state of the composites industry report. *Composites Manufacturing*, 19–23.

Mazumdar, S., Karthikeyan, D., Benevento, M., and R. Frassine. 2017. *State of the Composites Industry Report for 2017*. http://compositesmanufacturingmagazine.com/2017/01/composites-industry-report-2017/ (Accessed August 14, 2017).

Milbrandt, A., and S. Booth. 2016. *Carbon Fiber from Biomass*. Technical report, Clean Energy Manufacturing Analysis Center (CEMAC). https://www.nrel.gov/docs/fy16osti/66386.pdf (Accessed October 31, 2017).

Monolith. 2016. *Fact Sheet*. http://monolithmaterials.com/wp-content/uploads/2016/09/Monolith-Fact-Sheet-2016.pdf (Accessed March 1, 2017).

Monolith. 2017a. *Olive Creek Plant*. http://monolithmaterials.com/olive-creek/ (Accessed August 14, 2017).

Monolith. 2017b. *Nebraska Project Updates*. http://monolithmaterials.com/olive-creek/project-update-details/ (Accessed August 14, 2017).

Moshrefi, M.M., Rashidi, F., Bozorgzadeh, H.R., and S.M. Zekordi. 2012. Methane conversion to hydrogen and carbon black by DC-spark discharge. *Plasma Chem. Plasma Process.* 32:1157–1168.

Muradov, N. 2002. Thermocatalytic CO_2-free production of hydrogen from hydrocarbon fuels. *Proceedings of the 2002 U.S. Department of Energy Hydrogen Review*, National Renewable Energy Laboratory. https://www.nrel.gov/docs/fy02osti/32405a17.pdf.

Muradov, N., Chen, Z., and F. Smith. 2005a. Fossil hydrogen with reduced CO_2 emission: modeling thermocatalytic decomposition of methane in a fluidized bed of carbon particles. *Int. J. Hydrogen Energ.* 30:1149–1158.

Muradov, N., Smith, F., and A. T-Raissi. 2005b. Catalytic activity of carbons for methane decomposition reaction. *Catal. Today* 102–103:225–233.

Muradov, N., Smith, F., Bockerman, G., and K. Scammon. 2009. Thermocatalytic decomposition of natural gas over plasma-generated carbon aerosols for sustainable production of hydrogen and carbon. *Appl. Catal. A Gen.* 365:292–300.

Muradov, N.Z. 1993. How to produce hydrogen from fossil fuels without CO_2 emissions. *Int. J. Hydrogen Energ.* 18:211–215.

Muradov, N.Z. 1998. CO_2-free production of hydrogen by catalytic pyrolysis of hydrocarbon fuel. *Energ. Fuel* 12:41–48.

Muradov, N.Z. 2004. *Thermocatalytic Process for CO2-Free Production of Hydrogen and Carbon from Hydrocarbons*. Technical report, U.S. Department of Energy – Office of Scientific and Technical Information. https://www.osti.gov/scitech/servlets/purl/828215 (Accessed August 14, 2017).

Muradov, N.Z. 2005. Hydrogen production via catalytic reformation of fossil and renewable feedstocks. *Paper Presented at the Florida Universities Hydrogen Review 2005*, Florida Solar Energy Center. http://www.hydrogenresearch.org/NRM_Nov05/FSEC-Muradov-Catalytic%20Reformation-Nov05.pdf.

Muradov, N.Z., and T.N. Veziroğlu. 2005. From hydrocarbon to hydrogen–carbon to hydrogen economy. *Int. J. Hydrogen Energ.* 30:225–237.

Muradov, N.Z., and T.N. Veziroğlu. 2008. "Green" path from fossil-based to hydrogen economy: an overview of carbon-neutral technologies. *Int. J. Hydrogen Energ.* 33:6804–6839.

Olsvik, O., and F. Billaud. 1994. Thermal coupling of methane. A comparison between kinetic model data and experimental data. *Thermochim. Acta* 232:155–169.

Outotec Pori Research Center. 2002. *Outokumpu HSC Chemistry for Windows: User's Guide (Versions 5.1)*. https://www.scribd.com/document/81923599/00-HSC-Chemistry-5.

Park, S.-J., and G.-Y. Heo. 2015. Precursors and manufacturing of carbon fibers. In *Carbon Fibers*, ed. Parks, S.-J., 31–66. Dordrecht: Springer.

Park, S.-J., and S.-Y. Lee. 2015. History and structure of carbon fibers. In *Carbon Fibers*, ed. Parks, S.-J., 1–30. Dordrecht: Springer.

Pinilla, J.L., Suelves, I., Lazaro, M.J., and R. Moliner. 2008. Kinetic study of the thermal decomposition of methane using carbonaceous catalysts. *Chem. Eng. J.* 138:301–306.

Pinilla, J.L., Utrilla, R., Lazaro, M.J., Suelves, I., Moliner, R., and J.M. Palacios. 2009. A novel rotary reactor configuration for simultaneous production of hydrogen and carbon nanofibers. *Int. J. Hydrogen Energ.* 34:8016–8022.

Risby, J.P., and D. Pennington. 2012. Method for processing a gas and device for performing the method. *WO2012147054A1*. Published November 1, 2012.

Robertson, S.D. 1970. Carbon formation from methane pyrolysis over some transition metal surfaces - I. Nature and properties of the carbons formed. *Carbon* 8:395–400.

Rodríguez-Reinoso, F. 1998. The role of carbon materials in heterogeneous catalysis. *Carbon* 36:159–175.

Rostrup-Nielsen, J.R., Sehested, J., and J.K. Nørskov. 2002. Hydrogen and synthesis gas by steam and CO_2 reforming. *Adv. Catal.* 47:65–139.

Schwob, Y., Fischer, F., Fulcheri, L., and P. Willemez. 2000. Conversion of carbon or carbon-containing compounds in a plasma. *US20070183959A1*. Published August 8, 2000.

Sherman, L.M. 2007. *Carbon Nanotubes Lots of Potential - If the Price is Right.* Article, Plastics Technology. http://www.ptonline.com/articles/carbon-nanotubes-lots-of-po tentialif-the-price-is-right (Accessed August 14, 2017).

Snoeck, J.W., Froment, G.F., and M. Fowles. 1997. Kinetic study of the carbon filament formation by methane cracking on a nickel catalyst. *J. Catal.* 169:250–262.

Steinberg, M. 1999. Fossil fuel decarbonization technology for mitigating global warming. *Int. J. Hydrogen Energ.* 24:771–777.

Steinberg, M., and H.C. Cheng. 1989. Modern and prospective technologies for hydrogen production from fossil fuels. *Int. J. Hydrogen Energ.* 14:797–820.

Steinfeld, A. 2005. Solar thermochemical production of hydrogen––a review. *Sol. Energ.* 78:603–615.

Steinfeld, A., Kirillov, V., Kuvshinov, G., Mogilnykh, Y., and A. Reller. 1997. Production of filamentous carbon and hydrogen by solarthermal catalytic cracking of methane. *Chem. Eng. Sci.* 52:3599–3603.

Tao, X., Bai, M., Li, X., Long, H., Shang, S., Yin, Y., and X. Dai. 2011. CH_4–CO_2 reforming by plasma – challenges and opportunities. *Prog. Energ. Combust.* 37:113–124.

Tibbets, G. 1983. Carbon fibers produced by pyrolysis of natural gas in stainless steel tubes. *Appl. Phys. Lett.* 42:666–668.

T-Raissi, A., and D.L. Block. 2004. Hydrogen: automotive fuel of the future. *IEEE Power Energ. Mag.* 2:40–45.

Tsai, C.-H., and K.-T. Chen. 2009. Production of hydrogen and nano carbon powders from direct plasmalysis of methane. *Int. J. Hydrogen Energ.* 34:833–838.

U.S. Energy Information Administration. 2017a. *Natural Gas.* https://www.eia.gov/natural-gas/data.php (Accessed August 14, 2017).

U.S. Energy Information Administration. 2017b. *Annual Energy Outlook 2017.* https://www .eia.gov/outlooks/aeo/pdf/0383(2017).pdf.

U.S. Environmental Protection Agency. 1983. AP 42, *Fifth Edition Compilation of Air Pollutant Emission Factors*, Chapter 6 Organic Chemical Process Industry. https://ww w3.epa.gov/ttnchie1/ap42/ch06/final/c06s01.pdf (Accessed August 14, 2017).

Vinokurov, V., Sharafutdinov, R., and Y. Tychkov. 2005. Plasma-chemical processing of natural gas. *Chem. Technol. Fuel. Oil.* 41:112–115.

Wang, B., and G. Xu. 2003. Conversion of methane to C-2 hydrocarbons via cold plasma reactions. *J. Nat. Gas Chem.* 12:178–182.

Wang, M.-J., Reznek, S.A., Mahmud, K., and Y. Kutsovsky. 2003. Carbon black. In *Kirk-Othmer Encyclopedia of Chemical Technology*, ed. Kroschwitz, J.I., and A. Seidel, 761–803. Hoboken, NJ: John Wiley & Sons.

Weill, J., Cameron, C.J., and C. Raimbult. 1990. IFP processes for the direct conversion of methane into higher hydrocarbons. *Paper Presented at The European Oil and Gas Conference: A Multidisciplinary Approach in Exploration and Production R&D*, Palermo, Italy.

Wheeler, R.V., and W.L. Wood. 1928. The pyrolysis of methane. *Fuel* 7:535–539.

Wikipedia. 2017. *Graphene.* https://en.wikipedia.org/wiki/Graphene#Production (Accessed August 14, 2017).

Witten, E., Kraus, T., and M. Kühnel. 2016. *Composites Market Report 2016 - Market Developments, Trends, Outlook and Challenges.* Market report, Carbon Composites. http://www.eucia.eu/userfiles/files/20161128_market_report_2016_english.pdf (Accessed August 14, 2017).

Wulff, R.G. 1936. Process of producing acetylene gas. *US2037056.* Published April 14, 1936.

Xu, Y., Bao, X., and L. Lin. 2003. Direct conversion of methane under nonoxidative conditions. *J. Catal.* 216:386–395.

8 Methane Conversion on Single-Atom Catalysts

Xiaoyan Liu, Hua Liu, and Aiqin Wang

CONTENTS

8.1 INTRODUCTION

Methane, as the main component of natural gas, shale gas, coal bed methane, and methane hydrate, is a most promising raw material because of its high reserves and relatively inexpensive price; it also is a potent greenhouse gas (Rovik et al. 2009; Kim et al. 2008). Unfortunately, most of these methane reserves are located in remote and depopulated areas. The high transportation cost of methane over a long distance makes its usage economically unviable. In addition it is vital to convert methane on a large scale and on-site to transportable high-density energy sources or high value-added chemicals such as methanol, olefins, aromatics, and hydrogen.

The molecule of methane has a fully symmetrical tetrahedral geometry with four equivalent C–H bonds with the sp³ hybridization. Furthermore, methane exhibits the highest C–H bond strength and the first bond dissociation energy (BDE) is about 440 kJ·mol⁻¹, meaning that methane is the most stable alkane (Kim et al. 2009). In order to allow an electrophilic or nucleophilic attack, methane needs a relatively high local electric field to be polarized itself due to the absence of nondipole moment and extremely small polarizability (2.84×10^{-40} C²·m²·J⁻¹) (Amos 1979).

As a result of constant efforts during the last decades, various routes to activate methane have been developed. As shown in Figure 8.1, methane conversion processes can be divided into high-temperature and low-temperature processes

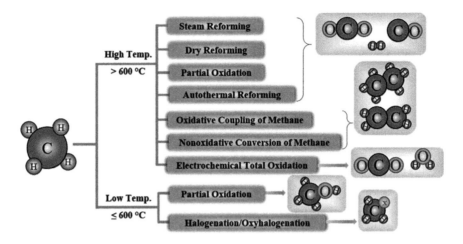

FIGURE 8.1 Schematic of selected methane conversion processes at high and low temperatures. (Reproduced with permission from Wang et al. 2017, Copyright 2017 Elsevier.)

(Wang et al. 2017). The former mainly involves the partial or total oxidation of methane (main products: CO, H_2, H_2O, and CO_2), oxidative coupling (Eichelbaum et al. 2015; Farrell et al. 2016; Ito et al. 1985; Keller and Bhasin 1982; Lunsford 1995; Pakhare et al. 2013; Takanabe and Iglesia 2009; Voskresenskaya et al. 1995; Zhang et al. 2006; Kondratenko and Baerns 2008), and nonoxidative conversion of methane that can obtain C_{2+} hydrocarbons (e.g., ethane, ethylene, and benzene) (Guo et al. 2014; Hu et al. 2009; Sakbodin et al. 2016; Wang et al. 1993; Zheng et al. 2008). Industrially, the methane can be converted to methanol via two steps: the first step is steam reforming of methane at high temperature to produce CO and H_2; the second step is to synthesize methanol by the Fischer–Tropsch (F–T) process. However, the high-temperature processes need more demanding requirements for the reactor, besides, they are energy intensive and high cost (Lunsford 2000). Hence, the development of one-step process for conversion of methane to high value-added chemicals at low temperatures is of great importance (Ogura and Takamagari 1986).

It has been reported that methane can be activated by halogen radicals through cleaving C–H bands under relatively mild conditions to produce alcohols, ethers, or olefins (Ogura and Takamagari 1986; Lorkovic et al. 2004; Weissman and Benson 1984; Olah et al. 1985; Olah 1987; Arvidsson et al. 2017). However, it makes the potential industrial application quite challenging due to the toxicity and high corrosiveness of the halogenation or oxy-halogenation reactions. Based on the above discussion, it is an ideal approach for utilization of CH_4 by directly converting CH_4 to oxygenated products such as HCHO, CH_3OH, and CH_3COOH under mild conditions (Osadchii et al. 2018; Tang et al. 2018; Shan et al. 2017; Kwon et al. 2017; Tomkins et al. 2019; Sushkevich et al. 2017). For example, the direct partial oxidation of methane to methanol (MTM) is thermodynamically feasible even at a low temperature; nevertheless, there is no industrial process for the MTM reaction even though it has been the subject of active research for almost a century. In addition to the high barriers in activating the methane, the C–H bond of the methanol is much active than that of the methane (Ravi et al. 2017).

Despite the difficulty for the direct partial oxidation of MTM, it has been investigated in the diverse fields (e.g., biological (Dalton et al. 1990), heterogeneous, homogeneous, etc.) by experimental and/or modeling (e.g., building models by density functional theory (DFT) calculation, synthesizing model catalyst, etc.) methods. For example, it can be realized in homogeneous catalysis by using strong acid to get methanol from methane directly. Periana et al. (1998) demonstrated, in concentrated sulfuric acid, platinum bipyrimidine complexes could catalyzed the selective oxidation of methane at temperatures around 200°C, making promising progress in homogeneous solution-phase catalysis. Recent researches indicated that the complexes of Pt (Periana et al. 1998), Hg (Periana et al. 1993), Pd (Periana et al. 2003), Rh (Lin and Sen 1994), Au (Shilov and Reidel 1984), Ru (Gunsalus et al. 2017), and V (Gunsalus et al. 2017) can promote selective oxidation of methane in a strong acid medium, such as oleum, trifloroacetic acid, hydrochloric acid, or hydrobromic acid at 80 ~ 500°C. In the homogeneous catalytic process, the active sites are coordinated metal complexes, which can reach 100% atom utilization efficiency. However, there are challenges in the separation, recycling, and corrosion for commercial application. It is highly desirable to develop heterogeneous catalyst systems for the MTM reaction.

The single-atom catalysts (SACs) have been emerging as an emerging technique since it was firstly proposed in 2011 by Qiao et al. (2011) for the CO oxidation over the Pt_1/FeO_x. The unique structure of the single atom active sites made the SACs promising for high selectivity for various reactions. The concept of the SAC was specified in 2018 in a review paper published by Wang et al. (2018). The isolated single sites were defined as the same atoms in each site with the same structures. Nevertheless, the coordination environment of the single atoms in the SACs do not have to be the same, but it must be a separated single atom, that is, the nearest neighbor must not be the same atom (Figure 8.2) (Liu 2017). Based on this concept, the SACs could be classified to the metal oxide supported SACs, single-atom alloys (SAAs), single atoms in the framework of the supports (e.g., nitrogen modified carbon materials, metal-organic frameworks (MOFs), zeolites, graphene, etc.). In addition, the SACs exhibited the advantages of the homogeneous and heterogeneous catalysis by getting high utilization of the active atoms and easy separation and recycling of the catalyst. Furthermore, the SACs are much stable than the enzymes, as pictured in Figure 8.3. The SACs can integrate the good points of the homogeneous, heterogeneous, and enzyme catalysts. They have been applied to various fields including catalytic synthesis of fine chemicals (Tomkins et al. 2017), electrocatalysis, photocatalysis, and so on. Therefore, the SACs are potential to be promising catalysts for the activation of methane. The direct MTM reaction is a hot topic in the past few years. There were many review papers published on this topic (Ravi et al. 2017, 2019; Latimer et al. 2018; Meng et al. 2019). But the SACs used on the MTM reaction has not been summarized yet. In this chapter, we will review the applications of the SACs to the activation of C–H bond of methane. In addition

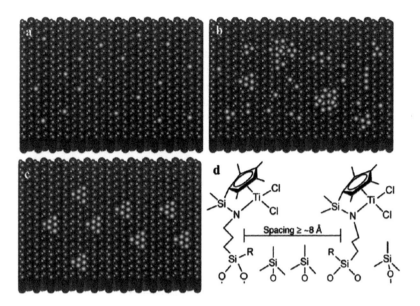

FIGURE 8.2   Schematic illustrations of (a) a supported single metal-atom catalyst (SAC), (b) an atomically dispersed supported metal catalyst (ADSMC), (c) an extreme example of a single site heterogeneous catalyst (SSHC), and (d) a site-isolated heterogeneous catalyst (SIHC). (Reproduced with permission from Liu 2017. Copyright 2016 American Chemical Society.)

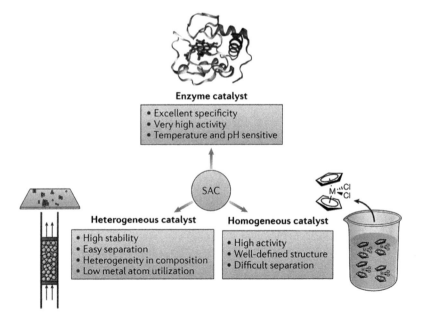

FIGURE 8.3   SACs integrated the advantageous features of enzyme, homogeneous, and heterogeneous catalysts. (Reproduced with permission from Wang et al. (2018). Copyright 2018, Springer Nature.)

to the traditional thermal heterogeneous catalysis, the nonthermal heterogeneous catalysis was also briefly summarized. The reaction conditions of the MTM in this chapter conducted under mild conditions at reaction temperatures lower than 200°C, unless specified.

8.2 METHANE TO METHANOL OVER THE SACS BY THERMAL CATALYSIS

The activation of methane over the SACs by heterogeneous thermal catalysis will be summarized by clarifying to the base metal and precious metal SACs. By literature review, it turned out that the base metal SACs for the MTM reaction were mainly focusing on Fe and Cu SACs in frameworks, while for the precious metal SACs, there were reports on different precious metals supported on metal oxides, in frameworks and SAAs. Therefore, the base metal SACs for the MTM reaction were classified according to different nature of the supports, at the same time, the precious metal SACs for this reaction were divided by different precious metals. The metals that were reported to be active for the direct MTM reaction were summarized in Figure 8.4.

8.2.1 BASE METAL SACs

The enzymes with iron or copper centers have been proved to be active for the direct oxidation of MTM under mild conditions. However, the stability of the enzymes is not good enough for application; in addition, they are sensitive to many factors including pH values, temperatures, and so on. Learning from the natural enzymes, several strategies have been developed, for example, to stabilize the enzymes in mesoporous silica nanospheres (Liu et al. 2016), to choose a support with the structure like enzymes (e.g., zeolites, graphene, MOFs, etc.), and so on. The synthesis of the SACs can be ascribed to the mimic of the enzymes in heterogeneous catalysis. Therefore, most of the SACs applied to the MTM reaction were those supported in the frameworks, especially for the base metal SACs. Here, the typical base metal Fe and Cu supported on three different supports (zeolite, MOF, and graphene) were chosen as examples of SACs for the MTM reaction.

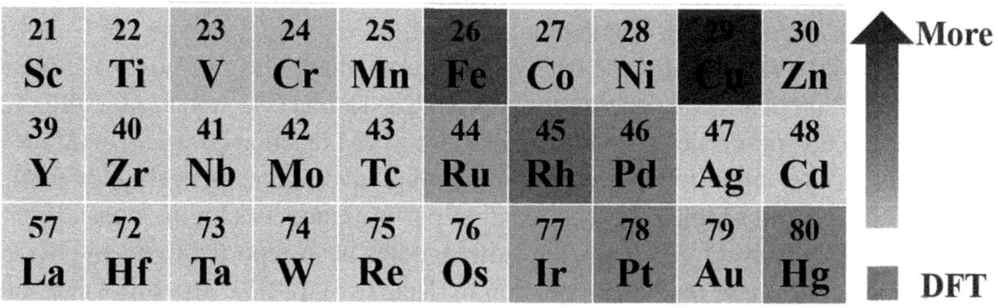

FIGURE 8.4 The metals reported to be active for the direct MTM reaction.

8.2.1.1   Zeolite-Supported Base Metal SACs

As microporous aluminosilicate materials, zeolites have a three-dimensional structure comprised of the units that mainly include $[SiO_4]^{4-}$ and $[AlO_4]^{5-}$ tetrahedra (T). Different arrangements of T form 20 secondary units and these subunits can be further combined to form a variety of zeolite frameworks which are classified small, medium, and large pores according to their pore sizes. In addition, the incorporation of trivalent Al^{3+} substituting for tetravalent Si^{4+} into the siliceous zeolite lead to formation of Brønsted acid sites (BAS). Therefore, the acidity of zeolite is directly related to the Si/Al ratio (the higher the Si/Al ratio, the lower the acidity) (Xing et al. 2017). For their unique three-dimensional structure and abundant BAS, zeolites are used as catalysts or supporting materials for many reactions.

Zeolite is a unique support for the catalysts applied to MTM reaction (Xing et al. 2017; Kidnay and Hiza 1963). Recently, several review papers have been published on the zeolite-supported catalysts for MTM reaction. Mahyuddin et al. (2019) reviewed the zeolite-supported base metal SACs (e.g., Fe (Hammond et al. 2012; Bols et al. 2018), Cu (Marijke et al. 2005; Sheppard et al. 2014; Narsimhan et al. 2015, 2017; Alayon et al. 2012; Grundner et al. 2015; Pappas et al. 2018, 2017; Beznis et al. 2010; Wulfers et al. 2015; Park et al. 2017) Ni (Shan et al. 2014), etc.) for MTM reaction based on the catalytic mechanism and active sites investigated by DFT calculations. They proposed the MTM occurs on the fixed bed in two typical processes: continuous reaction with the $O_2/N_2O/H_2O$ as the oxidants (Figure 8.5a) and chemical looping process (Figure 8.5b). The former mainly happens in liquid–solid phase reaction in batch reaction. The roles of the oxidant and solvent are very important. The later chemical looping process happened by a two-step reaction in gas–solid fixed-bed reaction, which is a stoichiometric reaction to some extent. Therefore, the catalytic performances of the catalysts cannot be compared if they tested under different reaction conditions. Generally, the reaction conditions were also listed when the reaction rate and/or selectivity to methanol were compared, as listed in Table 8.1.

Among the base metal-exchanged zeolite catalysts, the Cu was the most efficient one in fixed-bed gas-phase reaction (chemical looping process), which was also reviewed (Tomkins et al. 2017; Ravi et al. 2019; Park et al. 2019), respectively. Now, the zeolite-supported Cu SACs were among the most efficient systems for the MTM reaction, as shown in Table 8.1. Even using oxygen as the oxidant, it can show relatively higher catalytic performance. As to the active sites of the zeolite-supported Cu SACs, Sushkevich et al. (2017) reported that the yield of the Cu/Mordenite (MOR) of gas-phase methane oxidation with oxygen as the oxidant could be 0.204 mole of CH_3OH per mole of copper. The redox nature of the Cu (e.g., formation of CuII/CuI couples) played a key role for the catalysis process, which was proved by the in situ X-ray absorption near edge spectroscopy (XANES) results (Newton et al. 2018). The Cu without redox properties proved to be inert for MTM reaction (Brezicki et al. 2019).

For the zeolite-supported Fe catalysts, most were investigated in the liquid by using H_2O_2 as the oxidant (Table 8.1). The reaction performance was highly dependent on the nature of the solvent. Recently, the sulfolane has been reported to be an efficient solvent for liquid-phase methane oxidation by Xiao et al. (2019). The sulfolane/water mixture with an appropriate proportion could greatly enhance the methanol production and selectivity over the Fe–MFI zeolite catalyst with H_2O_2 as oxidant.

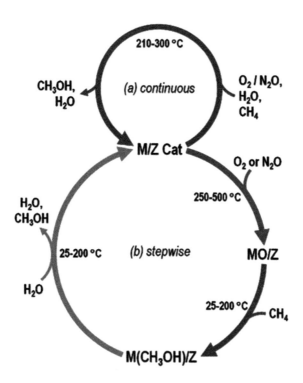

FIGURE 8.5 Schematic representation of (a) continuous and (b) stepwise processes of methane hydroxylation by metal-exchanged zeolite catalysts (M/Z Cat). (Reproduced with permission from Mahyuddin et al. 2019. Copyright 2019 Royal Society of Chemistry.)

Both continuous and stepwise processes were applied to the zeolite-supported Cu SACs, they showed higher performance in the chemical looping process than other base metal SACs. However, for the zeolite-supported Fe SACs in the chemical looping process, the reports were relatively fewer. In the liquid-phase reaction using H_2O_2 as the oxidant, the activity of the Fe-based SACs is higher than Cu. As reported by Hammond et al. (2012) mainly due to the production of the Fenton's reagent (Fe^{4+}= O) by the Fe^{2+} (aq) reacting H_2O_2. Although the conversion of methane over the Fe/ZSM-5 was high but the selectivity to methanol was low. The addition of Cu could greatly promote the selectivity of methanol, as shown in Table 8.1.

Currently, most research on zeolite-supported base metals focused on the Cu and Fe. It might be because the catalytic performances for the MTM over the zeolite-supported other d-block metals were not as good as those of the Fe and Cu based catalysts. But this does not mean that other metals have no chance to attain good catalytic performance. Other base metals in the d-block (e.g., Ni, Co, Zn, etc.) were also valuable for further fundamental research (Raynes et al. 2019). The types of the zeolites applied as the support of the SACs for the direct MTM reaction including ZSM-5, SSZ-13, MFI, CHA, and so on. It is certain that the acidity of the zeolites can influence the catalytic activity of the zeolite-supported SACs for the direct MTM reaction. By tuning the ratio of the Si/Al, the acidity of the zeolite support

TABLE 8.1
Selected Experimental Results of the MTM

Entry	Catalyst	Reaction temp.°C	Phase	Reaction time/min	Oxidant	Solvent	Oxy. umol/ g_{cat}	CH$_3$OH umol/ g_{cat}	Oxy. sel. [%]	CH$_3$OH sel. [%]	Ref.
1	Cu-ZSM-5	175	gas	--	O$_2$	--	--	8.2	--	--	Bols et al. 2018
2	Cu-ZSM-5	150	gas	60	N$_2$O	--	--	0.629	--	--	Marijke et al. 2005
3	Cu-ZSM-5	200	gas	30	O$_2$	--	<50	13	96	8	Sheppard et al. 2014
4	Cu-MOR	200	gas	20	O$_2$	--	--	13	--	--	Narsimhan et al. 2015
5	Cu-MOR	200	gas	240	O$_2$	--	--	160	--	80	Alayon et al. 2012
6	Cu-MOR	200	gas	360	O$_2$	--	--	170	--	90	Grundner et al. 2015
7	Cu-MOR	210	gas		O$_2$	--	--	82	--	--	Pappas et al. 2018
8	Cu-MOR	200	gas	30	H$_2$O	--	--	0.204/TON	--	97	Narsimhan et al. 2017
9	Cu-MOR	200	gas	360	H$_2$O	--	--	118.5	--	95	Grundner et al. 2015
10	Cu-SSZ-13	200	gas	--	O$_2$	--	--	31	--	--	Beznis et al. 2010
11	Cu-SSZ-16	200	gas	--	O$_2$	--	--	39	--	--	Beznis et al. 2010
12	Cu-SSZ-19	210	gas	--	O$_2$	--	--	36	--	--	Beznis et al. 2010
13	Cu-SSZ-13	200	gas	30	O$_2$	--	--	30	--	--	Wulfers et al. 2015
14	Cu-SSZ-13	200	gas	60	O$_2$	--	--	125	--	--	Park et al. 2017
15	Fe-SSZ-13	RT	gas	30	N$_2$O	--	--	26.8	--	--	Mahyuddin et al. 2019
16	Cu-NU-1000	200	gas	180	O$_2$	--	--	17	--	90	Zheng et al. 2019
17	Cu-NU-1000	200	gas	180	O$_2$	--	--	17.7	--	--	Ikuno et al. 2017
18	Fe-silicalite	70	liquid	30	H$_2$O$_2$	H$_2$O	6741	599	96	8	Hammond et al. 2012
19	Fe-ZSM-5	50	liquid	30	H$_2$O$_2$	H$_2$O	6967	826	83	12	Hammond et al. 2012

(Continued)

TABLE 8.1 (CONTINUED)
Selected Experimental Results of the MTM

Entry	Catalyst	Reaction temp.°C	Phase	Reaction time/min	Oxidant	Solvent	Oxy. umol/ g_{cat}	CH_3OH umol/ g_{cat}	Oxy. sel. [%]	CH_3OH sel. [%]	Ref.
20	Cu-ZSM-5	50	liquid	30	H_2O_2	H_2O	2648	2419	88	83	Hammond et al. 2012
21	Cu/ Fe-ZSM-5	50	liquid	30	H_2O_2	H_2O	7011	6993	85	85	Hammond et al. 2012
22	Fe-ZSM-5	50	liquid	120	H_2O_2	sulfolane		18800	--	--	Mahyuddin et al. 2019
23	Fe-MIL-53	40-60	liquid	120	H_2O_2	H_2O	~48	--	--	--	Osadchii et al. 2018
24	FeN$_4$- graphene	25	liquid	600	H_2O_2	H_2O	--	~115	--	--	Cui et al. 2018
25	Rh-ZSM-5	150	liquid	180	$CO+O_2$	H_2O	21259	--	--	--	Tang et al. 2018
26	Pd-ZSM-5	50	liquid	30	H_2O_2	H_2O	3846	264	--	--	Shan et al. 2017
27	Rh-ZrO$_2$	70	liquid	60	H_2O_2	H_2O	--	~36	--	--	Kwon et al. 2017

could be tuned. Thus, the catalytic activity could be influenced to some extent. The relationship between the structure of the zeolites and the catalytic performances could also affect the catalytic performance. It was proposed that the zeolites with smallpores exhibit higher activity than those with medium or largepores over the zeolite-supported Cu SACs (Park et al. 2019). The effect of the nature of the zeolites on different base metals still needs further investigation.

8.2.1.2 MOF-Supported Base Metal SACs

MOFs is an organic-inorganic hybrid material with intramolecular pores formed by self-assembly of organic ligands and inorganic metal ions or clusters through coordination bonds. In 1995, Yaghi et al. (1995) proposed the concept of MOF and initiated an alarming rate in the following two decades. Since then, various MOF materials were designed and synthesized, for example, the MIL-100 (Ferey et al. 2004), MIL-101 (Ferey 2005), ZIF-1 to ZIF-12 (Park et al. 2006), etc. The MOFs have the advantage of very high surface area (up to several thousand m^2/g), easy to be functionalized and well-designed porous structure; therefore, they were favored by many scientific researchers and developed in many applications.

MOF-supported SACs have recently emerged as a new material for MTM reaction. Recently, Liu et al. (2019) published a review on the implementation of MOF-supported SACs for the C–H bond activation. Generally, the MOF-supported SACs can be classified by the position of the active sites: (i) at metal nodes; (ii) in organic struts; and (iii) in the pores of MOFs. Most of them concerned the mimic of enzymes with metal ion centers. The investigation on the MOF-supported catalysts for MTM reaction developed step by step from cluster to diatoms and then single atom. In 2017, Ikuno et al. (2017) applied the NU-1000 confined Cu–oxo clusters with a few Cu atoms to the oxidation of MTM for the chemical looping process by using O_2 as the pretreatment oxidant. The selectivity for MTM was 45–60% with the byproducts of dimethyl ether and CO_2. Two years later, they further proved that the dicopper rather than the clusters of copper confined in the NU-1000 could get high selectivity from MTM (Zheng et al. 2019). The selectivity could be promoted from 70% to 90% when the methane pressure was increased from 1 to 40 bar. This is a typical case of the enzymes in the pores of MOF support. In 2018, the MOF stabilized bis (mu-oxo) dicopper species was reported to be active and highly selective for the methane oxidation to methanol reaction by using N_2O as the pretreatment oxidant (Baek et al. 2018). The structure of the MOF-supported Cu catalyst was claimed to mimic that of the particulate methane monooxygenase (pMMO) with a dicopper structure. The MOF-808 bearing imidazole units was chosen as the support to fix and stabilize the ligands. Although the dicopper sites were seem not form metallic state, the distance between the two copper atoms was around 2.50 Å, which is close to that of the Cu–Cu distance in the Cu foil (2.55 Å). Therefore, according to the above reports, the MOF stabilized SACs for MTM reaction seemed not solely ascribed to the single metal-atom center, the dimers could act as the efficient sites as that in the enzymes.

Recently, Vitillo et al. (2019) predicted the MOF (e.g., MIL-100, MIL-101, and MIL-808; MIL = materials of institute Lavoisier) supported Fe SACs was potential for the conversion of light alkanes to higher-value oxygenates at low temperatures with N_2O as the oxidant. As shown in Figure 8.6a, the activation of N_2O was calculated to be the

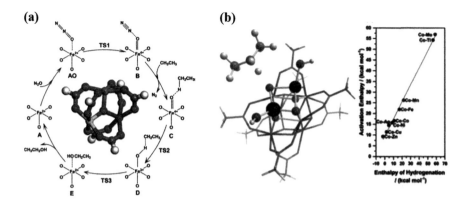

FIGURE 8.6 (a) Catalytic cycle for ethane to ethanol conversion on the divalent Fe (II) sites in MOFs. Reproduced with permission from Simons et al. (2018). Copyright 2019 American Chemical Society; (b) The cluster model containing a single node of NU-1000 and the Co–M oxide and the plot for the C–H bond activation energy on Co–M oxide clusters against enthalpy of hydrogenation of the cluster. (Reproduced with permission from Simons et al. 2018. Copyright 2018 American Chemical Society.)

rate determining step. The cycle is also surveyed to work for the MTM. Similar design was applied to the synthesis of the isolated Fe single sites for catalyzing the MTM reaction. According to the computational results, in the MIL-53 (Al) structure, both monomeric and dimeric Fe species can catalyze oxidation of methane with H_2O_2, thus, the overall activity of the catalysts might be determined by the coordination sphere of the Fe centers, including its geometry and quality of the ligands.

Simons et al. (2018) summarized the dopant metal effect with Co for the MOF (Zr-based nodes of NU-1000) supported SACs by calculations. The results showed that the barrier for the activation of the C–H bond of propane varies from 238.2 to 38.9 kJ/mol with different dopants (Figure 8.6b). These results could guide the design of efficient MOF-supported SACs for MTM. Further experiments are still needed to test the predictions obtained by the above-mentioned theoretical regulations.

Until now, many SACs supported on the MOFs have been synthesized. It proved that some of them showed activity in the direct oxidation of MTM reaction. The MOFs with different structures (NU-1000 and MIL) were applied as the support of the SACs for the MTM reaction. Besides the SACs, the dicopper sites were also synthesized. It indicated that the MOFs could be an ideal support for constructing different active structures, which could provide good models for the fundamental investigation for the MTM reaction. However, the MOFs are suffering from the poor stability during the reaction or storage. In addition, the MOFs supported SACs did not show higher activity than other materials such as the zeolites. Therefore, further study is still needed to enhance the catalytic performances of MOF-supported SACs for the direct MTM reaction.

8.2.1.3 Two-Dimensional Materials Supported Base Metal SACs

The two-dimensional materials show good potential as supports for SACs, which can ensure the attachment of the reactants to the active sites and the specific structure for fundamental research. Graphene is one of the typical two-dimensional materials.

It is a type of carbon nanomaterial arranged in a hexagonal honeycomb network composed of carbon monoatomic layer with sp² hybrid orbitals. The carbon atoms have four valence electrons, of which three electrons generate sp² bonds. That is, each carbon atom contributes an unbonded electron located in the p_z orbital. The p_z orbitals of neighboring atoms being perpendicular to the plane can form a π bond (similar to a benzene ring). The newly formed π bond is half-filled and the electrons can run through the entire layer. Therefore, it has excellent electrical properties, which makes it a good support for real and model catalysts.

The graphene-supported SACs were also reported to be active for the direct oxidation of MTM. Recently, by screening the graphene-supported 3D-transition metals (Mn, Fe, Co, Ni, and Cu), Cui et al. (2018) reported the O–FeN$_4$–O sites confined in the graphene showed activity for C–H bond activation at room temperature to produce methanol with H_2O_2, which is a good breakthrough for MTM reaction under mild conditions. Using the DFT calculations, TOF–MS, and ¹³C NMR, they proposed the reaction pathways of methane conversion. First, the C–H bond of methane is activated by the unique O–FeN$_4$–O structure to produce the methyl radical. Then methyl radical is converted into CH$_3$OOH and CH$_3$OH. The CH$_3$OH can be further oxidized to HOCH$_2$OOH and HCOOH over the O–FeN$_4$–O structure (Figure 8.7).

Before it was tested, several works had predicted that the graphene-supported base metal SACs might be potential for the direction oxidation of MTM by DFT calculations. For example, in 2014, Impeng et al. (2014) found the Fe–O active site could be generated on graphene by the decomposition of N$_2$O over the Fe-embedded graphene catalyst and the oxygen-centered radicals on the catalyst can activate the strong C–H bond of methane. The formation of methanol was even much facile than that over the zeolite-supported Fe SACs. Furthermore, compared with the Fe SACs supported on the BN nanosheets, Impeng et al. (2015) found the reaction barriers for the homolytic cleavage of the methane by the oxygen-center radical were close to each other (87.5 kJ//mol for the FeO/BN and 73.3 kJ/mol for FeO/graphene), which were similar to that of the enzymatic catalysts. Thus, the graphene-supported Fe SACs were predicted to be active for the oxidation of MTM by using N$_2$O as the oxidant. Yuan et al. (2018) proposed the single-atom Co-embedded graphene catalyst could also catalyze MTM reaction with the N$_2$O as the oxidant. They predicted that it might be active for the activation of C–H

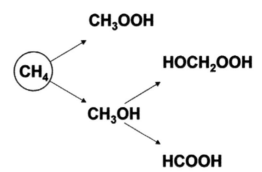

FIGURE 8.7 A possible reaction path for methane oxidation over FeN$_4$/GN catalyst. (Reproduced with permission from Cui et al. 2018. Copyright 2018 Elsevier.)

under mild conditions, which is the rate-limiting step for MTM reaction. The DFT calculations on the graphene-supported SACs for the direct MTM reaction were all applying N_2O as an oxidant, but it has not been validated by experiment yet.

Although the graphene-supported Fe SAC was proved to be active for MTM reaction by using H_2O_2 under mild conditions (Grundner et al. 2015), the reaction rate is very low, more efficient catalysts should be designed for further research. The two-dimension materials show good potential as supports for SACs, which can guarantee the attachment of the reactants to the active sites and the specific structure for fundamental research. Therefore, in addition to the graphene, other typical two-dimensional materials such as the BN nanosheets, MoS_2 nanosheets, and so on, can be the good candidates for the support of the SACs for MTM reaction.

8.2.1.4 Other Supports

In addition to the above-mentioned supports, there are other materials that can also synthesize catalysts for the activation of the C–H bond of methane. Guo et al. (2014) reported that the single Fe ions confined in the lattice of SiC were very stable against a high-temperature reaction. It can catalyze the nonoxidative conversion of methane to ethylene at 1363 K with the maximum conversion of methane at 48.1% and ethylene selectivity at 48.4%. Moreover, the catalytic performance even did not decay even after 60 h. The formation of the byproducts was prohibited over the Fe SAC. Recently, Sakbodin et al. (2016) further validated the single iron catalyst $Fe@SiO_2$ for nonoxidative conversion of methane. At 1323 K, a methane conversion of 23% is obtained, yielding C_2 hydrocarbons and aromatics without coke formation.

The catalysts based on Mo and V-based oxides were also studied in the partial oxidation of methane (Smith et al. 1993; Spencer and Pereira 1987, 1989; Spencer et al. 1990; Barbaux et al. 1988). Besides, an inversely supported $CeO_2/Cu_2O/Cu$ (111) catalyst was also reported to be able to activate methane at room temperature, over which the C, CH_x fragments, and CO_x species could be produced on the oxide surface (Zuo et al. 2016). Due to low cost of base metals and the efficient performances for the activation of methane under certain conditions, the metal oxides or the supported base metal SACs are getting increased attention.

For the base metal SACs in frameworks, the zeolite-supported Cu SACs exhibited the highest catalytic performances for the direct MTM reaction for the chemical looping process, while the zeolite-supported Fe SACs exhibited the highest activity in liquid MTM reaction by using H_2O_2 as oxidant. Among the base metal SACs, the zeolite-supported base metal SACs are the most popular. Although the MOF and graphene-supported base metal SACs also showed catalytic activity, their performances were relatively low. Further study is still needed to explore the efficient base metal SACs for the direct MTM reaction.

8.2.2 PRECIOUS METAL SACs

8.2.2.1 Rh SACs

Until now, Rh SACs have been one of the most important precious metal systems for the activation of MTM. The Rh-based organic metallic compounds in homogeneous catalysis (Hristov and Ziegler 2003; Chepaikin et al. 2002) and/or the Rh (111) in the model systems (Fratesi and de Gironcoli 2006) had been studied for the activation of methane long before the concept of SACs was proposed. Therefore, the supported Rh SACs might be one of the most potential candidates for the catalytic conversion of methane under mild conditions. Furthermore, it was predicted that the activation of methane was not easy to occur on bulk Rh surfaces by DFT calculations. For example, in 2003, Liu and Hu (2003) reported that the energy barrier for the activation of the C–H bond of CH_4 was much higher on the flat surfaces of the Rh and Pd than on the metal atoms at the steps and edges. This had been taken as a general rule for predicting the active sites on the metal surfaces. Later, in 2006, Fratesi and de Gironcoli (2006) reported that the catalytic reaction of MTM would not happen on the clean Rh (111) surfaces and the Rh adatoms on Rh (111). The adsorption of oxygen on the adatoms Rh on the different Rh surfaces (e.g., (111), (100), (110), (211), kink, etc.) was calculated to be active for the activation of methane to produce syngas. According to the electronic analysis results, the presence of oxygen atom might be originated from the strong interaction of acid-base pair sites on oxygen-metal systems, which could lead to a low dehydrogenation barrier (Baek et al. 2018). The above DFT calculation results showed that methane could not be adsorbed over the flat Rh surfaces but could be activated on the Rh atoms positioned at the kinks, edges or after being oxidized. These features happened to be the advantages of the SACs. In 2014, Duarte et al. (2014) synthesized the Rh SAC supported on the $xSm_2O_3–yCeO_2–Al_2O_3$ and applied it to the steam reforming of methane. The coking effect was greatly prohibited over the Rh SAC compared with the Rh nanoparticles. Therefore, the Rh SACs hold great promise for the direct oxidation of MTM.

The first work reported on the application of Rh SAC experimentally was published in 2017 by Kwon et al. (2017), who found the ZrO_2 supported Rh SAC was active for the direct oxidation of MTM reaction and recyclable by using H_2O_2 as the oxidant under mild conditions (70 °C) (Figure 8.8a and b). They synthesized the Rh_1/ZrO_2 catalyst by decreasing the loading of Rh from 5% to 0.3%. The formation of the Rh single atom on the support was proved by the Extended X-ray Absorption Fine Structure (EXAFS) and in situ CO-Diffuse Fourier Transform Infrared Spectrum (DRIFTS) (Figure 8.8c and d). The products of methanol over the Rh_1/ZrO_2 (the loading of Rh was 0.3%) catalyst could attain 1.25 mol/mol_{Rh} within 60 min, which is much higher than those of the other Rh catalysts with the loadings of 2% and 5% and also much higher than the Pd, Pt, and Ir with the loading of 0.3%. That is, the Rh_1/ZrO_2 is much efficient for the activation of C–H bond of methane under mild conditions than the Rh nanoparticles and the single atoms of other precious metals including Pd, Pt, and Ir. However, there were also other products in addition to methanol over the Rh_1/ZrO_2 including MeOOH and CO_2 (Figure 8.8a). In addition, only three different loadings of Rh were studied: 5%, 2% and 0.3%. The intrinsic activity on the single Rh atoms was not clearly demonstrated. That is, the plateau of the activity with the loading of Rh single atoms was not obtained. Therefore, the active sites of the Rh_1/ZrO_2 catalyst might not be unique, as reported in our previous work on the Fe–N–C single-atom catalyst (Liu et al. 2017), which might result in the relatively lower selectivity. When the O_2 was selected as the oxidant (at 260 °C), ethane and CO_2 could be obtained over the Rh SAC and Rh nanoparticles, respectively. The DFT calculation results indicated that the Rh single atoms preferred to couple the CH_3^* instead of the deep oxidation of CH_4.

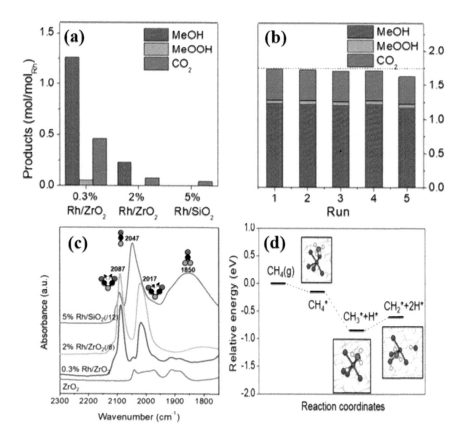

FIGURE 8.8 (a) Direct methane oxidation results using H_2O_2 as an oxidant in an aqueous solution and 0.3 wt%Rh/ZrO$_2$, 2 wt%Rh/ZrO$_2$ and 5 wt%Rh/SiO$_2$ catalysts; (b) Recyclability test results performed with the 0.3 wt% Rh/ZrO$_2$ catalyst (reaction condition: 30 bar of 95% CH$_4$/He, 70°C, 1 h, 0.5 M H$_2$O$_2$ and catalyst 30 mg); (c) DRIFT spectra of CO molecules on the ZrO$_2$, 0.3 wt%Rh/ZrO$_2$, 2 wt%Rh/ZrO$_2$ and 5 wt%Rh/SiO$_2$ catalysts; (d) Energy diagram of methane activation on the RhZr-Hyd (4.3 OH/nm^2) model with optimized geometries of the intermediates. (Reproduced with permission from Kwon et al. 2017. Copyright 2017 American Chemical Society.)

Recently, Harrath et al. (2019) reported that the formation of methanol over the Rh$_1$/ZrO$_2$ catalyst was energetically preferred than that of the MeOOH and CO$_2$, which led to the high selectivity to methanol. Furthermore, they systematically calculated the relationship between the activation of the reactants and the coordination structure of the M$_1$/ZrO$_2$ (M = Rh, Pd, Ir, Pt) catalysts and found that the activation of oxygen and methane were essentially different. The spontaneous dissociative adsorption of H$_2$O$_2$ could happen on the Rh$_1$/ZrO$_2$ catalyst, which is a key factor that initiates the radical reaction of the active sites and the hydrogenating of the support surface surrounding the Rh single atoms. Nevertheless, this process could not occur over the Pd$_1$/ZrO$_2$, Ir$_1$/ZrO$_2$ and Pt$_1$/ZrO$_2$ catalysts. By further screening the SACs of the transition metals and other precious metals on ZrO$_2$ (M$_1$/ZrO$_2$, M = Fe, Co, Ni, Cu, Ru, Ag, Os, Au) by considering the two key factors that controlled the activity and selectivity (namely, the spontaneous dissociative adsorption of H_2O_2 and the hydrogenation of the oxide support surface), they found the Ru_1/ZrO_2 and Fe_1/ZrO_2 catalysts were most potential for the direction conversion of MTM. Based on the DFT results, further experiment is still needed to synthesize the efficient catalyst for MTM reaction.

Generally, the H_2O_2 or O_2 is used as the oxidant for the direct conversion of MTM in autoclave and gas–solid chemical looping process, respectively. In 2017, Shan et al. (2017) for the first time introduced CO and O_2 for thSe catalytic oxidation and carbonylation of CH_4 to acetic acid over the zeolite-supported or the titanium-dioxide-supported Rh SAC. The conversion of CH_4 at 150°C after 6 h of reaction can attain a record of 4%, which is a great breakthrough for the activation of CH_4 under mild conditions. The reaction pathway was proposed in Figure 8.9. The CH_3OH was proved not to be the intermediate of the acetic acid. That is, the final product did not come from the carbonylation of methanol and CO, but the methane directly participated in the formation of acetic acid via CO-insertion reaction. Similar process was also reported by Tang et al. (2018) who further proposed the structure the Rh single atom was Rh_1O_5 and manifested that the introducing of CO instead of CO_2 could make the aimed product. These works not only manifested that the activation of the C–H bond of methane over the Rh SACs, but also created a new way for synthesiz-ing fine chemicals from methane, which was quite inspiring for further designing for new reactions involving the utilization of methane.

8.2.2.2 Other Precious Metal SACs

In addition, the Rh SACs, the Pd_1O_4 SAC supported on ZSM-5 was also proved to be active for the direct conversion of methane to formic acid (Guo et al. 2014). The addition of CuO could promote the selectivity to methanol by catalyzing the decom-position of the extra H_2O_2. Zhao et al. (2014) reported that the anion $PtAl_2O_4^-$ can activate methane to formaldehyde with high selectivity. It was the single Pt atom in the anion that activated methane instead of oxygen. This result indicated that the alumina supported Pt SAC might be active for the activation of C–H bond, although it has not been experimentally proved yet. The Pt SAA that obtained by diluting Pt to single atom with Cu (denoted as Pt/Cu SAA) was reported to be active for the activa-tion of C–H bond of the methane, methyl groups and butane by Marcinkowski et al. (2018). By studying the real catalysis, building model system in ultrahigh vacuum (UHV) conditions and doing simulations by DFT, they proposed that the formation of the Pt/Cu SAA can enhance the activation of C–H bond compared with monome-tallic Cu catalyst. In addition, the stability of the catalysts under realistic operating

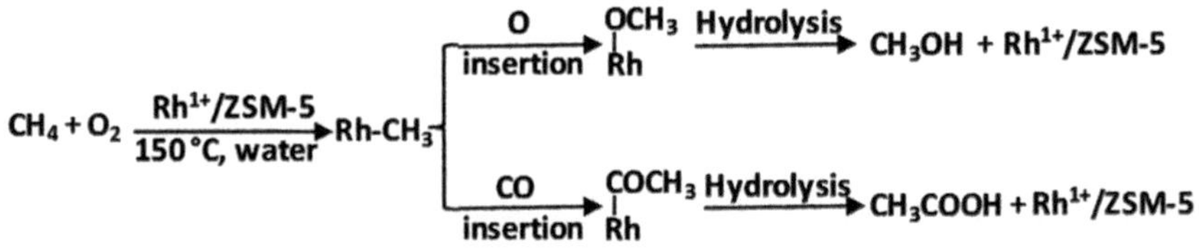

FIGURE 8.9 Possible reaction pathways of the catalytic conversion of MTM and acetic acid on Rh_1/ZSM-5. (Reproduced with permission from Shan et al. 2017. Copyright 2017 Springer Nature.)

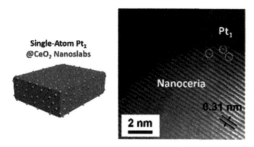
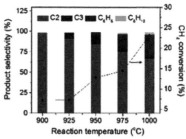

FIGURE 8.10 (a) The schematic of sintering resistant Pt_1/CeO_2 catalyst, (b) high-resolution HAADF-STEM images of the 0.5% $Pt_1@CeO_2$ catalyst and (c) Catalytic performance for the nonoxidative CH_4 conversion on the 0.5% $Pt_1@CeO_2$ catalyst as functions of the reaction temperature. (Reproduced with permission from Xie et al. 2018. Copyright 2018 American Chemical Society.)

conditions could also be greatly promoted and the coking problem could be avoided. Therefore, the design of the Pt/Cu SAA catalysts could provide a new way to synthesize a coke-resistant and cost-effective catalyst for industrial applications.

In addition to the activation of C–H bond over the Pt SACs under mild conditions, the sintering resistant Pt_1/CeO_2 was synthesized and applied to the methane conversion to C_2 hydrocarbons (ethane, ethylene, and acetylene) at 975 °C. The methane conversion could attain 14.4% with selectivity toward C_2 products at 74.6% (Figure 8.10) (Xie et al. 2018b). Unlike the direct oxidation of MTM at mild conditions, if the sintering resistant SACs could be synthesized, the activation of methane could be done at high temperatures by leading to different products instead of methanol.

Until now, the precious SACs of Rh, Pd, and Pt were reported to be active for the direct conversion of methane under mild conditions. The nature of the SACs highly depends on the support, which must be considered while designing the catalysts. Other precious metals – for example, the colloids of Au–Pd nanoparticles – could catalyze the direct conversion of methane by using O_2 as an oxidant in the presence of H_2O_2. By using the isotopically labeled oxygen, one can see that 70% of oxygen of the methanol was from gas-phase oxygen (Smith et al. 2019). Therefore, it was predicted that the reaction went through a radical process. That is, the colloid of Au–Pd nanoparticles might catalyze the formation of the methyl radicals, then it reacted with the oxygen molecule to produce methanol. Therefore, for the bimetallic catalysts, they do not have to be supported. The most efficient structures of the Au–Pd nanoparticles for the activation of methane were not clear. In addition, the role of the protecting agents was not well understood. In addition, Sirijaraensre and Limtrakul (2015) found there was synergy effect between the Au_4Pd clusters and the graphene for the oxidation of MTM according to their DFT calculation results. Without the graphene, the Au_4Pd nanoclusters were inert for MTM reaction.

Although the precious metals showed catalytic activity for the direct partial oxidation of MTM, the catalytic performance was not the most efficient. The supported base metal catalysts have been more interesting than the precious metal-based systems for MTM. It is interesting to know the catalytic performances of the precious metals from the fundamental research point of view.

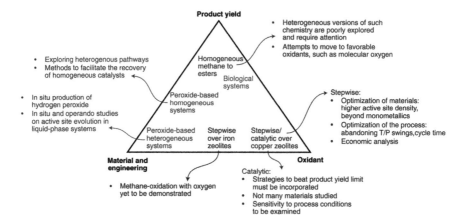

FIGURE 8.11 Evaluation of the established approaches for methane oxidation using a triangular model. The different approaches are ranked based on three criteria: product yield, choice of oxidant, material and engineering aspects, and future research directions in these topics are shown. (Reproduced with permission from Ravi et al. 2019. Copyright 2019 Springer Nature.)

According to the earlier mentioned summary of the literature, although some precious metals were reported to be active for the direct MTM reaction, it turned out that the zeolite-supported SACs exhibited the highest potential for the direct MTM reaction (as summarized in Table 8.1). However, the thermal heterogeneous catalysis still has a very long way to go. As proposed by Ravi et al. (2019) (Figure 8.11), high product yield is highly desirable. Currently, the homogenous catalysis can get to a relatively high point for the methane to esters, but applying oxygen as the oxidant for this process is still challenging. The yield of methane conversion for the heterogeneous catalysis by supported catalysts is still very low. In addition, the prevention of the over oxidation of methanol is difficult. Although some progress has been made on recognizing the real active structure, the performance is still far from application level. The SACs in the frameworks might be a possible way to realize the direct MTM reaction; yet, further investigation is needed to get new breakthrough.

8.3 METHANE TO METHANOL OVER THE SACS BY NONTHERMAL CATALYSIS

The catalytic activation of the C–H bond in methane in traditional thermal catalysis has been making progress although the activity is not industrially satisfactory yet. The widely accepted mechanism for the activation of the CH_4 is the formation of $CH_3{}^*$ radicals. Besides the traditional thermal catalysis, as mentioned in a recent review paper on the methane activation (Meng et al. 2019), other methods that could help to stimulate the molecules under mild conditions were also developed, including photocatalysis (Taylor et al. 1993), spark discharge (Kado et al. 2003), electrocatalysis (Spinner and Mustain 2013; Ma et al. 2017), and so on. One of the most popular ways is the photocatalysis, which has been found to be active for the activation of methane in the 1990s (Taylor et al 1993; Taylor 2003). Typically, the metal salt (Hu et al. 2018) semiconductors (Villa et al. 2015; Xie et al. 2018a) were applied as the catalysts for the photocatalysis process under gas-/liquid-phase reaction. The supported SACs began to be used after the concept proposed. For example, recently, Zhou et al. (2019) found that the methane could be converted to ethanol by introducing Cu species into polymeric carbon nitride (PCN) by using H_2O_2 as the oxidant under mild photocatalysis conditions. The isolated Cu sites could activate the adjacent C atom in PCN to participate in the formation of the C_2 product. Li et al. (2011) synthesized a Zn^+-modified ZSM-5 zeolite catalyst with superior photocatalytic activity for selectively activating C–H bond of an alkane molecule upon irradiation of a high-pressure mercury lamp or sunlight at room temperature. Mechanistic studies suggest there is a two-stage photoexcitation process. The first stage is electrons are transferred to the Zn^{2+} centers from the zeolite framework, which requires minimum energy corresponding to an about 390 nm-wavelength to form Zn^+ centers. The second stage is the Zn^+ centers adsorb methane driven by visible light of wavelengths shorter than 700 nm. A schematic energy diagram for the whole processes involved in the photocatalytic reaction is given in Figure 8.12.

As to the photocatalysis of the SACs, the reaction route is different from that of the thermal catalysis process, but it could provide a way to realize the methane activation under mild conditions. Although the sparkle discharge and electrocatalysis were also reported to be active for the activation of methane, the SACs have not been reported to show activity by these two methods. The SACs were thought to be valuable to be investigated in the electrocatalysis (Xie et al. 2018c), but there was report showed that the Pt SAC showed no activity for the methane oxidation reaction (Kamiya et al. 2014; Kim et al. 2018). The SACs might not be efficient in all of

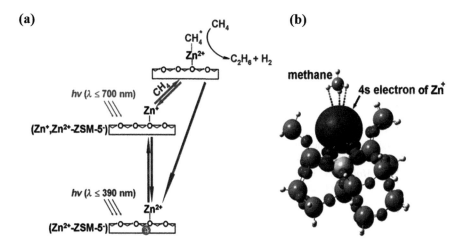

FIGURE 8.12 (a) Schematic energy diagram for the processes of the photocatalytic reaction. (b) The B3LYP hybrid exchange-correlation optimized geometry of the adsorbed methane molecule attracted by the Zn^+ active site (red: O, blue: Si, pink: Al, gray: C, white: H, and green: the 4s electron of Zn^+). (Reproduced with permission from Li et al. 2011. Copyright 2011 Wiley-VCH.)

the nonthermal catalysis, probably due to the fundamental differences among them. Further study is still needed to find the potential application of the SACs by different nonthermal methods.

8.4 CONCLUSIONS

The SACs applied to the C–H bond activation of methane has been summarized. Compared with the supported nanoparticles, the SACs have the following advantages: (i) single atom anchored in a rich chemical environment exhibiting structures similar to the enzymes can activate methane under mild conditions; (ii) the selectivity to methanol could be enhanced; (iii) the coking effect could be efficiently suppressed; (iv) the utilization of active metals could be maximized, which significantly reduce production costs, especially in precious metal catalysts; and (v) the structure-performance relationship could be built to get deep insight into the catalytic mechanism. But there is still a long way to go for the industrial application of the SACs to the direct MTM reaction because the reactivity, selectivity and stability are all needed to be enhanced. However, the existing knowledge about methane conversion and the recent promising progress in both designing innovative catalysts and reactor design have already provided significant guidance for further promoting development of direct methane conversion to value-added chemicals on the SACs.

ACKNOWLEDGEMENTS

The authors are grateful for the support from the National Key Projects for Fundamental Research and Development of China (2018YFB1501602 and 2016YFA0202801), the National Natural Science Foundation of China (21690080, 21690084, 21673228, 21776271, 21721004, and 21606227), the Strategic Priority Research Program of the Chinese Academy of Sciences (XDB17020100), and the Dalian National Laboratory for Clean Energy (DNL180303). We also thank the BL 14W beamline at the Shanghai Synchrotron Radiation Facility (SSRF). This chapter is dedicated to the 70th anniversary of Dalian Institute of Chemical Physics, CAS.

REFERENCES

Alayon, E. M., M. Nachtegaal, M. Ranocchiari, and J. A. van Bokhoven. 2012. Catalytic conversion of methane to methanol over Cu-mordenite. *Chem. Commun.* 48: 404–406.
Amos, R. D. 1979. An accurate*ab initio*study of the multipole moments and polarizabilities of methane. *Mol. Phys.* 38: 33–45.
Arvidsson, A. A., V. P. Zhdanov, P. A. Carlsson, H. Grönbeck, and A. Hellman. 2017. Metal dimer sites in ZSM-5 zeolite for methane-to-methanol conversion from first-principles kinetic modelling: is the $[Cu-O-Cu]^{2+}$ motif relevant for Ni, Co, Fe, Ag, and Au? *Catal. Sci. Technol.* 27 (7): 1470–1477.
Baek, J., B. Rungtaweevoranit, X. Pei, et al. 2018. Bioinspired metal-organic framework catalysts for selective methane oxidation to methanol. *J. Am. Chem. Soc.* 140 (51): 18208–18216.
Barbaux, Y., A. R. Elamrani, E. Payen, L. Gengembre, J. P. Bonnelle, and B. Grzybowska. 1988. Silica supported molybdena catalysts characterization and methane oxidation. *Appl. Catal.* 44: 117–132.

Beznis, N. V., B. M. Weckhuysen, and J. H. Bitter. 2010. Cu-ZSM-5 zeolites for the formation of methanol from methane and oxygen: probing the active sites and spectator species. *Catal. Lett.* 138 (1–2):14–22.

Bols, M. L., S. D. Hallaert, B. E. R. Snyder, et al. 2018. Spectroscopic identification of the α-Fe/α-O active site in Fe-CHA zeolite for the low-temperature activation of the methane C–H Bond. *J. Am. Chem. Soc.* 140: 12021–12032.

Brezicki, G., J. D. Kammert, T. B. Gunnoe, C. Paolucci, and R. J. Davis. 2019. Insights into the speciation of Cu in the Cu-H-mordenite catalyst for the oxidation of methane to methanol. *ACS Catal.* 9 (6): 5308–5319.

Chepaikin, E. G., A. P. Bezruchenko, and A. A. Leshcheva. 2002. Homogeneous rhodium-copper-halide catalytic systems for the oxidation and oxidative carbonylation of methane. *Kinet. Catal.* 43 (4): 507–513.

Cui, X., H. Li, Y. Wang, et al. 2018. Room-temperature methane conversion by graphene-confined single iron atoms. *Chemistry* 4 (8): 1902–1910.

Dalton, H., D. D. S. Smith, and S. J. Pilkington. 1990. Towards a unified mechanism of biological methane oxidation. *FEMS Microbiol. Lett.* 87 (3–4): 201–207.

Duarte, R. B., F. Krumeich, and J. A. van Bokhoven. 2014. Structure, activity, and stability of atomically dispersed Rh in methane steam reforming. *ACS Catal.* 4 (5): 1279–1286.

Eichelbaum, M., M. Havecker, C. Heine, et al. 2015. The electronic factor in alkane oxidation catalysis. *Angew. Chem. Int. Ed.* 54: 2922–2926.

Farrell, B. L., V. O. Igenegbai, and S. Linic. 2016. A viewpoint on direct methane conversion to ethane and ethylene using oxidative coupling on solid catalysts. *ACS Catal.* 6: 4340–4346.

Ferey, G. 2005. A chromium terephthalate-based solid with unusually large pore volumes and surface area. *Science* 309 (5743): 2040–2042.

Ferey, G., C. Serre, and C. M. Draznieks. 2004. A hybrid solid with giant pores prepared by a combination of targeted chemistry, simulation, and powder diffraction. *Angew. Chem. Int. Ed.* 116 (46): 6456–6461.

Fratesi, G., and S. de Gironcoli. 2006. Analysis of methane-to-methanol conversion on clean and defective Rh surfaces. *J. Chem. Phys.* 125 (4): 210–243.

Grundner, S., M. A. Markovits, G. Li, et al. 2015. Single-site trinuclear copper oxygen clusters in mordenite for selective conversion. *Nat. Commun.* 6: 7546–7554.

Gunsalus, N. J., A. Koppaka, S. H. Park, S. M. Bischof, B. G. Hashiguchi, and R. A. Periana. 2017. Homogeneous functionalization of methane. *Chem. Rev.* 117 (13): 8521–8573.

Guo, X. G., G. Z. Fang, X. Li, and J. Yang. 2014. Direct, nonoxidative conversion of methane to ethylene, aromatics, and hydrogen. *Science* 344 (6184): 616–619.

Hammond, C., M. M. Forde, M. H. Ab Rahim, et al. 2012. Direct catalytic conversion of methane to methanol in an aqueous medium by using copper-promoted Fe-ZSM-5. *Angew. Chem. Int. Ed.* 51: 5129–5133.

Harrath, K., X. Yu, H. Xiao, and J. Li. 2019. The key role of support surface hydrogenation in the CH_4 to CH_3OH selective oxidation by a ZrO_2-supported single-atom catalyst. *ACS Catal.* 9: 8903–8909.

Hristov, I. H., and T. Ziegler. 2003. Density functional theory study of the direct conversion of methane to acetic acid by $RhCl_3$. *Organometallics* 22 (17): 3513–3525.

Hu, A. H., J. J. Guo, H. Pan, and Z. Zuo. 2018. Selective functionalization of methane, ethane, and higher alkanes by cerium photocatalysis. *Science* 361 (6403): 668–672.

Hu, J. Z., J. H. Kwak, Y. Wang, et al. 2009. Studies of the active sites for methane dehydroaromatization using ultrahigh-field solid-state ^{95}mo NMR spectroscopy. *J. Phys. Chem. C* 113: 2936–2942.

Huang, W., S. Zhang, Y. Tang, et al. 2016. Low-temperature transformation of methane to methanol on Pd_1O_4 single sites anchored on the internal surface of microporous silicate. *Angew. Chem. Int. Ed.* 55 (43): 13441–13445.

Ikuno, T., J. Zheng, A. Vjunov, et al. 2017. Methane oxidation to methanol catalyzed by Cu-oxo clusters stabilized in NU-1000 metal-organic framework. *J. Am. Chem. Soc.* 139 (30): 10294–10301.

Impeng, S., P. Khongpracha, C. Warakulwit, et al. 2014. Direct oxidation of methane to methanol on Fe-O modified graphene. *RSC Adv.* 4 (24): 12572–12578.

Impeng, S., P. Khongpracha, J. Sirijaraensre, B. Jansang, M. Ehara, and J. Limtrakul. 2015. Methane activation on Fe- and FeO-embedded graphene and boron nitride sheet, role of atomic defects in catalytic activities. *RSC Adv.* 5 (119): 97918–97927.

Ito, T., J. Wang, C. H. Lin, and J. H. Lunsford. 1985. Oxidative dimerization of methane over a lithium-promoted. *J. Am. Chem. Soc.* 107: 5062–5068.

Kado, S., K. Urasaki, Y. Sekine, and K. Fujimoto. 2003. Direct conversion of methane to acetylene or syngas at room temperature using non-equilibrium pulsed discharge. *Fuel* 82 (11): 1377–1385.

Kamiya, K., R. Kamai, K. Hashimoto, et al. 2014. Platinum-modified covalent triazine frameworks hybridized with carbon nanoparticles as methanol-tolerant oxygen reduction electrocatalysts. *Nat. Commun.* 5: 5040.

Keller, G. E., and M. M. Bhasin. 1982. Synthesis of ethylene via oxidative coupling of methane. *J. Catal.* 73: 9–19.

Kidnay, A. J., and M. J. Hiza. 1963. Adsorption of methane and nitrogen on silica gel, synthetic zeolite, and charcoal. *J. Phys. Chem.* 67 (8): 1725–1727.

Kim, D. H., H. Y. Kim, J. H. Ryu, and H. M. Lee. 2009. Phase diagram of Ag-Pd bimetallic nanoclusters by molecular dynamics simulations, solid-to-liquid transition and size-dependent behavior. *Phys. Chem. Chem. Phys.* 11: 5079–5085.

Kim, H. Y., H. G. Kim, H. K. Da, and H. M. Lee. 2008. Overstabilization of the metastable structure of isolated ag-pd bimetallic clusters. *J. Phys. Chem. C* 112: 17138–17142.

Kim, J., H. E. Kim, and H. Lee. 2018. Single-atom catalysts of precious metals for electrochemical reactions. *ChemSusChem* 11 (1): 104–113.

Kondratenko, E. V., and M. Baerns. 2008. Handbook of heterogeneous catalysis. *Book* 6: 3010–3023.

Kwon, Y., T. Y. Kim, G. Kwon, J. Yi, and H. Lee. 2017. Selective activation of methane on single-atom catalyst of rhodium dispersed on zirconia for direct conversion. *J. Am. Chem. Soc.* 139 (48): 17694–17699.

Latimer, A. A., A. Kakekhani, A. R. Kulkarni, and J. K. Norskov. 2018. Direct methane to methanol, the selectivity–conversion limit and design strategies. *ACS Catal.* 8 (8): 6894–6907.

Li, L., G. D. Li, C. Yan, et al. 2011. Efficient sunlight-driven dehydrogenative coupling of methane to ethane over a Zn^+-modified zeolite. *Angew. Chem. Int. Ed.* 50 (36): 8299–8303.

Lin, M., and A. Sen. 1994. Direct catalytic conversion of methane to acetic-acid in an aqueous-medium. *Nature* 368: 613–615.

Liu, C. C., C. Y. Mou, S. S. F. Yu, and S. I. Chen. 2016. Heterogeneous formulation of the tri-copper complex for efficient catalytic conversion of methane into methanol at ambient temperature and pressure. *Energy Environ. Sci.* 9 (4): 1361–1374.

Liu, J. 2017. Catalysis by supported single metal atoms. *ACS Catal.* 7 (1): 34–59.

Liu, M. J., J. Wu, and H. W. Hou. 2019. Metal-organic framework (MOF)-based materials as heterogeneous catalysts for C-H bond activation. *Chem. Eur. J.* 25 (12): 2935–2948.

Liu, W. G., L. L. Zhang, X. Liu, et al. 2017. Discriminating catalytically active FeN_x species of atomically dispersed Fe-N-C catalyst for selective oxidation of the C-H bond. *J. Am. Chem. Soc.* 139 (31): 10790–10798.

Liu, Z. P., and P. Hu. 2003. General rules for predicting where a catalytic reaction should occur on metal surfaces: a density functional theory study of C-H and C-O bond breaking/making on flat, stepped, and kinked metal surfaces. *J. Am. Chem. Soc.* 125 (7): 1958–1967.

Lorkovic, I. M., A. Yilmaz, G. A. Yilmaz, et al. 2004. A novel integrated process for the functionalization of methane and ethane, bromine as mediator. *Catal. Today* 98 (1–2): 317–322.

Lunsford, J. H. 1995. The catalytic oxidative coupling of methane. *Angew. Chem. Int. Ed. Engl.* 34: 970–980.

Lunsford, J. H. 2000. Catalytic conversion of methane to more useful chemicals and fuels a challenge for the 21st century. *Catal. Today* 63: 165–174.

Ma, M., B. J. Jin, P. Li, et al. 2017. Ultrahigh electrocatalytic conversion of methane at room temperature. *Adv. Sci.* 4 (12): 379–387.

Mahyuddin, M. H., Y. Shiota, and K. Yoshizawa. 2019. Methane selective oxidation to methanol by metal-exchanged zeolites: a review of active sites and their reactivity. *Catal. Sci. Technol.* 9 (8): 1744–1768.

Marcinkowski, M. D., M. T. Darby, J. Liu, et al. 2018. Pt/Cu single-atom alloys as coke-resistant catalysts for efficient C-H activation. *Nat. Chem.* 10 (3): 325–332.

Marijke, H. G., P. J. Smeets, B. F. Sels, P. A. Jacobs, and R. A. Schoonheydt. 2005. Selective oxidation of methane by the bis (μ-oxo)dicopper core stabilized on ZSM-5 and mordenite zeolites. *J. Am. Chem. Soc.* 127: 1394–1395.

Meng, X. G., X. J. Cui, N. P. Rajan, L. Yu, D. H. Deng, and X. H. Bao. 2019. Direct methane conversion under mild condition by thermo-, electro-, or photocatalysis. *Chemistry* 5 (9): 2296–2325.

Narsimhan, K., K. Iyoki, K. Dinh, and Y. Román-Leshkov. 2017. Catalytic oxidation of methane into methanol over copper-exchanged zeolites with oxygen at low temperature. *ACS Centr. Sci.* 2 (6): 424–429.

Narsimhan, K., V. K. Michaelis, G. Mathies, W. R. Gunther, R. G. Griffin, and Y. Román-Leshkov. 2015. Methane to acetic acid over Cu-exchanged zeolites: mechanistic insights from a site-specific carbonylation reaction. *J. Am. Chem. Soc.* 137: 1825–1832.

Newton, M. A., A. J. Knorpp, A. B. Pinar, V. L. Sushkevich, D. Palagin, and J. A. van Bokhoven. 2018. On the mechanism underlying the direct conversion of methane to methanol by copper hosted in zeolites, braiding Cu K-Edge XANES and reactivity studies. *J. Am. Chem. Soc.* 140 (32): 10090–10093.

Ogura, K., and K. Takamagari.1986. Direct conversion of methane to methanol, chloromethane and dichloromethane at room-temperature. *Nature* 319 (6051): 308–308.

Olah, G. A. 1987. Electrophilic methane conversion. *Acc. Chem. Res.* 20 (11): 422–428.

Olah, G. A., B. Gupta, M. Farina, et al. 1985. Selective monohalogenation of methane over supported acidic or platinum metal catalysts and hydrolysis of methyl halides over gamma-alumina-supported metal oxide/hydroxide catalysts: a feasible path for the oxidative conversion of methane into methyl alcohol/dimethyl ether. *J. Am. Chem. Soc.* 107 (24): 7097–7105.

Osadchii, D. Y., A. I. Olivos-Suarez, A. Szécsényi, et al. 2018. Isolated Fe sites in metal organic frameworks catalyze the direct conversion of methane to methanol. *ACS Catal.* 8 (6): 5542–5548.

Pakhare, D., C. Shaw, D. Haynes, D. Shekhawat, and J. Spivey. 2013. Effect of reaction temperature on activity of Pt- and Ru-substituted lanthanum zirconate pyrochlores ($La_2Zr_2O_7$) for dry (CO_2) reforming of methane (DRM). *J. CO_2 Util.* 1: 37–42.

Pappas, D. K., A. Martini, M. Dyballa, et al. 2018. The nuclearity of the active site for methane to methanol conversion in Cu-mordenite: a quantitative assessment. *J. Am. Chem. Soc.* 140: 15270–15278.

Pappas, D. K., E. Borfecchia, and M. Dyballa. 2017. Methane to methanol: structure-activity relationships for Cu-CHA. *J. Am. Chem. Soc.* 139: 14961–14975.

Park, K. S., Z. Ni, A. P. Cote, et al. 2006. Exceptional chemical and thermal stability of zeolitic imidazolate frameworks. *Proc. Natl. Acad. Sci.* 103 (27): 10186–10191.

Park, M. B., E. D. Park, and W. S. Ahn. 2019. Recent progress in direct conversion of methane to methanol over copper-exchanged zeolites. *Front. Chem.* 7: 514–520.

Park, M. B., S. H. Ahn, A. Mansouri, M. Ranocchiari, and J. A. van-Bokhoven. 2017. Comparative study of diverse copper zeolites for the conversion of methane into methanol. *ChemCatChem* 9: 3705–3713.

Periana, R. A., D. J. Taube, and E. R. Evitt. 1993. A mercury-catalyzed, high-yield system for the oxidation of methane to methanol. *Science* 259: 340–343.

Periana, R. A., D. J. Taube, S. Gamble, H. Taube, T. Satoh, and H. Fujii. 1998. Platinum catalysts for the high-yield oxidation of methane to a methanol derivative. *Science* 280 (5363): 560–564.

Periana, R. A., O. Mironov, D. Taube, G. Bhalla, and C. J. Jones. 2003. Catalytic, oxidative condensation of CH_4 to CH_3COOH in one step via C-H activation. *Science* 301 (5634): 814–818.

Qiao, B., A. Wang, X. Yang, et al. 2011. Single-atom catalysis of CO oxidation using Pt_1/FeO_x. *Nat. Chem.* 3 (8): 634–641.

Ravi, M., M. Ranocchiari, and J. A. van Bokhoven. 2017. The direct catalytic oxidation of methane to methanol-a critical assessment. *Angew. Chem. Int. Ed.* 56 (52): 16464–16483.

Ravi, M., V. L. Sushkevich, A. J. Knorpp, et al. 2019. Misconceptions and challenges in methane-to-methanol over transition-metal-exchanged zeolites. *Nat. Catal.* 2 (6): 485–494.

Raynes, S., M. A. Shah, and R. A. Taylor. 2019. Direct conversion of methane to methanol with zeolites: towards understanding the role of extra-framework d-block metal and zeolite framework type. *Dalton Trans.* 48 (28): 10364–10384.

Rovik, A. K., S. Klitgaard, S. K. Dahl, C. H. Christensen, and I. Chorkendorff. 2009. Effect of alloying on carbon formation during ethane dehydrogenation. *Appl. Catal. A Gen.* 358: 269–278.

Sakbodin, M., Y. Wu, S. C. Oh, E. D. Wachsman, and D. Liu. 2016. Hydrogen-permeable tubular membrane reactor, promoting conversion and product selectivity for non-oxidative activation of methane over an Fe (c)SiO_2 catalyst. *Angew. Chem. Int. Ed. Engl.* 55: 16149–16152.

Shan, J., W. Huang, L. Nguyen, et al. 2014. Conversion of methane to methanol with a bent mono (μ-oxo)dinickel anchored on the internal surfaces of micropores. *Langmuir* 30 (28): 8558–8569.

Shan, J. J., M. W. Li, L. Allard, et al. 2017. Mild oxidation of methane to methanol or acetic acid on supported isolated rhodium catalysts. *Nature* 551 (7682): 605–608.

Sheppard, T., C. D. Hamill, A. Goguet, D. W. Rooney, and J. M. Thompson. 2014. A low temperature, isothermal gas-phase system for conversion of methane to methanol over Cu–ZSM-5. *Chem. Commun.* 50: 11053–11055.

Shilov, A. E., and D. Reidel. 1984. Activation of saturated hydrocarbons by transition metal complexes. *ChemInform* 284 (2): C40–C40.

Simons, M. C., M. A. Ortuño, V. Bernales, et al. 2018. C-H bond activation on bimetallic two-atom Co-M oxide clusters deposited on Zr-based MOF nodes, effects of doping at the molecular level. *ACS Catal.* 8: 2864–2869.

Sirijaraensre, J., and J. Limtrakul. 2015. Modification of the catalytic properties of the Au_4 nanocluster for the conversion of methane-to-methanol: synergistic effects of metallic adatoms and a defective graphene support. *Phys. Chem. Chem. Phys.* 17 (15): 9706–9715.

Smith, M. R., L. Zhang, S. A. Driscoll, and U. S. Ozkan. 1993. Effect of surface species on activity and selectivity of MoO_3/SiO_2 catalysts in partial oxidation of methane to formaldehyde. *Catal. Lett.* 19 (1): 1–15.

Smith, P. J., L. Smith, N. F. Dummer, et al. 2019. Investigating the Influence of reaction conditions and the properties of ceria for the valorisation of glycerol. *Energies* 12 (7):1359.

Spencer, N. D., and A. J. Pereira. 1989. V_2O_5-SiO_2-catalyzed methane partial oxidation with molecular oxygen. *J. Catal.* 116: 399–406.

Spencer, N. D., and C. J. Pereira. 1987. Partial oxidation of CH_4 to HCHO over a MoO_3-SiO_2 catalyst A kinetic study. *AIChE J.* 33 (11): 1808–1812.

Spencer, N. D., C. J. Pereira, and R. K. Grasselli. 1990. The effect of sodium on the MoO_3/SiO_2-catalyzed partial oxidation of methane. *J. Catal.* 126: 546–554.

Spinner, N., and W. E. Mustain. 2013. Electrochemical methane activation and conversion to oxygenates at room temperature. *J. Electrochem. Soc.* 160 (11): F1275–F1281.

Sushkevich, V. L., D. Palagin, M. Ranocchiari, and J. A. van Bokhoven. 2017. Selective anaerobic oxidation of methane enables direct synthesis of methanol. *Science* 356 (6337): 523–527.

Takanabe, K., and E. Iglesia. 2009. Mechanistic aspects and reaction pathways for oxidative coupling of methane on $MnNa_2WO_4SiO_2$ catalysts. *J. Phys. Chem. C* 113: 10131–10145.

Tang, Y., Y. Li, V. Fung, et al. 2018. Single rhodium atoms anchored in micropores for efficient transformation of methane under mild conditions. *Nat. Commun.* 9 (1): 1231.

Taylor, C. E. 2003. Methane conversion via photocatalytic reactions. *Catal. Today* 84 (1–2): 9–15.

Taylor, C. E., R. P. Noceti, J. R. D'Este, and D. V. Martello. 1993. Photocatalytic production of methanol and hydrogen from methane and water. *Stud. Surf. Sci. Catal.* 101: 407–416.

Tomkins, P., A. Mansouri, V. L. Sushkevich, et al. 2019. Increasing the activity of copper exchanged mordenite in the direct isothermal conversion of methane to methanol by Pt and Pd doping. *Chem. Sci.* 10 (1): 167–171.

Tomkins, P., M. Ranocchiari, and J. A. van Bokhoven. 2017. Direct conversion of methane to methanol under mild conditions over Cu-zeolites and beyond. *Acc. Chem. Res.* 50 (2): 418–425.

Villa, K., S. Murcia-Lopez, T. Andreu, and J. R. Morante. 2015. Mesoporous WO_3 photocatalyst for the partial oxidation of methane to methanol using electron scavengers. *Appl. Catal. B Environ.* 163: 150–155.

Vitillo, J. G., A. Bhan, C. J. Cramer, C. C. Lu, and L. Gagliardi. 2019. Quantum chemical characterization of structural single Fe (II) sites in MIL-type metal-organic frameworks for the oxidation of methane to methanol and ethane to ethanol. *ACS Catal.* 9 (4): 2870–2879.

Voskresenskaya, E. N., V. G. Roguleva, and S. G. Anshits. 1995. Oxidant activation over structural defects of oxide catalysts in oxidative methane coupling. *Catal. Rev.* 37: 101–143.

Wang, A., J. Li, and T. Zhang. 2018. Heterogeneous single-atom catalysis. *Nat. Rev. Chem.* 2 (6): 65–81.

Wang, B., S. Albarracín-Suazo, Y. Pagán-Torres, and E. Nikolla. 2017. Advances in methane conversion processes. *Catal. Today* 285: 147–158.

Wang, L., M. Xie, G. Xu, J. Huang, and Y. Xu. 1993. Dehydrogenation and aromatization of methane under non-oxidizing conditions. *Catal. Lett.* 21: 35–41.

Weissman, M., and S. W. Benson. 1984. Pyrolysis of methyl chloride, a pathway in the chlorine-catalyzed polymerization of methane. *Int. J. Chem. Kinet.* 16 (4): 307–333.

Wulfers, M. J., S. Teketel, B. Ipek, and R. F. Lobo. 2015. Conversion of methane to methanol on copper-containing small-pore zeolites and zeotypes. *Chem. Commun.* 51: 4447–4450.

Xiao, P. P., Y. Wang, T. Nishitoba, J. N. Kondo, and T. Yokoi. 2019. Selective oxidation of methane to methanol with H_2O_2 over an Fe-MFI zeolite catalyst using sulfolane solvent. *Chem. Commun.* 55 (20): 2896–2899.

Xie, J. J., R. X. Jin, A. Li, et al. 2018a. Highly selective oxidation of methane to methanol at ambient conditions by titanium dioxide-supported iron species. *Nat. Catal.* 1 (11): 889–896.

Xie, P. F., T. C. Pu, A. Nie, et al. 2018b. Nanoceria-supported single-atom platinum catalysts for direct methane conversion. *ACS Catal.* 8 (5): 4044–4048.

Xie, S. J., S. Lin, Q. Zhang, Z. Tian, and Y. Wang. 2018c. Selective electrocatalytic conversion of methane to fuels and chemicals. *J. Energy Chem.* 27 (6): 1629–1636.

Xing, B., J. Ma, R. Li, and H. Jiao. 2017. Location, distribution and acidity of Al substitution in ZSM-5 with different Si/Al ratios—a periodic DFT computation. *Catal. Sci. Technol.* 7: 5694–5708.

Yaghi, O. M., G. Li, and H. Li. 1995. Selective binding and removal of guests in a microporous metal-organic framework. *Nature* 378 (6558): 703–706.

Yuan, J. Y., W. Zhang, W. H. Li, et al. 2018. A high performance catalyst for methane conversion to methanol, graphene supported single atom Co. *Chem. Commun.* 54 (18): 2284–2287.

Zhang, L. L., Y. J. Ren, W. Liu, A. Wang, and T. Zhang. 2018. Single-atom catalyst, a rising star for green synthesis of fine chemicals. *Natl. Sci. Rev.* 5 (5): 653–672.

Zhang, Z., X. E. Verykios, and M. Baerns. 2006. Effect of electronic properties of catalysts for the oxidative coupling of methane on their selectivity and activity. *Catal. Rev.* 36: 507–556.

Zhao, Y.-X., Z.-Y. Li, Z. Yuan, X.-N. Li, and S.-G. He. 2014. Thermal methane conversion to formaldehyde promoted by single platinum atoms in $PtAl_2O_4$-cluster anions. *Angew. Chem. Int. Ed.* 53 (36): 9482–9486.

Zheng, H., D. Ma, X. Bao, et al. 2008. Direct observation of the active center for methane dehydroaromatization using an ultrahigh field ^{95}Mo NMR spectroscopy. *J. Am. Chem. Soc.* 130: 3722–3723.

Zheng, J., J. Ye, M. A. Ortuno, et al. 2019. Selective methane oxidation to methanol on cu-oxo dimers stabilized by zirconia nodes of an NU-1000 metal-organic framework. *J. Am. Chem. Soc.* 141 (23): 9292–9304.

Zhou, Y., L. Zhang, and W. Wang. 2019. Direct functionalization of methane into ethanol over copper modified polymeric carbon nitride via photocatalysis. *Nat. Commun.* 10 (1): 506–513.

Zuo, Z. J., P. J. Ramírez, S. D. Senanayake, P. Liu, and J. A. Rodriguez. 2016. Low-temperature conversion of methane to methanol on CeO_x/Cu_2O catalysts: water controlled activation of the C-H bond. *J. Am. Chem. Soc.* 138 (42): 13810–13813.

9 Active Sites in Mo/HZSM-5 Catalysts for Nonoxidative Methane Dehydroaromatization

Sonit Balyan, Puneet Gupta, Tuhin S. Khan, K.K. Pant, and M. Ali Haider

CONTENTS

9.1 INTRODUCTION

Methane dehydroaromatization (MDA) in nonoxidative environments introduces a single step conversion of methane to higher hydrocarbon, yielding high selectivity toward the formation of aromatic compounds (Spivey and Hutchings 2014; Majhi et al. 2013; Ma et al. 2013). The catalyst used in the process is desirable to be bifunctional having a metal site for methane activation and Brønsted acid site for aromatization. In 1993, the first evidence for direct conversion of methane to aromatic compounds was demonstrated by Wang et al. via the MDA reaction conducted in a fixed bed reactor over the Mo/HZSM-5 catalyst at around 973 K and atmospheric pressure. While the reaction is known to yield low methane conversion (< 11.2% at 973 K), the aromatic product selectivity in this reaction is generally higher (> 80%) at 973 K (Kikuchi et al. 2006; Matus et al. 2007; Majhi and Pant 2014), which has generated considerable interest in the research community to come up with innovative ideas on reactor design. For example by introducing recycle and membrane operations to increase the overall conversion and product yield (Qiu et al. 1997; Morejudo et al. 2016; Cook et al. 2009). A detailed review (Spivey and Hutchings 2014) may be referred for discussions on reactor design and operations of MDA reaction. Another concern about this reaction is catalyst stability. Molybdenum catalyst supported on zeolite structure is prone for deactivation by carbon deposition in the form of graphite and polyaromatic hydrocarbons (Ma et al. 2002; Karakaya et al.

2016; Tessonnier et al. 2008; Li et al. 2006). For example, at 6 hr time-on-stream, conversion values are observed to reduce from 13% to 8% at 973 K (Tshabalala et al. 2015; Majhi, Dalai, and Pant 2015; Mishra et al. 2017). Attempts to regenerate the catalyst on stream have shown limited success with the loss of active sites (Portilla, Llopis, and Martínez 2015; Ohnishi 1999; Xu et al. 2011; Ma et al. 2004; Gimeno et al. 2010). The nature of the active sites itself forms an interesting discussion in research, spanning over the contributions made in research during the past 25+ years. This chapter is focused on elaborating on the discussions on active site for the MDA process, which are explored via experiments and quantum mechanical ab-initio density functional theory (DFT) calculations in numerous studies (Gao et al. 2014, 2015; Lezcano-González et al. 2016). Understandings on the nature of active sites are expected to provide a molecular level mechanistic insight of the MDA reaction with a possibility to engineer the catalyst for increased stability and higher product yields. Following discussion elaborates on the thermodynamics, reactivity, and catalyst structure leading to an in-depth analysis of the nature of the active sites.

9.2 THERMODYNAMICS AND REACTIVITY

Direct conversion of methane to aromatic compounds are favorable at high temperatures since at standard conditions the reaction is highly endothermic with positive values of ΔH (Skutil and Taniewski 2006; Moghimpour, Sohrabi, and Sahebdelfar 2012) as given in the following equation to produce benzene:

$$6CH_4 \rightarrow C_6H_6 + 9H_2 \quad \Delta H_r^\circ = 523\,\frac{kJ}{mol}, \quad \Delta G_r^\circ = 435\,\frac{kJ}{mol} \quad (9.1)$$

In addition, two other reactions compete in nonoxidative conditions to produce carbon and ethylene as follows:

$$CH_4 \rightarrow C + 2H_2 \quad \Delta H_r^\circ = 74.6\,\frac{kJ}{mol}, \quad \Delta G_r^\circ = 52.5\,\frac{kJ}{mol} \quad (9.2)$$

$$2CH_4 \rightarrow C_2H_4 + 2H_2 \quad \Delta H_r^\circ = 201.5\,\frac{kJ}{mol}, \quad \Delta G_r^\circ = 169.6\,\frac{kJ}{mol} \quad (9.3)$$

Figure 9.1 shows the equilibrium conversion profile of methane with increasing temperature to produce carbon, ethylene, and benzene. It is clearly evident that methane activation to produce carbon dominates at all temperatures and only at temperatures above 973 K, an appreciable conversion (11.6%) may be achieved toward the formations of aromatics (benzene). At this temperature, equilibrium conversion toward carbon formation is significantly high [~90%] (Saraswat and Pant 2011), so the reaction is likely to be marked by high degree of catalyst deactivation. Carbon formation could become severe in cases where the reaction temperature is further increased, above 973 K (Xu, Bao, and Lin 2003). On comparing methane conversion to benzene versus ethylene (Figure 9.1), it can be asserted that aromatics formation is more favored as compared to olefins in nonoxidative environments.

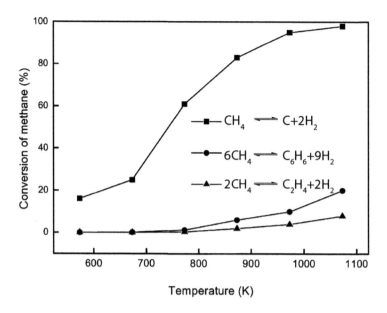

FIGURE 9.1 Thermodynamic equilibrium conversion with increasing temperature for methane conversion in a nonoxidative environment. (Reprinted from Xu, Bao, and Lin 2003. Copyright 2003 with permission from Elsevier.)

Figure 9.2 is showing time-on-stream conversion profile of methane conversion over a typical Mo_2C/HZSM-5 catalyst tested (Solymosi et al. 1997) in a nonoxidative environment. Product selectivity toward the formation of benzene, toluene, ethane, and ethylene are shown in the same plot. As observed, selectivity of benzene is increased with time initially and after 40 min of the reaction remained constant around 80%. However, methane conversion is marked with significant reduction starting from 7.5% to 2% after 2 hr of the reaction. Catalyst deactivation is caused by the deposition of polyaromatic hydrocarbons inside the zeolite pores (Kosinov et al. 2019). As compared to benzene; ethylene, ethane, and toluene selectivity remained lower than 10% during the reaction.

Catalyst regeneration is suggested to be a direct strategy to improve upon the conversion values. Martínez and co-workers (Portilla, Llopis, and Martínez 2015) have proposed a reaction-regeneration protocol for MDA reaction over Mo/HZSM-5 catalyst. After 6 hr time-on-stream, under methane environment at 973 K, in situ regeneration using air is performed over the catalyst bed at 813 K for 6 hr. Coke removal after first cycle is confirmed by thermo gravimetric and elemental analysis. This reaction-regeneration protocol is repeated for six times as shown in Figure 9.3. As observed in Figure 9.3(a) catalyst activity is regained after the regeneration cycle and the magnitude of methane conversion is negligibly decreased in consecutive cycles. Improvement in benzene selectivity during the first 2 hr of the reaction is consistently observed in all of the cyclic operations, Figure 9.3(b). On prolong operations (> 2 hr), benzene selectivity tends to decrease. This loss in benzene selectivity is more pronounced on the higher number of cycles, Figure 9.3 (b).

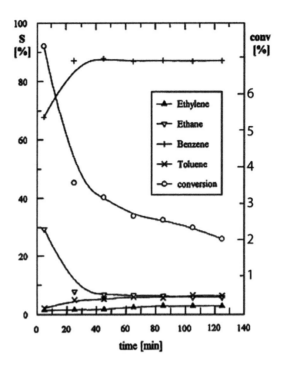

FIGURE 9.2 Methane conversion and product selectivity trend over time on a Mo/HZSM-5 catalyst at 973K. (Reprinted from Solymosi et al. 1997. Copyright 1997 with permission from Elsevier.)

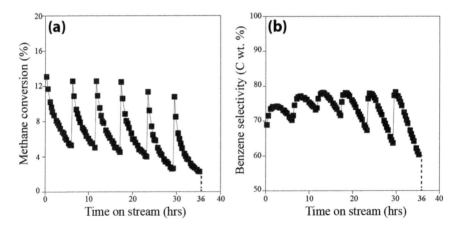

FIGURE 9.3 (a) Methane conversion and (b) benzene selectivity over time for six consecutive reaction-regeneration cycle. (Reprinted from Portilla, Llopis, and Martínez 2015. Copyright 2015 with permission from Royal Society of Chemistry.)

9.3   GENERATION OF THE ACTIVE SITES

As-synthesized Mo/HZSM-5 catalyst contains the oxide form of Mo species in +6 oxidation state (MoO_2). From operando Raman, infra-red (IR) and in situ UV-visible spectroscopy, Gao et al. (2015) have identified three different structures of molybdenum oxide species, depending upon the anchoring site of the zeolite structure. The anchoring site in turn is dependent on the Si/Al ratio of the zeolite. The three structures are listed as:

(i) $Mo(= O)_2$ anchored on double Al-atom site (Figure 9.4 (a)) inside the zeolite framework. Spectra of such species are more pronounced in zeolites of low Si/Al ratio (~15).

(ii) $Mo(= O_2)OH$ on single Al-atom sites (Figure 9.4 (b)) inside the zeolite framework. This structure is visible on increasing the Si/Al ratio to a range of 25 to 40.

(iii) $Mo(= O)_2$ anchored on double SiO_2 sites (Figure 9.4 (c)) on the outer surface of the zeolite framework. These species tend to originate from the migration of the molybdenum species to the external surface of the zeolites having significantly high (~140) Si/Al ratio, where hardly any Al-sites are available for anchoring.

From these experimental studies, the authors have further elaborated on a molecular level engineering strategy to control the distribution of Mo species anchored on the double Al-sites inside the zeolite framework, which are responsible for higher catalyst activity and aromatic selectivity. In essence, it is required to eliminate all Mo species anchored on the extra framework double SiO_2 sites (Gao et al. 2015), so as to obtain higher methane conversions and benzene yield.

In order to generate the active site for a dehydroaromatization reaction, Mo/HZSM-5 catalyst is pretreated with methane along with H_2. In this induction period, negligible product selectivity is observed, since methane is expected to react with molybdenum oxide species to form molybdenum oxycarbide and carbide species. Hydrogen

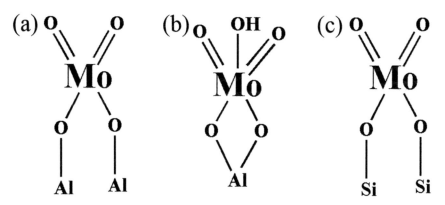

FIGURE 9.4   Anchoring sites of molybdenum species on the zeolite support via (a) double Al atom site, (b) single Al atom site, and (c) extra framework double SiO_2 site.

is required to clean the surface from graphitic carbon deposits (Wang, Lunsford, and Rosynek 1997). Generally the oxycarbide and carbide forms of molybdenum are ascribed for aromatic formation during MDA reaction (Lezcano-González et al. 2016).

Liu et al. (1999) have utilized X-ray absorption near edge structure (XANES) studies of as-synthesized and methane-treated Mo/HZSM-5 catalysts, and observed distinct changes in the Mo K-edge XAFS spectra. As shown in Figure 9.5, the Mo K-edge Fourier transform function of the as-synthesized Mo/HZSM-5 is corresponding to MoO_3 crystal. However, on treating the catalyst in methane environment for 1 hr of the induction period, a significant shift in the Mo K-edge spectra is observed, which is aligned more with the Mo_2C powder spectra, Figure 9.5(e). Thus,

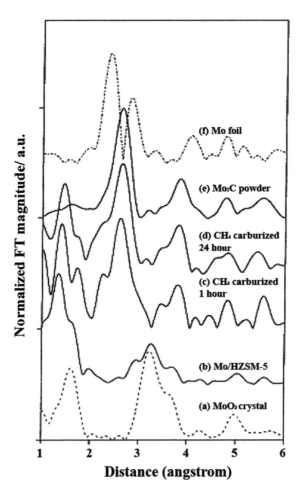

FIGURE 9.5 Mo K-edge Fourier transform spectra showing the oxidation states of Mo under different environment (a) MoO_3 crystal, (b) Mo/HZSM-5, (c) Mo/HZSM-5 carburized for 1 hr under CH_4 environment, (d) Mo/HZSM-5 carburized for 24 hr under CH_4 environment, (e) Mo_2C powder, and (f) Mo foil. (Reprinted from Liu et al. 1999. Copyright 1999 with permission form Elsevier.)

on carburization for 1 hr, in presence of methane, molybdenum oxide species are converted into molybdenum carbide species. Interestingly, in the Mo K-edge spectra of 1 hr carburized catalyst, a distinct peak at 2.23 Å is observed, which is attribute to molybdenum oxycarbide species. On prolong carburization, in presence of methane for 24 hr, the oxycarbide peak at 2.23 Å disappeared (Figure 9.5(d)) and spectra is more aligned with the Mo_2C structure.

The evolution of molybdenum carbide species in presence of methane and hydrogen-methane mixtures is elucidated from the X-ray photoelectron spectroscopy (XPS) of the as-synthesized Mo/HZSM-5 catalyst and carburized catalysts. Figure 9.6 shows representative spectra taken by Wang, Lunsford, and Rosynek (1997). In as-synthesized catalyst, XPS peak at 233.1 eV of the binding energy corresponds to the core electrons of Mo^{6+} oxidation state. On carburization in pure CH_4 for 2 hr, the overall intensity of XPS peaks is reduced, which is likely due to carbon deposition on the catalyst. More specifically, the reduced intensity peaks in the carburized catalyst are shifted from the position of as-synthesized catalyst to the positions of the XPS peaks obtained for molybdenum carbide (Figure 9.6(b)), indicating the formation of carbide species. In a separate experiment, in order to reduce the amount of carbon deposits, carburization of as-synthesized Mo/HZSM-5 is carried out under a mixture of CH_4 and H_2. The pre-treatment is carried out for 16 hr and in this case the intensity of XPS spectra is significantly higher than that of the catalyst carburized in pure methane. The binding energy of XPS peaks are corresponding to that of Mo_2C, i.e., 227.9 eV (Figure 9.6(c)), which is consistent with pure Mo_2C, where Mo is in +2 oxidation state.

9.4 NATURE OF THE ACTIVE SITES

Activity and product selectivity obtained in MDA process greatly depends upon the active molybdenum carbide species. Numerous studies have explored and elaborated on the structure of Mo carbide species obtained after the induction period, trying to link it with the reactivity of the zeolite catalyst for MDA (Lezcano-González et al. 2016). In addition, the carbide species is expected to evolve with time-on-stream, eventually leading to catalyst deactivation.

In a recent study, Lezcano-González et al. (2016) have utilized fluorescence detection with XANES along with X-ray emission and diffraction spectra to investigate in an operando manner the time-on-stream evolution of molybdenum carbide species and the structure of the carbide species responsible for aromatic formation. In confirmation to the earlier reports on the generation of the active site, molybdenum oxide species on the double Al-sites, are first converted to the oxycarbide species (MoC_xO_y) in the presence of methane. In this step, Mo^{6+} is subsequently reduced to Mo^{4+}. Methane and lower hydrocarbons (C_2H_x/C_3H_x) are formed in this step as shown in Figure 9.7. This initial transformation of the oxide species to oxycarbide species is estimated to occur in 4 min time-on-stream. On prolong operations, the oxycarbide species formed on the double Al-sites are shifted to single Al-sites, which corresponds to the reduction of Mo^{4+} to Mo^{3+} and generation of a Brønsted acid site, Figure 9.7. Slowly the oxycarbide is detached from the zeolite framework to form the carbide MoC_3 species (Figure 9.7) with subsequent reduction of Mo^{3+} to Mo^{2+}. Authors have simultaneously analyzed the products during the evolution of

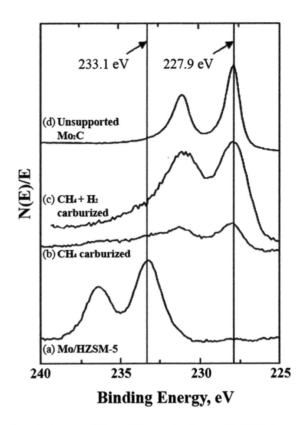

FIGURE 9.6 Chemical states of Mo carbide species observed in XPS characterization of the Mo/HZSM-5 catalyst carburized in different environment (a) Mo/HZSM-5, (b) Mo/HZSM-5 carburized under CH₄ environment, (c) Mo/HZSM-5 carburized under CH₄/H₂ environment, and (d) unsupported Mo₂C. (Reprinted from Wang, Lunsford, and Rosynek 1997. Copyright 1997 with permission from Elsevier.)

molybdenum species and have ascribed high yield of benzene formation to the generation of MoC_3 species as the active sites. On prolong operation in methane environment, the detached clusters of molybdenum carbide (MoC_3) species are observed to migrate to the outer surface of the zeolite where they tend to agglomerate and form larger size Mo_2C-like nanoparticles, which are responsible for the reduction in active surface area and carbon deposition. The key to molecular level engineering of the Mo/HZSM-5 catalyst for the MDA process lies in the stabilization of MoC_x species inside the zeolite framework to achieve high aromatic yield by simultaneously avoiding the sintering of the clusters and subsequent agglomeration to form Mo_2C like nanoparticles. This was evident in the synthesis of the zeolite support with low Silica to Alumina support ratio (SAR) by Kosinov et al. (2017) wherein at low SAR, the resultant zeolite with higher Brønsted acidity facilitated anchoring of the molybdenum oxide precursors in the pore of zeolite on double Al site.

In situ IR experiments by Gao et al. (2014) on the Mo/HZSM-5 catalyst synthesized with an optimum Si/Al ratio of 15, are further revealing upon the generation

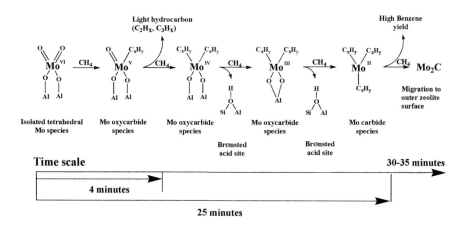

FIGURE 9.7 Evolution of molybdenum species on zeolite support during MDA reaction. (Reprinted from Lezcano-González et al. 2016. Copyright 2016 with permission from Wiley.)

and the nature of active sites. Authors have prepared catalyst samples with different Mo loadings ranging from 0 wt. % to 3.3 wt. % on the zeolite support. At lower Mo loading the framework Al site are available for anchoring of the Mo species. However, at higher Mo leading external Si sites are also available for anchoring. DFT simulations combined with quantum mechanical/molecular mechanical (QM/MM) calculations further elaborated on the structure and nature of the molybdenum carbide species formed after the induction period. From theoretical calculations, it is asserted that the carbide species are expected to anchor on the double Al-atom site in the zeolite framework (pores) only when the C/Mo ratio of the carbide cluster remains lower than 1.5. On the contrary, an increased C/Mo ratio of the carbide cluster is expected to facilitate the migration of the carbide cluster to the outer surface for anchoring on the extra framework double SiO_2 sites. Since at the outer surface the carbide species are expected to agglomerate and produce carbonaceous deposits, a molecular level engineering approach is proposed here to maintain the value of C/Mo ratio below 1.5 by using intermittent hydrogen or oxygen treatment (Gao et al. 2014). Following this argument an apt geometry of the active cluster for MDA is suggested to be Mo_2C in which C/Mo ratio is 0.5. However, this stands in contrast to the aforementioned observation by Lezcano-González et al. (2016) where the active sites for MDA is suggested to be MoC_3 species where C/Mo ratio is 3.

Methane activation and subsequent formation of ethylene on the two different types of carbide clusters (Mo_4C_2 and Mo_2C_6) proposed as the active sites is studied by Khan et al. (2018) in detail using DFT calculations. Geometry of the two carbide clusters is obtained from genetic algorithm-based optimization of various forms of Mo_xC_y cluster search combined with DFT calculations on the zeolite framework by Gao et al. (2014). The two different forms of clusters, Mo_4C_2 and Mo_2C_6 are shown in Figure 9.8.

Khan et al. (2018) have studied methane activations at different available sites on the cluster. In summary, at the Mo_4C_2 cluster methane dissociates with an activation energy of 116 kJ/mole and reaction energy of −32 kJ/mole as represented in the

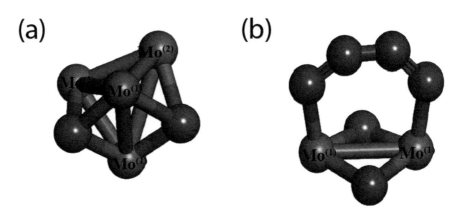

FIGURE 9.8 Structures of (a) Mo_4C_2 and (b) Mo_2C_6 clusters reported in literature. (Reprinted from Khan et al. 2018. Copyright 2018 with permission from American Chemical Society.)

reaction diagram, Figure 9.9. On the first methane activation, CH_3 group of the disso-ciated methane molecule is observed to bind on the Mo site, while the H atom binds to the neighboring C atom of the Mo_4C_2 cluster (Khan et al. 2018). Interestingly, first methane activation on Mo_2C_6 cluster proceeds in a similar fashion calculating a similar value of activation energy (119 kJ/mol) as shown in Figure 9.9. However, a significant difference in the activation energies is observed for the second methane activation on the two clusters. The calculated value of intrinsic activation barrier for the second methane activation is 182 kJ/mol on Mo_2C_6, which is significantly higher than that of Mo_4C_2 ($E_a = 117$ kJ/mol), Figure 9.9.

More interesting and significant difference in the reactivity of the two clusters is observed in C–C coupling of the adsorbed CH_3 species. Activation energies for the coupling reaction of the CH_3 intermediates on the Mo_2C_6 cluster is negligible in

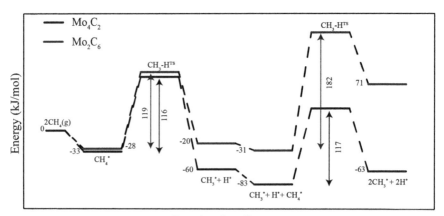

FIGURE 9.9 A comparison of reaction energy diagram for first and second methane dehy-drogenation on Mo_4C_2 and Mo_2C_6 clusters. (Reprinted from Khan et al. 2018. Copyright 2018 with permission from American Chemical Society.)

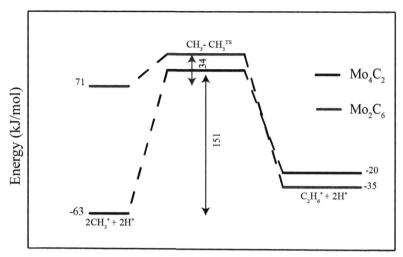

FIGURE 9.10 A comparison of reactions energetics for CH_3–CH_3 coupling on Mo_4C_2 and Mo_2C_6 clusters. (Reprinted from Khan et al. 2018. Copyright 2018 with permission from American Chemical Society.)

value ($E_a = 34$ kJ/mol) as compared to that of Mo_4C_2 cluster (151 kJ/mol) as shown in Figure 9.10. Overall, methane activation appears to be rate limiting on Mo_2C_6 clusters, while C–C coupling reaction is facilitated with significantly reduced barrier. In contrast, on the Mo_2C_4 cluster, both methane activation and C–C coupling reactions could exercise significant degree of rate control.

Adsorbed ethane molecule on respective carbide clusters undergo subsequent C–H bond activation to form ethylene. For Mo_4C_2, activation barriers for C–H activation are calculated to be 70 kJ/mole and 95 kJ/mol as shown in Figure 9.11.

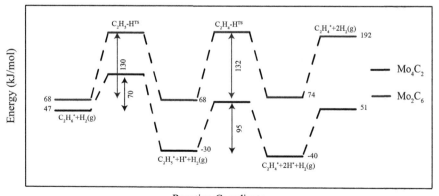

FIGURE 9.11 A comparison of reaction energy diagram for ethane dehydrogenation to form ethylene on Mo_4C_2 and Mo_2C_6 clusters. (Reprinted from Khan et al. 2018. Copyright 2018 with permission from American Chemical Society.)

As compared to Mo_4C_2, subsequent C–H bond activation to form ethylene on the Mo_2C_6 cluster is calculated to be relatively higher (E_a = 130 kJ/mol and 131 kJ/mol, Figure 9.11. Overall methane dehydrogenation on the Mo_2C_6 cluster appears to be a challenge in terms of calculated activation barriers for C–H activations. The presence of less number of Mo atoms in the Mo_2C_6 cluster makes the cluster less electron dense hence the metal-adsorbate binding are weak, which is likely to cause a higher C–H activation barrier with more endothermicity, however the coupling reaction could simultaneously become more favorable for the same cluster as observed in DFT calculations. Ethylene thus produced is expected to diffuse to the Brønsted sites of the zeolite. At the Brønsted sites, ethylene undergoes desired aromatization to form benzene and other aromatic products.

From the aforementioned mechanistic calculations on different carbide clusters of molybdenum, one thinks of a molecular strategy, in which depending on the anchoring site and dopants in the structure, the reactivity of the cluster is fine-tuned. Nature of the anchoring site of the zeolite framework is likely to alter the residual charge of the structure. A separate study is therefore undertaken by Khan et al. (2018) to study the effect of residual charge on C–H activation and C–C coupling reactions. The authors calculated a significant increase in the activation energy of first methane dehydrogenation step on increasing the cluster charge from neutral to +2, on both the clusters as mentioned in Table 9.1. This implies that reduced carbide clusters are facilitating methane activation. A comparative analysis of C–H bond activation in different steps of the reaction is presented at neutral and +1 residual charge of the two clusters, Table 9.2. In general, the trend follows that more reduced clusters are helpful in methane activation. The desired reduced state of the carbide cluster may be obtained in reactions by a careful choice of process conditions, catalyst dopants and Si/Al ratio of the zeolite framework. An interesting trend for C–C coupling reaction is observed on increasing the residual charge of the carbide cluster from neutral to +1. In contrast to C–H activation steps, the activation barrier for C–C coupling reaction on the two carbide clusters are observed to be significantly reduced

TABLE 9.1

Effect of Cluster Residual Charge on First Methane Dehydrogenation over Mo_4C_2 and Mo_2C_6 Clusters

| | Activation energy (kJ/mol) | |
	Mo_4C_2	Mo_2C_6
Neutral	116	119
Charge +1	142	133
Charge +1.5	163	134
Charge +2	174	135

TABLE 9.2

Comparison of Activation Energy for First and Second Methane Dehydrogenation and Ethane Dehydrogenation Steps on Respective Clusters with Increasing Charge

	Activation energy (kJ/mol)	
	Neutral	Charge +1
Mo_4C_2		
First CH_4 dehydrogenation	116	142
Second CH_4 dehydrogenation	117	131
Ethane dehydrogenation	70	183
Mo_2C_6		
First CH_4 dehydrogenation	119	133
Second CH_4 dehydrogenation	182	181
Ethane dehydrogenation	130	140

Reprinted from Khan et al. (2018) Copyright (2018) with permission from American Chemical Society.

on increasing the residual charge to +1 as listed in Table 9.3. While on the Mo_4C_2 cluster, the activation barrier for the C–C coupling step is reduced from +151 kJ/mol to +63 kJ/mol on increasing the charge to +1. On the Mo_2C_6 cluster the C–C coupling reaction effectively is barrier less with the +1 residual charge, as shown in Table 9.3. Thus, on both the carbide clusters, residual charge on the two clusters play an effective control on altering the activation energies of the C–H activation and C–C coupling steps. In essence, an optimum charge on the cluster is expected to facilitate both C–H activation and C–C coupling reactions, which may be obtained by molecular level engineering of the cluster in reaction environments.

TABLE 9.3

Comparison of Activation Energy of CH_3–CH_3 Coupling Steps on Respective Clusters with Increasing Charge

	Activation energy (kJ/mol)	
	Neutral	Charge +1
Mo_4C_2		
CH_3–CH_3 coupling	151	63
Mo_2C_6		
CH_3–CH_3 coupling	34	~0

Reprinted from Khan et al. (2018) Copyright (2018) with permission from American Chemical Society.

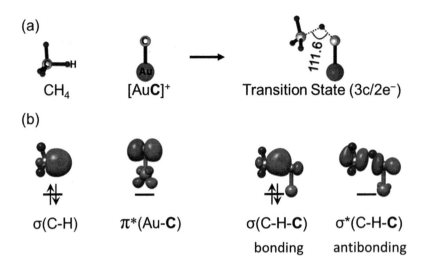

FIGURE 9.12   (a) [AuC]⁺ mediated methane C–H activation via a three-centered/two-electron transition state. (b) Quasi-restricted orbitals of the reactants and the transition state. (Reprinted from Li et al. 2016. Copyright 2016 with permission from Wiley.)

A deeper understanding on the reactivity of the Mo_2C_6 and Mo_4C_2 clusters for methane C–H activation and subsequent C–C coupling steps may be developed by acquiring the knowledge of how the molecular orbitals of a molybdenum carbide unit interacts with C–H bonds. In fact, in a recent study by Schwarz and co-workers (Li et al. 2016), quasi-restricted orbitals have been used to explore the reactivity of a diatomic monocationic gold carbide species ([AuC]⁺) with a methane C–H bond as shown in Figure 9.12. Based on their detailed analysis, the authors have pinpointed evidently that the hydrogen in the C–H bond migrates as a hydride, and the carbon of [AuC]⁺ serves as a site to accommodate the hydride. More specifically, it has been noted that the C–H activation takes place due to the interaction between a doubly-occupied σ(C–H) lobe of CH_4 and an empty π*-orbital of [AuC]⁺ as shown in Figure 9.12(b). This vacant π*-orbital localized mostly on the carbon of [AuC]⁺ acts as an electrophilic site for receiving the pair of σ(C–H) electrons from methane, thus resulting in the formation of a three-centered/two-electron (C–H–C) transition state for C–H activation. Such an overlap of the orbitals produces a doubly-filled bonding σ(C–H–C) orbital and a vacant antibonding σ*(C–H–C) orbital. Moreover, the 3c/2e⁻ transition state is nonlinear (∡ C–H–C = 112°), which increases the interaction between the carbon of CH_4 and the carbon of [AuC]⁺, leading to C–C coupling.

In another study, Li et al. (2014) have investigated methane activation by a diatomic monocationic molybdenum carbide species ([MoC]⁺). On the basis of their computational-based mechanistic study, the authors have proposed that the methane activation is initiated by the insertion of the [MoC]⁺ moiety into the C–H bond, leading to the formation of a three-centered (C–H–C) transition state as shown in Figure 9.13. However, unlike in the case of [AuC]⁺, here, the three atoms (C–H–C) have been found much like in a linear fashion, thus a hydrogen atom transfer process has been proposed. The 3c/2e⁻ transition state dissociates to yield the intermediate

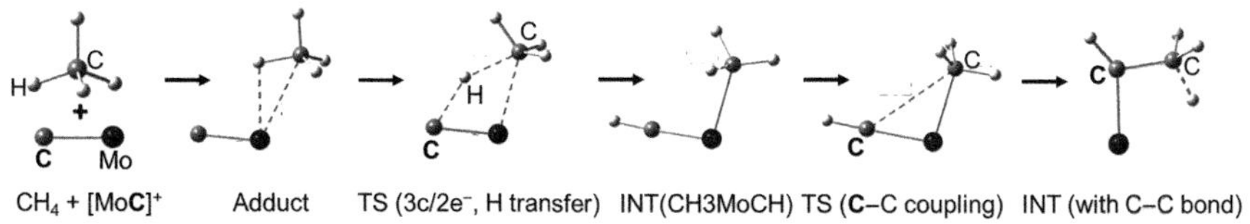

| CH₄ + [MoC]⁺ | Adduct | TS (3c/2e⁻, H transfer) | INT(CH3MoCH) | TS (C–C coupling) | INT (with C–C bond) |

FIGURE 9.13 [MoC]⁺ mediated methane C–H activation via a three-centered/two-electron transition state with a subsequent C–C coupling step. (Reprinted from Li et al. 2014. Copyright 2014 with permission from Wiley.)

CH₃MoCH, which in a subsequent step leads to a transition step featuring the C–C coupling step. Clearly, the C–H activation by [MoC]⁺ proceeds differently than the C–H activation by [AuC]⁺, We therefore speculate that the σ(C–H) orbital would interact differently with the two carbides. Thus, a quasi-restricted orbital analysis would be useful to shed light on the reactivity of [MoC]⁺ with CH₄. Such analysis would further be useful to fine tune the reactivity of their parent clusters, e.g., Mo₂C₆ and Mo₄C₂ with CH₄.

Based on aforementioned understandings on the reactivity and structure of molybdenum carbide clusters formed on the zeolite framework in the MDA process, a molecular level approach in catalyst regeneration is envisaged. Gao et al. (2015) have demonstrated a reaction-regeneration cycle for optimizing catalyst performance. Using operando Raman spectroscopy a clear evidence for the formation of molybdenum oxide species anchored on dual Al-sites is elucidated on catalyst regeneration for short duration (< 100 min) in oxygen environment at 753 K. Authors, further suggested to optimize the regeneration time, since on prolong regeneration (> 100 min), molybdenum oxide species are expected to migrate to external surface double SiO₂ site which may lead to the formation of carbon deposits during reaction. Figure 9.14 is clearly showing the effect of regeneration time on aromatic formation rate. On treating the catalyst with oxygen for a short duration (~100 min) in the regeneration cycle, more aromatics are formed in the reaction cycle, since the molybdenum carbide species formed after the induction period are directed to anchor on the framework double Al-sites. In contrast, on treating the catalyst for a relatively long duration (~120 min) in the regeneration cycle (Figure 9.14), lesser aromatics are formed in the reaction cycle, since the precursor molybdenum oxide species are expected to migrate to external double SiO₂ sites, where agglomeration of the carbide clusters formed during the induction and the reaction period is consequently leading to deactivation. The experimental observations on the reactivity of the two catalysts regenerated for different time periods are supported by DFT calculations by the authors (Gao et al. 2015). C–H bond activation is preferred over the Mo₄C₂ cluster anchored on the double Al-atom site with an activation energy of 112 kJ/mol as compared to the Mo₄C₂ cluster anchored on the external double SiO₂ sites (Eₐ = 140 kJ/mol). This is demonstrating, how reaction conditions at the macroscopic level are rationally determined from microscopic understandings developed on the structural evolution and nature of the active sites in the MDA process. In addition, the molecular level insights obtained from operando experiments and aforementioned DFT calculations, are elucidating a rational for catalyst design, wherein the formation

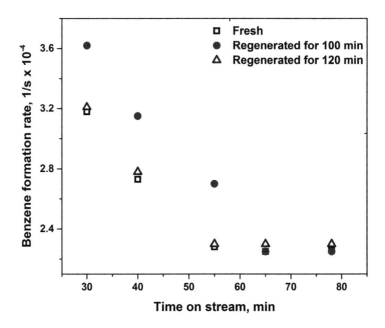

FIGURE 9.14 Benzene formation rate versus time-on-stream comparison for fresh and regenerated catalyst. (Reprinted from Gao et al. 2015. Copyright 2015 with permission from Sciencemag.)

and stability of active sites can be maintained by controlling the reaction conditions, choice of promoters and smart design of zeolite supports.

9.5 CONCLUSIONS

Activation of methane and the formation of aromatic products over a Mo/HZSM-5 catalyst in nonoxidative conditions is marred by severe catalyst deactivation due to carbon deposition and coke formation. In order to circumvent this issue, understanding the nature and time-on-stream evolution of catalytic active sites is of paramount importance. The active site in essence is bifunctional in nature where methane activation and ethylene formation occur at active molybdenum species and aromatization proceeds on the Brønsted acid sites of the framework zeolite. The active molybdenum species in the form of carbide clusters anchored on the framework zeolite are produced on catalyst pre-treatment under carburizing environment. The evolution of active carbide clusters via migration and agglomeration to the outer surface of the zeolite, forming larger size molybdenum carbide nanoparticles are suggested to cause catalyst deactivation on prolong treatment at the reaction conditions. The strategy to stabilize the catalyst, therefore essentially lies in controlling the stability of active molybdenum carbide clusters anchored on the framework zeolite by tuning the carburizing and reaction conditions and choosing an optimum Si/Al ratio in zeolite synthesizing. In addition, catalyst regeneration protocols are developed from mechanistic insights to achieve higher conversion and desired catalyst stability.

Furthermore, DFT calculations are suggesting toward optimizing the charge on the carbide cluster, which may provide energetically favored routes for C–H bond activation and C–C coupling reactions on the active carbide clusters. This may further be achieved in experiments by fine tuning the anchoring site via processing condition or using an external dopant.

REFERENCES

Cook, B., D. Mousko, W. Hoelderich, and R. Zennaro. 2009. Conversion of Methane to Aromatics over Mo$_2$C/ZSM-5 Catalyst in Different Reactor Types. *Appl. Catal. A Gen.* 365(1): 34–41.

Gao, J., Y. T. Zheng, J. M. Jehng, Y. D. Tang, I. E. Wachs, and S. G. Podkolzin. 2015. Identification of Molybdenum Oxide Nanostructures on Zeolites for Natural Gas Conversion. *Science* 348(6235): 686–690.

Gao, J., Y. Zheng, G. B. Fitzgerald, et al. 2014. Structure of Mo$_2$C$_x$ and Mo$_4$C$_x$ Molybdenum Carbide Nanoparticles and Their Anchoring Sites on ZSM-5 Zeolites. *J. Phys. Chem. C* 118(9): 4670–4679.

Gimeno, M. P., J. Soler, J. Herguido, and M. Menéndez. 2010. Counteracting Catalyst Deactivation in Methane Aromatization with a Two Zone Fluidized Bed Reactor. *Ind. Eng. Chem. Res.* 49: 996–1000.

Karakaya, C., S. H. Morejudo, H. Zhu, and R. J. Kee. 2016. Catalytic Chemistry for Methane Dehydroaromatization (MDA) on a Bifunctional Mo/HZSM-5 Catalyst in a Packed Bed. *Ind. Eng. Chem. Res.* 55(37): 9895–9906.

Khan, T. S., S. Balyan, S. Mishra, K. K. Pant, and M. A. Haider. 2018. Mechanistic Insights into the Activity of Mo-Carbide Clusters for Methane Dehydrogenation and Carbon-Carbon Coupling Reactions to Form Ethylene in Methane Dehydroaromatization. *J. Phys. Chem. C* 122(22): 11754–11764.

Kikuchi, S., R. Kojima, H. Ma, J. Bai, and M. Ichikawa. 2006. Study on Mo/HZSM-5 Catalysts Modified by Bulky Aminoalkyl-Substituted Silyl Compounds for the Selective Methane-to-Benzene (MTB) Reaction. *J. Catal.* 242(2): 349–356.

Kosinov, N., E. A. Uslamin, L. Meng, A. Parastaev, Y. Liu, and E. J. M. Hensen. 2019. Reversible Nature of Coke Formation on Mo/ZSM-5 Methane Dehydroaromatization Catalysts. *Angew. Chem. Int. Ed.* 58(21): 7068–7072.

Kosinov, N., F. J. A. G. Coumans, E. A. Uslamin, et al. 2017. Methane Dehydroaromatization by Mo/HZSM-5: Mono- or Bifunctional Catalysis? *ACS Catal.* 7(1): 520–529.

Lezcano-González, I., R. Oord, M. Rovezzi, et al. 2016. Molybdenum Speciation and Its Impact on Catalytic Activity during Methane Dehydroaromatization in Zeolite ZSM-5 as Revealed by Operando X-Ray Methods. *Angew. Chem. Int. Ed.* 55(17): 5215–5219.

Li, B., S. Li, N. Li, et al. 2006. Structure and Acidity of Mo/ZSM-5 Synthesized by Solid State Reaction for Methane Dehydrogenation and Aromatization. *Micropor. Mesopor. Mat.* 88: 244–253.

Li, J., S. Zhou, M. Schlangen, T. Weiske, and H. Schwarz. 2016. Hidden Hydride Transfer as a Decisive Mechanistic Step in the Reactions of the Unligated Gold Carbide [AuC]$^+$ with Methane under Ambient Conditions. *Angew. Chem. Int. Ed.* 55(42): 13072–13075.

Li, Z. Y., Z. Yuan, Y. X. Zhao, and S. G. He. 2014. Methane Activation by Diatomic Molybdenum Carbide Cations. *Chem. A Eur. J.* 20(14): 4163–4169.

Liu, S., L. Wang, R. Ohnishi, and M. Ichikawa. 1999. Bifunctional Catalysis of Mo/HZSM-5 in the Dehydroaromatization of Methane to Benzene and Naphthalene XAFS/TG/ DTA/ MASS/FTIR Characterization and Supporting Effects. *J. Catal.* 181(2): 175–188.

Ma, D., D. Wang, L. Su, Y. Shu, Y. Xu, and X. Bao. 2002. Carbonaceous Deposition on Mo/HMCM-22 Catalysts for Methane Aromatization: A TP Technique Investigation. *J. Catal.* 208(2): 260–269.

Ma, H., R. Kojima, R. Ohnishi, and M. Ichikawa. 2004. Efficient Regeneration of Mo/HZSM-5 Catalyst by Using Air with NO in Methane Dehydro-Aromatization Reaction. *Appl. Catal. A Gen.* 275(1–2): 183–187.

Ma, S., X. Guo, L. Zhao, and S. Scott. 2013. Recent Progress in Methane Dehydroaromatization: From Laboratory Curiosities to Promising Technology. *J. Energy Chem.* 22(1): 1–20.

Majhi, S. and K. K. Pant. 2014. Direct Conversion of Methane with Methanol toward Higher Hydrocarbon over Ga Modified Mo/H-ZSM-5 Catalyst. *J. Ind. Eng. Chem.* 20(4): 2364–2369.

Majhi, S., A. K. Dalai, and K. K. Pant. 2015. Methanol Assisted Methane Conversion for Higher Hydrocarbon over Bifunctional Zn-Modified Mo/HZSM-5 Catalyst. *J. Mol. Catal. A Chem.* 398: 368–375.

Majhi, S., P. Mohanty, H. Wang, and K. K. Pant. 2013. Direct Conversion of Natural Gas to Higher Hydrocarbons: A Review. *J. Energy Chem.* 22(4): 543–554.

Matus, E. V., I. Z. Ismagilov, O. B. Sukhova, et al. 2007. Study of Methane Dehydroaromatization on Impregnated Mo/ZSM-5 Catalysts and Characterization of Nanostructured Molybdenum Phases and Carbonaceous Deposits. *Ind. Eng. Chem. Res.* 46(12): 4063–4074.

Mishra, S., S. Balyan, K. K. Pant, and M. A. Haider. 2017. Non-Oxidative Conversion of Methane into Higher Hydrocarbons over Mo/MCM-22 Catalyst. *J. Chem. Sci.* 129(11): 1705–1711.

Moghimpour, B. P., M. Sohrabi, and S. Sahebdelfar. 2012. Thermodynamic Analysis of Nonoxidative Dehydroaromatization of Methane. *Chem. Eng. Technol.* 35(10): 1825–1832.

Morejudo, S. H., R. Zanon, S. Escolastico, et al. 2016. Direct Conversion of Methane to Aromatics in a Catalytic Co-Ionic Membrane Reactor. *Science* 353(6299): 563–566.

Ohnishi, R. 1999. Catalytic Dehydrocondensation of Methane with CO and CO_2 toward Benzene and Naphthalene on Mo/HZSM-5 and Fe/Co-Modified Mo/HZSM-5. *J. Catal.* 182(1): 92–103.

Portilla, M. T., F. J. Llopis, and C. Martínez. 2015. Non-Oxidative Dehydroaromatization of Methane: An Effective Reaction-Regeneration Cyclic Operation for Catalyst Life Extension. *Catal. Sci. Technol.* 5(7): 3806–3821.

Qiu, P., J. H. Lunsford, and M. P. Rosynek. 1997. Steady-State Conversion of Methane to Aromatics in High Yields Using an Integrated Recycle Reactor System. *Catal. Lett.* 48(1/2): 11–15.

Saraswat, S. K. and K. K. Pant. 2011. Ni-Cu-Zn/MCM-22 Catalysts for Simultaneous Production of Hydrogen and Multiwall Carbon Nanotubes via Thermo-Catalytic Decomposition of Methane. *Int. J. Hydrogen Energy* 36(21): 13352–13360.

Skutil, K. and M. Taniewski. 2006. Some Technological Aspects of Methane Aromatization (Direct and via Oxidative Coupling). *Fuel Proc. Technol.* 87: 511–521.

Solymosi, F., J. Cserényi, A. Szöke, T. Bánsági, and A. Oszkó. 1997. Aromatization of Methane over Supported and Unsupported Mo-Based Catalysts. *J. Catal.* 165: 150–161.

Spivey, J. J. and G. Hutchings. 2014. Catalytic Aromatization of Methane. *Chem. Soc. Rev.* 43(3): 792–803.

Tessonnier, J. P., B. Louis, S. Rigolet, M. J. Ledoux, and C. Pham-Huu. 2008. Methane Dehydro-Aromatization on Mo/ZSM-5: About the Hidden Role of Brønsted Acid Sites. *Appl. Catal. A Gen.* 336(1–2): 79–88.

Tshabalala, T. E., N. J. Coville, J. A. Anderson, and M. S. Scurrell. 2015. Dehydroaromatization of Methane over Sn-Pt Modified Mo/H-ZSM-5 Zeolite Catalysts: Effect of Preparation Method. *Appl. Catal. A Gen.* 503: 218–226.

Wang, D., J. H. Lunsford, and M. P. Rosynek. 1997. Characterization of a Mo/ZSM-5 Catalyst for the Conversion of Methane to Benzene. *J. Catal.* 169(1): 347–358.

Wang, L., L. Tao, M. Xie, G. Xu, J. Huang, and Y. Xu. 1993. Dehydrogenation and Aromatization of Methane under Non-Oxidizing Conditions. *Catal. Lett.* 21(1–2): 35–41.

Xu, Y., J. Lu, J. Wang, Y. Suzuki, and Z. G. Zhang. 2011. The Catalytic Stability of Mo/HZSM-5 in Methane Dehydroaromatization at Severe and Periodic CH4-H2 Switch Operating Conditions. *Chem. Eng. J.* 168(1): 390–402.

Xu, Y., X. Bao, and L. Lin. 2003. Direct Conversion of Methane under Nonoxidative Conditions. *J. Catal.* 216(1–2): 386–395.

10 Natural Gas Dehydroaromatization

Rekha Yadav and Sreedevi Upadhyayula

CONTENTS

10.1 INTRODUCTION

Conversion of natural gas (consist of 90% methane) to value-added hydrocarbon products can serve as a key to bridge the fossil fuel gap in forthcoming years (Schwach et al. 2017). With advancement in technology; i.e., the horizontal drilling and hydraulic fracturing techniques, more and more shale gas reserves are being discovered and upgraded (Schwach et al. 2017). The world natural gas reserves were reported 196.9 trillion cubic meters for the year 2018 as shown in Figure 10.1 (BP 2019). The abundance of natural gases makes it easily available and cheap feedstock. However, due to technical, regulatory, and economic constraints natural gas are flared in atmosphere and, hence, wasted (Monster et al. 2019; Fox et al. 2019). In 2018, about 3% increase in flaring of natural gas was recorded, i.e. about 145 billion cubic meters (Bamji 2019). Methane, major constituent of natural gas, is a

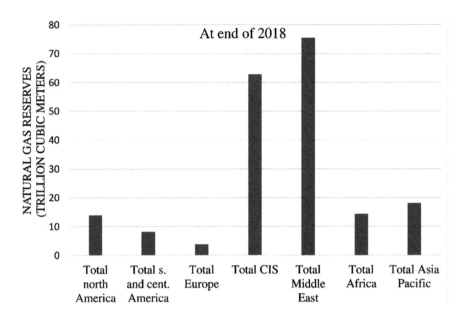

FIGURE 10.1 Distribution of total natural gas reserves of word in trillion cubic meters at end of year 2018.

greenhouse gas and possess a global warming potential 21 times that of carbon dioxide. Hence, flaring of natural gas during oil production and petroleum refining and petrochemical processes respectively should be avoided.

The catalytic conversion of methane to easily transferrable higher hydrocarbons of gasoline range at production sites is of high commercial interest (Segers 1998). In this regard, different possible catalytic routes including direct and indirect methane conversions have been explored. Among these, one of attractive route is direct conversion of methane to aromatics in the absence of oxidants. This method is known as methane dehydroaromatization (MDA) and possess higher probability for aromatic production compared to oxidative methods (Choudhary et al. 2006; Galadima and Muraza 2019; Vollmer et al. 2019a; Kanitkar and Spivey 2018). In the last decade, tremendous effort has been made to develop catalysts for one step nonoxidative conversion of natural gas. However, due to rapid deactivation and thermodynamic limitations, economically feasible catalytic systems for direct conversion of methane to hydrocarbons are not available.

Methane is a stable and symmetric organic molecule consisting of four equivalent C–H covalence bonds due to the sp³ hybridization of the central carbon atom with bond energy of 435 kJ/mol. Methane molecule lacks dipole moment and possesses very small polarizability (2.84×10^{-40} $C^2m^2J^{-1}$). Hence, economical, direct utilization of methane is of great challenge (Xu and Lin 1999).

Different approaches are made and explored for methane conversion to economically efficient chemicals and fuels (Scurrell 1987; Abdulrasheed et al. 2019). A few of the methods are Fischer–Tropsch approach on steam reforming or partial oxidized methane product (Bahri et al. 2019), oxidation of methane to methanol and formaldehyde, coupling of methane to other hydrocarbons and direct conversion of methane to aromatics in nonoxidative conditions (Ma et al. 2013; Galadima and Muraza 2019). However, most of these processes occurs at high temperatures and that leads to poor economics. Also, the catalyst shows severe deactivation at higher operation temperature and with low yield of the desired hydrocarbon products. In this chapter, we will cover the fundamental aspects of direct utilization of methane including the thermodynamics features, catalysis, as well as kinetics involved for nonoxidative coupling of methane.

10.2 THERMODYNAMICS OF METHANE ACTIVATION

Methane conversion to benzene is known as "dehydrocyclization", "aromatization", and "dehydroaromatization". The conversion of methane to benzene or other valuable products is an extremely unfavorable thermodynamic reaction. Higher positive value of $\Delta G_{reaction} = +433$ kJ mol^{-1} (Xu et al. 2003) for direct conversion of methane to benzene implies that very high temperatures are required for benzene formation. For theoretical calculation involving CH_4 as reactant and benzene and hydrogen as product (with no solid carbon, C(s)). The study showed that benzene formation at equilibrium can only be attained at very high temperatures (Spivey and Hutchings 2014).

An analogous equilibrium study with C(s) in the calculation, shows virtually no benzene formation at equilibrium. Thus, implying that the reaction goes to solid carbon [C(s)] and hydrogen, and not to benzene. The solid carbon, C(s), in the reaction corresponds to "coke" (Lunsford 2000).

The inference of following studies is, conversion of methane to benzene is thermodynamically favorable at higher temperatures and is subjected to be kinetically driven to undesired products such as carbon.

10.3 STRUCTURE AND PROPERTIES OF ZEOLITES USED FOR METHANE ACTIVATION

Zeolite and zeolite like molecular sieves have been extensively explored for methane conversion to hydrocarbons. Zeolites are highly crystalline water-containing aluminosilicates of natural or synthetic origin with highly ordered structures. It is three dimensional structures contains alternating SiO_4 and AlO_4 tetrahedrals interconnected through common oxygen atoms. The excess negative charge in zeolite resulting from aluminium content is compensated by alkali or alkaline earth metals and these metal ions are occupied in channels and voids. The general formula of zeolites is

$$M_{x/n}\left[\left(AlO_2\right)_x \left(SiO_2\right)_y\right] \cdot zH_2O$$

M = alkali and alkaline earth metals
n = valency of metal
x, y = oxides of alumina and silica variables
z = number of molecules of water of hydration

The composition is characterized by the Si/Al atomic ratio or by the molar ratio $M = SiO_2/Al_2O_3$. The Si and Al tetrahedrals combine in different shapes to form secondary building units, which further arranges to form different zeolitic structures. Zeolites are characterized by the framework structure based on distribution of different tetrahedral atoms (Si and Al), cell dimensions and symmetry. The entrances to the cavities of the zeolites are formed by 6-, 8-, 10-, and 12-ring apertures small, medium, and wide pore zeolites (Szostak 1989). Zeolites with different morphologies are distinguished by three letter code. Due to its excellent thermal and hydrothermal stability attributable to their crystalline structure, they act as potential catalysts in various petrochemical processes.

The pore size properties, acidity, and shape selectivity of zeolites and its analogous are important in refining and petrochemicals (Tanabe and Hölderich 1999; Marcilly 2001; Vermeiren and Gilson 2009; Degnan 2003; Shamzhy et al. 2019). Not only zeolites, but different solid acid catalyst plays a significant role in industrial applications. Tanabe and Hölderich (1999) reviewed the application of zeolites, oxides, complex oxides, phosphates, ion-exchange resins, clays, etc. as industrial catalysts and discussed their contribution. Impact and importance of zeolites in petrochemical industries have been reviewed by Vermeiren and Gilson (2009). The review discussed the role of zeolites as fluid catalytic cracking (FCC), as a gasoline making process; and hydro-cracking, as a middle distillate making process. Zeolites with topology MFI, FAU, and to a minor extend MOR are very versatile materials studied for the above mentioned processes (Vermeiren and Gilson 2009). Degnan (2003) specifically discussed the role of shape selectivity's of zeolites in context with petrochemical processes. Based on the reaction need the active sites of different zeolites can also be tailored (Shamzhy et al. 2019). ZSM-5 (Zeolite Socony Mobil-5) is one of the most studied zeolites for methane dehydroaromatization. ZSM-5 belongs to the pentasil family and possesses MFI topology.

10.4 DEHYDROAROMATIZATION OF METHANE

Several metal ions over different catalysts (zeolitic and nonzeolitic) have been evaluated as nonoxidative coupling catalysts for converting methane to C_2 and higher hydrocarbons (Olah et al. 1971).

The first report on methane activation appeared in the 1970s and showed that methane can carry out a homologation reaction in the presence of superacid medium (Olah and Olah 1971). This result accelerated research into direct conversion of methane using different solid-based catalysts. In recent years, the direct conversion of methane to higher hydrocarbons has been extensively studied over metal assisted heterogeneous catalysts. The different steps involved in methane gas dehydroaromatization over a catalyst system or support are shown in Figure 10.2.

10.4.1 NONOXIDATIVE CONVERSION OF METHANE OVER METAL-BASED CATALYST

The nonoxidative conversion of methane is a very important reaction as it can avoid production of carbon dioxide with high selectivity towards benzene. Since the direct conversion of methane to hydrocarbons is not thermodynamically favorable, different metal-based heterogeneous catalysts were explored for the same.

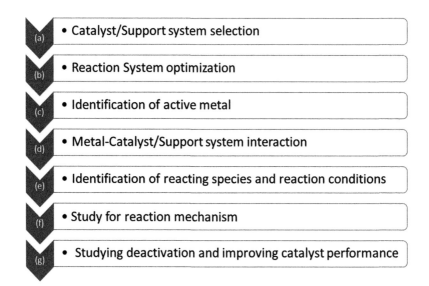

(a) • Catalyst/Support system selection

(b) • Reaction System optimization

(c) • Identification of active metal

(d) • Metal-Catalyst/Support system interaction

(e) • Identification of reacting species and reaction conditions

(f) • Study for reaction mechanism

(g) • Studying deactivation and improving catalyst performance

FIGURE 10.2   General steps involved in methane gas dehydroaromatization.

10.4.1.1 Mo-Based Catalysts

Molybdenum-based MFI zeolite catalysts are the most studied catalyst for methane dehydroaromatization. Table 10.1 summarizes the literature review on Mo/ZSM-5 catalyst. Wang et al. (1993) studied the Mo/H-ZSM-5 catalyst in a continuous flow model for methane dehydroaromatization and explored the active state of Mo, responsible for reaction. These catalysts showed good benzene selectivity of 100% with low (7.2%) conversion. The studies showed that Mo_2C is the active state for the desired reaction, which may be formed by initially exposing the catalyst to CH_4/H_2 mixtures. Kosinov et al. (2016) used spectroscopy and microscopy to study the interaction of active sites in Mo/ZSM-5 catalysts. They reported that the active centers in Mo/ZSM-5 are partially reduced single-atom Mo sites which are stabilized over zeolite framework. To find out the mechanism involved in formation of benzene, the authors used isotopic methane and applied it in a pulse reaction technique. The MDA was reported to be carried by hydrocarbon pool mechanism where initial methane activation products combine with polyaromatic carbon species to form benzene via secondary reactions. The effect of Mo loading, reaction temperature, and space velocity on the catalytic performance over Mo/H-ZSM-5 were studied by Xu et al. (1994). The higher loading of Mo species in H-ZSM-5 zeolite resulted in decrease in its crystallinity. The nature, amount and assess ability of acidic sites plays a crucial role in the catalytic performance. The 2% Mo/H-ZSM-5 calcined at 500°C showed aromatization activity with 5.6% conversion and > 90% selectivity to aromatics at 700°C. The authors concluded ethylene as initial product from experimental results based on variation of space velocity. On the basis of these results a possible mechanism for methane dehydrogenation and aromatization has been proposed by authors in which both the heterolytic splitting of methane in a solid acid environment and a molybdenum carbene-like complex as an intermediate are proposed. Han et al. (2019) studied

TABLE 10.1

Literature Review on Methane Dehydroaromatization over Mo/ZSM-5

Reference	T (°C)	GHSV (mL/ g_{cat} h)	Catalyst (Si/Al)	wt% Mo	Methane conversion (%)	Benzene selectivity (%)
(Kosinov et al. 2016)	700	1680	ZSM-5 (13)	5	–	8.8 (5 h)
(Wang et al. 1993)	700	1440	H-ZSM-5 (25)	2	7.2	100
(Wang et al. 1993)	700	1440	NaZSM-5 (25)	2	0	0
(Xu et al. 1994)	700	–	H-ZSM-5 (50)	2	5.6	90.4
(Solymosi et al. 1995)	700	–	H-ZSM-5 (55)	2	3.68[a]	–
(Wong et al. 1996a)	700	1440	H-ZSM-5 (50)	2	8.2	100[b]
(Szoke and Solymosi 1996)	700	–	H-ZSM-5 (55)	2	5.7	–
(Natesakhawat et al. 2015)	700	–	H-ZSM-5 (25)	4	7.4	78.5
(Sedel'nikova et al. 2017)	750	1000	H-ZSM-5 (40)	4	16	61.6
(Julian et al. 2020)	700	1500	H-ZSM-5 (23)	5	11.3	6.1 (5 h)
(Julian et al. 2020)	700	1500	H-ZSM-5 (23)[c]	5	11.7	8.5 (5 h)

[a] Maximum conversion, [b] All aromatic, [c] Synthesized using solvothermal method under supercritical conditions and reducing atmosphere (SC-STSE).

the effect of different metal loadings of Mo (1,3,5,7 wt%) on ZSM-5 (SiO$_2$/Al$_2$O$_3$ = 30) for MDA. At 700°C and total weight hourly space velocity (WHSV) of 1500 mL g_{cat}^{-1} h^{-1}, 5 wt% Mo/H-ZSM-5 showed better conversion then rest of catalyst which was due to high Mo dispersion leading to accessibility of Mo$_2$C and Brønsted acid sites. The selectivity toward benzene followed the order 5Mo/H-ZSM-5 > 7Mo/H-ZSM-5 > 3Mo/H-ZSM-5 > 1Mo/H-ZSM-5 after 10 h of reaction.

Vollmer et al. (2019b) reported that at higher Mo loadings, the selectivity toward coke increases. Theoretically for zeolite Si/Al = 25, 5.8 wt% of Mo is maximum amount of Mo that can be anchored on zeolite framework with Al/Mo = 1. However, at this dispersion, Mo nanoparticles are also formed which prolongs the induction period and catalyst deactivates. The authors found that there exists a synergetic effect between Mo mono- or dimeric species and zeolite acidity for aromatization. The selectivity toward aromatics i.e. benzene and naphthalene linearly increase with the amount of Mo mono- or dimeric species present. However, with higher Mo loadings, the selectivity to coke also increases due to diffusion limitation of products.

Liu and Xu (1999) studied in-situ pyridine probed FT–IR spectroscopy to determine the interaction between Mo and ZSM-5. The authors reported decrease in stretching modes at 3740 and 3618 cm^{-1} for 6 wt% Mo/ZSM-5 corresponding to silanol groups and bridged hydroxyl groups, Si(OH)Al respectively. The study showed that Mo species are preferentially located on Brønsted acidic sites after calcination at 500°C. Also, the total acidity and number of Brønsted acidic sites decrease with increase Mo loading. Since Mo/H-ZSM-5 is a bifunctional catalyst-free Brønsted acid sites on H-ZSM-5 are needed for good catalytic performance.

Solymosi et al. (1995) studied dehydrogenation of methane on MoO_3 supported on various oxides in a fixed bed, continuous flow reactor. The different oxides studied were SiO_2, Al_2O_3, TiO_2, MgO, and H-ZSM-5. The highest conversion of CH_4 (~5.4%) was reached for MoO_3/SiO_2, and the lowest (0.44%) for MoO_3/MgO. The study concluded that formation of hydrocarbons required the pre-reduction of Mo^{6+} to Mo^{4+}.

To distinguish the role of Mo and H-ZSM-5, different metal ions were explored over H-ZSM-5. Wong et al. (1996a) used Mo/H-ZSM-5, W/H-ZSM-5 and Mo-W/H-ZSM-5 for ethane and methane dehydroaromatization. W/H-ZSM-5 catalyst exhibits excellent selectivity of 80% for ethane aromatization; however, it remains inactive for methane conversion. Mo/HZM-5 showed 8.2% conversion with 100% selectivity toward aromatics. The effect of tungsten in Mo-W/H-ZSM-5 catalyst showed increase in conversion (11%) and lower coke formation, which was related to the formation of Mo-W mixed oxide. Since Mo-W mixed oxides showed lower coke formation, which is major reason for catalyst deactivation.

Szöke and Solymosi (1996) used K_2MoO_4/ZSM-5 catalyst under non-oxidizing conditions on continuous flow fixed-bed reactor. The reaction was reported at 700°C, after a partial reduction of the catalyst, products like benzene, ethylene and ethane were reported with benzene selectivity in the range of 60–70%.

Natesakhawat et al. (2015) studied the MDA over hydrogen-permselective palladium membrane reactors using 4 wt% Mo/H-ZSM-5 catalysts. Different WHSV from 750 to 9000 cm^3 g_{cat}^{-1} h^{-1} has been reported. Lower WHSVs (at and below 3000 cm^3 g_{cat}^{-1} h^{-1}) favored benzene as major product whereas at higher WHSVs, C_2H_4 dominated. Due to selective removal of H_2 from the reaction products in catalytic membrane reactors benzene yield was reported to be improved. The effect of molybdenum precursor on the physicochemical properties of the Mo/ZSM-5 catalyst was studied by Sedel'nikova et al. (2017). The activity and on-stream stability of the Mo-containing zeolite catalyst depend on the type of molybdenum compound that was used in catalyst preparation.

Julian et al. (2020) synthesised 5 wt% Mo/ZSM-5 (Si/Al = 23) by a solvothermal method under supercritical conditions and reducing atmosphere (SC-STSE). SC-STSE at 700°C, and GHSV of 1500 $mL/g_{cat}h$ showed 11.5% conversion with 8.8% selectivity toward benzene after 5 hours of reaction. The catalytic results were better compared to impregnated 5 wt% Mo/ZSM-5 catalyst, it might be due to improved metal dispersion in case of solvothermal method.

Hence, Mo/ZSM-5 is an active catalyst for methane dehydroaromatization. The active sites for the reactions are Mo_2C species formed inside the zeolite. The Mo species present on the outer surface of catalyst attracts coke formation and hence blocks the active zeolite sites.

10.4.1.2 Zn-Based Catalysts

The second most studied transition metal over ZSM-5 after Mo is Zn. Zn/ZSM-5 as catalyst for methane dehydroaromatization literature review are summarized in Table 10.2. Lai and Veser (2016) investigated the nature of Zn-containing H-ZSM-5 catalysts by preparing catalysts via three different synthetic approaches, which

TABLE 10.2

Literature Review on Methane Dehydroaromatization over Zn/ZSM-5

Reference	T (°C)	GHSV (mL/ g_{cat}h)	Catalyst (Si`/Al)	wt% Zn	Methane conversion (%)	Benzene selectivity (%)
(Lai and Veser 2016)	700	3750	ZSM-5 (31)	0.75[a]	~ 1	72
(Lai and Veser 2016)	700	3750	ZSM-5 (31)	6.4[b]	~ 1	61
(Lai and Veser 2016)	700	3750	ZSM-5 (40)	6.2[c]	~ 1	64
(Liu et al. 2011)	700	1680	ZSM-5 (25)	2	5.5	–
(He et al. 2019)[d]	400	–	ZSM-5 (23)	5	2.2	4.1
(He et al. 2019)[d]	400	–	ZSM-5 (23)[e]	5	1.7	3.2
(He et al. 2019)[d]	400	–	ZSM-5 (23)[f]	5	1.9	4.5
(He et al. 2019)[d]	400	–	ZSM-5 (80)	5	1.5	3.0
(He et al. 2019)[d]	400	–	ZSM-5 (80)[e]	5	1.4	1.9
(He et al. 2019)[d]	400	–	ZSM-5 (80)[f]	5	1.7	3.1

[a] Wet ion exchange for 24 h, [b]Wet impregnation, [c]Core shell synthesis, [d]Performed in batch reactor with 5.0 MPa with methane, [e]Treated with hexafluorosilicate, [f]Steam treated.

included wet ion exchange, wet impregnation, and core–shell synthesis. The metallic zinc (Zn^0)nanoclusters inside the zeolite micropores were found to be the most active species for the reaction; however, zinc could be vaporized at elevated temperatures The same is confirmed by loss of Zn active sites after reaction. The more stable active Zn^{2+} site (anchored at exchange site of zeolite) do not show aromatization.

Wet impregnation Zn-based/H-ZSM-5 catalysts were investigated for the MDA reaction under the conditions of atmospheric pressure and supersonic jet expansion (SJE) by Liu et al. (2011). The results revealed that under an SJE condition, the Zn/H-ZSM-5 catalyst exhibited high catalytic activity. The authors suggested heterolytic dissociative of methane over nanosized ZnO cluster present inside the H-ZSM-5 channels. The conclusion was based on the Zn LMM Auger signal detected on fresh catalyst splits after the catalyst was exposed to CH_4, implying the presence of two Zn species that are different in coordination environment. Hence the catalytic activity of 2 wt% Zn/H-ZSM-5 was related to the formation of methoxy and methyl species.

Catalytic activation of methane using n-pentane as co-reactant over Zn/H-ZSM-11 zeolite possessing MEL topology was studied (Anunziata et al. 2003). The catalyst showed an aromatics yield of 40% at 500°C and w/f = 30 g h mol^{-1} with a $C_1 = (C_1 + C_5)$ molar fraction (XC_1) = 0.30. The Zn-containing H-ZSM-11 catalyst acts as a hydride acceptor generating C_2, C_3, C_4, and their carbenium species (highly reactive intermediates) on strong Lewis sites, through an EDA adduct, which enables C_1 activation with high productivity.

He et al. (2019) investigates the impact of different Al sites on co-aromatization of methane and propane over 5 wt% Zn/H-ZSM-5 catalysts. The framework Al and extra-framework aluminium (EFAL) sites, were varied using different techniques. EFAL sites were removed from the catalyst by treating it with hexafluorosilicate and steam treatment is used to convert framework Al sites to EFAL. To find the Al effect the authors studied different parameters like average carbon number (molar fraction of product molecules with variable carbon numbers), substitution index (fraction of carbon atoms attached to substituent groups among the phenyl ring carbon atoms in the product molecules), and aromatic carbon fraction (carbon atoms in the phenyl rings/all the carbon atoms in the product matrix). The study was carried out in a batch reactor, with propane and methane at 5.0 MPa pressure. 5 wt% Zn/H-ZSM-5 (Si/Al = 23) has higher average carbon number of 10.07 compared to 5 wt% Zn/H-ZSM-5 (Si/Al = 80) and 5 wt% Zn/H-ZSM-5 (Si/Al = 280) with 9.10 and 8.98, respectively, under methane environment. The corresponding increase in aromatic carbon fraction was reported with increase in Al ratio. The results indicated that framework Al atoms are essential for methane aromatization in ZSM. The NH_3-TPD and DRIFTS studies demonstrated that framework Al and EFAL species correspond to Brønsted and Lewis acid sites of the catalyst, respectively. Therefore, high concentration of Al in catalyst increases methyl aromatization.

He et al. (2019b) studied effect of Zn distribution and the reaction pathway for the co-aromatization of methane with propane. The reaction was carried out in a batch reactor at 400°C and 5.0 MPa pressure with isotopic methane ($^{13}CH_4$). The higher concentration of Zn in the inner pores of zeolites is confirmed with Scanning Transmission X-ray microscopy images. The XANES and XPS spectra analysis of spent catalyst showed that Zn concentration increases on the external surface. This movement of Zn to the external framework increases the Brønsted acidity inside the pores disturbing the active catalytic sites. The authors proposed that during of methane activation in presence of propane, first C_4 species are formed which undergoes dimerization to form C_8 species at the catalytic sites, with formation of single-ring aromatics. The Zn species located inside the inner pores then facilitates the methane as methyl groups to phenyl rings.

Jarvis et al. (2019) studied the co-aromatization of methane and light straight run naphtha (LSR; see Figure 10.3) over 5 wt% Zn/H-ZSM-5, 1 wt% Pt/H-ZSM-5, and 5 wt% Zn-1 wt% Pt/H-ZSM-5 (Si/Al = 280). The NH_3-TPD revealed that all the catalysts have same number of total acidic sites, with 1 wt% Pt/H-ZSM-5 having no medium acidic sites and highest strong acidic sites. 1 wt% Pt-Zn/H-ZSM-5 showed 70% LSR conversion with 19% xylene selectivity with 8.3 ($\mu g/g_{cat} \cdot h$) coke yield. The 5 wt% Zn-1 wt% Pt/H-ZSM-5, on other hand, showed 62% conversion with 36% selectivity toward xylene with 1.7($\mu g/g_{cat} \cdot h$) coke yield. Hence implying that 5 wt% Zn-1 wt% Pt/H-ZSM-5 contains appropriate acidic sites for the reaction. The addition of Zn in the catalyst reduces the coke formation and hence also reduces catalyst deactivation.

Kazansky and Serykh (2004) prepared different Zn catalysts with wet impregnation, ion exchange and chemical reaction of hydrogen of zeolite with Zn vapours. Two MFI with Si/Al ratio 25 and 40 were compared with 2 wt% Zn. The DRIFT spectra analysis showed complete substitution of acidic protons by zinc ions for MFI (Si/Al = 25) when prepared using a chemical reaction. The complete substitution of protons indicates that if a Zn^{2+} cation is localized at the aluminium sites, the corresponding Al site will lack a proton and the negative charge will be partially

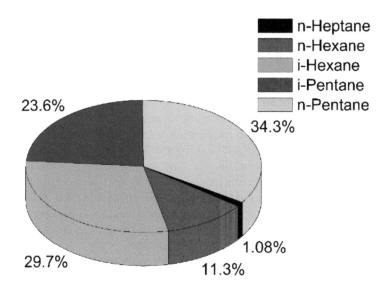

FIGURE 10.3   Percentage distribution of components in LSR.

compensated excessive positive charges on Zn^{2+}. Hence, these partially compensated positive charges sites exhibit high chemical activity.

Anunziata and Mercado (2006) explored methane aromatization with light gasoline as co-reactor over Zn/H-ZSM-11 with 0.54, 1.33, and 2.13 Zn^{2+}/cell units. Zn/H-ZSM-11 (2.13) showed the highest conversion of 23.2 mol% and 898 mol% for methane and light gasoline, respectively, at 500°C and 1 atm pressure (methane (w/f) = 15 g h mol^{-1} and light gasoline (w/f) = 98 g h mol^{-1}). The high conversion of methane was attributed to high number of weak, moderate and strong Lewis acidic sites over Zn/H-ZSM-11 (shown by pyridine FTIR studies). Also lack of Brønsted sites on catalyst results in decreased cracking reaction.

Luzgin et al. (2008) explored Zn/H-BEA zeolite (Si/Al = 18, 7.2 wt% Zn) for methane aromatization using ^{13}C methane during conversion of methane and propane. The author denied the presence of surface methyl species on un-protonated zeolites. The methoxy and zinc-methyl species were claimed to be formed on ZnO inside the zeolite channels.

The direct aromatization of methane was not observed over Zn-modified zeolite under the reaction conditions (Luzgin et al. 2009). However, alkylation of ^{13}C methane in aromatic products was observed with benzene, toluene and xylenes molecules, which are produced from propane. Similar to Mo/ZSM-5, Zn/ZSM-5 is active catalyst for methane dehydroaromatization. However, it shows quick deactivation on time on stream. Also, the methane conversion is less on Zn/ZSM-5 compared to Mo/ZSM-5.

10.4.1.3   Transitional Metal Catalysts

The effect of different transition metal ions (TMI) over H-ZSM-5 zeolites (with TMI = Mo, Fe, V, W, and Cr) on nonoxidative methane conversion were explored (Weckhuysen et al. 1998a; Jiang et al. 1999). Before catalytic performance, all of the

TABLE 10.3

Effect of Different Transition Metal (2 wt %) over ZSM-5 (Si/Al = 25) on Methane Dehydroaromatization at GHSV = 800 h⁻¹, 750°C for 3 h Reaction (Weckhuysen et al. 1998a; Jiang et al. 1999)

Metal (M)	Methane conversion (%)	Benzene selectivity (%)
Mo	7.5	71.2
Mo	7.9	72.2
Fe	3.9	45.5
Fe	4.1	73.4
V	0.6	35.5
V	3.2	31.6
Cr	0.5	28.2
Cr	1.1	72.0
W	2.3	40.6
W	2.4	50.8

catalysts were reduced in CO at high temperatures. 2 wt% TMI/H-ZSM-5 was used for nonoxidative conversion and the activities decreased as follows: Mo (18.3) > W (10.8) > Fe (5.7) > V (3.9) > Cr (1.5), where the numbers in parentheses are the rates of methane reaction (molecule reacted/metal atom/h) obtained after 3 h on-stream. The conversion and benzene selectivity are summarized in Table 10.3.

Denardin and Perez-Lopez (2019) studied transitions metals (Cu, Ni, Co, and Zn) as promoters over Fe/ZSM-5 promoted catalysts. Among all transition metals, a shorter induction period was reported on Cu-promoted and it also showed high rate of benzene formation. Ca-promoted catalyst produced less coke and significant amounts of aromatics.

Metal cation-acidic proton bifunctional (Ag/H-ZSM-5) catalyst was reported to have heterolytic dissociation of a C–H bond in methane to form $CH_3^{\delta+}$ and metal hydride species. The $CH_3^{\delta+}$ species reacts with ethylene to form propylene and acidic protons (Baba and Abe 2003). The metal halide species reacts with the acidic proton and form hydrogen. Ethylene reacted with methane to give propylene and hydrogen over Ag/H-ZSM-5.

Lim and Kim (2019) studied Mn/H-ZSM-5 catalysts with various loading amounts of Mn for methane dehydroaromatization (MDA) reaction. Combined various analytical and spectroscopic analysis indicated the presence of various manganese species, such as isolated Mn^{2+} ion, isolated Mn^{3+} ion, agglomerated MnO, and agglomerated Mn_2O_3 species, depends on the loading amount of Mn. The authors reported that 5 wt% Mn loading is optimum to have isolated Mn^{2+} and Mn^{3+} ion species. Among different species Mn^{3+} serves as main active site in MDA. With increase in Mn concentration the relative crystallinity of the ZSM-5 decreases. The methane conversion decreases in the order: 5 wt% Mn/H-ZSM-5 > 7.5 wt% Mn/H-ZSM-5 > 10 wt% Mn/H-ZSM-5 > 2.5 wt% Mn/H-ZSM-5. The high conversion on 5 wt% Mn might be due to isolated Mn^{2+} ion, isolated Mn^{3+} ion present at acidic sites.

Li et al. (2018) studied co-aromatization of methane and paraffin-rich raffinate oil on zinc and gallium modified ZSM-5 catalysts. 5% Zn-Ga/ZSM-5 (80:1) showed a conversion of about 95% with about 70% of xylene in the compositional yield. The excellent performance is attributed to highly distributed Zn, Ga species, which enhanced the Lewis acidity and synergistic influence of aromatic products.

Stepanov et al. (2019) studied the effect of 5 wt% Re/ZSM-5 on methane aromatization. The catalytic results showed that 5 wt% Re/ZSM-5 catalyst exhibited a conversion of less than 4%. However, 0.5 wt% Re–4.0%Mo/ZSM-5 showed a conversion of 14% with 7% benzene yield. The high conversion in the catalyst might be due to well-distributed Re nanoparticles and high amount of strong acidic sites.

Vollmer et al. (2019c) compared Fe and Mo active metals over ZSM-5 (Si/Al = 24). After 240 minutes on time of stream, 3.37 wt% Mo/H-ZSM-5 and 2 wt% Fe/H-ZSM-5 showed a conversion of 9% and 7%, respectively. The longer activation period of Fe resulted in decrease of methane yield with increased coke. The active sites were monomer in case of Mo and dimer in case of Fe.

Yaghinirada et al. (2019) studied effect of Ag, Cd, Cr, Mo, Zn, and Mn (3 wt% of each metal) H-ZSM-5 for methane aromatization. The conversion decreases in the order: Mo/H-ZSM-5 > Zn/H-ZSM-5 > Mn/H-ZSM-5 > Ag/H-ZSM-5 > Cd/H-ZSM-5 > Cr/H-ZSM-5 and the corresponding benzene yields are 6.44, 3.99, 3.23, 2.36, 2.29, and 1.37%, respectively.

Razdan et al. (2019) studied Zr as H_2 absorbent to MoCx/H-ZSM-5 in five different configurations including staged-bed, stratified-bed, and interpellet physical mixtures. The configuration used were as following (i) MoCx/H-ZSM-5, (ii) Zr packed upstream of catalyst, (iii) Zr packed downstream of catalyst, (iii) Zr packed in both upstream and downstream, i.e., the catalyst is sandwiched between Zr, and (iv) physical mixture of Zr and catalyst. The H_2 formed on Mo/H-ZSM-5 during dehydrogenation and cyclization inhibits methane aromatization. Use of Zr scavenger resulted in 14% conversion to benzene, which is near to its kinetic limit. The study showed that benzene synthesis rate decreases with increase in hydrogen effluent rate. The two catalyst configurations, Zr packed downstream and physical mixture, showed reversible effects on catalyst deactivation.

Various transition metal ions have been studied over ZSM-5 zeolite. However, among all Mo containing ZSM-5 showed the best conversion of methane and selectivity toward benzene.

10.4.1.4 Effect of Promoters

The role of promoters has been explored for nonoxidative methane dehydroaromatization. Sridhar et al. (2020) explored the effect of Co and Ni additive on 6 wt% Mo/H-ZSM-5. Both Co and Ni provided high structural stability by enhancing the Mo retention in the zeolitic framework.

Wang et al. [61] studied the dehydro-oligomerization of methane to ethylene and aromatics over Mo/ZSM-5 catalyst. They observed that catalyst with 2 wt% Mo exhibited optimum activity with 6.7% conversion and 80.8% selectivity toward benzene. Further modification of the catalyst with the addition of Li or P showed a decrease in activity.

Since CO pretreated catalysts showed good conversion, the influence of pretreatment conditions on methane aromatization performance over Mo/H-ZSM-5 and

Mo–Cu/H-ZSM-5 was studied by Zhang et al. (2003). Before catalysis, the catalysts were pretreated in N_2 and CH_4 flow at 750°C, catalyst pretreated with CH_4 showed more coke formation. The study indicated that the addition of Cu promoted the conversion of methane, which was much higher than over Mo/H-ZSM-5. After reacting for 1 h, the methane conversion over Mo-Cu/H-ZSM-5 and Mo/H-ZSM-5 was 10.5% and 7.5%, respectively, with benzene selectivity of 90.0% and 89.8%, respectively. Also, a benzene yield of 9.45% at 700°C was reported, which is near the thermodynamic equilibrium conversion (ca. 11.4%). Furthermore, more deactivation was observed in Mo/H-ZSM-5 than over Mo-Cu/H-ZSM-5.

The promotional effect of Ru on Mo/H-ZSM-5 zeolite was reported by Shu et al. (1997). With increase in Ru content (0 to 0.7%) over Mo/ZSM-5 catalyst, the amount of strong acid sites decreases with increase in amount of moderate and weak acid sites. For a catalyst with a Mo/Ru ratio of less than 3 or higher than 7 (Mo loading is 2%), no promotional effect was observed. The influence of adding Fe, Cr, Co, and Ga into 3%Mo/H-ZSM-5 catalyst on methane aromatization and selectivity was studied (Dong et al. 2004). The experimental results reported promotional effect of Fe, Cr, Co, and Ga for the dehydrogenation and dissociation of methane. The methane conversion follows the order Ti > Cr > Co > Zn > Ni = Rh > Re > Ga > Au > Fe > Mo > Ag > Cu, and the increase of benzene yield on the catalysts modified with different elements follows the order Co > Cr > Ga > Ni > Zn > Fe > Rh > Re > Ga > Au > Mo. The combined metal catalysts that showed better selectivity to hydrocarbons were: Co–Mo, Fe–Mo, Ga–Mo, Cr–Mo, Ni–Mo, and Au–Mo/H-ZSM-5 catalysts. Table 10.4 summarizes the effect of addition of promotors on Mo/ZSM-5.

Coupling of methane and methanol in reactant feed over bifunctional Ga-, Zn-, In-, and/or Mo-Modified ZSM-5 Zeolites was reported by Choudhary et al. (2005). No methane conversion was reported in absence of methanol and/or bifunctional zeolite catalyst. The study is very beneficial as the methanol or dimethyl ether to be used as co-reactant in feed can be produced from methane itself through methane – syngas – methanol or dimethyl ether technology. Also, methane converted per mole of additive can be close to 1.0, hence only small amount of additive will be required.

Denardin and Perez-Lopez (2019) studied the effect of different transition metal promoters (Cu, Ni, Co, and Zn) on Fe/ZSM-5. The selectivity towards benzene followed the order 5 wt% Fe/ZSM-5 > 5 wt% Fe-1 wt% Cu/ZSM-5 > 5 wt% Fe-1 wt% Co/ZSM-5~5 wt% Fe-1 wt% Zn/ZSM-5 ≫ 5 wt% Fe-1 wt% Ni/ZSM-5. On the other hand, the induction time follows the order: 5 wt% Fe-1 wt% Cu/ZSM-5 < 5 wt% Fe/ZSM-5 < 5 wt% Fe-1 wt% Zn/ZSM-5~5 wt% Fe-1 wt% Co/ZSM-5. The catalyst promoted by Cu showed the maximum benzene selectivity of 41%, which may be due to the reduction of FeO to Fe^0 in the presence of Cu.

Tshabalala et al. (2014) studied the promotional effect of tin on 0.5 wt% Pt/2 wt% Mo/H-ZSM-5 catalyst. With the addition of Sn, the methane conversion decreases from 7.2 to 4.3%; however, an increase (64–83%) in the benzene selectivity was reported. Hence Sn as a promotor increased the benzene selectivity with a decrease in coke formation.

Abdelsayed et al. (2015) studied the effect of Fe and Zn promoters on Mo/H-ZSM-5. In case of Fe as promoter, most of the Fe remained on the external surface

TABLE 10.4

Literature Review of Nonoxidative Conversion of Methane with Different Promotors

Reference	T (°C)	ZSM-5 (Si/Al)	Promotor	wt% Mo	Methane conversion (%)	Benzene selectivity (%)
(Dong et al. 2004)	700	39.5	2 wt% Co	3	11.2	68.4
(Dong et al. 2004)	700	39.5	2 wt% Fe	3	9.2	63.1
(Dong et al. 2004)	700	39.5	2 wt% Ga	3	9.7	57.9
(Dong et al. 2004)	700	39.5	2 wt% Cr	3	12.1	63.0
(Dong et al. 2004)	700	39.5	2 wt% Ni	3	10.4	59.1
(Dong et al. 2004)	700	39.5	2 wt% Zn	3	11.1	60.8
(Dong et al. 2004)	700	39.5	2 wt% Ti	3	13.7	61.4
(Dong et al. 2004)	700	39.5	2 wt% Rh	3	10.4	60.6
(Dong et al. 2004)	700	39.5	2 wt% Re	3	10.1	64.7
(Dong et al. 2004)	700	39.5	2 wt% Au	3	9.5	62.0
(Dong et al. 2004)	700	39.5	2 wt% Ag	3	8.3	60.7
(Dong et al. 2004)	700	39.5	2 wt% Cu	3	8.2	61.4
(Choudhary et al. 2005)	550	49.6	2 wt% GaAl	2	0.96[a]	–
(Choudhary et al. 2005)	550	49.6	2 wt% Zn	2	0.78[b]	–
(Ramasubramanian et al. 2019)	750[c]	30	1 wt% K	10	28	50
(Ramasubramanian et al. 2019)	750[c]	30	1 wt% Rh	10	26	40
(Ramasubramanian et al. 2019)	750[c]	30	1 wt% Fe	10	25	20

Moles methane converted per mole methanol with CH_4/CH_3OH [a]15:1, [b]14.1:1, [c]Reaction performed in a batch reactor

of ZSM-5. 35% increase in the benzene formation was reported after 12 h of reaction compared to unpromoted catalyst which might be due increased Lewis acidic sites. Whereas in case of Zn promoted catalyst only slight increase in benzene formation was observed, which lasted for 3 h only. After that the catalyst behaved as unpromoted catalyst which might be due to loss of more than 50% Zn from catalyst confirmed with EDX results.

Ramasubramanian et al. (2019) studied effect of K, Rh, and Fe (1 wt%) as promoters over 10 wt% Mo/H-ZSM-5. The methane conversion decreases in the order 10 wt% Mo/H-ZSM-5 > 1 wt% K/10 wt% Mo/H-ZSM-5 > 1 wt% Rh/10 wt% Mo/H-ZSM-5~ 1 wt% Fe/10 wt% Mo/H-ZSM-5 with time on stream of 45 min. The selectivity toward benzene for corresponding time on stream is reported as 10 wt% Mo/H-ZSM-5 > 1 wt% K/10 wt% Mo/H-ZSM-5 > 1 wt% Fe/10 wt% Mo/H-ZSM-5 > 1 wt% Rh/10 wt% Mo/H-ZSM-5. After 255 min, the conversion for all the catalysts became similar with 1 wt% K/10 wt% Mo/H-ZSM-5 showing the highest selectivity towards benzene. The higher selectivity of benzene over 1 wt% K/10 wt%

Mo/H-ZSM-5 was reported due to less coke formation over catalyst which might be due to modification of acidic sites.

The addition of promotors affects the methane dehydroaromatization reaction in different ways. Some of the promotors showed positive effect on conversion but deactivates after some time, others show less deactivation at the cost of methane conversion and benzene selectivity.

10.4.1.5 Addition of Other Hydrocarbons

Choudhary et al. (1997) studied the conversion of methane to higher hydrocarbons over H-galloaluminosilicate ZSM-5 type (MFI) zeolite in the presence of alkenes. In combination with propene, 36.2% conversion of methane with 90.7% selectivity towards aromatics was observed at 550°C. The authors reported that methane activation resulted from a hydrogen-transfer reaction with alkenes.

The positive effect of coke formed on the catalysts was explored by Pierella et al. (1997), which suggested that it may act as an initiator for pure methane conversion, which have the same effect as the C_2^+ additive in the feed. The pure methane converted to aromatics over partially coke catalysts MoW/H-ZSM-5 at 600°C gave a yield of 6.8%, whereas the fresh Mo(VI)/H-ZSM-5 remained inactive at the temperature.

Baba et al. (2005) showed that methane reacts with ethene over the In-supported ZSM-5 catalyst at 400°C producing higher hydrocarbons such as propene and toluene. Under the same conditions, H-ZSM-5 showed no catalytic activity for methane conversion, only ethane conversion was reported. In/ZSM-5 pretreated with hydrogen at 450°C after calcining at 630°C exhibited the highest conversion of CH_4 of 11.8% at 400°C.

Propane as co-reactant over cobalt and zinc-impregnated H-ZSM-5 catalysts were evaluated using a fixed-bed microreactor (Liu et al. 2008b). At WHSV of 1.6 g $h^{-1}g_{cat}^{-1}$ and 600°C, C_1 conversion reached 36.7% with selectivity of aromatic products above 88.7% at atmospheric pressure. The incorporation of cobalt and zinc into H-ZSM-5 catalyst decreased the strength of strong Brønsted acidic sites and increased in Lewis acidic sites were reported. The micropore area of the catalyst was also decreased with increase in metal ions concentration which resulted in the decrease of the coke on the catalyst.

Wang et al. (2016) studied the technical feasibility of ethanol with methane for hydroaromatization over H-ZSM-5 in the presence of Zn and Ag loading (5% Zn-1% Ag/ZSM-5, 5% Zn/ZSM-5 and 1% Ag/ZSM-5). 1% Ag/ZSM-5 due to well-distributed Ag showed promising catalytic performance for ethanol aromatization under methane environment at 400°C and 1 atm pressure. High amount of coke was also formed on catalyst due to strong Brønsted and Lewis surface acid sites.

Chu and Qiu (2003) coupled dehydroaromatization of methane over Mo/H-ZSM-5 with ethane. The addition of 6.3% of ethane to methane feed at 725°C and 3 atm pressure for 7 h. The rate of benzene formation was increased to 1930 nmolC/g-cat/s (on carbon basis). Similar results have been reported by Matus et al. (2009). Also, the condensation of carbonaceous deposit increased in presence of ethane.

The addition of hydrocarbons showed a positive effect on the methane conversion as well as selectivity toward benzene and higher hydrocarbons. The aromatic products produced by additive hydrocarbons on zeolite sites facilitates methane substituted on the aromatic rings.

10.4.2 Interaction of Methane with Different Catalyst Supports (Silica, Alumina, and H-ZSM-5)

In 1992, Koerts et al. (1992), reported hydroaromatization of methane over a bifunctional catalyst at lower temperatures. The C_2 to C_6 alkanes formation from methane at atmospheric pressure takes place in a two-step route in which methane is first thermally activated. In first step methane is decomposed by a reduced Group VIII metal catalyst (5% Ru, 3% Rh, and 10% Co over silica) into hydrogen and adsorbed surface carbonaceous species at 177–527°C. In a second reaction step a particular surface carbonaceous intermediate produces hydrocarbons upon hydrogenation at 27–127°C.

Kanitkar et al. (2019) explored sulfated zirconia (SZ) as catalyst support for Mo/SZ. Raman spectra and DRIFTS confirmed difference in acid sites when Mo was doped on SZ compared to Mo/ZSM-5, except at higher Mo loadings. All these catalysts showed methane conversions of 5–20% at 600–700°C. Also, Mo/SZ showed higher selectivity toward the heavier aromatics such as naphthalene and coke as compared to Mo/H-ZSM-5 at 700°C, which was more selective towards benzene. Mo oxide supported on ZSM-5/MCM-22 has been studied and reported that Mo carbide activates methane to form CH_x species. These species dimerized into C_2H_y and then oligomerized on the strong ZSM-5/MCM-22 Brønsted acid sites to form benzene (Abedin et al. 2019).

Guo et al. (2014) explained that presence of adjacent sites promotes C–C coupling resulting in oligomerization and deactivation of catalyst by carbonaceous deposits. The authors prepared single iron sites embedded in a nonacidic silica matrix and used for nonoxidative conversion of methane. At 1090°C, maximum conversion of 48.1% with ethylene selectivity of 53%, and total hydrocarbon selectivity of 99% were observed. The catalyst showed no deactivation up to 60 h time on stream.

Chen et al. (1995) investigated the interaction of methane with silica, alumina's (η, γ, and α) and H-ZSM-via FTIR studies. The surface OH groups played an important role in methane adsorption. The interaction sequence of methane with OH group follows the order Si–OH–Al > Al–OH > Si–OH, which agrees well with the strength of their acidities. The hydroxyls on η-alumina could be exchanged with deuterated methane (CD_4) at 300°C, whereas γ and α-alumina required higher temperature. This indicated that adsorption of methane occurs via formation of hydrogen bond with hydroxyl group and it follows the sequence: $\eta > \gamma > \alpha$-Al_2O_3, which was reported in accordance with the order of the numbers of their surface hydroxyl groups.

Chen et al. (1996) studied the interaction of methane with Mo/H-ZSM-5 zeolite at −100°C. The results revealed that methane is adsorbed at bridging hydroxyl groups (band at 2890 and 3002 cm^{-1}) and surface oxygen species (band at 2900 cm^{-1}). Thus, they provide active sites that causes C–H bond weakness which is first step in aromatization.

Wong et al. (1996b) studied Mo containing different catalytic supports like H-ZSM-5, HSAPO-34, and HY. Among all H-ZSM-5 catalyst showed the highest catalytic activity and stability. The catalytic activity and benzene selectivity over

Mo/HSAPO-34 catalyst decreased with increasing time on stream with high selectivity to ethene. The increased high selectivity of ethane with benzene expanse is due to nonavailability of active sites to aromatize it. Mo/HY showed negligible activity, which might be due to the rapid blocking of the pore system and the Mo active sites.

Julian et al. (2019) studied the effect acidic sites in ZSM-5 by comparing it with MCM-22. The authors used polyoxometalates (POMs) i.e. $[(n-C_4H_9)_4N]_2[Mo_6O_{19}]$ (denoted as "Mo$_6$") and $[(n-C_4H_9)_4N]_4[Mo_8O_{26}]$ (denoted as "Mo$_8$") as precursors of Mo for the catalyst. For comparison, ammonium molybdate tetrahydrate $(NH_4)_6Mo_7O_{24}\cdot4H_2O$, impregnated catalyst was also prepared (denoted as "Mo$_7$"). MCM-22 showed better stability with lower coke formation and high benzene yield. Also, 5% Mo$_6$/MCM-22 catalyst displays superior stability due to uniform distribution of active sites.

Abedin et al. (2019) studied Mo-sulfated hafnia (SH) for methane hydroaromatization. At 650°C, 1200 ml/(g$_{cat}$hr), 1 atm, 5% Mo/SH showed 9.5% conversion with aromatic selectivity of 50.2%, whereas 5% Mo/H-ZSM-5 showed a conversion of 7.5% with 12.9% selectivity toward aromatics after 120 minutes on time of stream. Pyridine DRIFTS studies showed that SH has both Brønsted and Lewis acidic sites. The NH$_3$-TPD results showed that the addition of Mo oxides decreased the total number of acidic sites, which might be due to exchange between protonated sulfate sites on hafnia and Mo, however slight increase in Lewis acidic intensity of the catalyst was reported.

10.4.3 Formation and Nature of Active Sites

For dehydroaromatization of methane, different reaction mechanism has been proposed. Among those a few reaction mechanisms have been discussed in detail. Wong et al. (1996a) proposed the following mechanism. Reduction of Mo^{6+} occurred during the reaction as demonstrated by the XPS measurement. In that case, hydrocarbon chain initiation involved Mo active sites in either 4+ or 5+ oxidation states. Since the Mo^{4+} concentration was higher than that of Mo^{5+}, molybdenum-carbene intermediate was formed on Mo^{4+}.

Wang et al. (1997) proposed a new Mo phase formed under methane atmosphere, which is a prerequisite of the formation of aromatics C$_2$ hydrocarbons. The new Mo phase most possibly is molybdenum carbide (MoC). The formation of molybdenum carbide active phase is responsible for methane nonoxidative transformation to aromatics and C$_2$ hydrocarbons. The reactions taking place are suggested as:

$$CH_4 \rightarrow C + 2H_2$$

$$MoO_3 + H_2 \rightarrow MoO_2 + H_2O$$

$$MoO_3 + C \rightarrow MoC + CO + CO_2$$

$$MoO_2 + 3C \rightarrow MoC + 2CO$$

Zheng et al. (2008) studied active center of methane dehydroaromatization by ^{95}Mo NMR spectroscopy. The study confirmed that the Mo species which migrates to the zeolite Brønsted acidic sites upon calcination serves as active sites for methane dehydroaromatization.

Zaikovskii et al. (2006) studied the active sites of 4 wt% Mo/H-ZSM-5 catalyst using electron microscopic and ESR studies. Mo was present as clusters in the zeolitic channels and as Mo_2C crystallite on zeolite surface. For all the catalyst active (time on stream = 10 min) and deactivated with coke, 2–3 nm carbon layer in form of graphite was observed on Mo_2C. The results suggested that primary activation of methane occurs on Mo clusters present in zeolite cavities but proceeds over zeolite active sites.

The nature of active phase on the Co–Ga/H-ZSM-5 catalyst was discussed by Liu et al. (2008). Less amount of coke was observed due to decrease in activity of acidic sites by addition of Co and Ga. Xu and Lin (1999) compared the different reaction mechanism given in literature for bifunctional catalysts and concluded that there are quite a lot of similarities between short chain alkane aromatization and methane aromatization. The authors believed the methane aromatization as an extension of short chain alkane aromatization.

10.4.4 Induction Period of Methane Dehydroaromatization Reaction

Since Mo/ZSM-5 is the most studied catalyst for methane dehydroaromatization, several researches are focused on the reaction kinetics based on that catalyst. Wang et al. (1997b) and Ma et al. (2000c) reported that Mo^{6+} is converted to Mo_2C during pretreatment of Mo/ZSM-5 at 700°C in CH_4/H_2 mixture and this transformation is referred to as the induction period. In methane reaction over molybdenum modified catalysts, methane activation takes place over Mo_2C. Hence Mo_2C acts as active site in methane aromatization.

The temperature-programmed surface reaction (TPSR) techniques for 6Mo/HMCM-22 showed that ethylene and benzene were formed after induction period. Also, it revealed that the catalyst remained active even when molybdenum carbide was covered with carbonaceous deposits (Ma et al. 2000c). The study also concluded that the pretreatment of MoO_3 into molybdenum carbide before the reaction, reduces the reaction temperature to 574°C from 700°C. The presence of Mo_2C species was confirmed using XPS studies. Based on XPS studies the authors reported the transformation of molybdenum oxide to molybdenum carbide takes place by a two-step reaction (Ma et al. 2000c; Weckhuysen et al. 1998b) as follows:

$$MoO_3 + CH_4 \rightarrow MoO_2 + CO_2 + H_2O$$

$$MoO_2 + 2CH_4 \rightarrow Mo_2C + CO_2 + H_2 + CO + H_2O$$

At first, MoO_3 is transformed to MoO_2 in the presence methane and thus producing carbon dioxide and water as products. The second stage is however reported as vigorous and results in phase transformation of MoO_2 to Mo_2C (Xu et al. 2003). During reaction, the oxygen deficiency or elimination allows the carbonaceous deposits to interact with Mo species to form Mo_2C. Xu et al. (2003) discussed the synergistic effect of Mo_2C and zeolite acidic sites and suggested that ethylene formation takes place over Mo_2C active sites, whereas zeolite acidic sites catalyze the subsequent conversion to benzene. Masiero et al. (2009) studied the effect of metal loading with different Mo/Fe ratios for Mo–Fe/ZSM-5 catalysts. Up to 10 wt% metal loading with an Mo/Fe ratio between 2 and 4, the catalyst showed good conversion and selectivity towards benzene. Also, the authors reported an increase in induction time as Fe content was increased. However, Men et al. (2001) in their studies detected the formation of benzene during the induction period and ascribed it to the reduction of Mo^{6+} species, which are considered to be slightly active methane aromatization species. Jiang et al. (1999) studied pulse reaction over Mo/ZSM-5 and reported two kinds of Mo species and reported that formation of carbonaceous deposit could shorten the induction period.

Liu et al. (2007) studies over Mo/H-ZSM-5 showed that the conversion of methane and selectivity towards hydrocarbons stabilizes after induction period of 80 min on time of stream. No active site change was observed by Lacheen and Iglesia (2005) when the catalyst was pre-reduced with CO, H_2, or in CH_4–H_2 mixtures and resulted in similar product formation rates as directly activated in CH_4 reactant mixtures.

Rahman et al. (2019) studied the various species formed during induction period with different pretreatment. The four types of treatment were performed and studied using temperature-programmed reduction and carburization profiles: (1) heating the catalyst in a reducing gas, H_2, up to reaction temperature and switching to CH_4; (2) heating the catalyst in a reducing gas, H_2, mixed with dilute CH_4; (3) heating the catalyst in CH_4 up to reaction temperature; and (4) heating the catalyst in an inert gas (commonly He) up to reaction temperature and then switching to CH_4 or to H_2 followed by CH_4 or to H_2/CH_4 mixture. All the pretreated samples (with CH_4) showed the presence of Mo_2C species on H-ZSM-5 (see Table 10.5). The mechanism involved in Mo_2C varied with pretreatment conditions. However, catalyst treated

---

TABLE 10.5

Treatment Condition Applied to 10 wt% MoO_x/H-ZSM-5 and the Corresponding Mo Species Evolved

Treatment	Mo Species Evolved
$H_2 \rightarrow CH_4$[a]	$MoO_3 \rightarrow MoO_2 \rightarrow Mo \rightarrow Mo_2C$
CH_4/H_2[b]	$MoO_3 \rightarrow MoO_2 \rightarrow Mo + MoO_2 + Mo_2C \rightarrow Mo_2C$
CH_4[b]	$MoO_3 \rightarrow MoO_2 \rightarrow Mo_2C$
$He \rightarrow CH_4$[a]	$MoO_3 \rightarrow$ disperesed $MoO_x \rightarrow MoO_2 \rightarrow Mo_2C$
$He \rightarrow CH_4/H_2$[a]	$MoO_3 \rightarrow MoO_2 \rightarrow Mo_2C$
$He \rightarrow H_2$[a]	$MoO_3 \rightarrow MoO_2 \rightarrow Mo$

Where [a] $_X \rightarrow Y$ samples were treated in X and then gas was switched to Y. [b] $_{X/Y}$ samples were heated in X or a mixture of X/Y and maintained in that gas all of the time. (Reprinted from Journal of Catalysis, Vol. 375, Rahman, M., Infantes-Molina, A., Boubnov, A., Bare, S. R., Stavitski, E. Sridhar, A., Khatib, S. J. Increasing the catalytic stability by optimizing the formation of zeolite-supported Mo carbide species ex situ for methane dehydroaromatization, Pages No. 314–328, Copyright (2019), with permission from Elsevier.)

with H_2 followed by CH_4, or in a dilute mixture of CH_4/H_2 showed higher stability, which might be due to higher Mo dispersion.

10.4.5 THE ROLE OF BRØNSTED ACID SITES

Mo/ZSM-5 is one of the most studied catalyst for MDA. To investigate the role of Brønsted acidic sites Zhao et al. (2019) studied the effect of Si/Al ratio on Mo/H-ZSM-5. In general, methane dehydrogenation reactions take place in presence of strong Brønsted acid sites. In case of zeolite catalyst, as discussed in previous section, oligomerization of C_2 intermediates occurs on the acidic sites of the zeolite. However, the zeolites possess Lewis as well as Brønsted acidic sites. Classically, it was generally accepted that the molybdenum is anchored on the Brønsted acid sites of the zeolite as $(Mo_2O_5)^{2+}$ species. The formation of dimer structure $(Mo_2O_5)^{2+}$ was confirmed using Raman spectroscopy (Li et al. 2000).

Tessonnier et al. (2008) studied the role of Brønsted acid sites over Mo/ZSM-5 for methane dehydroaromatization reaction. The authors compared the classical scheme for which Mo is believed to perform the dehydrogenation and coupling of methane to ethylene and compared it with their results. They reported that the dispersion of molybdenum species in zeolite channels as monomer or dimer depends upon Si/Al ratio of H-ZSM-5. NH_3-TPD of H-ZSM-5 before and after Mo loading showed decrease in acidic sites after Mo loading indicating that acidic sites act as anchoring sites for molybdenum inside the zeolite channels.

The effect of Brønsted acid strength in alkane activations has been studied by density functional theory (DFT) calculations (Chu et al. 2012). The authors used a solid acid models with varying acid strengths (weak-, strong-, to superacid) and evaluated reactions including methane hydrogen exchange (C–H bond activation), propane dehydrogenation (C–H bond dissociation and H–H bond formation), and propane cracking (C–C cleavage). The calculation showed that activation barriers (E_{act}) of all the reactions decrease linearly, with increasing the acid strength. Also, the propane cracking was reported to be most sensitive to acid strength, while the methane hydrogen exchange reaction showed least sensitivity to acid strength. Ionic character in transition state of organic framework played an important role in the susceptibility for the reactivity to acid strength. The ionic transition states of (0.913–0.964 (Mulliken charge) |e|), while the methane hydrogen exchange has a less ionic transition and (0.646–0.773 |e|), were reported for propane cracking reaction and methane hydrogen exchange respectively. As discussed earlier, the SiO_2/Al_2O_3 ratio is one of the determining factors for Mo species cleavage with zeolitic framework. The synergistic effect of Mo and acidic sites was observed for methane aromatization (Ma et al. 2005). For the catalytic activity of Mo/HMCM-22 catalysts for methane aromatization in the presence of oxygen. A special focus was made on the carbonaceous deposits of catalysts upon deactivation. With different coke deposits on molybdenum species and Brønsted acid sites different temperature-programmed oxidation profile was reported. Three different kinds of carbonaceous deposits were reported in the oxidation for Mo/HMCM-22 catalysts. The coke included carbidic carbon in molybdenum carbide (both from internal Mo ions and external Mo particles), molybdenum-associated coke, and (possibly) two aromatic-type cokes on acid sites. The TPO profile revelled the deactivation of molybdenum carbide, the TPO of catalyst showed absence of carbidic carbon indicating the catalyst covered by molybdenum-associated coke. Also, it was concluded from observations that the coke formation decreases the catalytic performances.

Dong et al. (2004) found maximum benzene yields and good catalytic activity with a 3% Mo/H-ZSM-5 catalyst with a SiO_2/Al_2O_3 ratio of 39.5 in the presence of promoters. The NH_3-TPD revelled that addition of Fe and Co additives to the Mo/H-ZSM-5 catalysts decreases the acidity of catalyst for strongly acidic site. Lu et al. (1999, 2001) suggested that a greater number of Brønsted acidic sites results in large amount of carbonaceous depositions in the methane aromatization reaction hence decrease in activity. In the study, the authors dealuminated H-ZSM-5 (Si/Al = 2 5) to remove excess framework aluminium (Brønsted acidic sites) before loading Mo. The temperature-programmed oxidation (TPO) of coked Mo/H-ZSM-5 suggested less coke over dealuminated catalyst.

Su et al. (2002) showed that as SiO_2/Al_2O_3 ratio increases from 24 to 55, the number of Brønsted acidic sites becomes almost half. Since Brønsted acidic sites are needed to anchor the Mo species, a smaller number of MoO_x species were found with zeolite having Si/Al higher ratio. Fewer Brønsted sites results in lower migration of Mo into the zeolite channels. Also, higher Si/Al ratio resulted in less methane conversion with high amount of coke. Ding et al. (2002) demonstrated that Brønsted acidic sites and molybdenum species located on the external surface of the zeolite over Mo/H-ZSM-5 catalysts when prepared by mechanical mixing are responsible for unselective catalytic reactions. The authors studied nonoxidative methane aromatization on silica covered external Mo and acidic sites. The catalytic result showed increased one ring aromatics with slow deactivation.

To study the generation of Brønsted acidic sites, a physical mixture of MoO_3 and H-ZSM-5 are mixed followed by calcination. The heating of sample resulted in migration of MoO_x which was calculated by the rate of H_2O evolution (Borry et al. 1999). ^{27}Al NMR revealed that with increase in Mo loading decrease in Brønsted acidic sites was observed. 4–5wt% Mo loading is optimum to generate well distribution of active sites. However, higher Mo resulted in $(MoO_3)_n$ oligomers or unreducible and inactive $Al_2(MoO_4)_3$ domains via reactions with framework Al atoms and ultimately decreasing the Brønsted acidic sites (Borry et al. 1999). The detailed XPS study over Mo/ZSM-5 revealed that the carbonaceous deposits during catalytic activity are formed on the Mo-supported external surface of ZSM-5 (Shu et al. 1999). Tan et al. (2002) studied the role of Brønsted acidity by introducing Cs^+ to Mo/ZSM-5 catalyst. The addition of Cs+ replaced the Bronsted acid sites and the catalytic activity of Cs/Mo/H-ZSM-5 was decreased after 5 min on time on stream. In situ ^1H MAS NMR spectroscopic observation of proton species on a Mo-modified H-ZSM-5 zeolite catalyst suggested that as the reaction temperature was increased to 700°C, the blue-white color of the catalyst was changed to black due to Mo_2C formation. Also Brønsted acid signal at $\delta = 4.1$ decreases with appearance of two new peaks at $\delta = 6.8, 7.9$ corresponding to water molecule formed and adsorbed on Lewis acidic site (Ma et al. 2000a). The EPR measurements showed two types of Mo species inside the zeolite channels in the form of $Al(I)...MoO_x$ or $Al(II)...MoO_x$, based on their reducibilities by CH_4. Also, on the zeolite external surface, Mo species was present as MoO_3 crystallite with octahedral (oct) coordination or MoO_x phase with a square-pyramidal (squ) coordination. The reducibility for all the Mo species was represented as $MoO_3oct \approx MoO_xsqu > Al(I)...MoO_x > Al(II)...MoO_x$ (Ma et al. 2000b). For nanosized H-ZSM-5 the impregnated Mo remains on the external surface and interact with Brønsted sites. whereas for micro sized H-ZSM-5 it migrates zeolitic channels which are more beneficial for methane dehydroaromatization.

10.4.6 THE NATURE OF THE CARBONACEOUS DEPOSIT AND ITS ROLE IN THE REACTION

The catalyst deactivation occurs due to deposition of carbonaceous deposits or coke on the active sites of catalysts (Sahoo et al. 2003; He et al. 2010). The major problem encountered during methane dehydroaromatization is catalyst deactivation. The main reasons that lead to deactivation of catalyst are as following: (i) pore blockage, i.e., carbonaceous deposits formed within the zeolite cavity, (ii) active sites coverage, i.e., poisoning of the active sites by heavy carbonaceous products, and (iii) structure alterations which may be caused by reaction conditions. In the present case, the catalyst deactivation occurs as a result of coke formation. Over zeolites two different kinds of coke were reported viz soft coke and hard coke. Hard coke results when higher reaction temperatures and extended time on stream are used in bulky reactants (Karge 1991; Mann 1997). The soft coke can be removed from catalyst via calcination, hence regenerating the catalyst. Over Mo/MCM-22 the carbonaceous deposits increase with the Mo loading from 2 to 10 wt%, the increase in carbonaceous deposits are due to increased acidic sites. Also, the increased contact time during reaction leads to increase in carbonaceous deposits (Shu et al. 2000). When the acidity is high, the coke formation is polycyclic aromatic with three or four fused rings (Lee et al. 2012).

Guisnet et al. (2009) discussed in detail about the coke formation and the methods to minimize it. The important factors to consider for coke reduction are pore structure, acidity of catalyst and the reaction conditions. It is important to note that coke formation and catalyst deactivations are also result of shape selective catalysis of zeolite (Ma et al. 2002a). The different amounts of coke under similar reaction conditions were reported which was due to different acidities possessed by the catalysts. The catalyst with Mo/H-ZSM-5 contained weak Brønsted acidic sites compared to parent zeolite. The steam treatment of catalyst removes the tetrahedral Al hence decreasing number of Brønsted acidic sites and hence less coke formation was reported.

Shu et al. (1997) reported that when Ru was added as promoter over Mo/ZSM-5, the splitting of C–H bond was increased. Furthermore, the DTA (differential thermal analysis) suggested that carbonaceous deposit helped in the reaction. Weckhuysen et al. [50] studied the nature of carbonaceous deposits on Mo/H-ZSM-5 catalysts using ^{13}C CPMAS NMR and reported two types of carbonaceous deposits one located on the acidic sites and other on the Mo species. The carbonaceous deposit on Mo species helps in reduction of induction time of reaction. Jiang et al. (1999) used ^{13}C solid state NMR spectroscopy to distinguish the carbonaceous deposits. They concluded that there were two kinds of carbonaceous deposits; one associated with Brønsted acid sites and the other was associated with molybdenum carbide or partially reduced Mo species.

Different carbonaceous species and their amounts are distinguished and studied in detail using temperature-programmed techniques. During aromatization different carbon species are formed over catalyst active sites. Based on the intermediates formed during the reaction, some groups have divided carbonaceous deposits as carbidic carbon in molybdenum carbide, molybdenum-associated coke, and aromatic-type cokes on acid sites (Ma et al. 2002b; Liu et al. 2002). Among all, carbidic carbon over molybdenum species helps in reducing the induction period (Liu et al. 2005, 2001).

Kosinov et al. (2019) reported that an increase in pressure has positive effect on methane conversion with increased selectivity toward xylene and toluene. The increased selectivity was independent of Mo loading, methane space velocity as well as temperature. The increased pressure increases the product diffusion from catalyst hence increasing conversion and selectivity.

10.4.7 Plasma Catalysis for Methane Activation

Various nonconventional systems such as plasma, microwaves, photocatalysis, and non-Faradaic reactions have been investigated and studied by numerous groups to resolve the deactivation of catalysts which occurs in conventional systems. Electrons with high energy are introduced to reactants for the dissociation of the gas feed molecules into radicals. The electron discharge systems can be classified into two types: equilibrium discharge and non-equilibrium discharge (Oshima et al. 2013).

Park et al. (2017) studied plasma assisted methane dehydroaromatization using dielectric barrier discharge (DBD) reactor. Before the reaction the catalyst was applied with CH_4 feed with argon at 700°C for 1 hour to convert Mo oxide to Mo_2C. For reaction, 500°C temperature with continuous sinusoidal voltage wave function with 3.5 kV and 3 kHz was applied. In plasma condition good conversion of methane was observed with decrease in benzene selectivity with high amount of ethane in reaction product.

Heintze and Magureanu (2002) reported feasibility of benzene formation from methane into benzene by combining plasma and surface catalysis. A detailed study indicated that acetylene production was constant and the selectivity towards ethylene increases in parallel to benzene over time on stream implying that product formation does not take place by C_2 intermediate mechanism.

Li et al. (2007) explored oxygen free methane conversion over two-stage plasma-followed-by-catalyst (PFC) (Ni/H-ZSM-5) reactor at atmospheric pressure and low temperature. H_2–CH_4 feed was introduced to PFC containing Fe, Co, Ni, Cu, and Zn metal-loaded H-ZSM-5. Ni-loaded H-ZSM-5 catalysts showed the best catalytic activity at low temperatures.

Nonoxidative aromatization of methane over a Mo–Fe/H-ZSM-5 catalyst via TGA was studied by Zhu et al. (2005). Like conventional systems, two types of carbonaceous species were reported over catalysts. The activation energy for coke burning was reported higher for the low-temperature peak in the plasma-treated catalyst (treated for 1 hour) at 600°C, 550°C, and 699°C, whereas the high-temperature peak the activation energy was higher in the conventional catalyst at 600°C, 550°C, and 699°C. Also, the DTG peak temperature (high-temperature peak) for plasma-treated catalysts was shifted to a lower value. The observations suggest that carbonaceous species in the low-temperature region were higher in plasma-treated catalysts than in the conventional catalyst.

10.5 CONCLUSIONS

The fossil fuels are precious to mankind and must be utilized wisely. Methane has higher global warming potential than CO_2, hence flaring of methane to atmosphere must be avoided. Methane has potential to be converted to higher hydrocarbons of gasoline range under nonoxidative conditions. In the current chapter, we have discussed the C–H bond activation, and reaction pathways involved in MDA over different catalytic systems. The studies show that significant progress has been made in conventional and nonconventional routes to increase the higher hydrocarbon conversion and selectivity. A significant positive progress has been observed in the process, which seems very promising. It also suggests that this field still needs new technologies and catalyst systems that are coke-resistant.

REFERENCES

Abdelsayed, V., Shekhawat, D., and Smith, M. W. 2015. Effect of Fe and Zn promoters on Mo/HZSM-5 catalyst for methane dehydroaromatization. *Fuel* 139:401–410.
Abdulrasheed, A., Jalil, A. A., Gambo, Y., Ibrahim, M., Hambali, H. U., and Hamid, M. Y. S. 2019. A review on catalyst development for dry reforming of methane to syngas: recent advances. *Renew. Sust. Energy Rev.* 108:175–193.
Abedin, M. A., Kanitkar, S., Bhattar, S., and Spivey, J. J. 2019. Sulfated hafnia as a support for Mo oxide: a novel catalyst for methane dehydroaromatization. *Catal. Today.* doi:10.1016/j.cattod.2019.02.021.
Anunziata, O. A., and Mercado, G. G. 2006. Methane transformation using light gasoline as co-reactant over Zn/H-ZSM11. *Catal. Lett.* 107(1–2):111–116.
Anunziata, O. A., Mercado, G. V. G., and Pierella, L. B. 2003. Catalytic activation of methane using n-pentane as co-reactant over Zn/H-ZSM-11 zeolite. *Catal. Lett.* 87(3–4):167–171.
Baba, T., and Abe, Y. 2003. Metal cation–acidic proton bifunctional catalyst for methane activation: conversion of $^{13}CH_4$ in the presence of ethylene over metal cations-loaded H-ZSM-5. *Appl. Catal. A Gen.* 250(2):265–270.
Baba, T., Abe, Y., Nomoto, K., et al. 2005. Catalytic transformation of methane over in-loaded ZSM-5 zeolite in the presence of ethene. *J. Phys. Chem. B* 109(9):4263–4268.
Bahri, S., Patra, T., and Upadhyayula, S. 2019. Synergistic effect of bifunctional mesoporous ZSM-5 supported Fe-Co catalyst for selective conversion of syngas with low Ribblet ratio into synthetic fuel. *Micro. Meso. Mater.* 275:1–13.
Bamji, Z. 2019. *Global Gas Flaring Inches Higher for the First Time in Five Years.* World Bank Blog. https://blogs.worldbank.org/opendata/global-gas-flaring-inches-higher-first-time-five-years (accessed December 29, 2019).
Borry, R. W., Kim, Y. H., Huffsmith, A., Reimer, J. A., and Iglesia, E. 1999. Structure and density of Mo and acid sites in Mo-exchanged H-ZSM5 catalysts for nonoxidative methane conversion. *J. Phys. Chem. B* 103(28):5787–5796.
BP. 2019. *BP Statistical Review of World Energy 2019*, 19th ed.
Chen, L., Lin, L., Xu, Z., et al. 1995. Interaction of methane with surfaces of silica, aluminas and HZSM-5 zeolite. A comparative FT-IR study. *Catal. Lett.* 35(3–4):245–258.

Chen, L., Lin, L., Xu, Z., et al. 1996. Fourier transform-infrared investigation of adsorption of methane and carbon monoxide on HZSM-5 and Mo/HZSM-5 zeolites at low temperature. *J. Catal.* 161(1):107–114.

Choudhary, V. R., Kinage, A. K., and Choudhary, T. V. 1997. Low-temperature nonoxidative activation of methane over H-galloaluminosilicate (MFI) zeolite. *Science* 275(5304):1286–1288.

Choudhary, V. R., Mondal, K. C., and Mulla, S. A. 2005. Simultaneous conversion of methane and methanol into gasoline over bifunctional Ga-, Zn-, In-, and/or Mo-modified ZSM-5 zeolites. *Angew. Chem. Int. Ed.* 44(28):4381–4385.

Choudhary, V. R., Mondal, K. C., and Mulla, S. A. R. 2006. Process for the simultaneous conversion of methane and organic oxygenate to C2 to C10 hydrocarbons United States Patent, *Patent No. 7022888*.

Chu, W., and Qiu, F. 2003. Remarkable promotion of benzene formation in methane aromatization with ethane addition. *Topics Catal.* 22(1–2):131–134.

Chu, Y., Han, B., Fang, H., Zheng, A., and Deng, F. 2012. Influence of acid strength on the reactivity of alkane activation on solid acid catalysts: a theoretical calculation study. *Micro. Meso. Mater.* 151:241–249.

Degnan, Jr. T. F. 2003. The implications of the fundamentals of shape selectivity for the development of catalysts for the petroleum and petrochemical industries. *J. Catal.* 216(1–2):32–46.

Denardin, F., and Perez-Lopez, O. W. 2019. Tuning the acidity and reducibility of Fe/ZSM-5 catalysts for methane dehydroaromatization. *Fuel* 236:1293–1300.

Ding, W., Meitzner, G. D., and Iglesia, E. 2002. The effects of silanation of external acid sites on the structure and catalytic behavior of Mo/H–ZSM5. *J. Catal.* 206(1):14–22.

Dong, Q., Zhao, X., Wang, J., and Ichikawa, M. 2004. Studies on Mo/HZSM-5 complex catalyst for methane aromatization. *J. Nat. Gas Chem.* 13(1):36–40.

Fox, T. A., Barchyn, T. E., Risk, D., Ravikumar, A. P., and Hugenholtz, C. H. 2019. A review of close-range and screening technologies for measuring fugitive methane emissions in upstream oil and gas. *Environ. Res. Lett.* 14:069601.

Galadima, A., and Muraza, O. 2019. Advances in catalyst design for the conversion of methane to aromatics: a critical review. *Catal. Surv. Asia* 23:149–170.

Guisnet, M., Costa, L., and Ribeiro, F. R. 2009. Prevention of zeolite deactivation by coking. *J. Mol. Catal. A Chem.* 305(1–2):69–83.

Guo, X., Fang, G., Li, G., et al. 2014. Direct, nonoxidative conversion of methane to ethylene, aromatics, and hydrogen. *Science* 344(6184):616–619.

Han, S. J., Kim, S. K., Hwang, A., et al. 2019. Non-oxidative dehydroaromatization of methane over Mo/H-ZSM-5 catalysts: a detailed analysis of the reaction-regeneration cycle. *Appl. Catal. B Environ.* 241:305–318.

He, P., Jarvis, J. S., Meng, S., et al. 2019b. Co-aromatization of methane with propane over Zn/HZSM-5: the methane reaction pathway and the effect of Zn distribution. *Appl. Catal. B Environ.* 250:99–111.

He, P., Wang, A., Meng, S., et al. 2019. Impact of Al sites on the methane co-aromatization with alkanes over Zn/HZSM-5. *Catal. Today* 323:94–104.

He, S., Sun, C., Yang, X., Wang, B., Dai, X., and Bai, Z. 2010. Characterization of coke deposited on spent catalysts for long-chain-paraffin dehydrogenation. *Chem. Eng. J.* 163(3):389–394.

Heintze, M., and Magureanu, M. 2002. Methane conversion into aromatics in a direct plasma-catalytic process. *J. Catal.* 206(1):91–97.

Jarvis, J., He, P., Wang, A., and Song, H. 2019. Pt-Zn/HZSM-5 as a highly selective catalyst for the Co-aromatization of methane and light straight run naphtha. *Fuel* 236:1301–1310.

Jiang, H., Wang, L., Cui, W., and Xu, Y. 1999. Study on the induction period of methane aromatization over Mo/HZSM-5: partial reduction of Mo species and formation of carbonaceous deposit. *Catal. Lett.* 57(3):95–102.

Julian, I., Hueso, J. L., Lara, N., et al. 2019. Polyoxometalates as alternative Mo precursors for methane dehydroaromatization on Mo/ZSM-5 and Mo/MCM-22 catalysts. *Catal. Sci. Technol.* 9:5927–5942.

Julian, I., Roedern, M. B., Hueso, J. L., et al. 2020. Supercritical solvothermal synthesis under reducing conditions to increase stability and durability of Mo/ZSM-5 catalysts in methane dehydroaromatization. *Appl. Catal. B Environ.* 263:118360.

Kanitkar, S., Abedin, M. A., Bhattar, S., and Spivey, J. J. 2019. Methane dehydroaromatization over molybdenum supported on sulfated zirconia catalysts. *Appl. Catal. A Gen.* 575:25–37.

Kanitkar, S. R., and Spivey, J. J. 2018. Light alkane aromatization: efficient use of natural gas. In *Natural Gas Proc. in Midstream Downstream*, eds. Elbashir, N. O., El-Halwagi, M. M., Economou, I. G., and Hall, K. R., 379–401. John Wiley & Sons, Hoboken, NJ, USA.

Karge, H. G. 1991. Coke formation on zeolites. *Stud. Surf. Sci. Catal.* 58:531–570. Elsevier.

Kazansky, V. B., and Serykh, A. I. 2004. Unusual localization of zinc cations in MFI zeolites modified by different ways of preparation. *Phys. Chem. Chem. Phys.* 6(13):3760–3764.

Koerts, T., Deelen, M. J., and Van Santen, R. A. 1992. Hydrocarbon formation from methane by a low-temperature two-step reaction sequence. *J. Catal.* 138(1):101–114.

Kosinov, N., Coumans, F. J., Uslamin, E., Kapteijn, F., and Hensen, E. J. 2016. Selective coke combustion by oxygen pulsing during Mo/ZSM-5-catalyzed methane dehydroaromatization. *Angew. Chem. Int. Ed.* 55(48):15086–15090.

Kosinov, N., Uslamin, E. A., Meng, L., Parastaev, A., Liu, Y., and Hensen, E. J. 2019. Reversible nature of coke formation on Mo/ZSM-5 methane dehydroaromatization catalysts. *Angew. Chem. Int. Ed.* 58(21):7068–7072.

Lacheen, H. S., and Iglesia, E. 2005. Isothermal activation of $Mo_2O_5^{2+}$–ZSM-5 precursors during methane reactions: effects of reaction products on structural evolution and catalytic properties. *Phys. Chem. Chem. Phys.* 7(3):538–547.

Lai, Y., and Veser, G. 2016. Zn-HZSM-5 catalysts for methane dehydroaromatization. *Environ. Prog. Sust. Energy* 35(2):334–344.

Lee, K. Y., Kang, M. Y., and Ihm, S. K. 2012. Deactivation by coke deposition on the HZSM-5 catalysts in the methanol-to-hydrocarbon conversion. *J. Phys. Chem. Solids* 73(12):1542–1545.

Li, Q., Zhang, F., Jarvis, J., et al. 2018. Investigation on the light alkanes aromatization over Zn and Ga modified HZSM-5 catalysts in the presence of methane. *Fuel* 219:331–339.

Li, W., Meitzner, G. D., Borry III, R. W., and Iglesia, E. 2000. Raman and X-ray absorption studies of Mo species in Mo/H-ZSM5 catalysts for non-oxidative CH_4 reactions. *J. Catal.* 191(2):373–383.

Li, X. S., Shi, C., Xu, Y., Wang, K. J., and Zhu, A. M. 2007. A process for a high yield of aromatics from the oxygen-free conversion of methane: combining plasma with Ni/ HZSM-5 catalysts. *Green Chem.* 9(6):647–653.

Lim, T. H., and Kim, D. H. 2019. Characteristics of Mn/H-ZSM-5 catalysts for methane dehydroaromatization. *Appl. Catal. A Gen.* 577:10–19.

Liu, B. S., Jiang, L., Sun, H., and Au, C. T. 2007. XPS, XAES, and TG/DTA characterization of deposited carbon in methane dehydroaromatization over Ga–Mo/ZSM-5 catalyst. *Appl. Surf. Sci.* 253(11):5092–5100.

Liu, B. S., Zhang, Y., Liu, J. F., et al. 2011. Characteristic and mechanism of methane dehydroaromatization over Zn-based/HZSM-5 catalysts under conditions of atmospheric pressure and supersonic jet expansion. *J. Phys. Chem. C* 115(34):16954–16962.

Liu, H., Li, T., Tian, B., and Xu, Y. 2001. Study of the carbonaceous deposits formed on a Mo/HZSM-5 catalyst in methane dehydro-aromatization by using TG and temperature-programmed techniques. *Appl. Catal. A Gen.* 213(1):103–112.

Liu, H., Shen, W., Bao, X., and Xu, Y. 2005. Methane dehydroaromatization over Mo/HZSM-5 catalysts: the reactivity of MoCx species formed from MoOx associated and non-associated with Brönsted acid sites. *Appl. Catal. A Gen.* 295(1):79–88.

Liu, H., Su, L., Wang, H., Shen, W., Bao, X., and Xu, Y. 2002. The chemical nature of carbonaceous deposits and their role in methane dehydro-aromatization on Mo/MCM-22 catalysts. *Appl. Catal. A Gen.* 236(1–2):263–280.

Liu, J. F., Jin, L., Liu, Y., and Yun-Shi, Q. I. 2008. Methane aromatization over cobalt and gallium-impregnated HZSM-5 catalysts. *Catal. Lett.* 125(3–4):352–358.

Liu, J. F., Liu, Y., and Peng, L. F. 2008b. Aromatization of methane by using propane as co-reactant over cobalt and zinc-impregnated HZSM-5 catalysts. *J. Mol. Catal. A Chem.* 280(1–2):7–15.

Liu, W., and Xu, Y. 1999. Methane dehydrogenation and aromatization over Mo/HZSM-5: in situ FT–IR characterization of its acidity and the interaction between Mo species and HZSM-5. *J. Catal.* 185(2):386–392.

Lu, Y., Ma, D., Xu, Z., Tian, Z., Bao, X., and Lin, L. 2001. A high coking-resistance catalyst for methane aromatization. *Chem. Commun.* (20):2048–2049.

Lu, Y., Xu, Z., Tian, Z., Zhang, T., and Lin, L. 1999. Methane aromatization in the absence of an added oxidant and the bench scale reaction test. *Catal. Lett.* 62(2–4):215–220.

Lunsford, J. H. 2000. Catalytic conversion of methane to more useful chemicals and fuels: a challenge for the 21st century. *Catal. Today* 63(2–4):165–174.

Luzgin, M. V., Rogov, V. A., Arzumanov, S. S., Toktarev, A. V., Stepanov, A. G., and Parmon, V. N. 2008. Understanding methane aromatization on a Zn-modified high-silica zeolite. *Angew. Chem. Int. Ed.* 47(24):4559–4562.

Luzgin, M. V., Rogov, V. A., Arzumanov, S. S., Toktarev, A. V., Stepanov, A. G., and Parmon, V. N. 2009. Methane aromatization on Zn-modified zeolite in the presence of a co-reactant higher alkane: how does it occur? *Catal. Today* 144(3–4):265–272.

Ma, D., Lu, Y., Su, L., et al. 2002a. Remarkable improvement on the methane aromatization reaction: a highly selective and coking-resistant catalyst. *J. Phys. Chem. B* 106(34):8524–8530.

Ma, D., Shu, Y., Bao, X., and Xu, Y. 2000b. Methane dehydro-aromatization under nonoxidative conditions over Mo/HZSM-5 catalysts: EPR study of the Mo species on/in the HZSM-5 zeolite. *J. Catal.* 189(2):314–325.

Ma, D., Shu, Y., Cheng, M., Xu, Y., and Bao, X. 2000c. On the induction period of methane aromatization over Mo-based catalysts. *J. Catal.* 194(1):105–114.

Ma, D., Shu, Y., Zhang, W., Han, X., Xu, Y., and Bao, X. 2000a. In Situ ^{1}H MAS NMR spectroscopic observation of proton species on a Mo-modified HZSM-5 zeolite catalyst for the dehydroaromatization of methane. *Angew. Chem. Int. Ed.* 39(16):2928–2931.

Ma, D., Wang, D., Su, L., Shu, Y., Xu, Y., and Bao, X. 2002b. Carbonaceous deposition on Mo/HMCM-22 catalysts for methane aromatization: a TP technique investigation. *J. Catal.* 208(2):260–269.

Ma, D., Zhu, Q., Wu, Z., et al. 2005. The synergic effect between Mo species and acid sites in Mo/HMCM-22 catalysts for methane aromatization. *Phys. Chem. Chem. Phys.* 7(16):3102–3109.

Ma, S., Guo, X., Zhao, L., Scott, S., and Bao, X. 2013. Recent progress in methane dehydroaromatization: from laboratory curiosities to promising technology. *J. Energy Chem.* 22(1):1–20.

Mann, R. 1997. Catalyst deactivation by coke deposition: approaches based on interactions of coke laydown with pore structure. *Catal. Today* 37(3):331–349.

Marcilly, C. 2001. Evolution of refining and petrochemicals. What is the place of zeolites? *Stud. Surf. Sci. Catal.* 135:37–60.

Masiero, S. S., Marcilio, N. R., and Perez-Lopez, O. W. 2009. Aromatization of methane over Mo-Fe/ZSM-5 catalysts. *Catal. Lett.* 131(1–2):194–202.

Matus, E. V., Sukhova, O. B., Ismagilov, I. Z., Tsikoza, L. T., and Ismagilov, Z. R. 2009. Peculiarities of dehydroaromatization of CH_4–C_2H_6 and CH_4 over Mo/ZSM-5 catalysts. *React. Kin. Catal. Lett.* 98(1):59–67.

Men, L., Zhang, L., Xie, Y., et al. 2001. A new phenomenon in the induction period of the methane dehydroaromatization reaction. *Chem. Commun.* (18):1750–1751.

Mønster, J., Kjeldsen, P., and Scheutz, C. 2019. Methodologies for measuring fugitive methane emissions from landfills – a review. *Waste Manage.* 87:835–859.

Natesakhawat, S., Means, N. C., Howard, B. H., et al. 2015. Improved benzene production from methane dehydroaromatization over Mo/HZSM-5 catalysts via hydrogen-permselective palladium membrane reactors. *Catal. Sci. Technol.* 5(11):5023–5036.

Olah, G. A., and Olah, J. A. 1971. Electrophilic reactions at single bonds. IV. Hydrogen transfer from, alkylation of, and alkylolysis of alkanes by alkylcarbenium fluoroantimonates. *J. Am. Chem. Soc.* 93(5):1256–1259.

Olah, G. A., Halpern, Y., Shen, J., and Mo, Y. K. 1971. Electrophilic reactions at single bonds. III. HD exchange and protolysis (deuterolysis) of alkanes with superacids. The mechanism of acid-catalyzed hydrocarbon transformation reactions involving the. sigma. electron pair donor ability of single bonds via three-center bond formation. *J. Am. Chem. Soc.* 93(5):1251–1256.

Oshima, K., Shinagawa, T., and Sekine, Y. 2013. Methane conversion assisted by plasma or electric field. *J. Japan Petro. Inst.* 56(1):11–21.

Park, S., Lee, M., Bae, J., et al. 2017. Plasma-assisted non-oxidative conversion of methane over Mo/HZSM-5 catalyst in DBD reactor. *Topics Catal.* 60(9–11):735–742.

Pierella, L. B., Wang, L., and Anunziata, O. A. 1997. Methane direct conversion to aromatic hydrocarbons at low reaction temperature. *React. Kin. Catal. Lett.* 60(1):101–106.

Rahman, M., Infantes-Molina, A., Boubnov, A., et al. 2019. Increasing the catalytic stability by optimizing the formation of zeolite-supported Mo carbide species ex situ for methane dehydroaromatization. *J. Catal.* 375:314–328.

Ramasubramanian, V., Lienhard, D. J., Ramsurn, H., and Price, G. L. 2019. Effect of addition of K, Rh and Fe over Mo/HZSM-5 on methane dehydroaromatization under non-oxidative conditions. *Catal. Lett.* 149(4):950–964.

Razdan, N. K., Kumar, A., and Bhan, A. 2019. Controlling kinetic and diffusive length-scales during absorptive hydrogen removal in methane dehydroaromatization on MoCx/H-ZSM-5 catalysts. *J. Catal.* 372:370–381.

Sahoo, S. K., Rao, P. V. C., Rajeshwer, D., Krishnamurthy, K. R., and Singh, I. D. 2003. Structural characterization of coke deposits on industrial spent paraffin dehydrogenation catalysts. *Appl. Catal. A Gen.* 244(2):311–321.

Schwach, P., Pan, X., and Bao, X. 2017. Direct conversion of methane to value-added chemicals over heterogeneous catalysts: challenges and prospects. *Chem. Rev.* 117(13): 8497–8520.

Scurrell, M. S. 1987. Prospects for the direct conversion of light alkanes to petrochemical feedstocks and liquid fuels-a review. *Appl. Catal.* 32:1–22.

Sedel'nikova, O. V., Stepanov, A. A., Zaikovskii, V. I., Korobitsyna, L. L., and Vosmerikov, A. V. 2017. Preparation method effect on the physicochemical and catalytic properties of a methane dehydroaromatization catalyst. *Kin. Catal.* 58(1):51–57.

Segers, R. 1998. Methane production and methane consumption: a review of processes underlying wetland methane fluxes. *Biogeochemistry* 41(1):23–51.

Shamzhy, M., Opanasenko, M., Concepción, P., and Martínez, A. 2019. New trends in tailoring active sites in zeolite-based catalysts. *Chem. Soc. Rev.* 48(4):1095–1149.

Shu, J., Adnot, A., and Grandjean, B. P. 1999. Bifunctional behavior of Mo/HZSM-5 catalysts in methane aromatization. *Ind. Eng. Chem. Res.* 38(10):3860–3867.

Shu, Y., Ma, D., Xu, L., Xu, Y., and Bao, X. 2000. Methane dehydro-aromatization over Mo/ MCM-22 catalysts: a highly selective catalyst for the formation of benzene. *Catal. Lett.* 70(1–2):67–73.

Shu, Y., Xu, Y., Wong, S. T., Wang, L., and Guo, X. 1997. Promotional effect of Ru on the dehydrogenation and aromatization of methane in the absence of oxygen over Mo/ HZSM-5 catalysts. *J. Catal.* 170(1):11–19.

Solymosi, F., Erdöhelyi, A., and Szöke, A. 1995. Dehydrogenation of methane on supported molybdenum oxides. Formation of benzene from methane. *Catal. Lett.* 32(1–2):43–53.

Spivey, J. J., and Hutchings, G. 2014. Catalytic aromatization of methane. *Chem. Soc. Rev.* 43(3):792–803.

Sridhar, A., Rahman, M., Infantes-Molina, A., Wylie, B. J., Borcik, C. G., and Khatib, S. J. 2020. Bimetallic Mo-Co/ZSM-5 and Mo-Ni/ZSM-5 catalysts for methane dehydroaromatization: a study of the effect of pretreatment and metal loadings on the catalytic behavior. *Appl. Catal. A Gen.* 589:117247.

Stepanov, A. A., Zaikovskii, V. I., Korobitsyna, L. L., and Vosmerikov, A. V. 2019. Nonoxidative conversion of methane to aromatic hydrocarbons in the presence of ZSM-5 zeolites modified with molybdenum and rhenium. *Petro. Chem.* 59(1):91–98.

Su, L., Xu, Y., and Bao, X. 2002. Study on bifunctionality of Mo/HZSM-5 catalysts for methane dehydro-aromatization under non-oxidative condition. *J. Nat. Gas Chem.* 11(1/2):8–27.

Szöke, A., and Solymosi, F. 1996. Selective oxidation of methane to benzene over K_2MoO_4/ ZSM-5 catalysts. *Appl. Catal. A Gen.* 142(2):361–374.

Szostak, R. 1989. *Molecular Sieves Principles of Synthesis and Identification.* VanNostrand Reinhold, New York, NY.

Tan, P. L., Leung, Y. L., Lai, S. Y., and Au, C. T. 2002. The effect of calcination temperature on the catalytic performance of 2 wt.% Mo/HZSM-5 in methane aromatization. *Appl. Catal. A Gen.* 228(1–2):115–125.

Tanabe, K., and Hölderich, W. F. 1999. Industrial application of solid acid–base catalysts. *Appl. Catal. A Gen.* 181(2):399–434.

Tessonnier, J. P., Louis, B., Rigolet, S., Ledoux, M. J., and Pham-Huu, C. 2008. Methane dehydro-aromatization on Mo/ZSM-5: about the hidden role of Brønsted acid sites. *Appl. Catal. A Gen.* 336(1–2):79–88.

Tshabalala, T. E., Coville, N. J., and Scurrel, M. S. 2014. Dehydroaromatization of methane over doped Pt/Mo/H-ZSM-5 zeolite catalysts: the promotional effect of tin. *Appl. Catal. A Gen.* 485(5): 238–244.

Vermeiren, W., and Gilson, J. P. 2009. Impact of zeolites on the petroleum and petrochemical industry. *Topics Catal.* 52(9):1131–1161.

Vollmer, I., Mondal, A., Yarulina, I., Abou-Hamad, E., Kapteijn, F., and Gascon, J. 2019b. Quantifying the impact of dispersion, acidity and porosity of Mo/HZSM-5 on the performance in methane dehydroaromatization. *Appl. Catal. A Gen.* 574:144–150.

Vollmer, I., Ould-Chikh, S., Aguilar-Tapia, A., et al. 2019c. Activity descriptors derived from comparison of Mo and Fe as active metal for methane conversion to aromatics. *J. Am. Chem. Soc.* 141(47):11814–11824.

Vollmer, I., Yarulina, I., Kapteijn, F., and Gascon, J. 2019a. Progress in developing a structure-activity relationship for the direct aromatization of methane. *ChemCatChem* 11(1):39–52.

Wang, A., He, P., Yung, M., Zeng, H., Qian, H., and Song, H. 2016. Catalytic co-aromatization of ethanol and methane. *Appl. Catal. B Environ.* 198:480–492.

Wang, D., Lunsford, J. H., and Rosynek, M. P. 1997b. Characterization of a Mo/ZSM-5 catalyst for the conversion of methane to benzene. *J. Catal.* 169(1):347–358.

Wang, L., Tao, L., Xie, M., Xu, G., Huang, J., and Xu, Y. 1993. Dehydrogenation and aromatization of methane under non-oxidizing conditions. *Catal. Lett.* 21(1–2):35–41.

Wang, L., Xu, Y., Wong, S. T., Cui, W., and Guo, X. 1997. Activity and stability enhancement of MoHZSM-5-based catalysts for methane non-oxidative transformation to aromatics and C2 hydrocarbons: effect of additives and pretreatment conditions. *Appl. Catal. A Gen.* 152(2):173–182.

Weckhuysen, B. M., Rosynek, M. P., and Lunsford, J. H. 1998b. Characterization of surface carbon formed during the conversion of methane to benzene over Mo/H-ZSM-5 catalysts. *Catal. Lett.* 52(1–2):31–36.

Weckhuysen, B. M., Wang, D., Rosynek, M. P., and Lunsford, J. H. 1998a. Conversion of methane to benzene over transition metal ion ZSM-5 zeolites: I. Catalytic characterization. *J. Catal.* 175(2):338–346.

Wong, S. T., Xu, Y., Liu, W., Wang, L., and Guo, X. 1996b. Methane activation without using oxidants over supported Mo catalysts. *Appl. Catal. A Gen.* 136(1):7–17.

Wong, S. T., Xu, Y., Wang, L., et al. 1996a. Methane and ethane activation without adding oxygen: promotional effect of W in Mo-W/HZSM-5. *Catal. Lett.* 38(1–2):39–43.

Xu, Y., and Lin, L. 1999. Recent advances in methane dehydro-aromatization over transition metal ion-modified zeolite catalysts under non-oxidative conditions. *Appl. Catal. A Gen.* 188(1–2):53–67.

Xu, Y., Bao, X., and Lin, L. 2003. Direct conversion of methane under nonoxidative conditions. *J. Catal.* 216(1–2):386–395.

Xu, Y., Liu, S., Guo, X., Wang, L., and Xie, M. 1994. Methane activation without using oxidants over Mo/HZSM-5 zeolite catalysts. *Catal. Lett.* 30(1–4):135–149.

Yaghinirada, E., Aghdasiniaa, H., Naghizadeha, A., and Niaeia, A. 2019. Non-oxidative conversion of methane to aromatics over modified zeolite catalysts by transitional metals. *Iranian J. Catal.* 9(2):147–154.

Zaikovskii, V. I., Vosmerikov, A. V., Anufrienko, V. F., et al. 2006. Properties and deactivation of the active sites of an MoZSM-5 catalyst for methane dehydroaromatization: electron microscopic and EPR studies. *Kin. Catal.* 47(3):389–394.

Zhang, Y., Wang, D., Fei, J., and Zheng, X. 2003. Influence of pretreatment conditions on methane aromatization performance of Mo/HZSM-5 and Mo-Cu/HZSM-5 catalysts. *J. Nat. Gas Chem.* 12(2):145–149.

Zhao, K., Jia, L., Wang, J., Hou, B., and Li, D. 2019. The influence of the Si/Al ratio of Mo/HZSM-5 on methane non-oxidative dehydroaromatization. *New J. Chem.* 43(10):4130–4136.

Zheng, H., Ma, D., Bao, X., et al. 2008. Direct observation of the active center for methane dehydroaromatization using an ultrahigh field ^{95}Mo NMR spectroscopy. *J. Am. Chem. Soc.* 130(12):3722–3723.

Zhu, X., Yu, K., Li, J., Zhang, Y. P., Xia, Q., and Liu, C. J. 2005. Thermogravimetric analysis of coke formation on plasma treated Mo-Fe/HZSM-5 catalyst during non-oxidative aromatization of methane. *React. Kin. Catal. Lett.* 87(1):93–99.

11 Multifunctional Reactors for Direct Nonoxidative Methane Conversion

Su Cheun Oh, Sichao Cheng, Ying Pan,
Emily Schulman, and Dongxia Liu

CONTENTS

11.1 INTRODUCTION

11.1.1 NATURAL GAS AS AN ALTERNATIVE TO CRUDE OIL

Presently, the fuel and chemical industries depend almost solely on crude oil as a feedstock (Lee 2019). The increasing pressure to reduce the carbon dioxide (CO_2) footprint and the volatility of the crude oil market drives the movement to replace the feedstock base for these industries (Fiorentino, Zucaro, and Ulgiati 2019). In addition, the sharp rise of wealth and mobility in new economies within South and East Asia pushes the fuel and chemical markets to seek feed material alternatives to crude oil to produce gasoline, diesel, jet fuel, and specialty chemicals such as base oils, waxes, and solvents (Abdel-Aal and Alsahlawi 2013). Coal, biomass and natural gas provide alternatives to crude oil (Chu and Majumdar 2012). The use of coal increases the CO_2 footprint, reducing its viability as a cleaner alternative fuel/chemical feedstock (Boden, Marland,

305

and Andres 2009; Hong and Slatick 1994). Even with significant research efforts throughout the past decades, renewable biomass-fueled processes are mostly operational solely at the research scale and cannot fulfill the requirements of fuel/chemical markets (Winandy et al. 2008; Biddy et al. 2016). Therefore, natural gas, which does not experience the severe drawbacks of using coal or biomass, is the most promising replacement for crude oil. The discovery of shale gas basins in North America and other countries around the world further promotes natural gas as an abundant, accessible, and inexpensive energy source and raw material for the fuel, chemical and polymer industries (Boyer et al. 2011; Wood 2012; Wenrui, Jingwei, and Bin 2013).

11.1.2 Gas-to-Liquid Conversion Pathways in Natural Gas Upgrading

Currently, ~40% of natural gas resources are considered to be stranded or uneconomical due to a combination of resource size and distance from the market, (*e.g.* smaller and more remote resources are less economical) (Khalilpour and Karimi 2012). This is a result of the difficulties associated with storage and transport of natural gas (Khalilpour and Karimi 2012). By breaking C–H bonds in methane (CH_4), the main component in natural gas, and forming C–C bonds to produce hydrocarbon liquids via the gas-to-liquid (GTL) process, the value of the products increases. Transporting the liquid products is significantly easier, as the density of the liquid phase products is 1000 times greater than that of the gas-phase feed (e.g. methane vs. benzene (C_6H_6)). This density increase is dramatic, requiring a much smaller pipeline for transport, or perhaps one rather than 1000 tanker trucks. Therefore, GTL technologies directly influence the ability to use global natural gas resources to address the continual rise in demand for new and existing fuel/chemical markets.

The incorporation of GTL technologies into methane conversion processes has resulted in three pathways (Bao, El-Halwagi, and Elbashir 2010; Wood, Nwaoha, and Towler 2012). The first is a multistep conversion process, in which natural gas is first converted into synthesis gas (i.e. a mixture of carbon monoxide (CO) and hydrogen (H_2)) followed by Fischer–Tropsch (FT) synthesis of higher hydrocarbons. This process, like most hydrocarbon processing (i.e. oil refining), benefits economically from its scalability, with cost savings arising from the ability to develop extremely large plants. This requires large capital investments, and the plants must be located in close proximity to natural gas resources. Therefore, the FT synthesis is only economically viable for a handful of natural gas reserves and requires well-capitalized corporations and state-run enterprises to confront the high initial cost. The second GTL process involves the direct conversion of CH_4 to ethylene (C_2H_4) through the oxidative coupling of methane (OCM). This process is being explored by startup companies such as Siluria, but it suffers from low yields (< 25%) of desired products and substantial waste resulting from the complete and partial oxidation of CH_4 (i.e. CO_2, CO) exiting the system. The third process converts methane into C_2 (i.e. ethane (C_2H_6), ethylene (C_2H_4) and acetylene (C_2H_2)) and aromatics via direct nonoxidative methane coupling (DNMC), as indicated by the exemplary reaction equation: $2\ CH_4 = 1/2\ C_2H_4 + 1/6\ C_6H_6 + 5/2\ H_2$. The DNMC products (e.g. C_2H_4 and C_6H_6) are considered to be very valuable feedstocks for the chemical and polymer industries. The DNMC pathway is significantly simpler than the multi-step FT process, which means lower investment costs are needed and stranded gas can potentially be utilized. This gives natural gas investigators access to higher value markets, which allows them to partially diversify oil pricing and potentially increase revenues.

11.1.3 CHALLENGES IN DIRECT NONOXIDATIVE METHANE CONVERSION

Past research efforts have investigated noncatalytic and catalytic DNMC processes (Choudhary, Aksoylu, and Wayne Goodman 2003; Holmen, Olsvik, and Rokstad 1995). Noncatalytic DNMC pyrolyzes CH_4 to achieve high C_2H_2 yield but requires very high reaction temperatures (> 1573 K) (Horn and Schlögl 2015; Khan and Crynes 1970; Holmen, Rokstad, and Solbakken 1976; Holmen, Olsvik, and Rokstad 1995; Olsvik and Billaud 1993). The catalytic DNMC process was originally discovered by using a bifunctional molybdenum-loaded Zeolite Socony Mobil-5 (Mo/ZSM5) catalyst in 1993 (Wang et al. 1993). Thousands of catalysts have been reported for use in DNMC since, but Mo/ZSM5 is still considered the most active and promising metal/zeolite system for this reaction (Spivey and Hutchings 2014; Alvarez-Galvan et al. 2011). Although substantial enhancement in hydrocarbon product selectivity has been realized using Mo/ZSM5 compared to that of noncatalytic methane pyrolysis, DNMC typically occurs at < 1023 K due to catalyst stability and is limited by low CH_4 conversion and fast catalyst deactivation (<~10 h) due to coke deposition (Alvarez-Galvan et al. 2011; Karakaya and Kee 2016). The iron/silica (Fe/SiO_2, single Fe site embedded in a SiO_2 matrix reported in 2014) (Guo et al. 2014) is a recently-developed, effective catalyst for DNMC chemistry. Compared to Mo/ZSM5, Fe/SiO_2 requires higher reaction temperatures (1223–1363 K), produces high C_2 and higher hydrocarbons (collectively referred to as C_{2+}) and results in less coking.

The DNMC reaction stoichiometry shows the gas volume increases significantly with increasing conversion. This limits methane equilibrium conversion, with ~11% conversion at ~973 K and ~23% conversion at ~1073 K, as illustrated in Figure 11.1(a). High selectivity toward C_{2+} hydrocarbons is limited to the kinetically controlled regime, since thermodynamics tends to decompose methane to carbon (C) and H_2.

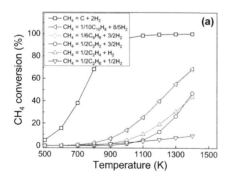
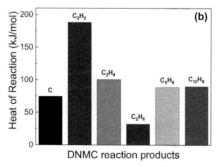

FIGURE 11.1 (a) Equilibrium conversion of CH_4 and (b) enthalpy of reaction (ΔH_{rxn}) at standard temperature and pressure conditions for the formation of C_2H_2, C_2H_4, C_2H_6, C_6H_6, naphthalene ($C_{10}H_8$), and carbon, from DNMC reactions. (Calculated using HSC Chemistry 6.0 Software.)

Additionally, the DNMC reaction is highly endothermic (see Figure 11.1(b)), and heat supply becomes a technical challenge. These barriers have placed limitations on the advances of DNMC research in the field of reaction engineering (Karakaya and Kee 2016). Scale up, reactor design, process development, and economics have rarely been studied for DNMC in the past; thus limited content in open literature focuses on the DNMC process at the reactor scale. In this chapter, we discuss the reactor design and development for DNMC in methane upgrading, focusing on the multifunctional reactors that concurrently tackle two or more barriers present in the DNMC process. It is promising to realize high carbon and thermal efficiencies as well as low catalyst deactivation in these reactors to achieve technoeconomic process viability of DNMC.

11.2 MEMBRANE REACTOR FOR NONOXIDATIVE METHANE CONVERSION

According to the reaction stoichiometry of DNMC reactions, the methane conversion to C_{2+} hydrocarbons or C produces hydrogen as a co-product in large quantities. As shown in Figure 11.1(a), the H_2 quantity stays in the range of 0.5 moles (formation of C_2H_6) to 2.0 moles (formation of C) when considering the conversion of 1 mole of CH_4. The DNMC reactions are favored at low pressure and high temperature conditions (refer to Figure 11.2(a)). Based on Le Châtelier's principle, the removal of hydrogen from the product stream will shift the thermodynamic equilibrium of DNMC to higher CH_4 conversions (see Figure 11.2(b)). This can potentially achieve improved CH_4 conversion at relatively lower temperatures. Membrane reactors (MRs) can be used to remove H_2 from DNMC in situ, and thus hold the promise of circumventing limitations of thermodynamic equilibrium, increasing product yields and producing pure H_2. Moreover, the process efficiency obtained by combining the catalytic reaction and H_2 separation steps into a single MR unit reduces capital cost and improves system modularity resulting, in the potential to efficiently and cost-effectively produce value-added C_{2+} products from nonoxidative methane upgrading.

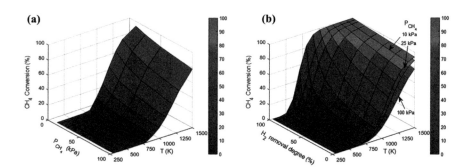

FIGURE 11.2 Thermodynamic analysis for the reaction of $CH_4 = 1/6\ C_6H_6 + 3/2\ H_2$ in DNMC: (a) CH_4 conversion as a function of reaction temperature and pressure and (b) enhancement in CH_4 conversion due to equilibrium shift through H_2 removal according to Le Châtelier's principle.

11.2.1 HYDROGEN-PERMEABLE MEMBRANE REACTOR

Research efforts toward the development of hydrogen-permeable MRs for DNMC were initiated during the 1990s. Both simulation and experimental studies were conducted to understand the effects of H_2 removal on overcoming the thermodynamic limitation and improving the C_{2+} product yield in DNMC. A substantial enhancement of the methane conversion was predicted by the simulation work in the Iglesia group using a reactor system containing Mo/ZSM5 catalyst and a mixed ionic-electronic conducting (MIEC) membrane (Borry et al. 1998). The initial attempts at parallel experimental studies, however, did not demonstrate favorable results, mainly due to the lack of membranes with sufficient H_2 permeation flux and the rapid catalyst deactivation upon H_2 removal. Continual research on DNMC has been focused on employing different membrane materials and/or membrane geometries to increase H_2 flux and conquer fast catalyst deactivation in the MRs. There are a great variety of membrane materials available for the separation of H_2 from a mixture of other gases, namely polymer membranes (Phair and Badwal 2006; Perry, Nagai, and Koros 2006; Ockwig and Nenoff 2007), nanoporous ceramic membranes (Gallucci et al. 2013; Ockwig and Nenoff 2007), palladium (Pd)-based membranes (Ockwig and Nenoff 2007; Barelli et al. 2008; Yun and Oyama 2011; Karakaya and Kee 2016) and dense MIEC membranes (Zhang et al. 2003; Li et al. 2015; Karakaya and Kee 2016). Given that the DNMC process requires high activation temperatures (typically > 873 K), only Pd-based and MIEC ceramic membranes were incorporated into the systems for DNMC reactions. Figure 11.3 demonstrates the schematics of a hydrogen-permeable tubular MR during DNMC, in which H_2 gas is removed in situ to result in the co-production of pure H_2 and C_{2+} (i.e. C_6H_6) products on opposite sides of the membrane. Other membrane geometries, such as pellet and hollow fiber, have also been used in hydrogen permeation membrane reactor systems for DNMC.

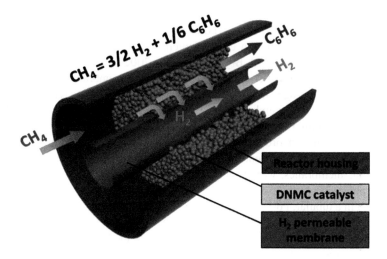

FIGURE 11.3 Schematics of a hydrogen-permeable MR for methane upgrading to benzene and hydrogen in DNMC reactions.

11.2.1.1 DNMC in Pd-Based Membrane Reactors

Pd-based membranes have been widely studied for hydrogen separation since the 1800s (Karakaya and Kee 2016). They can produce high-purity hydrogen during the separation process. Generally, hydrogen transports through the membrane in the following mechanism: (i) adsorption and dissociation of H_2 on the membrane surface to form atomic H_1; (ii) dissolution of H_1 into the Pd-based membrane matrix; (iii) diffusion of the H_1 through the Pd lattice in the membrane; and (iv) recombination of H_1 to form H_2 and desorption of H_2 molecules from the opposite membrane surface. The effective H_2 flux $\left(J_{H_2}\right)$ through a Pd-based membrane is governed by Sieverts' Law, which can be expressed as

$$J_{H_2} = \frac{P}{L}\left(p_{H_2,1}^n - p_{H_2,2}^n\right) \tag{11.1}$$

where P is the permeance (mol m^{-2} s^{-1} Pa^{-n}), L is the membrane thickness, and $p_{H_2,1}^n$ and $p_{H_2,2}^n$ are the hydrogen partial pressure of the feed and permeate sides, respectively. Ideally, n should be equal to 0.5. Pd-based membranes encounter several limitations despite their unique capability to permeate hydrogen. First, two different phases, α and β, are formed when the absorption of hydrogen occurs below its critical point (571 K and 2 MPa). This results in different crystal unit cell lattice parameters for the two phases, although both phases retain the pure Face Centered Cubic (FCC) structure. Defects tend to form on the membrane as a result of change in volume due to strain and recrystallization during the phase transformation (Yun and Oyama 2011). Second, Pd loses its ductility after exposure to hydrogen, leading to a process called hydrogen embrittlement (Rogers 1968). Membrane cracking can also happen during this process. Last, Pd tends to react strongly with carbon-containing species (Collins et al. 1996), sulfur (Paglieri and Way 2002), and carbon monoxide (CO) (Paglieri and Way 2002), which deactivates the surface of the membrane. In order to overcome these challenges, Pd can be alloyed with other metals to prevent phase transformation, hydrogen embrittlement, and metal poisoning in the membrane (Cheng and Yeung 1999; Bryden and Ying 2002; Fort, Farr, and Harris 1975; Jun and Lee 2000; Uemiya et al. 2007). Furthermore, membrane cost, resistance to coking, and high hydrogen flux under extreme reaction conditions remain challenges in the application of Pd membranes for H_2 separation in DNMC processes.

Table 11.1 summarizes the attempts to couple DNMC catalysts with a Pd or Pd-alloy membrane in order to circumvent thermodynamically limited methane conversion thus far. Different membrane configurations, including disk-shaped and tubular membranes, were tested, with tubular Pd-based membranes being the most popular due to their commercial availability. The DNMC reaction was first attempted in a tubular, Pd-based MR in 1989 using a Pt-Sn/Al$_2$O$_3$ catalyst (Andersen et al. 1989). However, poor DNMC performance was obtained due to severe coke formation within the reactor system. The discovery of the Mo/ZSM5 catalyst in 1993 reignited interest in Pd-membrane reactors once again in the 2000s. Since then, the coupling of Mo/ZSM5 catalysts with Pd or Pd-alloy membranes of different geometrical configurations (disk-shaped and tubular) to overcome the thermodynamic

TABLE 11.1

Overview of the Catalyst and Pd-Based H_2-Permeable Membrane Reactor Types Used for DNMC Reactions in Literature (Oh, Sakbodin, and Liu 2019)

Membrane type	Catalyst and reaction condition	Membrane geometry/dimension	Performance	Year/Reference
Pd membrane	Temperature: 400 K Pressure: ~ 101 kPa Catalyst: Pt–Sn/Al$_2$O$_3$	Tubular (5 mm outer diameter × 0.3 mm thickness × 1000 mm length)	CH$_4$ conv.: < 1.0% Selectivity: coke and trace of C$_2$ H$_2$ removal: N/A	1989 (Andersen et al. 1989)
Pd-coated alloy membrane	Temperature: 773–873 K Pressure: ~ 1 atm Catalyst: 0.5% Ru–3% Mo/ZSM5	Tubular (2.38 mm outer diameter × 178 mm length, unknown thickness)	CH$_4$ conv.: 2.5% Selectivity: highly selective toward benzene H$_2$ removal: up to ~210 mL STP min^{-1} H$_2$ permeation rate	2001 (Rival et al. 2001)
Pd–Ag alloy membrane	Temperature: 858 K Pressure: ~ 101 kPa Catalyst: Re/ZSM5 and Mo/ZSM5	Tubular (3.175 mm outer diameter × 177.8 mm length, unknown thickness)	CH$_4$ conv.: 7.5% Selectivity: > 95% aromatics H$_2$ removal: 75% removed	2001 (Wang et al. 2001)
Pd–Ag alloy membrane	Temperature: 873–973 K Pressure: ~ 101 kPa Catalyst: 0.5% Ru–3% Mo/ZSM5	Tubular (9.5 mm outer diameter × 70 mm length, unknown thickness)	CH$_4$ conv.: ~17% Selectivity: increased coke H$_2$ removal: N/A	2002 (Larachi et al. 2002)
Pd-coated Ta/Nb alloy membrane	Temperature: 873 K Pressure: ~ 101 kPa Catalyst: 0.5% Ru–3% Mo/ZSM5	Tubular (2.38 mm outer diameter × 178 mm length, unknown thickness)	CH$_4$ conv.: < 12% Selectivity: increased coke, highly selective toward benzene H$_2$ removal: up to ~200 mL STP min^{-1} H$_2$ permeation rate	2002 (Iliuta et al. 2002)

(Continued)

TABLE 11.1 (CONTINUED)

Overview of the Catalyst and Pd-Based H_2-Permeable Membrane Reactor Types Used for DNMC Reactions in Literature (Oh, Sakbodin, and Liu 2019)

Membrane type	Catalyst and reaction condition	Membrane geometry/dimension	Performance	Year/ Reference
Pd–Ag alloy membrane on porous stainless-steel support	Temperature: up to 973 K Pressure: ~ 101 kPa Catalyst: 0.5% Ru–3% Mo/ZSM5	Tubular (9.5 mm outer diameter × 0.005 mm thickness × 70 mm length)	CH_4 conv.: ~17% Selectivity: increased coke, highly selective toward benzene H_2 removal: up to ~0.0016 m^3 STP m^{-2} s^{-1} H_2 permeation rate	2003 (Iliuta, Grandjean, and Larachi 2003b)
Pd membrane	Temperature: 883 K Pressure: ~ 101 kPa Catalyst: 3 wt% Mo/ZSM5	Disk (20 μm thickness)	CH_4 conv.: 5–8% Selectivity: increased by 2–10 times for aromatics H_2 removal: 50–60%	2003 (Kinage, Ohnishi, and Ichikawa 2003)
Pd tubular membrane	Temperature: 973 K Pressure: ~ 101 kPa Catalyst: 4 wt% Mo/ZSM5	Tubular (6.6 mm outer diameter × 0.125 mm thickness × 46.4 mm length)	CH_4 conv.: ~10% Selectivity: ~80% benzene H_2 removal: 6.8–48.3% H_2 recovery	2015 (Natesakhawat et al. 2015)

Reproduced with permission from Oh, Sakbodin, and Liu 2019. Copyright 2019 The Royal Society of Chemistry limitations of DNMC has been an ongoing research topic. Despite different membrane configurations and metal/zeolite catalyst formulations across numerous studies, three major conclusions were drawn: (i) methane conversion and C_{2+} yield were improved with the use of a Pd-based membrane reactor compared to those of a fixed-bed reactor (FR) without in situ hydrogen removal; (ii) up to ~75% of hydrogen generated from the DNMC reaction was selectively removed by the Pd-based membranes; and (iii) hydrogen flux across the Pd-based membranes was significantly reduced due to carbon deposition on the membrane surfaces.

Soon after demonstrating the ability of Pd-based membrane reactors to remove H_2 in the DNMC reaction (Andersen et al. 1989), research on Pd-based membrane reactors in DNMC shifted gears to focus on understanding the mechanism and process control of DNMC in membrane reactors to optimize performances. In 2001, Larachi et al. examined the types of coke formed on a 3 wt% Mo-0.5wt% Ru/ZSM5 catalyst in a Pd–Ag alloy membrane reactor at a temperature range of 873 K to 973 K (Larachi et al. 2002). Three types of carbon, namely carbidic, graphite-like, and aromatic-aliphatic species, were formed in the membrane reactor. The group also discovered that, under continuous hydrogen removal conditions, the H/C ratio of the carbon deposits in the membrane reactor was lower than that in the fixed-bed reactors (FR), but the chemical states of carbon and molybdenum on the catalyst remained the same. When exposed to a mixture of CH_4/H_2 for a short period of time, the catalyst was regenerated, and the coke-deactivated sites were recovered. Next, the same group placed the Pd–Ag membrane on porous stainless steel via electroless plating as a strategy improve the membrane permeability and mechanical stability (Larachi et al. 2002; Rival et al. 2001; Iliuta, Grandjean, and Larachi 2003a). The DNMC performance showed methane conversion well beyond the thermodynamic limitation at 973 K, but it was followed by a sharp decrease after a certain period of reaction time. The carbonaceous species that formed under low hydrogen pressure and high reaction temperature adversely affected the catalyst activity. Later on, Wang et al. studied the method of hydrogen removal on the permeate side of the membrane reactor by introducing inert gas flow to sweep away the permeated hydrogen (Wang et al. 2001). The group applied a vacuum pump to remove the permeated hydrogen and successfully obtained pure hydrogen from the DNMC reaction using this membrane reactor system.

Natesakhawat et al., on the other hand, examined the effects of weight hourly space velocity (WHSV) on DNMC in a Pd-based membrane reactor in order to maximize methane conversion, product selectivity, and hydrogen removal while reducing coke formation (Natesakhawat et al. 2015). The membrane reactor performance was examined as a function of WHSV, which ranged from 750 to 9000 mL g_{cat}^{-1} h^{-1} at 973 K and 101 KPa pressure over a 4wt% Mo/ZSM5 catalyst. The group concluded that methane conversion and aromatic product yield decreased with increasing WHSV. The hydrogen removal dropped from 48.3% to 6.8% when WHSV was high. The optimized catalyst performance and hydrogen recovery were obtained when an intermediate WHSV of 3000 mL g_{cat}^{-1} h^{-1} was used. Benzene yield enhancement as high as ~360% was also achieved. In addition, the hydrogen found in the retentate stream helped alleviate coke accumulation. However, the carbon deposited on the inner surface of the membrane reactor in contact with the catalyst bed. This had caused a decrease in hydrogen permeability over the time-on-stream (TOS) of 15 hours.

The major challenge when using Pd-based membrane reactors for the DNMC reaction is the rapid deactivation of the catalyst due to coke formation. Therefore, catalyst/membrane system regeneration is crucial in the commercialization of the process. Several groups have carried out studies on the regeneration of the catalyst/membrane systems. For example, Larachi et al. performed regeneration studies on the Pd–Ag alloy tubular membrane over Ru–Mo/HZSM5 catalyst in DNMC by switching back and forth between membrane and FR conditions (Larachi et al. 2002). CH_4/H_2 gas mixtures were also added to the reaction to facilitate regeneration of the system. The group concluded that the coke formation was reversible when switching from the membrane reactor to a FR, since more H_2 was present under FR conditions. The additional CH_4/H_2 regeneration step also helped reduce coke formation in the catalyst/membrane system as the H_2 gas assisted in the rehydrogenation and restoration of the active sties. Next, Natesakhawat et al. demonstrated that coke accumulation for Mo/ZSM5 catalysts coupled with a Pd tubular membrane reactor could be alleviated if the concentration of H_2 in the retentate stream was maximized (Natesakhawat et al. 2015). Similarly, Kinage et al. found that the addition of H_2 to the reaction using Mo/ZSM5 catalyst and a Pd membrane mitigated the blockage of the Pd membrane caused by carbon deposition (Kinage, Ohnishi, and Ichikawa 2003). Regeneration studies, including the effects of hydrogen sulfide (H_2S) (Tarditi et al. 2015a), CO (Li et al. 2007), and hydrocarbons on the Pd-based membranes (Jung et al. 2000) as well as the dehydrogenation of hydrocarbons via Pd-based membrane reactors (Sheintuch and Dessau 1996; Weyten et al. 2000), have also been conducted. The results showed that Pd-based membrane reactors can be regenerated by H_2 (Tarditi et al. 2015b; Sheintuch and Dessau 1996; Li et al. 2007) or O_2 (Li et al. 2007; Jung et al. 2000; Weyten et al. 2000) to completely restore the initial H_2 flux through the membrane. Nevertheless, a few studies found that the H_2 selectivity on Pd-based membranes decreased after a few cycles of regeneration treatment (Li et al. 2007).

11.2.1.2 DNMC in Ceramic Membrane Reactors

Dense MIEC ceramic membranes are promising candidates for hydrogen removal in DNMC given their thermal and chemical stabilities at high temperature conditions. The representative MIEC ceramic materials are partially substituted perovskite-type oxides such as calcium zirconate ($CaZrO_3$), strontium cerate ($SrCeO_3$), and barium cerate ($BaCeO_3$) (Guan 1998). The hydrogen transport mechanism in MIEC consists of the following steps: (i) dissociation of hydrogen and ionization of hydrogen to from hydroxide defects (proton defects) and electrons; (ii) migration of hydroxide defects and electrons through the membrane; (iii) reduction of defects to hydrogen molecules with electrons at the permeate side; and (iv) desorption of hydrogen molecules from the membrane surface. According to the Wagner equation,

$$J_{OH_O^\bullet} = -\frac{1}{L}\left[\frac{RT}{2F^2}\int_{P_{H_2}^r}^{P_{H_2}^f}\sigma_{amb}d\ln P_{H_2}\right]. \tag{11.2}$$

the hydrogen permeation ($J_{OH_0^-}$ represents the flux hydroxide ions through the membrane) is affected by four different parameters: hydrogen partial pressure gradient (P'_{H_2} and P''_{H_2} represent the partial pressure on both sides of the membrane), membrane thickness (L), ambipolar conductivity of the membrane material (σ_{amb}), and operating temperature (T). R and F symbolize the gas constant and the Faraday constant in the equation, respectively. The major advantage of the dense MIEC membrane is that it has nearly 100% selectivity for hydrogen permeation. However, to be suitable for hydrogen separation, the membrane materials must possess both ionic and electronic conductivity in order to maintain electroneutrality. Furthermore, the ionic and electronic transference numbers should also be comparable to maximize hydrogen permeation.

Table 11.2 summarizes the MIEC ceramic membrane reactors developed for in situ hydrogen removal to improve methane conversion in DNMC. The first exploration into the use of MIEC for methane activation was reported by Hamakawa et al. in 1994 (Hamakawa, Hibino, and Iwahara 1994). More detailed simulations (Borry et al. 1998) and (Borry et al. 1998; Liu, Li, and Iglesia 2002) experimental studies were performed in the 2000s by Iglesia's group using a MIEC membrane in combination with a Mo/ZSM5 catalyst. Despite the promising simulation results, the experimental results using a dense $SrZr_{0.95}Y_{0.05}O_3$ membrane disk with dimensions of 16 mm dimeter and 1 mm thickness demonstrated unfavourable performance due to insufficient hydrogen flux and rapid catalyst deactivation under hydrogen removal conditions (Borry et al. 1998). The low surface-to-volume ratio of the disk-shaped membranes resulted in insufficient effective surface area, thus reducing hydrogen flux through the membranes. Therefore, Liu et al. fabricated a significantly thinner $SrZr_{0.95}Y_{0.05}O_3$ membrane of ~2 μm in thickness supported on a 1 mm thick porous support for the DNMC reaction (Liu, Li, and Iglesia 2002). Methane conversion slightly increased from 950 K to 993 K due to a higher hydrogen removal efficiency (16% removal) compared to that at 950 K (< 7% removal). Nevertheless, the higher rate of hydrogen removal also led to a faster catalyst deactivation. The addition of small amounts of CO_2 has shown a reduction in the deactivation rate and prolonged the lifetime of the catalyst, but the advancement of MIEC membrane reactors for DNMC is still limited by low hydrogen flux. Therefore, design and fabrication of more efficient membranes as well as coke resistant catalysts for the DNMC reaction are necessary.

In recent years, Xue et al. designed a MIEC hollow fiber membrane reactor for the DNMC reaction over a 6wt% Mo/ZSM5 catalyst (Xue et al. 2016). The innovative "U"-shaped $La_{5.5}W_{0.6}Mo_{0.4}O_{11.25-\delta}$ (LWM0.4) hollow fiber membrane reactor resulted in a higher surface-to-volume ratio compared to that of the disk and tubular membrane configurations. The methane conversion in the hollow fiber membrane reactor with constant hydrogen removal was higher than that in the FR without hydrogen removal. Furthermore, the methane conversion in the membrane reactor exceeds the thermodynamic value of 12% at 973 K. As for product yield, an aromatic hydrocarbon yield of ~4.7% was achieved in the membrane reactor, which is higher than that in the FR. However, the differences between the membrane and FR become less significant with time-on-stream (TOS) due to accelerated catalyst deactivation under hydrogen removal conditions.

TABLE 11.2

Overview of H_2-Permeable Ceramic Membrane Reactors Used for DNMC Reactions in Literature

Membrane type	Catalyst and reaction condition	Membrane geometry/dimension	Performance	Year/Reference
$SrCe_{0.95}Yb_{0.05}O_{3-\delta}$	Temperature: 1173 K Pressure: ~ 101 kPa Catalyst: Ag	Disk (12 mm diameter × 1 mm thickness)	CH_4 conv.: < 1% Selectivity: observed C_2 product, no analysis for other products. H_2 removal: N/A	1994 (Hamakawa, Hibino, and Iwahara 1994)
$SrZr_{0.95}Y_{0.05}O_3$	Temperature: 950 K Pressure: ~ 101 kPa Catalyst: 4 wt% Mo/ZSM5	Disk (16 mm diameter × 1 mm thickness)	CH_4 conv.: 10–12% Selectivity: > 90% C_{2+} H_2 removal: < 5%	1998 (Borry et al. 1998)
$SrCe_{0.95}Yb_{0.05}O_{3\delta}$	Temperature: 950–1000 K Pressure: ~ 101 kPa Catalyst: 4 wt% Mo/ZSM5 Co-feed: CO_2	Disk (2 μm thickness on 1 mm thickness porous substrate, unknown diameter)	CH_4 conv.: < 13.3% Selectivity: ~73% C_2–C_{10} and 6% C_{12+} H_2 removal: 7–16%	2002 (Liu, Li, and Iglesia 2002)
$SrCe_{0.95}Yb_{0.05}O_{3-\delta}$	Temperature: 1023–1273 K Pressure: ~ 101 kPa Catalyst: No catalyst	Hollow fiber (1.45 mm outer diameter × 0.65 mm inner diameter × 260 mm length)	CH_4 conv.: ~80% Yield: ~13.4% C_2 H_2 removal: N/A	2006 (Liu, Tan, and Li 2006)
$SrCe_{0.7}Zr_{0.2}Eu_{0.1}O_{3-\delta}$	Temperature: > 1223 K Pressure: ~ 101 kPa Catalyst: Fe/SiO_2	Tubular (6 mm outer diameter × 0.02 mm thickness × 150 mm length)	CH_4 conv.: > 30% Selectivity: ~99% C_2 and aromatics H_2 removal: 12–22%	2016 (Sakbodin et al. 2016a)
$BaZr_{0.7}Ce_{0.2}Y_{0.1}O_{3-\delta}$ *(external circuit)*	Temperature: 983 K Pressure: ~ 101 kPa Catalyst: 6 wt% Mo/MCM22	Tubular (10 mm outer diameter × 7 mm inner diameter × 250–300 mm length × 0.025–0.030 mm electrolyte thickness)	CH_4 conv.: < 12% Selectivity: > 80% aromatics H_2 removal: varies depending on applied current density	2016 (Morejudo et al. 2016a)
$La_{5.5}W_{0.6}Mo_{0.4}O_{11.25-\delta}$	Temperature: 973 K Pressure: ~ 101 kPa Catalyst: 6 wt % Mo/ZSM5	Hollow fiber U-shape (~0.2 mm thickness)	CH_4 conv.: ~10% Selectivity: 30–40% aromatics, 60–70% coke H_2 removal: 40–60%	2016 (Xue et al. 2016)

Reproduced with permission from Oh, Sakbodin, and Liu 2019. Copyright 2019 The Royal Society of Chemistry

The development of a high temperature and coke resistant catalyst (i.e. Fe/SiO_2) has provided new research opportunities for the MIEC membranes used in DNMC reactions. For example, Sakbodin et al. has shown the improved DNMC performance over this catalyst in a H_2-permeable tubular membrane reactor consisting of a $SrCe_{0.7}Zr_{0.2}Eu_{0.1}O_{3-\delta}$ thin-film (~20 μm) supported on the outer surface of a porous $SrCe_{0.8}Zr_{0.2}O_{3-\delta}$ tube (Sakbodin et al. 2016b) capped on one end. The reaction was performed at high temperature (1223 K to 1323 K) under constant hydrogen removal conditions; appreciable improvement in methane conversion was achieved using this reactor configuration. Figure 11.4 compares the methane conversion and product selectivity of the DNMC reaction between a fixed-bed reactor and a membrane reactor at different temperatures. Higher methane conversion is observed in the membrane reactor relative to that in the FR at all tested temperatures. The selectivity shifts toward larger products as temperature increases in both the FR and membrane reactors. Higher C_{2+} yield was also obtained at all tested temperatures when the membrane reactor was used. The effects of sweep gas flow rates on the conversion and selectivity in DNMC were also studied, since hydrogen partial pressure across the membrane reactor impacts the hydrogen permeation. Higher methane conversion (e.g. two times increase in conversion compared to that in the FR) was achieved when the sweep gas flow rate was increased. Also, the membrane reactor showed a stable performance with increasing TOS of the reaction.

11.2.2 OXYGEN-PERMEABLE MEMBRANE REACTORS

The most widely accepted mechanism of methane activation in DNMC literature (for metal/zeolite catalyst) is a bifunctional pathway requiring participation of both metal sites and Brønsted acid sites. Methane is activated on metal sites, releasing hydrogen

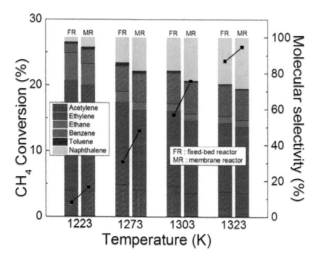

FIGURE 11.4 CH_4 conversion and product selectivity over Fe/SiO_2 catalysts in a fixed-bed reactor (FR) and packed-bed membrane reactor (MR) at different temperatures in the DNMC reaction. (Adapted with permission from Sakbodin et al. 2016b. Copyright 2016 Wiley.)

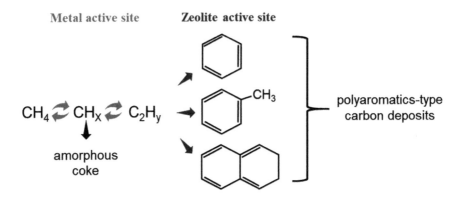

FIGURE 11.5 The most accepted pathway for methane activation over a metal/zeolite catalyst to form hydrocarbon products and coke in DNMC.

into the gas phase and forming surface CH_x species. Then, products of their dimerization into C_2H_y undergo oligomerization on the zeolite Brønsted acid sites to form benzene and other aromatics. The continuous carbon chain growth leads to the formation of polynuclear aromatics and eventually coke, which results in catalyst deactivation in the DNMC reaction. Figure 11.5 shows that coke can also form from the direct dehydrogenation of CH_x species in parallel with C–C bond formation steps. The addition of an oxidant, such as CO, CO_2, and oxygen (O_2), in the methane feed has been practiced in the FR to slow down the deactivation process in DNMC (Liu, Nutt, and Iglesia 2002). In the membrane reactor operation, the oxidant addition via an oxygen-permeable membrane has also been considered for mitigating coke formation. Figure 11.6 shows the working scheme of this type of membrane reactor in DNMC reactions.

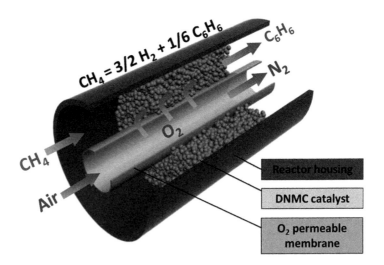

FIGURE 11.6 Schematics of an oxygen-permeable membrane reactor for methane upgrading to benzene and hydrogen in DNMC reactions. Oxidant is added in situ to mitigate catalyst deactivation.

In practice, Cao et al. (2013) developed the oxygen-permeable membrane reactor to suppress coke formation in the DNMC reaction in 2013. Their pellet-shaped membrane reactor consists of an asymmetric oxygen-transporting $Ba_{0.5}Sr_{0.5}Co_{0.8}Fe_{0.2}O_{3-\delta}$ (BSCF) perovskite membrane on a porous BSCF support. The membrane was placed beneath the catalyst at the end of a tubular FR. Methane was fed to the reactor via a smaller tube, which was inserted inside the catalyst bed. The oxygen permeated through the membrane from the sweep side to the reaction side. The equilibrium of the DNMC reaction shifted to the product side as the permeated oxygen oxidized the hydrogen. Then, the newly formed stream reacted with deposited coke on the catalyst to create the final products, CO and CO_2. This in turn reduced the formation of coke and prolonged the lifetime of the catalyst. Their study also showed that the distribution of oxygen through this type of MR could enhance the catalyst stability and selectivity toward aromatics. The key concept lies in the uniform and consistent distribution of oxygen throughout the reactor. The side reaction caused by partial or total oxidation of methane could potentially decrease the selectivity and yield of desired products if the oxygen is distributed unevenly along the reactor. Furthermore, the CH_4 conversion could be hindered by inhibiting the formation of the catalyst site (Mo_2C) in the Mo/ZSM5 catalyst.

In both FRs and oxygen-permeable MRs, the conversion of CH_4 and selectivity to aromatics decreased dramatically, but the oxygen-permeable MR showed generally higher CH_4 conversion and aromatics selectivity over a 15-hour reaction period. After a 6-hour induction period, the aromatics selectivity in the MR was able to reach a plateau and stabilize, but that of the FR continued declining. Further study of co-feeding oxygen in both FRs and MRs has proven that exceeding the critical O_2/CH_4 molar ratio causes the oxidative destruction and ultimate deactivation of the active site (Mo_2C). Additionally, the catalytic performance has shown significant improvement in the MR from upshifting the critical ratio from 0.02 to 0.04 at a given reaction temperature of 1023 K.

11.2.3 MEMBRANE REACTOR COUPLED WITH EXTERNAL CIRCUIT

Although the MIEC ceramic membrane has shown improved performance in DNMC for methane upgrading, the relatively lower ionic and/or electronic conductivity of the membrane materials limits the capability of H_2 removal from the reactor and thus further improvement in methane conversion. Alternatively, ion-conducting membranes were used to allow for the selective removal of H_2 driven by external current to overcome the thermodynamic limitations. Figure 11.7 shows the conceptual reactor design of an external circuit-driven, gas-permeable MR for the DNMC reaction. The reactor is separated into a sweep side and a reaction side. The oxygen is transported from the sweep side to the reaction side while hydrogen is transferred in the opposite direction within the electrolyte membrane simultaneously. The advantage of this type of electrolyte MR over a hydrogen/oxygen-permeable MR lies in the efficient gas removal or addition driven by the external current.

The application of external current on the ion-conducting membrane for methane activation can be traced back to 1990s. Stoukides et al. first studied the DNMC reaction on an ionic conducting electrolyte (yttria-doped strontium-ceria (YSC)) in

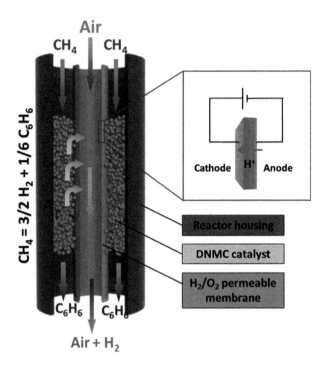

FIGURE 11.7 Schematics of a gas-permeable, ion-conducting MR driven by an external circuit for methane upgrading to benzene and hydrogen in DNMC reactions.

combination with Ag electrodes under the application of an external current (Chiang, Eng, and Stoukides 1991). The entire Ag/YSC/Ag assembly was an electrochemical cell for electrocatalytic DNMC. The Ag metal served as both the anode for the electrolyte and catalyst for methane activation, while the solid YSC electrolyte acted as a proton conductor. By applying a current to the cell, protons were pumped to or from the Ag electrodes. Under a closed circuit, the rate of DNMC increased up to eight times higher than the open-circuit rate. This study, however, only focused on the promoting effects of external current on methane activation.

In 2016, an innovative design of a co-ionic catalytic membrane reactor (CMR) together with an external circuit by Morejudo et al. solved the issues of both thermodynamic limitations and coke formation in DNMC. The MR was made from a 25-micron-thick $BaZr_{0.7}Ce_{0.2}Y_{0.1}O_{3-\delta}$ (BZCY72) electrolyte film on a porous BZCY72-Ni support with a layer of Cu-based anode applied on the electrolyte film on the opposite side of the film (Morejudo et al. 2016b). This special "sandwich" film design allowed for the electrochemical oxidation of H_2 into protons while preventing secondary conversion of hydrocarbons into coke by co-permeation of O_2. The ability to transfer hydrogen and oxygen across a membrane has a direct correlation to the applied current density. An immediate application of current will initiate the H_2 permeation process, which leads to higher aromatic yields (Figure 11.8a). The DNMC reaction in this ionic MR was catalyzed by 6wt% Mo/MCM-22 (Mobil Composition of Matter number 22) catalyst at approximately 973 K. By comparing

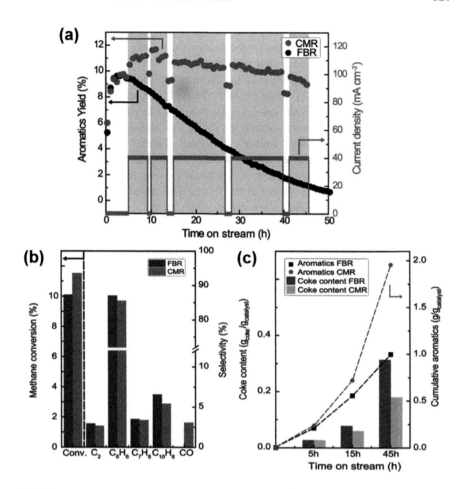

FIGURE 11.8 FR and CMR performance in DNMC using 6Mo/MCM-22 catalyst. (a) Aromatics yield with time-on-stream. Gray regions are hydrogen extraction period. (b) Methane conversion and product selectivity after 5 hrs (FR) and 8 hrs (CMR). (c) Cumulative aromatics and coke content selectivity in FR and CMR for DNMC reaction at 5 h, 15 h and 45 h. Reaction Conditions: 710°C, 1500 ml/g/h, 1 bar, and current density of 40 mA/cm². (Reproduced from Morejudo et al. 2016b with permission from *Science*.)

with the conventional FR design, the CMR offered a higher methane conversion and similar product selectivity (Figure 11.8b). Furthermore, the rate of coke formation in the CMR was significantly lower than that in FR at 45 h reaction time (Figure 11.8c).

11.3 CATALYTIC WALL REACTOR FOR DIRECT NONOXIDATIVE METHANE CONVERSION

Among all available methane conversion pathways, DNMC is evidenced as the most endothermic, requiring high temperatures to reach decent conversion and product yields. The high endothermicity indicates a high energy input for DNMC and thus creates large temperature gradients in either the reactor or the catalyst interior. In DNMC literature, studies have mainly focused on the development of catalyst materials to enhance methane conversion and catalyst durability. The design and operation of reactors can also contribute to the thermodynamic and kinetic controls of DNMC reactions to produce desired products. The millisecond catalytic wall reactor (Oh et al. 2019) made of DNMC catalyst on a support tube is an example of enhancing DNMC performance via innovation in reactor engineering.

The principle of the catalytic wall reactor for autothermal operation is to provide an improved method of heat exchange between the exothermic and endothermic reactions. By coupling the two reactions, it is possible to use recovery heat in order to operate desired reaction at autothermal conditions. The operation of a reaction with a residence time of approximately a few milliseconds means a very short contact time between the reactant and catalyst. It is a viable solution for chemical syntheses from alkanes, such as oxidative dehydrogenation of hydrocarbons for olefin or hydrogen production, (Venkataraman, Redenius, and Schmidt 2002; Kaminsky et al. 1997) because thermodynamically undesirable products with less or no carbon formation are achievable. In DNMC reactions, high selectivity to C_{2+} hydrocarbons is limited to the kinetically controlled regime since thermodynamics dictates the formation of carbon and hydrogen in DNMC (refer to Figure 11.1). Therefore, the millisecond catalytic wall reactor could run DNMC autothermally under kinetically controlled regions to produce C_{2+} hydrocarbons and protect the reaction system from deactivation via coking. This in turns lower the energy and capital costs.

11.3.1 CATALYTIC WALL REACTOR FOR AUTOTHERMAL OPERATION

Numerous groups have attempted to design catalytic wall reactors for coupling exothermic and endothermic reactions. The initial attempt to design a catalytic wall reactor dates back to the 1980s by Hunter et al., who combined a catalytic plate reactor for heating air and exchanging heat between the two reactions (Hunter and McGuire 1980). Later on, Arntz's group carried out detailed simulation and experimental studies on a catalytic wall reactor by coupling methane combustion with methane steam reforming endothermic reactions (Frauhammer et al. 1999). Schmidt's group combined exothermic methane catalytic combustion and endothermic reactions such as ethane cracking and ethanol steam reforming in a catalytic wall reactor (Venkataraman, Redenius, and Schmidt 2002). Not only were the heat transfer boundary layers was eliminated, > 99% ethanol conversion was achieved when the steam/carbon ratio of 3/1 was used. The reactor demonstrated some excellent characteristics: elimination of thermal boundary layers, high conversion rate, and stable reactor performance. Later on, the same group performed the simulation for the partial oxidation of methane to syngas in noble-metal (Rh and Pt)-coated monoliths (Wanat, Venkataraman, and Schmidt 2004). This simulation demonstrated that the reaction under higher pressure was dominated by gas-phase chemistry instead of surface chemistry. More recently, Bordes–Richard studied the oxidative dehydrogenation of propane and Fischer–Tropsch synthesis of clean fuels in a metallic plate-like catalytic wall reactor coated with a thin layer of VO_x/TiO_2 and Co/SiO_2 catalysts (Giornelli et al. 2007). Even though catalytic wall reactors have been studied for various catalytic reactions, it has never been attempted for the DNMC reaction.

11.3.2 DIRECT NONOXIDATIVE METHANE CONVERSION IN CATALYTIC WALL REACTOR

Recently, Oh et al. demonstrated the performance of the DNMC reaction in a millisecond catalytic wall reactor comprised of Fe/SiO_2 catalyst coated on the wall of the reactor with the exothermic coke combustion reaction and the endothermic catalytic DNMC reaction operating in autothermal conditions (Oh et al. 2019). The catalytic wall reactor not only manipulated methane conversion and product selectivity but also regulated coke formation, which is deemed detrimental to the reaction. Furthermore, the autothermal operation of the millisecond catalytic wall reactor was able to overcome the highly endothermic nature of the DNMC reaction by manipulating the heat supply and recovery during the reaction. In this work, stable methane conversion, C_{2+} selectivity, coke yield and long-term durability were achieved by using the millisecond catalytic wall reactor.

The concept of DNMC in a millisecond catalytic wall reactor reported by Oh et al. originated from the need for the critically small surface area of Fe/SiO_2 catalyst used in the DNMC reaction in a FR. The requirement of small amounts of Fe/SiO_2 catalyst for DNMC suggests the potential to develop millisecond catalytic wall reactors comprised of a catalyst coating layer on the reactor wall that offers equivalent catalyst surface area to that in the FR. The reactant flows through the reactor channel, and the reaction initiates on the wall surface, occurring on both the reactor wall and in the reactor channel due to gas-phase reactions. As shown in Figure 11.9, the catalytic wall reactor configuration consists of the smaller diameter catalytic wall reactor enclosed in a larger diameter housing tube, which operates autothermally by coupling and periodical swapping of endothermic DNMC (methane feed) and exothermic oxidative coke removal (air feed) on opposite sides of the reactor.

By comparing to the catalytic wall reactor performance in DNMC with that of the noncatalytic quartz reactor, higher CH_4 conversion in the catalytic wall reactor confirmed that the Fe/SiO_2 catalyst was successfully incorporated into the quartz tube wall. The catalytic wall reactor also showed stable performance for up to 50 hours with constant ~ 11.3% methane conversion and selectivities of 30.3% C_2, 21.2% benzene, 6.6% toluene, and 32.4% naphthalene, and the total selectivity toward these products was maintained at >91% with ~ 9.5% of coke selectivity (Figure 11.10). With this DNMC catalytic wall reactor design, the methane conversion was able to reach 33.9% at 1323 K with C_{2+} yield as high as 25.4% at 20 mL min^{-1} CH_4 feed flow rates.

The technological feasibility of the fabricated autothermal catalytic wall reactor was further proved by energy balance analysis. The DNMC reaction in a catalytic wall reactor sufficiently aligned with a recent agreement framework analysis, which suggested that economically feasible DNMC is achievable at > 25% methane conversion, < 20% coke formation, and low catalyst cost (Huang et al. 2018). The economic calculations exhibited a reduction in supplied energy requirements and cost by over six times with the incorporation of heat integration simulating the

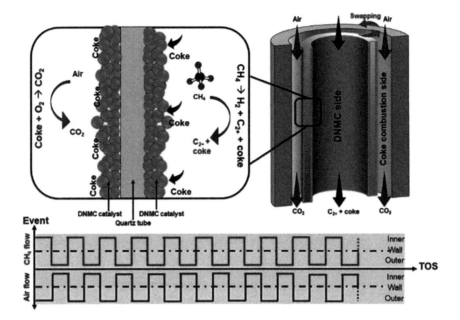

FIGURE 11.9 Schematic of DNMC and coke combustion on opposite sides of a catalytic wall reactor for autothermal operation in the process simulation. (Reproduced from Oh et al. 2019 with permission from Wiley.)

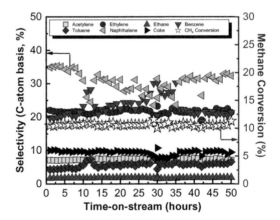

FIGURE 11.10 Long-term stability test of DNMC reaction in catalytic wall reactor. (1273 K, 20 mL min^{-1} gas flow rate, CH$_4$:N$_2$ = 9:1 (N$_2$: internal standard), 1 atm pressure, 0.075wt% Fe in Fe/SiO$_2$, 0.375 g catalyst.). (Reproduced from Oh et al. 2019 with permission from Wiley.)

autothermal process. The DNMC reaction produces multiple industrially-valuable chemicals and fuels – hydrogen, ethylene, and benzene – whose production rates are converted into retrievable prices. Since the cost of the methane feedstock is low and energy costs are significantly reduced using heat integration, the feasibility of netting a profit is reasonable. In combination with the low cost of a Fe/SiO$_2$ catalyst, the DNMC reaction in a catalytic wall reactor is an economically feasible and transformative technology for shifting the petrochemical sector to a natural gas feedstock in industry.

11.4 SUMMARY

The DNMC pathway is promising to produce C_{2+} hydrocarbons from abundant and low-cost natural gas resources in a one-step operation. The process is significantly simpler than the state-of-the-art syngas followed by FT synthesis, suggesting a great potential to lower investment costs in natural gas utilization. The major challenges in DNMC are low methane equilibrium conversion, fast catalyst coking, and high reaction endothermicity. Design and development of innovative multifunctional reactors, in combination with advancements in DNMC catalyst technology, could provide viable solutions to concurrently circumvent these challenges. Membrane reactors, coupling DNMC catalysts with hydrogen removal and/or oxygen addition membranes in one unit, can achieve enhancement in methane conversion and C_{2+} yield as well as mitigating catalyst deactivation. The further advancement of membrane reactors for DNMC, however, is constrained by lack of active/selective catalysts and membranes with high hydrogen/oxygen flux under high reaction temperature conditions.

The catalytic wall reactor is an emerging technology for DNMC. This technology benefits from the robust Fe/SiO_2 catalyst and the unique mixture of heterogeneous surface and homogenous gas-phase reactions at the medium-high temperature regimes. In the millisecond catalytic wall reactor, a thin layer of catalyst is deposited directly onto both sides of the reactor. The millisecond catalytic wall reactor flows reactant through the reactor channel, and the reaction occurs on the wall and in the gas phase. An exothermic reaction takes place on one side of the reactor and provides heat to the endothermic reaction occurring catalytically on the other side. Such a configuration permits the reactions to be run autothermally and adiabatically with a residence time of approximately a few milliseconds. Such short contact time is viable for synthesis of C_{2+} hydrocarbons in DNMC. The millisecond catalytic wall reactor, comprised of a methane activation catalyst and autothermal functionality is a potential viable solution to overcome the challenges in the DNMC process, especially the high endothermicity of the reaction. The catalytic wall reactor is also cost efficient since it requires relatively small amounts of catalyst but assures high throughput. Overall, membrane and catalytic wall reactors offer a unique reactor design opportunity to improve methane conversion, product selectivity and yield in the DNMC pathway. Therefore, the design and development of innovative multifunctional reactor systems could be the future of DNMC technology.

REFERENCES

Abdel-Aal, H. K., and M. A. Alsahlawi. 2013. *Petroleum Economics and Engineering*. CRC Press, Boca Raton, FL.

Alvarez-Galvan, M. C., Mota, N., Ojeda, M., Rojas, S., Navarro, R. M., and J. L. G. Fierro. 2011. Direct Methane Conversion Routes to Chemicals and Fuels. *Catalysis Today* 171 (1):15–23.

Andersen, A., Dahl, I. M., Jens, K. J., Rytter, E., Slagtern, A., and A. Solbakken. 1989. Hydrogen Acceptor and Membrane Concepts for Direct Methane Conversion. *Catalysis Today* 4 (3–4):389–397.

Bao, B., El-Halwagi, M. M., and N. O. Elbashir. 2010. Simulation, Integration, and Economic Analysis of Gas-to-Liquid Processes. *Fuel Processing Technology* 91 (7):703–713.

Barelli, L., Bidini, G., Gallorini, F., and S. Servili. 2008. Hydrogen Production through Sorption-Enhanced Steam Methane Reforming and Membrane Technology: A Review. *Energy* 33 (4):554–570.

Biddy, M. J., Davis, R., Humbird, D., Tao, L., Dowe, N., Guarnieri, M. T., Linger, J. G., Karp, E. M., Salvachúa, D., and D. R. Vardon. 2016. The Techno-Economic Basis for Coproduct Manufacturing to Enable Hydrocarbon Fuel Production from Lignocellulosic Biomass. *ACS Sustainable Chemistry & Engineering* 4 (6):3196–3211.

Boden, T. A., Gregg Marland, G., and R. J. Andres. 2009. *Global, Regional, and National Fossil-Fuel CO_2 Emissions*. Carbon Dioxide Information Analysis Center, Oak Ridge National Laboratory, US Department of Energy, Oak Ridge, TN, USA.

Borry, III, R. W., Lu, E. C., Kim, Y. H., and E. Iglesia. 1998. Non-oxidative Catalytic Conversion Of Methane With Continuous Hydrogen Removal. *Studies in Surface Science and Catalysis* 119:403–410.

Boyer, C., Clark, B., Jochen, V., Lewis, R., and C. K. Miller. 2011. Shale Gas: A Global Resource. *Oilfield Review* 23 (3):28–39.

Bryden, K. J., and J. Y. Ying. 2002. Nanostructured Palladium–Iron Membranes for Hydrogen Separation and Membrane Hydrogenation Reactions. *Journal of Membrane Science* 203 (1–2):29–42.

Cao, Z., Jiang, H., Luo, H., Baumann, S., Meulenberg, W. A., Assmann, J., Mleczko, L., Liu, Y., and J. Caro. 2013. Natural Gas to Fuels and Chemicals: Improved Methane Aromatization in an Oxygen-Permeable Membrane Reactor. *Angewandte Chemie International Edition* 52 (51):13794–13797. doi:10.1002/anie.201307935.

Cheng, Y. S., and K. L. Yeung. 1999. Palladium–Silver Composite Membranes By Electroless Plating Technique. *Journal of Membrane Science* 158 (1–2):127–141.

Chiang, P. H., Eng, D., and M. Stoukides. 1991. Electrocatalytic Methane Dimerization with a Yb-Doped $SrCeO_3$ Solid Electrolyte. *Journal of the Electrochemical Society* 138 (6):L11–L12.

Choudhary, T. V., Aksoylu, E., and D. W. Goodman. 2003. Nonoxidative Activation of Methane. *Catalysis Reviews* 45 (1):151–203.

Chu, S., and A. Majumdar. 2012. Opportunities and Challenges for a Sustainable Energy Future. *Nature* 488 (7411):294.

Collins, J. P., Schwartz, R. W., Sehgal, R., Ward, T. L., Brinker, C. J., Hagen, G. P., and C. A. Udovich. 1996. Catalytic Dehydrogenation of Propane in Hydrogen Permselective Membrane Reactors. *Industrial & Engineering Chemistry Research* 35 (12):4398–4405.

Fiorentino, G., Zucaro, A., and S. Ulgiati. 2019. Towards an Energy Efficient Chemistry. Switching from Fossil to Bio-Based Products in a Life Cycle Perspective. *Energy* 170:720–729. doi:10.1016/j.energy.2018.12.206.

Fort, D., Farr, J. P. G., and I. R. Harris. 1975. A Comparison of Palladium-Silver and Palladium-Yttrium Alloys as Hydrogen Separation Membranes. *Journal of the Less Common Metals* 39 (2):293–308. doi:10.1016/0022-5088(75)90204-0.

Frauhammer, J., Eigenberger, G., Hippel, L. V., and D. Arntz. 1999. A New Reactor Concept for Endothermic High-Temperature Reactions. *Chemical Engineering Science* 54 (15–16):3661–3670.

Gallucci, F., Fernandez, E., Corengia, P., and M. V. S. Annaland. 2013. Recent Advances on Membranes and Membrane Reactors for Hydrogen Production. *Chemical Engineering Science* 92:40–66.

Giornelli, T., Löfberg, A., Guillou, L., Paul, S., Courtois, V. L., and E. Bordes-Richard. 2007. Catalytic Wall Reactor: Catalytic Coatings of Stainless Steel By Vo_x/Tio_2 and Co/Sio_2 Catalysts. *Catalysis Today* 128 (3–4):201–207.

Guan, J. 1998. *Development of Mixed-Conducting Ceramic Membranes for Hydrogen Separation*. Argonne National Lab., IL, US.

Guo, X., Fang, G., Li, G., Ma, H., Fan, H., Yu, L., Ma, C., Wu, X., Deng, D., and M. Wei. 2014. Direct, Nonoxidative Conversion of Methane to Ethylene, Aromatics, and Hydrogen. *Science* 344 (6184):616–619.

Hamakawa, S., Hibino, T., and H. Iwahara. 1994. Electrochemical Hydrogen Permeation in a Proton-Hole Mixed Conductor and Its Application to a Membrane Reactor. *Journal of the Electrochemical Society* 141 (7):1720–1725.

Holmen, A., Olsvik, O., and O. A. Rokstad. 1995. Pyrolysis of Natural Gas: Chemistry and Process Concepts. *Fuel Processing Technology* 42 (2–3):249–267.

Holmen, A., Rokstad, O. A., and A. Solbakken. 1976. High-Temperature Pyrolysis Of Hydrocarbons. 1. Methane to Acetylene. *Industrial & Engineering Chemistry Process Design and Development* 15 (3):439–444.

Hong, B. D., and E. R. Slatick. 1994. Carbon Dioxide Emission Factors for Coal. *Quarterly Coal Report* 7:1–8.

Horn, R., and R. Schlögl. 2015. Methane Activation by Heterogeneous Catalysis. *Catalysis Letters* 145 (1):23–39.

Hu, W., Bao, J., and B. Hu. 2013. Trend and Progress in Global Oil and Gas Exploration. *Petroleum Exploration and Development* 40 (4):439–443.

Hunter, J. B., and G. McGuire. 1980. Method and Apparatus for Catalytic Heat Exchange. *US Patents US4214867A*, filed July 1978.

Iliuta, M. C., Grandjean, B. P. A., and F. Larachi. 2003a. Methane Nonoxidative Aromatization over Ru– Mo/HZSM-5 At Temperatures up to 973 K in a Palladium– Silver/Stainless Steel Membrane Reactor. *Industrial & Engineering Chemistry Research* 42 (2): 323–330.

Iliuta, M. C., Grandjean, B. P. A., and F. Larachi. 2003b. Methane Nonoxidative Aromatization over Ru–Mo/HZSM-5 at Temperatures up to 973 K in a Palladium–Silver/Stainless Steel Membrane Reactor. *Industrial & Engineering Chemistry Research* 42 (2):323–330. doi:10.1021/ie020486n.

Iliuta, M. C., Larachi, F., Grandjean, B. P. A., Iliuta, I., and A. Sayari. 2002. Methane Nonoxidative Aromatization over Ru– Mo/HZSM-5 in a Membrane Catalytic Reactor. *Industrial & Engineering Chemistry Research* 41 (10):2371–2378.

Jun, C. S., and K. H. Lee. 2000. Palladium and Palladium Alloy Composite Membranes Prepared by Metal-Organic Chemical Vapor Deposition Method (Cold-Wall). *Journal of Membrane Science* 176 (1):121–130. doi:10.1016/S0376-7388(00)00438-5.

Jung, S. H., Kusakabe, K., Morooka, S., and S. D. Kim. 2000. Effects of Co-Existing Hydrocarbons on Hydrogen Permeation through a Palladium Membrane. *Journal of Membrane Science* 170 (1):53–60.

Kaminsky, M. P., Huff, Jr. G. A., Calamur, N., and M. J. Spangler. 1997. Catalytic Wall Reactors and Use of Catalytic Wall Reactors for Methane Coupling and Hydrocarbon Cracking Reactions. *US Patent US5599510A*, filed December 1991.

Karakaya, C., and R. J. Kee. 2016. Progress in the Direct Catalytic Conversion of Methane to Fuels and Chemicals. *Progress in Energy and Combustion Science* 55:60–97.

Khalilpour, R., and I. A. Karimi. 2012. Evaluation of Utilization Alternatives for Stranded Natural Gas. *Energy* 40 (1):317–328. doi:10.1016/j.energy.2012.01.068.

Khan, M. S., and B. L. Crynes. 1970. Survey of Recent Methane Pyrolysis Literature. *Industrial & Engineering Chemistry* 62 (10):54–59.

Kinage, A. K., Ohnishi, R., and M. Ichikawa. 2003. Marked Enhancement of the Methane Dehydrocondensation toward Benzene using Effective Pd Catalytic Membrane Reactor with Mo/ZSM-5. *Catalysis Letters* 88 (3–4):199–202.

Larachi, F., Oudghiri-Hassani, H., Iliuta, M. C., Grandjean, B. P. A., and P. H. McBreen. 2002. Ru-Mo/HZSM-5 Catalyzed Methane Aromatization in Membrane Reactors. *Catalysis Letters* 84 (3–4):183–192.

Lee, R. P. 2019. Alternative Carbon Feedstock for The Chemical Industry? – Assessing the Challenges Posed by the Human Dimension in the Carbon Transition. *Journal of Cleaner Production* 219:786–796.

Li, H., Goldbach, A., Li, W., and H. Xu. 2007. PdC formation in Ultra-Thin Pd Membranes During Separation Of H2/CO Mixtures. *Journal of Membrane Science* 299 (1–2):130–137.

Li, W., Zhu, X., Cao, Z., Wang, W., and W. Yang. 2015. Mixed Ionic-Electronic Conducting (MIEC) Membranes for Hydrogen Production from Water Splitting. *International Journal of Hydrogen Energy* 40 (8):3452–3461.

Liu, Y., Tan, X., and K. Li. 2006. Nonoxidative Methane Coupling in a $SrCe_{0.95}Yb_{0.05}O_{3-A}$ (Scyb) Hollow Fiber Membrane Reactor. *Industrial & Engineering Chemistry Research* 45 (11):3782–3790.

Liu, Z., Li, L., and E. Iglesia. 2002. Catalytic Pyrolysis Of Methane on Mo/H-ZSM5 with Continuous Hydrogen Removal by Permeation through Dense Oxide Films. *Catalysis Letters* 82 (3–4):175–180.

Liu, Z., Nutt, M. A., and E. Iglesia. 2002. The Effects of CO_2, CO and H_2 Co-Reactants on Methane Reactions Catalyzed by Mo/H-ZSM-5. *Catalysis Letters* 81 (3):271–279. doi: 10.1023/a:1016553828814.

Morejudo, S. H., Zanón, R., Escolástico, S., Yuste-Tirados, I., Malerød-Fjeld, H., Vestre, P. K., Coors, W. G., Martínez, A., Norby, T., and J. M. Serra. 2016a. Direct Conversion Of Methane to Aromatics in a Catalytic Co-Ionic Membrane Reactor. *Science* 353 (6299):563–566.

Natesakhawat, S., Means, N. C., Howard, B. H., Smith, M., Abdelsayed, V., Baltrus, J. P., Cheng, Y., Lekse, J. W., Link, D., and B. D. Morreale. 2015. Improved Benzene Production from Methane Dehydroaromatization over Mo/HZSM-5 Catalysts via Hydrogen-Permselective Palladium Membrane Reactors. *Catalysis Science & Technology* 5 (11):5023–5036.

Ockwig, N. W., and T. M. Nenoff. 2007. Membranes for Hydrogen Separation. *Chemical Reviews* 107 (10):4078–4110.

Oh, S. C., Sakbodin, M., and D. Liu. 2019. Direct Non-Oxidative Methane Conversion In Membrane Reactor. *Catalysis* 31:127–165.

Oh, S. C., Schulman, E., Zhang, J., Fan, J., Pan, Y., Meng, J., and D. Liu. 2019. Direct Non-Oxidative Methane Conversion in a Millisecond Catalytic Wall Reactor. *Angewandte Chemie International Edition* 131 (21):7157–7160. doi:10.1002/ange.201903000.

Olsvik, O., and F. Billaud. 1993. Modelling of the Decomposition of Methane at 1273 K in a Plug Flow Reactor at Low Conversion. *Journal of Analytical and Applied Pyrolysis* 25:395–405.

Paglieri, S. N., and J. D. Way. 2002. Innovations in Palladium Membrane Research. *Separation and Purification Methods* 31 (1):1–169.

Perry, J. D., Nagai, K., and W. J. Koros. 2006. Polymer Membranes for Hydrogen Separations. *MRS Bulletin* 31 (10):745–749.

Phair, J. W., and S. P. S. Badwal. 2006. Review of Proton Conductors for Hydrogen Separation. *Ionics* 12 (2):103–115.

Rival, O., Grandjean, B. P. A., Guy, C., Sayari, A., and F. Larachi. 2001. Oxygen-free Methane Aromatization in a Catalytic Membrane Reactor. *Industrial & Engineering Chemistry Research* 40 (10):2212–2219.

Rogers, H. C. 1968. Hydrogen Embrittlement of Metals: Atomic Hydrogen from a Variety of Sources Reduces the Ductility of Many Metals. *Science* 159 (3819):1057–1064.

Sakbodin, M., Wu, Y., Oh, S. C., Wachsman, E. D., and D. Liu. 2016a. Hydrogen-Permeable Tubular Membrane Reactor: Promoting Conversion and Product Selectivity for Non-Oxidative Activation of Methane Over an Fe©Sio$_2$ Catalyst. *Angewandte Chemie International Edition* 128 (52):16383–16386.

Sheintuch, M., and R. M. Dessau. 1996. Observations, Modeling and Optimization of Yield, Selectivity and Activity during Dehydrogenation of Isobutane and Propane in a Pd Membrane Reactor. *Chemical Engineering Science* 51 (4):535–547.

Spivey, J. J., and G. Hutchings. 2014. Catalytic Aromatization of Methane. *Chemical Society Reviews* 43 (3):792–803.

Tarditi, A. M., Imhoff, C., Braun, F., Miller, J. B., Gellman, A. J., and L. Cornaglia. 2015a. PdCuAu Ternary Alloy Membranes: Hydrogen Permeation Properties in the Presence of H$_2$S. *Journal of Membrane Science* 479:246–255.

Uemiya, S., Endo, T., Yoshie, R., Katoh, W., and T. Kojima. 2007. Fabrication of Thin Palladium-Silver Alloy Film by Using Electroplating Technique. *Materials Transactions* 48 (5):1119–1123. doi:10.2320/matertrans.48.1119.

Venkataraman, K., Redenius, J. M., and L. D. Schmidt. 2002. Millisecond Catalytic Wall Reactors: Dehydrogenation of Ethane. *Chemical Engineering Science* 57 (13):2335–2343.

Wanat, E. C., Venkataraman, K., and L. D. Schmidt. 2004. Steam Reforming and Water–Gas Shift of Ethanol on Rh and Rh–Ce Catalysts in a Catalytic Wall Reactor. *Applied Catalysis A: General* 276 (1–2):155–162.

Wang, L., Murata, K., Sayari, A., Grandjean, B., and M. Inaba. 2001. Production of Ultra Highly Pure H$_2$ and Higher Hydrocarbons from Methane in One Step at Mild Temperatures and Development of the Catalyst under Non-Equilibrium Reaction Conditions. *Chemical Communications* (19):1952–1953.

Wang, L., Tao, L., Xie, M., Xu, G., Huang, J., and Y. Xu. 1993. Dehydrogenation and Aromatization of Methane under Non-Oxidizing Conditions. *Catalysis Letters* 21 (1–2):35–41.

Weyten, H., Luyten, J., Keizer, K., Willems, L., and R. Leysen. 2000. Membrane Performance: The Key Issues for Dehydrogenation Reactions in a Catalytic Membrane Reactor. *Catalysis Today* 56 (1–3):3–11.

Winandy, J. E., Rudie, A. W., Williams, R. S., and T. H. Wegner. 2008. A Future Vision for Optimally Using Wood and Biomass. *Forest Products Journal* 58 (6):6–16.

Wood, D. A. 2012. A Review and Outlook for the Global LNG Trade. *Journal of Natural Gas Science and Engineering* 9:16–27.

Wood, D. A., Nwaoha, C., and B. F. Towler. 2012. Gas-to-liquids (GTL): A Review of an Industry Offering Several Routes for Monetizing Natural Gas. *Journal of Natural Gas Science and Engineering* 9:196–208.

Xue, J., Chen, Y., Wei, Y., Feldhoff, A., Wang, H., and J. Caro. 2016. Gas to Liquids: Natural Gas Conversion to Aromatic Fuels and Chemicals in a Hydrogen-Permeable Ceramic Hollow Fiber Membrane Reactor. *ACS Catalysis* 6 (4):2448–2451. doi:10.1021/acscatal.6b00004.

Yun, S., and S. Ted Oyama. 2011. Correlations in Palladium Membranes for Hydrogen Separation: A Review. *Journal of Membrane Science* 375 (1–2):28–45.

Zhang, G., Dorris, S. E., Balachandran, U., and M. Liu. 2003. Interfacial Resistances of Ni–BCY Mixed-Conducting Membranes for Hydrogen Separation. *Solid State Ionics* 159 (1–2):121–134.

Anjaneyulu Koppaka, Niles Jensen
Gunsalus, and Roy A. Periana

CONTENTS

12.1 INTRODUCTION

Methane (CH_4), the principal component of natural gas (~90%) is the simplest, most abundant carbon-based material with an energy content per C–H bond that is comparable to the longer chain hydrocarbons (oil) (Olah et al. 2006; Caballero and Perez 2013). The development of next-generation processes that enable the clean and economical conversion of methane to all products that are currently derived from petroleum remains a long-standing, "Grand Challenge" within the field of chemistry (Schwarz 2011; Gunsalus et al. 2017). Importantly, utilization of methane as a source of fuel necessitates its chemical conversion to a product that is liquid under ambient conditions because of the expense associated with transporting a gas. Ideally, methane would be transformed into liquid fuels such as methanol, dimethyl ether, longer alkanes, diesel, or gasoline as these compounds retain most of the energy content that was present in the parent CH_4. Additionally, conversion of methane to liquid products (functionalization of methane) such as dimethyl ether or methanol would also facilitate the direct or indirect conversion of these precursors into the products such as olefins, aromatics, long chain hydrocarbons, etc. required by the commercial sector.

The key challenge with selective methane functionalization is that the oxidation products are much more reactive than methane by, essentially, every qualitative measure. As such, methane functionalization is often associated with substantial over oxidation to products such as CO_2 or coke which makes the selective methane conversion one of the most difficult, long-standing, "Grand Challenges" in all of chemistry. The reader can get a sense of this challenge by comparing various qualitative measures of reactivity of methane with those of methanol and ethylene (two potentially desired methane oxidation products), as shown in Figure 12.1. These challenges of relative reactivity, demonstrate why an economically feasible process

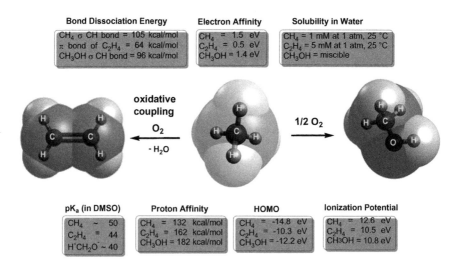

FIGURE 12.1 Comparison of reactivity of methane with ethylene and methanol through various quantitative measures (note: oxidative coupling reaction is not balanced).

for direct conversion of methane to methanol using O_2, which are often unselective radical processes has not been developed (Crabtree 1985, 1995). This also explains the low selectivity at high methane conversion in the thoroughly studied "oxidative coupling" of methane to ethylene (Labinger 1988; Lunsford 1995; Farrell et al. 2016; Galadima and Muraza 2016). Reactivity of methane with nucleophiles (qualitatively assessed by electron affinity), with electrophiles (HOMO energy), with protons (proton affinity), by electron loss (ionization potential), and by deprotonation (pK_a) all predict lower reactivity with methane as compared to its oxidation products, methanol and ethylene Figure 12.1 (Bischof et al. 2013).

Owing to the poor solubility of methane compared to products (i.e. methanol), for any practical system (where methane conversion is high) the solution phase concentration of product will far exceed that of the methane. Which necessitates that methane react much more rapidly than methanol ($k_{methane} \gg k_{methanol}$) to maintain high selectivity for the product. In fact, kinetic models indicated that in order to attain > 90% selectivity for methanol at greater than 15% methane conversion, the relative rate constants for the conversion of methane to methanol ($k_{methane}$) and methanol to CO_2 ($k_{methanol}$) must be at least 20:1 (this is considering a reactor with 1:1 liquid to gas ratio and with a methane pressure of 500 psig) (Hashiguchi et al. 2010).

Because of the aforementioned challenges, current commercial technology for methane conversion operates *via* multistep processes. Initially, methane is converted to syngas (a mixture of CO and H_2) at temperatures above 800°C. The generated syngas is then converted to methanol in a separate, subsequent, lower temperature process. The entire commercial methanol process has been highly optimized and operates with an overall yield of ~70% based on the amount of methane consumed, but the process is highly capital intensive which limits the use of methanol as a fuel (Olsbye 2016). Approximately 60% of the expense of current methanol production is related with capital costs for plant construction that arise from the high temperatures, large temperature variations and multiple steps that are required for the process (Periana et al. 1997; Lunsford 2000; Arakawa et al. 2001). In order to implement a new process for methane conversion, the key challenge is significant reduction of capital expense, ideally less energy consumption (reducing the operation temperatures to < 250°C would substantially lower the capital expenses) and lower emissions while maintaining the high selectivity of > 70% associated with the current process (or, ideally, improving selectivity).

Methane conversion to liquid products at lower temperatures requires a co-reactant (usually an oxidant) to provide a thermodynamic driving force for the reaction. The dehydrogenative coupling of methane to ethane and H_2 (Scheme 12.1) as well as of methane and H_2O to produce methanol and H_2 (Scheme 12.1) are not thermodynamically favorable processes at low temperatures and are only able to spontaneously progress at temperatures exceeding 800°C. However, both processes become highly thermodynamically favorable at lower temperatures with the inclusion of a co-reactant (in this case O_2) as shown in Scheme 12.1, respectively. It should be noted that O_2 (or atmospheric air, containing 21% O_2) is the only co-oxidant used in currently existing syngas processes. It is paramount that any next generation process for low-temperature methane conversion also uses only O_2 (or preferably air) as the only other consumed reagent as O_2 is the only economically viable oxidant for

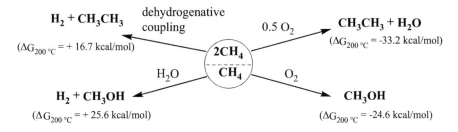

SCHEME 12.1 Thermodynamic values for various routes for conversion of methane to ethane or methanol.

the chemical conversion of methane to high volume, low value added, commodity chemicals and fuels such as methanol (Edwards and Foster 1986). The requirement for use of pure O_2 can also lead to unfavorable economics given the capital expense required for an air separation plant. It must be stated, however, that for higher value-added, lower volume products such as methane sulfonic acid (MSA), methyl amine, methyl chloride, etc. it could be plausible to utilize more expensive oxidants (i.e. hydrogen peroxide) and still develop chemical processes with favorable economics (McCoy 2016). Given the low cost of fuels and commodity chemicals relative to the higher cost of oxidants (i.e. H_2O_2), it is necessary that O_2, or preferably air, be the only co-reactant.

This chapter highlights scientific discoveries and studies that are homogeneous and utilizes organometallic and/or inorganic complexes for the net conversion of methane to stable, functionalized products. Discussion is limited to the systems that are well-defined, occurring solely in the liquid and/or gas phase. Systems involving solid phase reagents (metal oxides, nanotubes, MOFs, etc.), otherwise known as heterogenous systems, and biological systems will not be covered in any significant detail within this chapter. Within this chapter, the word "functionalization" is defined as the generation of a desired product through substitution of a methane hydrogen atom with various functional groups (hydroxy, amino, boryl, sulfonyl, etc.). The words "oxidation" and "activation" are defined differently from one another and are not interchangeably used with "functionalization."

12.2 ORGANIZATION STRUCTURE

Central to the organization of this chapter shown in Scheme 12.2 is the categorization of reported homogeneous reactions with methane. Emphasis is on practicality and, as discussed above, in order to meet the requirements of having methane augment or replace petroleum, O_2 (and more ideally, air) must be the only net co-reactant. For this reason, systems that use O_2 as the terminal oxidant or systems that use oxidants that can be regenerated from O_2 with favorable thermodynamics are only included in this chapter. Systems that utilize oxidants that cannot be regenerated from O_2 with favorable thermodynamics for the conversion of methane to methanol are not discussed herein. This classification of oxidants as O_2-regenerable (or non-O_2-regenerable) is based on reoxidation of the reduced species (from the reaction with methane) with O_2

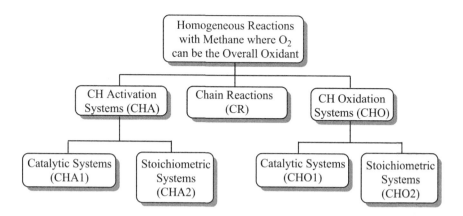

SCHEME 12.2 Organization structure for homogeneous methane functionalization systems (note: chain reactions are not discussed in this chapter).

to re-generate the oxidant and is based only on thermodynamic potentials. There are reports in the literature where methane is converted to smaller scale chemicals with potentially high value such as methanesulfonic acid (MSA) and other, more complex organic molecules (Sadow and Tilley 2003; Caballero et al. 2011; Cook et al. 2016; Smith et al. 2016; Diaz-Urrutia and Ott 2019; Gunsalus et al. 2019). These systems often require oxidants that are not O_2 regenerable but the products being generated are potentially high in value. For reasons of brevity, we have not included detailed discussion of these reports but wanted to state that this could be a potentially useful strategy for making small-scale, high value-added products from methane.

12.2.1 ORGANIZATION BASED ON PRACTICAL CONSIDERATIONS

Relatively few systems are known that consume O_2 directly, in the same reactor with methane. Most systems that are considered O_2-regenerable make use of stoichiometric oxidants that can subsequently be regenerated by O_2. In practical implementation, we envision these systems functioning in a two-stage, continuous process (Scheme 12.3)

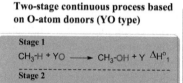

SCHEME 12.3 Functionalization of CH_4 to methanol by two-stage, O_2-regenerable: O-atom donors (YO) or electron-acceptor (Ox) type reactions.

TABLE 12.1

Thermodynamic Potentials for Some O-Atom Donors (Stage 2, Scheme 12.3)

	Y	YO	$Y + 1/2\ O_2 \rightarrow YO$ (ΔH_y (kcal/mol))
1	Me_2S	Me_2SO	−27
2	SO_2	SO_3	−24
3	NO_2^-	NO_3^-	−24
4	Na_2O	Na_2O_2	−23
5	MnO_2	MnO_4^{2-}	−23
6	VO^{2+}	VO_2^+	−19
7	NO	NO_2	−14
8	PbO	PbO_2	−14
9	C_5H_5N	C_5H_5NO	−13

where O-atom donors (YO), Table 12.1, or electron acceptors (Ox) can function as oxidizing species for methane oxidation in the first step and then subsequent regeneration of the oxidants from the reduced species (Y and H_2Ox, respectively) with O_2 in a second step. This process is modeled after the well-known Wacker process for oxidation of ethylene to acetaldehyde with stoichiometric amounts of Cu(II) as the oxidant, see Figure 12.2 (Eckert et al. 2005). While not immediately apparent, there are some distinct advantages to using O_2 indirectly rather than directly: (1) separation of O_2 from reactions with methane can aid in increasing overall reaction selectivity as many radical reactions are known to occur in the presence of triplet O_2; (2) Low-pressure air (containing 21% O_2) can be utilized in place of more expensive, pure oxygen thereby reducing operation costs; and (3) Separating methane and O_2 avoids the formation of potentially explosive mixtures.

For such a process to be viable for methane functionalization, the oxidant must be of sufficiently high thermodynamic potential to oxidize methane, but it must not be of so high potential that it cannot be regenerated with favorable thermodynamics using O_2 (hence O_2 regenerable). Using available data on thermodynamics, O_2-regenerable reagents have been identified, semi-quantitatively, by the propensity of the oxidant to donate an oxygen atom (O-atom donors, referred to as YO herein) or by the propensity of the oxidant to accept electrons (electron acceptors, referred to as Ox herein). Considering that the $\Delta H°_{tot}$ for converting methane to methanol is −30 kcal mol^{-1} (Scheme 12.3), it is necessary that the sum of the reaction steps, $\Delta H°_1$ and $\Delta H°_2$ (from Scheme 12.3), must be equal to −30 kcal mol^{-1}. In theory, $\Delta H°_2$ can be any value that satisfies the equation $\Delta H°_2 = -30 - \Delta H°_1$. However, in practicality there must be a driving force for the oxidation of methane, which necessitates that $\Delta H°_1 \leq 0$ kcal mol^{-1}. This means that, in practicality, $\Delta H°_2$ must lie between 0 and −30 kcal mol^{-1}. Holm and Donahue (1993) have reported a "thermodynamic scale for oxygen atom transfer reactions" using existing data on thermodynamics which can be used as a basis for identifying YO co-reactants that can be regenerated from O_2. Based on their work a list of potential O-atom donors YO that meet the requirements for methane functionalization are selected and presented in Table 12.1.

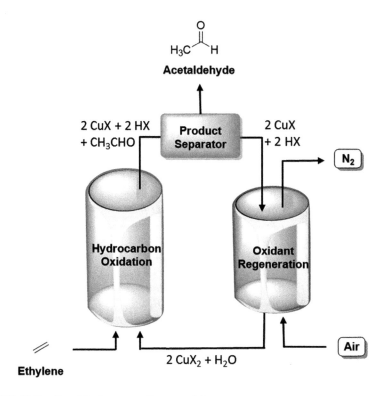

FIGURE 12.2 Simplified process diagram of the two-stage continuous Wacker process.

Analogously, appropriate electron acceptors can also be identified using similar analysis to that of O-atom donors with known electrochemical data and comparing them to the reduction potential of the two-component half reaction for methane oxidation to methanol. Using this method, we find that suitable air-regenerable electron acceptors will have a reduction potential within the range of 0.626 V and 1.229 V, versus the standard hydrogen electrode (SHE). Table 12.2 contains a selection of potential electron acceptors (Holm and Donahue 1993). While it must be stated that certain factors such as pH, solvent choice, ligands and/or other additives, etc. can drastically affect the O-atom donor and redox properties of these reagents, this analysis based on thermodynamic potentials provides a quantitative starting point for identifying practical oxidants for use in the conversion of methane to methanol (Ogura and Takamagari 1986). As this data show, powerful oxidants commonly used in methane functionalization (IO_4^-, H_2O_2, $S_2O_8^{2-}$) cannot be regenerated by O_2 and for this reason, these systems are not discussed within this chapter.

12.2.2 ORGANIZATION BASED ON MECHANISTIC CONSIDERATIONS

The discussion within this chapter has been narrowed down to include only those systems that have the thermodynamic potential to be regenerated using O_2 or air as this is a practical requirement for any system addressing the "Grand Challenge."

TABLE 12.2
Select Reduction Potentials within the
Range 0.626 V and 1.229 V (vs. SHE) (Lide
2007)

	Reaction	$E°$ (V)
1	$PtCl_6^{2-} + 2\,H^+ + 2\,e^- \rightarrow PtCl_4^{2-} + 2\,HCl$	0.73
2	$ReO_4^- + 2\,H^+ + e^- \rightarrow ReO_3 + H_2O$	0.77
3	$Fe^{3+} + e^- \rightarrow Fe^{2+}$	0.77
4	$Hg_2^{2+} + 2\,e^- \rightarrow 2\,Hg$	0.79
5	$Ag^+ + e^- \rightarrow Ag$	0.80
6	$ClO^- + H_2O + 2\,e^- \rightarrow Cl^- + 2\,OH^-$	0.84
7	$Pd^{2+} + 2\,e^- \rightarrow Pd$	0.91
8	$2\,Hg^{2+} + 2\,e^- \rightarrow Hg_2^{2+}$	0.92
9	$NO_3^- + 3\,H^+ + 2\,e^- \rightarrow HNO_2 + H_2O$	0.93
10	$V_2O_5 + 6\,H^+ + 2\,e^- \rightarrow 2\,VO^{2+} + 3\,H_2O$	0.96
11	$IO_3^- + 6\,H^+ + 6\,e^- \rightarrow I^- + 3\,H_2O$	1.08
12	$PtO_2 + 2\,H^+ + 2\,e^- \rightarrow PtO + H_2O$	1.01
13	$Cu^{2+} + 2\,CN^- + e^- \rightarrow [Cu(CN)_2]^-$	1.1
14	$MnO_2 + 4\,H^+ + 2\,e^- \rightarrow Mn^{2+} + 2\,H_2O$	1.22

In order to further categorize these systems, we have broken them down at a more fundamental level and attempted to classify them based on mechanistic consider-ations, as shown in Scheme 12.2. We have used the mechanistic proposals in the original reports, whenever possible, to classify these reactions. Where mechanis-tic information is not given, or is ambiguous, we have used our best judgement to attempt to place the systems in the categories shown in. *It is important to note that this classification scheme is based solely on the elementary reaction involving meth-ane; reactions in the mechanistic sequence that do not involve methane were not considered when classifying reactions.* We believe that adopting this "carbocentric" perspective is important because the reactivity of methane is the key challenge in developing systems for methane oxidation. Scheme 12.2 shows that we have catego-rized these reactions into three categories: (1) chain reactions (CR) which involve reaction between methane and a chain propagating species (i.e. carbocations, carb-anions, free radicals) generated from methane fragments; (2) CH activation (CHA) in which the carbon species generated from the elementary reaction with methane is not formally oxidized (i.e. the oxidation state of carbon remains at −IV); and (3) CH oxidation (CHO) in which the carbon species generated from the elementary reaction with methane has been formally oxidized (i.e. oxidation state of carbon > −IV). We acknowledge that chain reactions (CR) can formally be classified as CHO reactions because, for i.e. radicals, the carbon fragment is oxidized relative to methane, but because these systems are so different mechanistically from other CHO reactions, we decided to include them in their own, separate category and these are not discussed in this chapter. CHA and CHO reactions (nonchain reactions) are then further subdivided based on whether the species of constant composition that is undergoing the elementary reaction with methane is (A) a catalyst, present in sub-stoichiometric amounts relative to methane and the oxidant – these are classified as "catalytic reactions," or (B) present in stoichiometric amounts – these are classified as "stoichiometric reactions." It should be made clear that, although catalytic systems have stoichiometric amounts of an oxidant present, a reaction is considered a "catalytic reaction" because the stoichiometric oxidant is not the species undergoing the elementary reaction with methane; the catalyst is. For stoichiometric systems, the stoichiometric oxidant is the species that is undergoing the elementary reaction with methane and no catalyst is necessary. To assist in navigating this chapter, the various classifications in are keyed to the corresponding sections.

12.2.2.1 Classification Details of Mechanistic Considerations

12.2.2.1.1 Chain Reactions versus Catalytic Reactions

Chain reactions are often referred to as catalytic reactions. We have classified these two types of reactions differently to avoid confusion. Free-radical reactions typically utilize some type of radical initiator (Scheme 12.4), at times in the presence of a pro-moter, to initiate a radical process (Laidler 1996; McMurry 2004; Fink 2006). These initiators react with methane via free-radical substitution to generate methyl radicals and generate product in a chain reaction. Chain reactions are ended by termination sequences where the chain propagating species is quenched as in the coupling of two free-radical species. Although these reactions may appear to be catalytic in the sense that several moles of product are formed per mole of initiator consumed, they are not catalytic based on the strict definition of a catalyst as a permanent, unchanged species. As shown in Scheme 12.4, the methoxy radical, RO$^{\bullet}$ (the species undergoing the elementary reaction with methane) is a fragment of the reactants, methane and O_2 (step 1), and has been "permanently changed" in the reaction sequence. Although this methoxy radical is regenerated in the reaction sequence, the methoxy radical that is formed is composed of "new" atoms from the reactants, methane and O_2. This is unique to chain reactions and is fundamentally different from "true catalysis" as defined herein. In true catalysis, the core atoms that make up the catalytic motif are said to be unchanging and not composed, simply, of fragments of the reactants. It is often the case that a core catalytic motif can associate and dissociate with other atoms during the catalytic process, but the catalyst is strictly defined as that set of

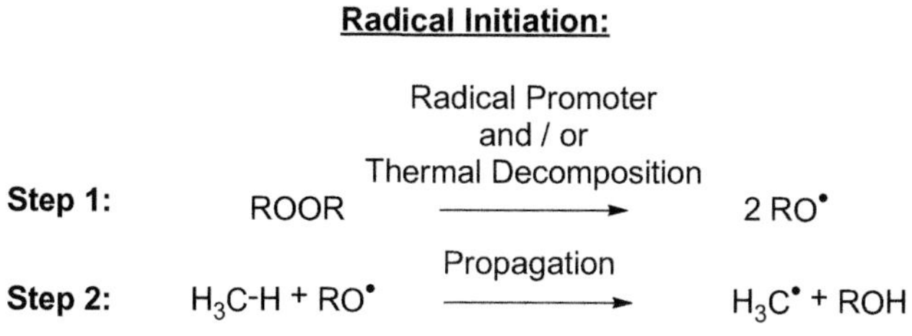

Radical Initiation:

Radical Promoter
and / or
Thermal Decomposition

Step 1: ROOR $\xrightarrow{\hspace{2cm}}$ 2 RO$^{\bullet}$

Propagation

Step 2: H$_3$C-H + RO$^{\bullet}$ $\xrightarrow{\hspace{2cm}}$ H$_3$C$^{\bullet}$ + ROH

SCHEME 12.4 Initiation of free-radical chain reactions for CH_4 functionalization by a per-oxide initiator.

core atoms that remains unchanged throughout the entire catalytic cycle. It is possible, and often the case, for radical processes (initiation, propagation, etc.) to be facilitated by a species that is present in catalytic amounts (Co(II), Cu(II), Fe(III), etc.), but if this species is not involved in the elementary CH cleavage reaction with methane then it does not fall under the classification of "catalytic," as defined above.

Free-radical processes for methane functionalization may seem problematic given the discussion above about weaker BDE in the product, methanol, than in methane. Contrarily, there have been many highly selective reactions that have been developed for methane oxidation that proceed by free-radical mechanisms. The explanation for this discrepancy is often attributed to the "polar effect" (otherwise known as "polarity effect") (Russell and Brown 1995; Russell and Williamson 1964a, b; Zavitsas and Pinto 1972; Tedder 1982; Parsons 2000; Heidbrink et al. 2001). This effect describes how polar forces influence the energetics of a transition state in a radical reaction. An attack on a CH bond by an electrophilic radical (i.e. halogen radicals, alkoxy radicals) results in a polar transition state where positive charge is built up on the carbon atom of the CH bond. Electron withdrawing groups adjacent to carbon will destabilize this transition state (increase the energetics) whereas electron donating groups will stabilize the transition state (decrease the energetics). In the functionalization of methyl radicals by SO_3, this is precisely what is occurring. The reaction is highly selective because the product, MSA, is effectively protected from forming a radical by H-atom abstraction because the electron withdrawing sulfonyl group increases the energetics of the transition state for reaction with electrophilic $CH_3SO_3\bullet$ radicals, thereby disfavoring the reaction pathway (over the more facile H-atom abstraction from methane). Nucleophilic, or even less electrophilic radicals would not be expected to show the same levels of selectivity against free-radical substitution of MSA (Fleming 1976). One other challenge associated with chain reactions is the relatively high cost of initiators that are often used in the processes. The key to overcoming this challenge is maximizing the number of times a reaction propagates (generating a molecule of product and a chain-carrying species to continue the reaction) relative to the initiation and termination sequences. This is a particularly difficult challenge given the high cost of initiators (and often times the oxidants employed) compared to the low value of methanol.

12.2.2.1.2   Classification Details of Nonchain Reactions

12.2.2.1.2.1   CH Activation and CH Oxidation

For all systems of methane oxidation that operate by "nonchain reactions," that is systems in which the elementary reaction with methane involves a reactant with a core motif that is not comprised only of methane fragments, we have subdivided them into two mechanistic categories; CH activation (CHA) and CH oxidation (CHO). These two reactions are shown in Scheme 12.5. CHA involves the reaction between a hypothetical species, LMX_2, and a CH bond to generate a L(X)M–C species containing an M–C bond. Although Shilov et al. (1997) showed that CHA has been used in the literature to describe reactions with alkanes that generate carbon radicals, carbocations, carbanions, and carbenes; we have adopted a stricter sense of the word "activation" and chosen not to classify these types of reactions as CHA events. (1) Activation events are reversible transitions: (2) the "activated species" is in a nearly identical physicochemical state; and (3) the "activated species" can more

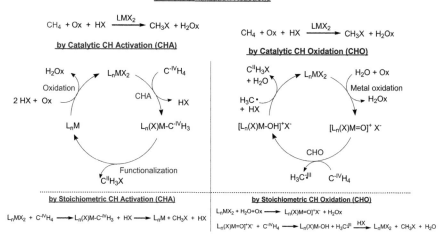

SCHEME 12.5 Methane functionalization by CH activation (CHA) and CH oxidation (CHO) in both catalytic and stoichiometric reactions.

facilely undergo a specific chemical transformation. In this chapter, the specific transformation is the conversion of methane to a desired product.

Quite possibly the most telling difference between CHA and CHO is that CHA leads to transition of a molecule (an alkane in our case) into a nearly identical physical or chemical state while CHO does not. This is because, by our definition of CHA, there is no formal change in oxidation state at the carbon in the elementary reaction with methane. Thus, the C in methane is assigned a formal oxidation state (FOS) of $-IV$ and the C in the L(X)M–C species generated by CHA is also assigned a FOS of $-IV$. Admittedly, FOS assignment is a formalism, but it is a generally accepted and useful method of assigning the approximate electronic state of a molecule. In a CHO event, in the elementary reaction with methane, the assigned FOS at C increases from $-IV$ in CH_4 to $> -IV$ in the obtained product. Thus, there has been substantial electronic change at C in CHO reactions and the new species is not in a "nearly identical chemical or physical state." We believe that the use of FOS assignments could help with facilely identifying CHA and CHO reactions and aid in discussion.

The basis for this classification scheme is that it could be expected that reactions proceeding without a change in FOS (CHA reactions) should be fundamentally different from those reactions that proceed with a change in FOS (CHO reactions). Effectively, CHA reactions are acid-base reactions while CHO reactions are redox reactions. In CHA reactions, as with proton transfer reactions, there is only little electronic reorganization in the transition state in moving from LMX_2 and methane to $L(X)M–CH_3$ and HX as compared to the electronic reorganization around the carbon when a CH bond is cleaved in CHO reactions. In CHO reactions there are more substantial electronic and configurational changes which is approximated by the change in FOS. Classification of reactions into these two categories will facilitate in highlighting commonalities and dissimilarities and help with the design of new catalysts. Furthermore, it is highly likely that the core motif for systems operating by CHA and CHO will be fundamentally different and require different design and reactivity characteristics. This seems to be valid as, in general, CHO catalysts and reagents are often strongly oxidizing, high oxidation state species while many systems that operate by CHA utilize low oxidation state, nonoxidizing species. In fact, the lower oxidation states of many transition metal CHA complexes can be more active than the higher oxidation state variants (Mironov et al. 2013).

12.2.2.1.2.2 Catalytic versus Stoichiometric Reactions
CHA and CHO reactions have been further divided into two sub-categories, those that are catalytic and those that are stoichiometric. Catalytic reactions, in the elementary reaction with methane, involve a reagent that is present in sub-stoichiometric amounts whose core composition remains "unchanged" throughout the course of the reaction as methane is consumed. Scheme 12.5, above the dotted lines, shows hypothetical catalytic cycles for CHA and CHO with a core "LM" motif that is constant through all the reactions within the catalytic cycle. In these catalytic reactions, an oxidant (either O_2 or an O_2-regenerable oxidant) is consumed in stoichiometric quantities, but the reactions are classified as catalytic because it is the catalytic species that is undergoing the elementary reaction with methane, not the stoichiometric oxidant. CHA and CHO reactions are classified as stoichiometric when the "LM" motif undergoing the elementary reaction with methane is present in stoichiometric quantities and is not reoxidized in situ by an oxidant. This is exemplified by the hypothetical reactions in Scheme 12.5 below the dotted line.

It is our hope that this categorization of methane reactions into chain reactions, CHA reactions and CHO reactions will help to add clarity to the large body of work that has been reported on homogeneous methane functionalization and facilitate the identification of new directions for research. CHA and CHO reactions have been further subdivided into stoichiometric and catalytic reactions with the hope that this further subcategorization could help in giving clarity to what systems may have the most potential for commercial development and guide further research in catalyst development. It is the intent of this chapter to give the reader a method to quickly find or classify various systems for homogeneous methane functionalization that have been reported.

12.3 H_2SO_4/SO_3 SYSTEMS FOR METHANE OXIDATION

One of the extensively studied system for oxidation of methane is based on H_2SO_4 and SO_3, either separately or as mixtures. These reagents are typically considered to be air-regenerable based on the assumption that the process can be integrated with the large-scale, commercial process for H_2SO_4 production which is prepared via SO_2 oxidation to SO_3 with O_2. Scheme 12.6 highlights the three general routes (**A, B, C**) for using H_2SO_4 and/or SO_3 as an air-regenerable oxidant that have been reported in the literature (Conley et al. 2006). Using these strategies, the objective is to design a continuous process, reminiscent of the well-established Wacker process, for conversion of methane to methanol using O_2 as the overall oxidant. Given the broad interest in this area of research, it is important to highlight some non-obvious disadvantages with these systems: (1) SO_3 is rapidly and irreversibly converted, in the presence

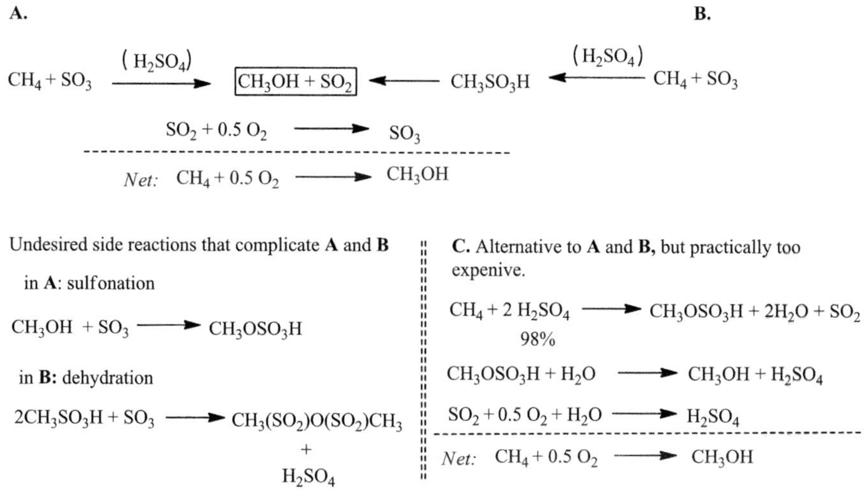

A.

$$CH_4 + SO_3 \xrightarrow{(H_2SO_4)} \boxed{CH_3OH + SO_2} \longleftarrow CH_3SO_3H$$

B.

$$CH_3SO_3H \xleftarrow{(H_2SO_4)} CH_4 + SO_3$$

$$SO_2 + 0.5\ O_2 \longrightarrow SO_3$$

$$Net:\quad CH_4 + 0.5\ O_2 \longrightarrow CH_3OH$$

Undesired side reactions that complicate **A** and **B**

in **A**: sulfonation

$$CH_3OH + SO_3 \longrightarrow CH_3OSO_3H$$

in **B**: dehydration

$$2CH_3SO_3H + SO_3 \longrightarrow CH_3(SO_2)O(SO_2)CH_3$$
$$+$$
$$H_2SO_4$$

C. Alternative to **A** and **B**, but practically too expenive.

$$CH_4 + 2\ H_2SO_4 \longrightarrow CH_3OSO_3H + 2H_2O + SO_2$$
$$98\%$$

$$CH_3OSO_3H + H_2O \longrightarrow CH_3OH + H_2SO_4$$

$$SO_2 + 0.5\ O_2 + H_2O \longrightarrow H_2SO_4$$

$$Net:\quad CH_4 + 0.5\ O_2 \longrightarrow CH_3OH$$

SCHEME 12.6 General routes and issues with the use of SO_3 and H_2SO_4 as a O_2-regenerable oxidant for the net functionalization of methane to methanol with O_2.

of water, to H_2SO_4 as it is both a potent dehydrating agent as well as a sulfonating agent; and (2) H_2SO_4 and water form an azeotrope at ~ 98% sulfuric acid that, for all practical purposes, prevents the conversion of H_2SO_4 into SO_3 and water. Due to these two properties of H_2SO_4/SO_3 systems, methanol formed in route **A** would likely be converted to methyl bisulfate (CH_3OSO_3H) by SO_3 in situ. Methane sulfonic acid, CH_3SO_3H (MSA) formed in route **B** would, most likely, be dehydrated by SO_3 to make methane sulfonic anhydride ($CH_3(SO_2)O(SO_2)\ CH_3$) and H_2SO_4. Given the properties of SO_3, these two side reactions are likely unavoidable. When these unavoidable side reactions are taken into account, in addition to the desired product methanol, undesired consumption of SO_3 and conversion to H_2SO_4 are also realized. While it is true that H_2SO_4 is a large-scale commodity chemical, it is unlikely that this process could be economically viable given the difference in scale of methanol production for use as a fuel and sulfuric acid production for chemical use. It could be possible to couple sulfuric acid production to methanol production for chemical use, but other factors such as the prohibitively high cost of separating methanol from concentrated sulfuric acid (*vide infra*) and high capital costs for plant construction (due to the corrosive nature of H_2SO_4) would likely lead to unfavorable economics.

Route **C** in Scheme 12.6 seems to address the issues highlighted in route **A** and **B** by utilizing H_2SO_4 that is $\leq$ 98%. The system is free of any SO_3 (so that SO_3 consumption is a non-issue) and any water formed in the chemical process can be subsequently removed by distillation, thus concentrating the sulfuric acid back to ~ 98%. Unfortunately, separating the product methanol, and reconcentratation of the solvent by removing water would be far too expensive for commercialization. One possibility to reduce separation costs would be to first separate methyl bisulfate from the reaction mixture and then hydrolyze the neat material, rather than performing a dilution of the entire sulfuric acid solvent. The challenge with this approach is that methyl bisulfate has a high boiling point (167°C at 0.013 bar) making separation by distillation too expensive. In order to make this approach viable, other methods of separation would need to be developed.

Another possibility for reducing separation costs would be to focus on developing systems that can operate in sulfuric acid that is $\leq 80\%$, making removal of the products much more facile and reconcentration of the solvent much less expensive. It should be noted, however, that many of the current catalysts that operate in H_2SO_4 have been found to be severely inhibited by water coordination when the concentration of sulfuric acid is $\sim < 90\%$. Additionally, it is questionable whether solutions of sulfuric acid that are $\leq 80\%$ are sufficiently oxidizing to act as a stoichiometric oxidant for methane conversion. Methane conversion to methanol under standard conditions (1 M H_2SO_4) is estimated to be unfavorable (Gunsalus et al. 2017).

An additional challenge associated with the use of sulfuric acid and/or SO_3 systems is that these systems react rapidly with higher alkanes (C > 1) to generate undesirable oxidation products (Konnick et al. 2014). This necessitates that methane must be separated from the additional components of natural gas (ethane, propane, etc.) prior to being subjected to the reaction conditions. The syngas process for methanol production does not mandate such separations because all of the hydrocarbons are first converted to CO/H_2. One possible area where the H_2SO_4/SO_3 systems could have utility is in the production of higher value, lower volume chemicals. A recent report by Grillo, a German firm, demonstrates this at pilot scale for the synthesis of MSA from methane and SO_3 (Diaz-Urrutia and Ott 2019). Because of the higher value of MSA (and the lower volume requirements relative to methanol), it is possible that the challenges addressed above are averted.

12.4 CHA1 CATALYTIC SYSTEMS THAT UTILIZE O_2/O_2-REGENERABLE OXIDANTS AND OPERATE BY NONCHAIN, CH ACTIVATION REACTIONS

To date, there have been many systems developed that interact with alkane C–H bonds and induce CHA. The principle challenge today is designing systems that integrate CHA with functionalization of the M–R (metal-alkyl) intermediates that are generated. It is often the case that systems that induce alkane CHA are incompatible with the conditions required for M–R functionalization, often due to species decomposition or inhibition. There are only a few known, facile pathways for functionalization of the M–CH₃ intermediates generated from CHA pathways and finding optimal conditions under which those functionalization pathways can operate in concert with CHA pathways is often difficult. Nevertheless, systems have been developed that achieve these transformations.

12.4.1 PLATINUM SYSTEMS

The first reported example for methane functionalization in which a CHA mechanism was proposed is the Shilov system (Gol'dshleger et al. 1972; Shilov 1984; Labinger and Bercaw 2002). In this system, K_2PtCl_4 was found to catalyze the conversion of methane to methanol and methyl chloride (CH_3Cl) at ~100°C in the presence of toichiometric amounts of K_2PtCl_6 using HCl/water as the solvent, Eq. (12.1). One remarkable feature of this system is that the Pt(IV) that is acting as a stoichiometric oxidant generates more of the active Pt(II) catalyst as the reaction proceeds. The reaction of K_2PtCl_4 with O_2 in HCl to generate Pt(IV) has favorable thermodynamics ($\Delta H = -29$ kcal/mol) and, as such, this reaction is considered O_2-regenerable. This system demonstrated that potentially useful products could be produced, with good selectivity, from methane using well-defined, soluble species that could be rationally tuned and designed. Shilov proposed a mechanism involving three key steps as shown in Scheme 12.7 (Shilov and Shul'pin 2000; Shilov and Shteinman 1977; Chen et al. 2007; Fekl and Goldberg 2003). Given the relative complexity of this system and competing off-cycle reactions, thorough mechanistic investigations were difficult. It was not until later, after the development and study of many model systems, that the initial mechanism put forth by Shilov gained traction and CHA was recognized as a new potential mechanism for alkane conversion (Rostovtsev et al. 1998, 2002; Sen 1998; Wick and Goldberg 1999; Butikofer et al. 2006; Vedernikov et al. 2006; Khusnutdinova et al. 2007; Pawlikowski et al. 2007; Vedernikov 2007; Yahav-Levi et al. 2008; Grice and Goldberg 2009).

$$CH_4 + K_2\left[PtCl_6\right] + H_2O \xrightarrow[\text{HCl/H}_2O]{[\text{Pt(H}_2\text{O)}_2\text{Cl}_2]} CH_3OH + K_2\left[PtCl_4\right] + 2HCl \quad (12.1)$$

This system suffered from low turnover frequency (TOF) ($< 10^{-5}$ s^{-1} at $< 100°C$) resulting in too low of volumetric productivity or Space–Time Yield (STY) for methane oxidation and the reaction generated a mixture of products. The catalyst lifetime was relatively short with a turnover number (TON) of < 20, primarily due to the

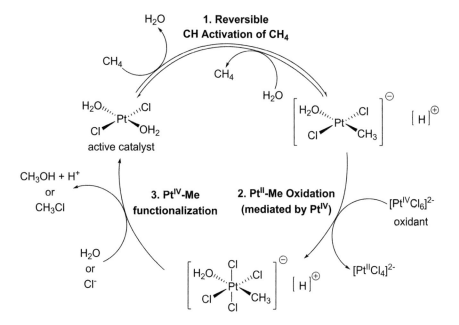

SCHEME 12.7   Platinum catalyzed oxidation of CH_4 to methanol with Pt^{IV}.

formation of Pt black (Pt black cannot be reoxidized with O_2 or air in any practical manner) and the requirement for stoichiometric Pt(IV) as an oxidant makes the process too expensive to implement in a hypothetical, two-step, air-regenerable process. These factors all prevented developing a commercial process based on this attractively simple system. The system is, undoubtedly, important because there have been substantial efforts to develop new and improved systems for methane functionalization by CHA based on platinum and other metals since its discovery.

Several research groups have investigated ligation of the platinum center to address the challenges associated with the Shilov system. This strategy was first reported by the Periana group (1998) in which they ligated Pt(II) with 2,2'-bipyrimidine. The resulting complex, η^2-(2,2'-bipyrimidyl)platinum dichloride ((bpym)PtCl$_2$), was found to be highly active and stable (catalyst TONs of 300 were reported with no observed loss in activity) for CHA/functionalization of methane in hot, concentrated sulfuric acid (Figure 12.3). The obtained methanol product was a mixture of methyl bisulfate (CH_3OSO_3H, the esterification product of methanol and H_2SO_4), protonated methanol ($CH_3OH_2^+$) and methanol (CH_3OH) with the ratio of three products dependent on the concentration of sulfuric acid. Sulfuric acid concentration decreases with increased methane conversion as water and product concentrations increase (Figure 12.3). In this system, sulfuric acid played three major roles. It acted as the reaction solvent and the stoichiometric oxidant. It also served as the in situ protecting reagent in the system via esterification of methanol to methyl bisulfate and/or via protonation of methanol. These electron deficient products were less prone to reactivity with the electrophilic Pt(II) catalyst, effectively rendering them "protected." Estimates have placed the reactivity of these products (CH_3OSO_3H and $CH_3OH_2^+$) at approximately 100 times less reactive than methane. Despite the products being present in concentrations of $\sim$ 1 M and methane having a solubility only in the millimolar range, Periana reported a selectivity for methanol-derived products of 81% at 90% methane consumption when reactions were run in oleum. STY of the system was found to be $\sim 10^{-6}$ mol mL^{-1} sec^{-1} at 220°C with a methane pressure of 500 psig and catalyst TOF was $\sim 10^{-3}$ sec^{-1}.

This system has gained attention from many different research groups since its disclosure in 1998. Mainly this is because of the relatively high STY, excellent selectivity for methanol-derived products, a catalyst motif that is robust and stable (even at temperatures up to 220°C in oleum), and the fact that SO_3 (or H_2SO_4) is used as the stoichiometric oxidant (given the well-established, commercial technology for oxidizing SO_2 to SO_3 with O_2). Unfortunately, (bpym)PtCl$_2$ is severely inhibited by water. Studies have shown that, at high methane concentrations where the buildup of water has been substantial (Figure 12.3), CHA of methane by the electrophilic

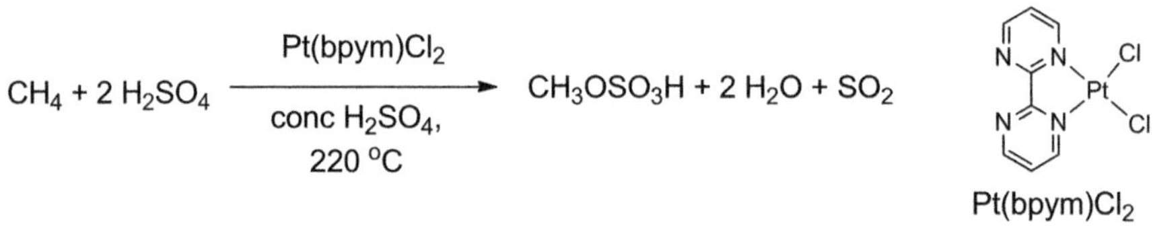

$$CH_4 + 2\,H_2SO_4 \xrightarrow[\substack{\text{conc } H_2SO_4, \\ 220\,^\circ C}]{\text{Pt(bpym)Cl}_2} CH_3OSO_3H + 2\,H_2O + SO_2$$

Pt(bpym)Cl$_2$

FIGURE 12.3 Functionalization of methane in Periana System with Pt(bpym)Cl$_2$.

species effectively stops when the concentration of sulfuric acid drops below ~ 90%. This corresponds to a methanol product concentration of ~ 1 M and gives an integrated catalyst rate of ~ 10^{-3} sec^{-1}. From a perspective of practicality for commercialization, such a slow rate necessitates an inventory of platinum that would be far too expensive for a competitive process and rate increases of 100–1000 times would likely be necessary (Conley et al. 2006).

To investigate the mechanism of this reaction and attempt to improve reaction characteristics, several research groups have conducted theoretical computational studies on the (bpym)PtCl$_2$ system, also known as the "Periana–Catalytica system" (Gilbert et al. 2001; Kua et al. 2002; Hristov and Ziegler 2003a; Xu et al. 2003; Paul and Musgrave 2007; Ahlquist et al. 2009). All these studies agree that, fundamentally, this reaction occurs in three key steps. First, methane coordinates and undergoes CHA with Pt(II) to make a Pt^{II}–Me. Next, the Pt^{II}–Me intermediate is oxidized to a Pt^{IV}–Me under the reaction conditions. Finally, reductive functionalization from the Pt^{IV}–Me gives the functionalized product, methyl bisulfate, and regenerates the active Pt(II) catalyst. These theoretical studies of the Periana–Catalytica system, though in agreement on the mechanistic basics, do not always agree on the exact identity of the chemical species involved in catalysis nor do they agree on the exact mechanism by which such a species activates methane.

Studies on the original Shilov system have shown that it is Pt(II) that does CHA, not Pt(IV). Interestingly, Periana and coworkers (Mironov et al. 2013) observed that (bpym)Pt(II) is rapidly oxidized to (bpym)Pt(IV) under the reaction conditions (200°C, concentrated H$_2$SO$_4$). This seemed to suggest that, within minutes, the system should be deactivated. This was inconsistent with the observed, high stability of the system. The initial hypothesis was that methane reactivity with (bpym)Pt(II) was substantially faster than the background oxidation of Pt(II) to Pt(IV). Upon careful examination, however, it was found that the background oxidation of (bpym) PtX$_2$ to (bpym)PtX$_4$ does in fact occur under reaction conditions with methane. Counterintuitively, this oxidation (originally thought to be highly undesirable) facilitates faster functionalization of methane. The oxidation of (bpym)Pt^{II}–Me to (bpym) Pt^{IV}–Me by H$_2$SO$_4$ (or SO$_3$) was found to be slow. Periana and coworkers found that **4** (formed via k$_4$, Scheme 12.8) acts as a highly effective oxidant for converting (bpym) Pt^{II}–Me (**2**) to (bpym)Pt^{IV}–Me (**3**) and regenerating active catalyst (bpym)PtX$_2$ (**1**, k$_5$, Scheme 12.8). This "self-repair" mechanism is highly favorable as **2** is on the order of parts per billion, **1** and **4** are between 10 and 100 mM and [H$_2$SO$_4$] is ~ 18 M. Despite the high concentration of H$_2$SO$_4$ leading to background oxidation of **1** to **4** (step k$_4$, Scheme 12.8), the generated (bpym)PtX$_4$ reacts with (bpym)Pt^{II}–Me (step k$_5$, Scheme 12.8) at enough of a sufficient rate to prevent catalyst deactivation. Calculations by Goddard show that reactions of (bpym)Pt^{II}–Me with (bpym)PtX$_4$ proceed with a barrier of just 5 kcal mol^{-1}. The oxidation would be predicted to proceed at diffusion-controlled rates. Essentially, if the off-cycle, background oxidation of Pt(II) to Pt(IV) is considered as a step in the catalytic cycle, oxidation of Pt^{II}–Me to Pt^{IV}–Me (step k$_2$) is no longer the rate determining step. Instead, formation of **4** (step k$_4$) becomes the rate determining step and oxidation of **2** to **3** (step k$_5$) is rapid.

As discussed previously in this chapter, one of the challenges with H$_2$SO$_4$ is the separation of the generated methanol from the reaction solvent. In this spirit, in an

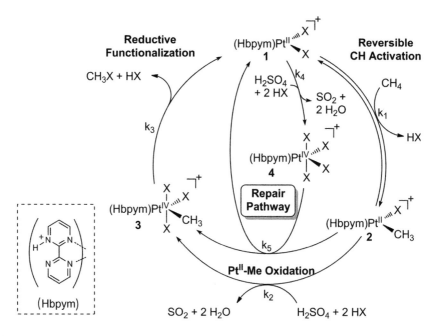

SCHEME 12.8 Revised mechanism in the Periana–Catalytica system. All species possess an outer-sphere anion (HSO_4^- or Cl^-), which have been omitted for clarity.

effort to improve the Periana–Catalytica system, the Michalkiewicz group (2006a) reported on the development of a technique for product separation based on low-pressure membrane distillation (LPMD). Using this technique, the authors first separated any remaining SO_3 from the post-reaction mixture and then were able to separate methylbisulfate from the reaction mixture, with close to 100% recovery, without the need for hydrolysis first. H_2SO_4 was retained in the reaction mixture and did not pass through the membrane. The collected product was subsequently hydrolyzed to give methanol. This report demonstrates an innovative, creative solution to the challenge of product separation from sulfuric acid. However, as discussed above, another challenge with the use of H_2SO_4/SO_3 systems is the irreversible conversion of SO_3 to H_2SO_4 with water that is co-generated in the reaction with methane. When methanol is produced at the scale required for fuels, the amount of H_2SO_4 produced would be much larger than global demand. It is not clear that this report addresses this challenge.

In another innovative report, Michalkiewicz and Kosowski (2007) were able to achieve methane conversions at ambient pressures (one atmosphere) using catalytic $PtCl_4$ in 25 wt% oleum in the temperature range of 130 to 220°C and achieved highest yield at 220°C. Through use of an absorption reactor, they were able to operate at very low pressures because the reactor maximized the surface area of the gas–liquid contact thus eliminating the mass-transport challenges that are typically associated with methane reactions. Lowering the pressure requirements for methane oxidation has the potential to be highly useful as the capital costs associated with plant construction could be drastically reduced without the requirement for high pressures.

Cheng et al. (2006) have shown that a number of platinum compounds, many of which were initially inactive in sulfuric acid, could be made to react with methane by the inclusion of ionic liquids in the reaction mixture (Eq. (12.2)). In comparison to many of the other reported platinum systems, these systems were reported to have a higher water tolerance and the selectivity for conversion of methane to methyl bisulfate was good. Despite the promising results, catalytic activity of the systems was found to be ~50% lower than the original Periana–Catalytica system and the systems were not overly robust with a reported TON of 3.5. Goddard and coworkers (Xu et al. (2008) studied these systems using computational methods and found that they operate through Shilov-like chemistry, but in an oxidizing media (concentrated H_2SO_4), by facilitating solubility of the otherwise insoluble species (such as $PtCl_2$):

$$CH_4 + SO_3 + H_2SO_4 \xrightarrow[\substack{H_2SO_4\,(96\%),\\220\,°C}]{\substack{PtCl_2\\ \text{Ionic Liquid}\,(6\,eq)}} CH_3OSO_3H + SO_2 + H_2O \qquad (12.2)$$

The groups of Yang et al. (1999) and Souma et al. (2002) have reported that Pt(II) can catalyze carboxylation of alcohols to carboxylic acids. In an extension of this, Zerella and Bell (2006a) found that platinum salts ($PtCl_2$, $(NH_3)_4PtCl_2$, (bipyridyl) $PtCl_2$, $Pt(acac)_2$) were competent to catalyze methane carbonylation to acetic acid (HOAc) in 96% sulfuric acid in the presence of CO. Inclusion of co-catalysts ($CuCl_2$, $VO(acac)_2$) and O_2 was found to increase the total product yield (both acetic acid and methylbisulfate). CO was found to be necessary for formation of HOAc as methyl-bisulfate was the only product formed in experiments run in the absence of CO. In a typical experiment, a reactor containing 3 mL of 96% H_2SO_4 and 20 mM Pt(II) cata-lyst is charged with equal pressure (400 psig of each) methane and CO and heated to 180°C for four hours. Under the reaction conditions, CO could also act as a reducing agent. The role of the co-catalysts was to reoxidize any Pt black (Pt^0) formed as a result of this reduction. Thus, the reaction with $PtCl_2$ and no co-catalyst gave 26.9 mM total products. The same reaction with the inclusion of $VO(acac)_2$ and 100 psig O_2 resulted in 55.6 mM total products. The highest reported yield of acetic acid was achieved by using the more soluble species, $Pt(acac)_2$ in the presence of $CuCl_2$ and O_2. In this case, acetic acid concentration was found to be 42 mM, corresponding to a TOF of $1.5 * 10^{-4}$ sec^{-1}.

Recently, Schuth and coworkers (Zimmermann et al. (2016) reported a system for conversion of methane to methylbisulfate with greater than 98% selectivity in 20% oleum. Using K_2PtCl_4 salts as catalysts, the authors reports a TON of greater than 16,000 (Periana–Catalytica system TON ~ 1000 under reported conditions). The experimental data suggest that methane oxidation in 20% oleum proceeds with > 90% selectivity independent of the oxidation state of the starting plati-num species. While the reported system does seem to give a better TON than the Periana–Catalytica system, the authors found that the original system proceeded with much better rates (46.1% methane conversion with Periana–Catalytica com-pared with 4.1% with K_2PtCl_4; catalyst = 50 mM, 215°C, 2.5 hours). The authors believe this rate difference is due to superior solubility properties of $(bpym)PtX_2$ as compared to K_2PtCl_4 in oleum.

12.4.2 MERCURY SYSTEMS

In 1950, Snyder and Grosse filed a patent (1950) for the conversion of methane to functionalized products facilitated by mercury salts as catalysts (Equation (12.3)). This report represents one of the earliest examples of homogeneous, catalytic methane functionalization. Methane and SO_3 were found to react in the presence of catalytic $HgSO_4$ salts over a temperature range of 100–450°C to give a mixture of oxidized products. Product composition was primarily CH_3OSO_3H, CH_3SO_3H and $CH_2(SO_3H)_2$ and the ratio of methane to SO_3 as well as the reaction temperature was found to be critical to the outcome of product selectivity. At 300°C, with a ratio of $CH_4:SO_3$ of 6.9 and a methane pressure of 1350 psi, the total conversion of methane to liquid products was found to be 44% with the ratio of methylbisulfate to MSA of 1.65: 1. System was heterogeneous and ill-defined, no mechanistic hypothesis was suggested:

$$2CH_4 \xrightarrow[\substack{SO_3 \\ 100-450\,°C}]{HgSO_4} CH_3SO_3H + CH_3OSO_3H \qquad (12.3)$$

Interestingly, it has long been known that Hg(II) can make stable bonds to carbon. The mercuration of benzene by HgX_2 salts has long been known to proceed to make C_6H_5-Hg-X species (Dimroth 1902). These reactions proceed under mild conditions (i.e. in HOAc) and provide a classic example of electrophilic aromatic substitution. It is true that CHA and electrophilic aromatic substitutions are, mechanistically, fundamentally different, but the mercuration of benzene does illustrate the propensity for Hg(II) to form strong bonds to carbon. Not surprisingly, methane will not react with Hg(II) under the mild conditions required for benzene mercuration as it is much less nucleophilic than benzene, but a report by Periana and coworkers (1993) did find that homogeneous, well-defined Hg(II) could be used for methane functionalization under strongly acidic conditions. The researchers were able to demonstrate that mercuric triflate reacted with methane stoichiometrically in trifluoromethanesulfonic acid (HOTf) to generate methyl triflate in quantitative yield based on added Hg(II) (Equation (12.4)). The nonoxidizing nature of triflic acid meant that the reduced Hg(II) could not be reoxidized and thus functioned only as a stoichiometric oxidant. However, by switching to the more oxidizing sulfuric acid, the group was able to realize catalysis with Hg(II). They reported selectivity of up to 85% for the conversion of methane to methylbisulfate in concentrated H_2SO_4, catalyzed by $HgSO_4$, with a yield of 50% based on added methane (Equation (12.5)). The major byproduct of the reaction was found to be CO_2. The authors state that sulfuric acid plays three roles in the system: (1) acts as the reaction solvent; (2) provides in situ protection of the generated methanol product via esterification; and (3) acts as the overall, stoichiometric oxidant for the reoxidation of Hg(0) (or Hg(I)) back to Hg(II):

$$Hg^{II}(CF_3SO_3)_2 + CH_4 \xrightarrow[CF_3SO_3H]{} CF_3SO_3CH_3 + CF_3SO_3H + Hg^0 \qquad (12.4)$$

$$CH_4 + 2H_2SO_4 \xrightarrow[conc.\,H_2SO_4]{HgSO_4} CH_3OSO_3H + 2H_2O + SO_2 \qquad (12.5)$$

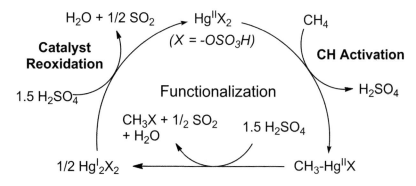

SCHEME 12.9 Catalytic cycle for the electrophilic CH activation of methane with $Hg^{II}(OSO_3H)_2$ to generate CH_3OSO_3H.

As shown in Scheme 12.9, the group hypothesized that this reaction proceeds by a Hg-facilitated, electrophilic CH activation of a methyl C–H bond to yield a Hg^{II}–Me intermediate. Functionalization of this species yields methylbisulfate and ½ an equivalent of reduced mercurous bisulfate, $Hg_2(OSO_3H)_2$. Hg(II) is regenerated by sulfuric acid oxidation of mercurous bisulfate. The authors provided evidence in support of this mechanism through the design of some key, fundamental experiments. First, heating methane with Hg(II) in D_2SO_4 resulted in small amounts of deuterium incorporation into methane, as determined by sampling of the gas phase. This suggests that a reversible CHA event between methane and Hg(II) is occurring to generate Hg^{II}–Me. This hypothesis is further supported by the observation that independently synthesized Hg^{II}–Me, when exposed to the reaction conditions, gave low yields of methane (microscopic reverse of CHA) and predominantly formed methylbisulfate as the major methyl product. The ratio of methylbisulfate to methane from the functionalization of Hg^{II}–Me matched well with the ratio of methylbisulfate to CH_3D observed in the reaction between methane and Hg(II) in D_2SO_4. In addition, $Hg-Me^{II}$ is observed by NMR in the post-reaction mixture. When reactions were run using ^{13}C-enriched CH_4, ^{13}C, ^{1}H and [199]Hg NMR confirmed that methyl mercury was a methane derived intermediate in the reaction as $Hg^{II}-^{13}CH_3$ was observed in the post-reaction mixture. Consistent with the initial report on Hg(II) functionalization of methane by Snyder and Grosse (1950), Periana et al. (1993) found that reactions had to be run in H_2SO_4 that was $\leq 98\%$. If the reaction was carried out in acid that was higher in concentration (i.e. containing some SO_3), there were issues with selectivity and other products such as MSA began to form. Similar to the previously described Periana–Catalytica system, this Hg(II) system for methane functionalization is inhibited by water and catalysis stops when the acid becomes $\sim \leq 80\%$ concentration.

Kataja, Song, and Huuska (1994) experimented with many different conditions and modifications to improve the efficiency/effectiveness of this Hg(II) system. They found that the rate of the reaction could be increased, slightly, when reactions were run under supercritical conditions. The largest influencer that they found on reaction rate, however, was temperature.

Gang et al. (2000) also conducted a detailed study on this Hg(II) system in sulfuric acid. The authors reported that the pressure drop in the reactors after equilibrium is reached is identical and independent of catalyst loading. The rate of reaction was seen to increase with increased loading of $HgSO_4$ up to the point where the solubility in the reaction media was reached (~0.8 wt% $HgSO_4$ in 65 wt% oleum). Additionally, they reported that stir rates needed to be > 720 rpm to ensure that the reactions were not mass-transfer limited. Stir rates above 720 rpm had no additional impact on reaction rates. Temperature profiles of the reaction indicated that increasing the temperature above 180°C actually has little effect on the overall rate of catalysis. This is because at elevated temperatures there is higher concentration of SO_3 in H_2SO_4 which causes reduced solubility of $HgSO_4$. The mechanism of the reaction was probed computationally by Cundari and coworkers (1998a) and was found to be in agreement with that initially proposed by the Periana group. Cundari (1998b) found that the exact activation barriers for the reaction were variable and highly dependent on the exact ligands coordinated to the Hg(II) center. They identified the transition state structure with bisulfate ligand displacement to be simultaneous with methane deprotonation and formation of the $Hg^{II}-CH_3$ bond (Figure 12.4). According to the authors, electron withdrawing groups on the "activating ligand" facilitate in CHA of methane, kinetically and thermodynamically. Electron donating groups on the "spectator ligand" facilitate in stabilizing the interactions between Hg(II) and methane in the transition state and lead to an ~ 6–12 kcal mol^{-1} reduction in energy requirements.

A report by Sen and coworkers (Sen et al. (1994), Basickes et al. (1996) utilized a similar system to that described by Periana, except the researchers also added in stoichiometric amounts of known radical initiators. In this system, they observed functionalization of both methane and ethane. The authors proposed a radical chain mechanism for the reaction (Scheme 12.10) that was in stark disagreement with the mechanism put forth by Periana involving Hg(II) facilitated CHA. They proposed that Hg(II) is able to act as a one electron oxidant due to the poor donor properties of the ligands and the highly acidic nature of the solvent. Via outer-sphere electron transfer followed by proton loss, this species interacts with methane to generate a

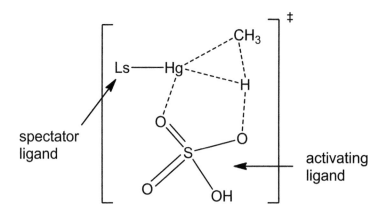

FIGURE 12.4 Ligand effects in electrophilic CH activation of methane with Hg.

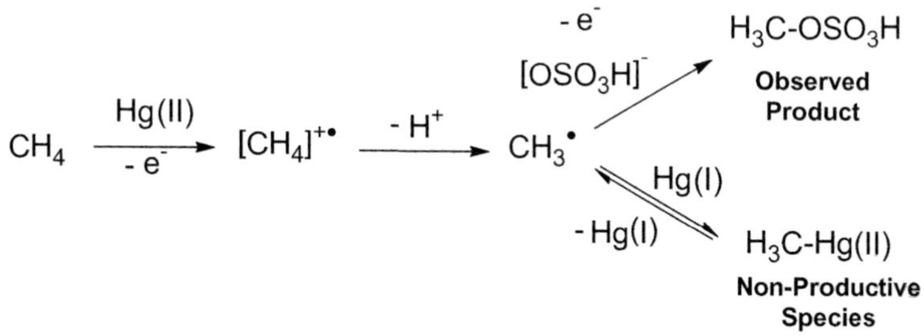

SCHEME 12.10 Radical mechanism for the Hg^{II} initiated functionalization of CH_4.

methyl radical. The observed methylmercury intermediate, according to Sen, could be formed from the reversible combination of a methyl radical with Hg(I), but it would not play a role in the propagation of the reaction mechanism. Oxidation of this methyl radical by Hg(II) leads to methylbisulfate.

Sen presented several points as evidence for a radical process over a CHA process: (1) using known radical initiators, under conditions where reoxidation of Hg(II) is known to not occur, and similar product yields are achieved; and (2) The microscopic reverse of functionalization (i.e. attempting to make $CH_3Hg(OSO_3H)$ from CH_3OSO_3H and Hg(II) or Hg(I)) was unsuccessful (Equations 12.6 and 12.3) C–C bond cleavage products were obtained, in up to 25% relative to added oxidant, when reactions were run with ethane. Similar yields of C–C bond cleavage products are observed in reactions between ethane and known radical initiators such as $K_2S_2O_8$ and Ce(IV). Precedence for such a mechanism is provided by the reaction of ethane with NO_2PF_6 as described by Olah and Lin (1971).

$$CH_3OSO_3H + Hg^I \text{ or } Hg^{II} \xrightarrow[\substack{180\ °C}]{\substack{H_2SO_4}} \times \quad CH_3HgOSO_3H \qquad (12.6)$$

A recent report by the Ess group (Fuller et al. 2016) examined these two alternative mechanisms using density functional theory. They reported that the radical mechanism proposed by Sen seems to be much less likely than the original CHA mechanism from Periana. The electron transfer pathways that Sen proposed were found to have a ΔG of ~77 kcal mol^{-1} while the alternative electrophilic substitution (CHA) mechanism was found to have a $\Delta G^{\ddagger}$ of 34 kcal mol^{-1}. Even when electron transfer reactions were coupled to proton transfer reactions (PCET), the ΔG of 41 kcal mol^{-1} still suggested that CHA was the clear low-energy pathway.

Mukhopadhyay and Bell (2004) found that Hg(II) catalysis could be utilized for the selective conversion of methane to MSA. Reactions were conducted in fuming sulfuric acid in the presence of O_2 and the temperature could not exceed 130°C or there was appreciable oxidation of MSA to methylbisulfate (CH_3OSO_3H). Under optimum conditions, it was found that $Hg(OTf)_2$, ($OTf = OSO_2CF_3$) gave the highest selectivity for MSA (CH_3SO_3H) over other functionalized products with 92% selectivity and a TON of 60. Other catalysts screened had higher catalytic activity, but they were all found to be less selective for the generation of MSA. These reactions could be further optimized to increase the yield of MSA by addition of SO_3 to the reactions up to 40 wt%, but increasing SO_3 above 30 wt% also gave an increase in product oxidation to methylbisulfate.

Interestingly, oxygen was found to be a necessary additive for increasing selectivity of MSA. Selectivity for reactions run in SO_3 for MSA were found to be 21% and 14% in the presence and absence of O_2, respectively. Additionally, higher yields of methylbisulfate were achieved when oxygen was excluded from the reaction mixture, increasing from ~2% to 10%. Lower O_2 pressures showed slower rates but increased selectivity for MSA whereas higher O_2 pressures gave the opposite effect; faster rates and lower selectivities. Study of this system led to the proposal that competing radical and electrophilic substitution pathways for methane sulfonation were both operating. The authors believe that the presence of O_2 helps to deactivate the radical pathway. The promotion of the electrophilic substitution (CHA) pathway leads to the observed higher selectivity (Scheme 12.11).

12.4.3 PALLADIUM SYSTEMS

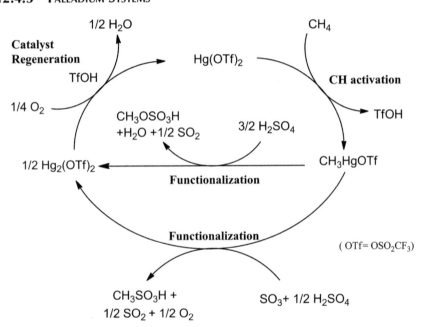

SCHEME 12.11   Catalytic cycle proposed by Mukhopadhyay and Bell for the functionalization of methane to CH_3OSO_3H and CH_3SO_3H with Hg(OTf).

Work by Sen and coworkers showed the viability of $PdSO_4$ for methane oxidation using H_2SO_4 containing 30% SO_3 as a solvent (Sen et al. 1994; Basickes et al. 1996). Under the reaction conditions (T = 160°C, Time = 1 day, CH_4 = 1000 psig, $PdSO_4$ = 0.1 mmol) the authors realized a TON of up 15 (TOF = 1.7×10^{-4} sec^{-1}) for the conversion of methane to methylbisulfate. Michalkiewicz, Kalucki, and Sošnicki (2003) were able to improve the rate and total TON of the reaction by replacing $PdSO_4$ with metallic palladium. Under the reaction conditions (T = 160°C, time = 2 hours, CH_4 = 500 psig, Pd^0 = 0.1–0.2 g, Solvent = H_2SO_4 w/30 wt% SO_3) the authors reported a TON of up to 47 (TOF = 6.5 × 10^{-3} sec^{-1}) for conversion of methane to methylbisulfate. The authors found that methylbisulfate selectivity was highly dependent on the amount of SO_3 and, although they achieved the highest TON with 30 wt% SO_3, more substantial amounts of CO_2 were also generated when $[SO_3] \geq 30$ wt% (CO_2 = ~ 7–8% of added CH_4). Catalyst recycling was demonstrated with this system for 10 cycles without any apparent loss in catalytic activity. The mechanism for the reaction is shown in Scheme 12.12:

$$2CH_4 + 4H_2SO_4 \xrightarrow[\substack{H_2SO_4\,(96\%) \\ 180\,°C}]{PdSO_4} CH_3CO_2H + 4SO_2 + 6H_2O \qquad (12.7)$$

Periana et al. (2003) reported an interesting variation of this system in 2003 in which they described "catalytic, oxidative condensation of CH_4 to CH_3CO_2H (HOAc) in one step via CH activation" using $PdSO_4$ in 96% sulfuric acid (Equation (12.7)). This system was unique because it did not require CO to generate acetic acid from methane. Control studies with $^{13}CH_4$ revealed that HOAc in this reaction was formed via oxidative condensation of two methane molecules as the formation of $^{13}CH_3{}^{13}CO_2H$ was observed (with both carbons being derived from $^{13}CH_4$). The reaction was found to occur with > 90% selectivity for the formation of HOAc and yields of up to 12% (of combined methanol/HOAc) based on added methane, were achieved. When reactions were run at 180°C for 1 hour a maximum TON of 18 was achieved (TOF = 5 × 10^{-3} sec^{-1}). Formation of acetic acid was calculated as resulting from an eight electron oxidation (TON = 4). The role of H_2SO_4 in this system was both reaction solvent and stoichiometric oxidant. The mechanism proposed for this reaction was a nonradical, two-step oxidation process involving CHA that is shown in Scheme 12.13. Evidence for the proposed mechanism was supported by experiments conducted with $^{13}CH_4$ and ^{12}CO that gave only $^{13}CH_3{}^{12}CO_2H$. It is important to note that

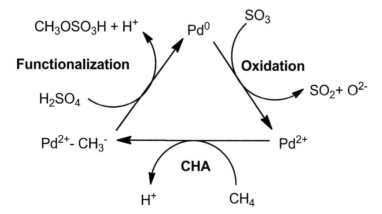

SCHEME 12.12 The catalytic cycle for Pd-catalyzed CH_4 functionalization to methylbisulfate with SO_3 and H_2SO_4.

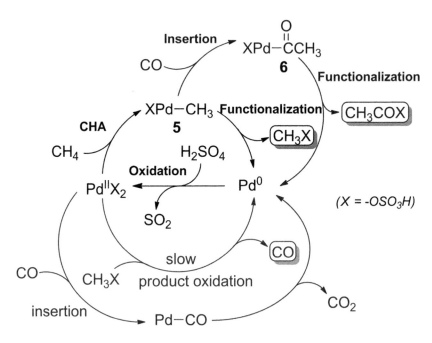

SCHEME 12.13   Catalytic cycle for Pd-catalyzed CH_4 functionalization to CH_3COOH in H_2SO_4.

the formation of CO from methyl bisulfate is slow (Scheme 12.13), but the insertion of CO into the Pd–Me species, **5**, to generate species **6** must proceed rapidly to prevent buildup of excess CO, which would be oxidized to CO_2. Controls suggest this is the case as formation of HOAc is proportional to added CO at low concentrations, but the reaction is inhibited at high CO concentrations, leading predominantly to CO_2 formation.

Bell and coworkers (Zerella et al. 2004, 2006b) studied this system to gain further insight. They found that they could integrate Cu(II) into the system and enable the use of O_2 as the terminal oxidant. The initial report utilized $CuCl_2$ as a co-catalyst (Zerella et al. 2004). Running reactions in 96% H_2SO_4 under an atmosphere of CH_4/O_2 resulted in the formation of 49 mM HOAc and 4 mM MeOH after 4 hours at 180°C with an overall selectivity for these two products of 47%. This corresponds to a TON of 10 (assuming HOAc as an 8 e^- oxidation and MeOH as a 2 e^- oxidation) and a TOF of 7×10^{-4} sec^{-1}. Consistent with the mechanism proposed by Periana, Bell confirmed that CO is an intermediate in the reaction and the authors proposed a stepwise mechanism (shown in Scheme 12.14) that is similar to the one proposed by Periana.

In a follow up report, the researchers investigated a similar reaction, but excluded the use of Cu(II) as a co-catalyst (but still contained O_2) (Zerella et al. 2006b). In this investigation, the authors report yields that were similar to the original system described by Periana (where O_2 was excluded from the reaction). When oxygen was included in the reaction, the authors found that yields of HOAc were dependent on the ratio of O_2 to methane as well as the overall partial pressures of the two gasses. Higher ratios of O_2: CH_4 and higher total pressures resulted in higher yields of HOAc

1. $Pd(OSO_3H)_2 + CH_4 \xrightarrow{\text{CHA}} CH_3\text{-}Pd(OSO_3H) + H_2SO_4$

2. $CH_3\text{-}Pd(OSO_3H) + CO \xrightarrow{\text{Insertion}} (CH_3CO)Pd(OSO_3H)$

3. $(CH_3CO)Pd(OSO_3H) + H_2O \xrightarrow{\text{Hydrolysis}} CH_3COOH + H_2SO_4 + Pd(0)$

4. $CH_3\text{-}Pd(OSO_3H) \xrightarrow{\text{Functionalization}} CH_3OSO_3H + Pd(0)$

5. $CH_3OSO_3H + H_2O \xrightarrow{\text{Hydrolysis}} CH_3OH + H_2SO_4$

SCHEME 12.14 Stepwise mechanism for Pd-catalyzed CH_4 functionalization to HOAc with H_2SO_4.

and a TON of up to 14 with selectivity for HOAc approaching 82%. The mechanism proposed for this reaction is analogous to the original mechanism proposed by Periana, Scheme 12.13 with the exception that O_2 is facilitating reoxidation of Pd in place of SO_3. The mechanistic details of these reactions have been investigated by Chempath and Bell (2006) using density functional theory. This study seems to support the previous mechanistic proposals.

In the original system disclosed by Periana, formation of palladium black (Pd^0) was found to be a major challenge that was exacerbated by high concentrations of CO. Reoxidation of Pd^0 by SO_3 is kinetically slow, but this rate of reoxidation seems to be increased in the systems described by Bell that include O_2 or O_2/Cu(II). As a result, higher TOF and total product concentrations were observed in these systems.

An et al. (2006) reported on the use of $Pd(OAc)_2$ in trifluoroacetic acid (a non-superacid media) to convert methane to methyl trifluoroacetate (MeTFA). The authors report that catalysis was achieved by employing a series of redox couples that enabled regeneration of the Pd(II) catalyst using O_2 as the terminal oxidant. It was necessary to couple three redox couples, Pd^0/Pd^{II}, p-benzoquinone (BQ)/hydroquinone (H_2Q) and $NO/NO_2/O_2$, in order to effectively allow the use of molecular oxygen as the terminal oxidant. Using this system, the authors report yields of MeTFA up to 106 µmol after 10 hours, corresponding to a TOF of 1.9×10^{-4} sec^{-1}. The overall redox couple and oxidation sequence is shown in Scheme 12.15.

$CH_4 + CF_3COOH \xrightarrow[- Pd^0]{Pd^{2+}} CF_3COOCH_3 + 2H^+$

$Pd^0 + 2H^+ + BQ \longrightarrow H_2Q + Pd^{2+}$

$H_2Q + NO_2 \longrightarrow BQ + NO + H_2O$

$NO + 0.5\,O_2 \longrightarrow NO_2$

regeneration of Pd^{2+} catalyst through coupled red-ox reaction

Net: $CH_4 + CF_3COOH + 0.5\,O_2 \longrightarrow CF_3COOCH_3 + H_2O$

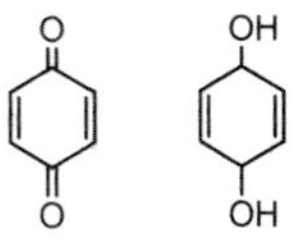

SCHEME 12.15 Pd^{II}-catalyzed methane functionalization to MeTFA.

Yuan et al. (2011, 2013a, b) also explored Pd catalysis for methane oxidation in trifluoroacetic acid. In the initial report, they utilized the molybdovanadophosphoric acid, $H_5PMo_{10}V_2O_{40}$, BQ and O_2 in their redox cycle (Yuan et al. 2011). The authors found that addition of perfluorooctane to the system resulted in a higher obtained yield of MeTFA (35 mM product, TON of 118) as compared to a system run in neat trifluoroacetic acid under otherwise identical conditions (0.91 mM product, TON of 3). The authors propose that C_8F_{18} increases the solubility of O_2 in the reaction medium which aids in regeneration of $H_5PMo_{10}V_2O_{40}$ and BQ. The authors state that they do not believe a free-radical mechanism is operating. It seems reasonable that, similar to the report above proposed by Periana, CHA could be operating.

Yuan group (2013a) extended this chemistry to show that methane could be oxidized to a mixture of HOAc and MeTFA in a system comprised of K_2PdCl_4, $H_5PMo_{10}V_2O_{40}$ and O_2 in trifluoroacetic acid. Acetic acid was observed to be the major product and only minor amounts of MeTFA were generated. The best reported run gave a TON of 4248 (4167 HOAc, 81 MeTFA) after 8 hours, corresponding to a TOF of 1.5×10^{-1} sec^{-1}. The mechanistic analysis of this system is complicated by a background reaction between methane and the molybdovanadophosphoric acid that gives functionalized product in the absence of palladium. Nevertheless, the authors propose a Pd(II)-mediated CHA event followed by reductive elimination to generate product. Reoxidation of Pd^0 to Pd^{II} by $H_5PMo_{10}V_2O_{40}$ was proposed to proceed by a radical mechanism. Unfortunately, insufficient data was provided in the report to speculate on the source of the carbon for the second methyl group in the HOAc product and further studies are required to ascertain whether it is derived from methane or from some other pathway, such as decarboxylation of the solvent.

The Yuan group also found that (2,2'-bipyridyl)PdCl$_2$ could also be used in trifluoroacetic acid to catalyze the conversion of methane to methanol with $H_5PMo_{10}V_2O_{40}$, BQ, and molecular oxygen. Similar to their original report, they found that the addition of perfluorooctane increased the yield of the reaction. Addition of water was also found to increase the yield of MeOH which the researchers proposed facilitated loss of chloride from the Pd center, which would facilitate reductive functionalization to generate methanol (Yuan et al. 2013b). This is shown in Scheme 12.16.

Mukhopadhyay and Bell (2003) performed selective sulfonation of methane to generate MSA in trifluoromethanesulfonic acid (TfOH) solvent with Pd(II) and Cu(II) salts as catalysts. SO_2 and O_2 were also employed as reagents in the system as shown in Equation (12.8). In order to obtain appreciable yields of MSA, it was found that the palladium and copper catalysts needed to be simultaneously present, with copper hypothesized to facilitate reoxidation of palladium by molecular oxygen. Under conditions of 1200 psig CH$_4$ and 30 psig SO$_2$ yields of MSA up to 20% (based on added SO$_2$) were obtained after 40 hours at 85°C. More work is needed to ascertain detailed mechanistic information about this system:

$$CH_4 + SO_2 + 0.5O_2 \xrightarrow[\substack{85°C, 12-20h, \\ TfOH}]{\text{Pd and Cu salts}} CH_3SO_3H \qquad (12.8)$$
$$\text{MSA}$$

Fujiwara and coworkers (Nishiguchi et al. 1992, Nakata et al. 1993, 1994, Fujiwara et al. 1996) developed systems for the conversion of methane and CO to HOAc using

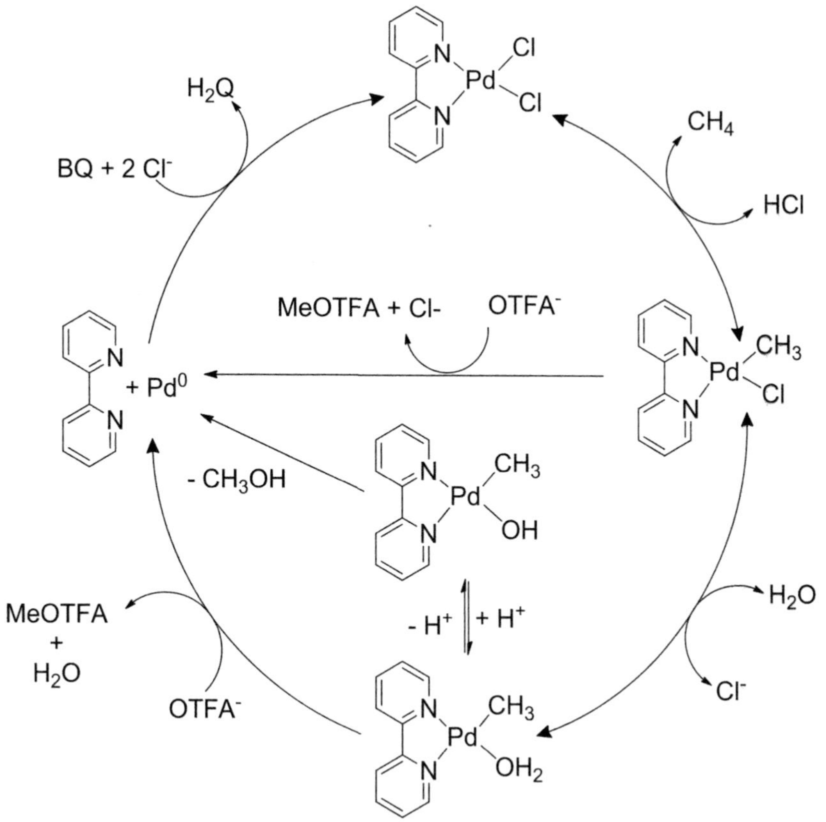

SCHEME 12.16 Oxidation of CH_4 mediated by a $PdCl_2(bpy)$ complex.

Pd(II) and/or Cu(II) catalysts in the presence of molecular oxygen in trifluoroacetic acid solvent. While palladium systems typically operate by electrophilic CHA, Cu(II) systems are known to promote radical reactions. The authors propose that the inclusion of Cu(II) in the reaction alters the catalytic mechanism (Scheme 12.17). The rate enhancement in the bimetallic system (both Pd and Cu) was attributed to formation of a 1:1 Pd(II):Cu(II) complex with increased electrophilicity. Switching from CO to CO_2 (Eq. (12.9)) was found to decrease the overall reaction rate, but a similar TON (~10–15) was achieved after 20 hours at 80°C:

$$CH_4 + CO_2 \xrightarrow[\substack{K_2S_2O_8,\text{HTFA} \\ 80\,°C,\,20\,h}]{\text{Pd(OAc)}_2 \text{ and/or Cu(OAc)}_2} \underset{\text{(AcOH)}}{CH_3COOH} \tag{12.9}$$

12.4.4 RHODIUM SYSTEMS

The Monsanto process, based on a rhodium catalyst, is used for the conversion of methanol to acetic acid (more recently it has been largely replaced by the greener and more efficient Cativa process, developed by BP Chemicals Ltd, which is a process

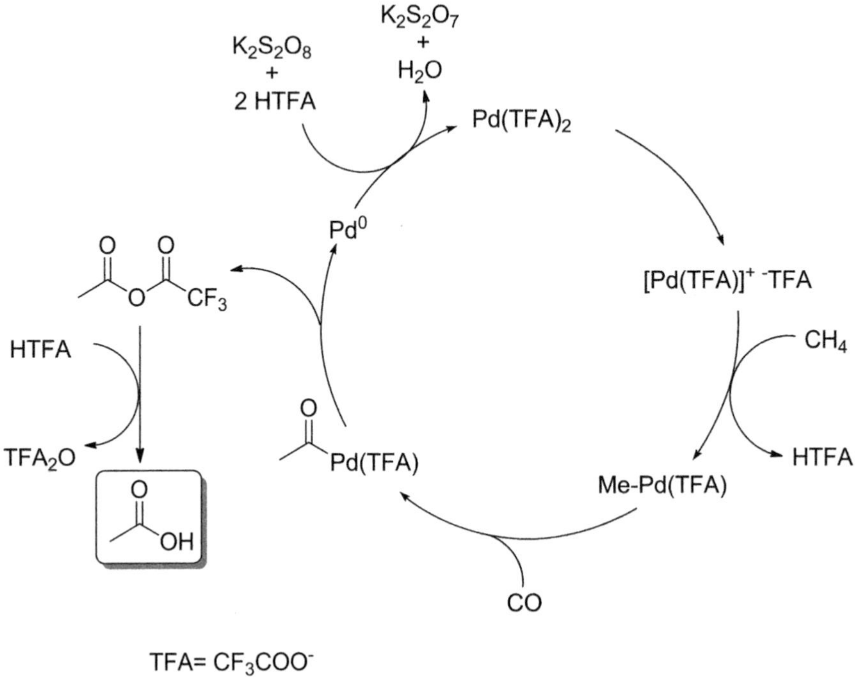

SCHEME 12.17 Pd-catalyzed CH_4 functionalization to HOAc with CO and O_2 or $K_2S_2O_8$ in CF_3CO_2H.

similar to the Monsanto process but uses an iridium-based catalyst) by carbonylation of methanol (Jones 2000; Sunley and Watson 2000). The required use of methanol in this process is less than ideal because the methanol is produced through the energy intensive, syngas process. The capability to directly convert methane to acetic acid using CO, and O_2 is an economically desirable process. Although the thermodynamics are not favorable for the functionalization of methane to acetaldehyde by reaction with CO ($\Delta G = +14$ kcal/mol), the reaction of methane with CO and O_2 to generate acetic acid is thermodynamically favorable (Chepaikin et al. 1998).

Lin and Sen (1994), using $RhCl_3$ as a catalyst, directly converted methane to acetic acid in the presence of CO and O_2 in an aqueous medium at a temperature of around 100°C. Methanol and formic acid are also generated in this system in significant yields. The proposed reaction mechanism involved CH activation of methane by Rh catalyst to make $Rh–CH_3$ intermediate, which then undergoes CO insertion to make $Rh–COCH_3$ and subsequently generates acetic acid after reaction with water. Addition of a promoter (Cl–, I–, or Pd/C) significantly increased the yields of acetic acid. Isotope labeling studies with $^{13}CH_3OH$ did not generate any $^{13}CH_3CO_2H$ post reaction, which suggested that the $Rh-CH_3$ species is derived from methane, not from methanol or methyl iodide (generated from $CH_3OH + HI$). In a striking dissimilarity, in the Monsanto process the in situ generated methyl iodide, generated from methanol is crucial, and adds to Rh catalyst in an oxidative addition fashion (Forster 1979).

The major problem in this system is low turnover rates (~ 0.1 turnover/h based on Rh) in water. The yields were significantly improved by adding a strong acid to the water which not only increased the reaction rates, but also changed the product distribution (Lin et al. 1996). In a 6:1(v/v) mixture of perfluorobutyric acid:water solvent system, the product distribution significantly shifted to favor the formation of methanol and its ester ($C_3F_7CO_2CH_3$) over acetic acid. The authors attributed the increase in catalytic rate to the increased nucleophilicity of the perfluorobutyrate anion as compared with water which can more easily capture the Rh–CH_3 species as the ester (Scheme 12.18). However, water is important in this reaction as no product formation was observed in pure perfluorobutyric acid.

When reaction was carried out in the presence of $^{13}CH_3OH$ and $^{12}CH_4$ the methyl products observed are $^{12}CH_3OH$, $C_3F_7CO_2^{12}CH_3$, and $^{12}CH_3CO_2H$, which are derived entirely from methane. Which suggests that methane is considerably more reactive toward Rh metal complex in this system than the corresponding oxyfunctionalized products, methanol and its ester ($C_3F_7CO_2CH_3$), despite lower homolytic bond strength (10 kcal mol^{-1} lower than the corresponding bond strength in methane). In this system product selectivities and yields are sensitive towards solvent type and the presence of additives. Addition of external radical sources was not observed to impact the product distribution suggesting that radical processes are not operating under these reaction conditions.

Chepaikin and coworkers (1998, 2001, 2002, 2006, 2010) developed a RhCl$_3$/(DCl)NaCl/KI system for the generation of acetic acid through oxidative carbonylation of methane in aqueous medium in the presence of CO and O_2. They also found that addition of acids (acetic acid and trifluoroacetic acid) to the aqueous medium accelerates the reaction rates. Less coordinating trifluoroacetic acid was found to be more effective. Selectivity for methanol-derived products, methanol and methyl trifluoroacetate, has been increased in the presence of CuO, which is attributed to the result of formation of CuCl-OOH, which helps with oxidation of the rhodium complex (Chepaikin et al. 2002). In this system the yields of methanol and methyltrifluoroacetate are found to be dependent on the on the ratio of Rh^{III}:Cu^{II}:Cl$^-$ and using a ratio of 1:20:6 was found be important for the higher yields. The yields of the other products observed in this system, acetic acid and formic acid, are found be fairly independent of Rh^{III}:Cu^{II}:Cl$^-$ ratio. A comparative kinetic isotope effect (KIE) study using CH_4 and CD_4 with RhCl$_3$/NaCl/KI ($k_H/k_D = 4.34 \pm 0.16$) and RhCl$_3$/CuO/NaCl (3.93 ± 0.18) catalytic systems in trifluoroacetic acid indicated a CH bond breaking mechanism and provided more support against the involvement of free radicals in

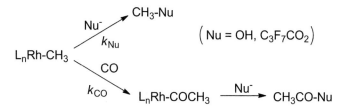

SCHEME 12.18 Suggested origin for product selectivity observed in Rh-catalyzed methane functionalization.

the reaction mechanism (Chepaikin et al. 2006). Studies conducted with isotopically labeled $^{18}O_2$, $H_2^{18}O$, and $CF_3CO^{18}OH$ indicated an O-atom transfer from oxygen molecule to Rh–CH$_3$ intermediate, which was believed to proceed either by direct insertion of O_2 in to the Rh–CH$_3$ bond (observed with Pt complexes and unknown with Rh) or generation of a Rh–CH$_3$-peroxo species and migration of the methyl group to one of the O-atoms of the peroxo (which seems more likely). Indeed, upon addition of titanium metal (used as a test for two-electron oxidizing agents such as H_2O_2) oxidation is observed under conditions similar to those of the parent reaction.

Hristov and Ziegler (2003b) performed DFT studies on Sen's system (Lin and Sen 1994) for Rh-catalyzed selective functionalization of methane with CO in the presence of O_2, and proposed the following mechanistic steps for the process (Scheme 12.19). The catalytic cycle mainly involved 4 steps for the generation of acetic acid: (1) C–H activation, (2) CO insertion, (3) reductive elimination, and (4) metal hydride oxidation. C–H activation step was studied by both oxidative addition (OA) and σ-bond metathesis (SBM) and the authors report that OA mechanism was seen to be favored with a significantly lower barrier of 24.8 kcal/mol compared to 29.8 kcal/mol for SBM. As shown in the Scheme 12.19, intermediate **9 (A or B)** was found to be responsible for the observed distribution of products through two different possible

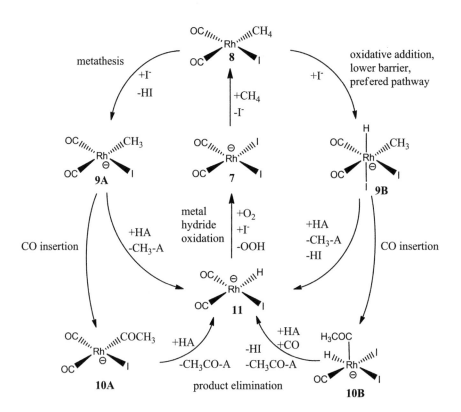

SCHEME 12.19 Proposed catalytic cycles for Rh-catalyzed methane functionalization with CO and O_2.

pathways in both the mechanisms. In the presence of more acidic solvent (i.e. in the presence of perfluorobutyric acid), attack by the anion is favored (to make methyl ester, CH_3–A) over CO insertion. This explains why acetic acid is the preferred product when only water, a low nucleophilic solvent, is used in the reaction.

Rhodium-based CH activation of methane has not received nearly the same level of attention as other systems. However, products derived from methane (methanol, its ester, and acetic acid) have been reported in yields up to 1.2 M. Although these systems do not display catalytic rates comparable to those observed in other homogeneous systems, the report by Sen's group, which was further investigated by the groups of Chepaikin and Ziegler, is an important discovery in the area of CH activation and functionalization chemistry.

12.4.5 IODINE SYSTEMS

Elemental iodine dissolved in H_2SO_4/SO_3 was found to be efficient for the selective catalytic oxidation of methane to methyl bisulfate (Periana et al. 2002). Product was not observed when < 98% H_2SO_4 was used, and oleum (2–3% SO_3) in H_2SO_4 was found to be critical for the catalytic oxidation. A 53% conversion of methane was achieved at 195°C with > 95% selectivity for C–O based products. Methyl bisulfate concentration of up to 1 M was achieved with a turnover frequency of 3.3×10^{-2} s^{-1} and a volumetric productivity of ~10^{-7} mol mL^{-1} s^{-1}. Other possible methane functionalized products CH_3OH, CH_3I, or CH_3SO_3H were not observed and only SO_2, and minor levels of CO_2, < 1% based on added CH_4, were observed in the gas phase. The H/D exchange (~50%) between gas phase CH_4 and D_2SO_4/SO_3 proceeds in just 30 minutes (Olah and Prakash 1985). Based on superacid chemistry, Olah and coworkers (1985) proposed CH_4D^+ as a plausible intermediate in this process. Interestingly H/D exchange was found to be suppressed in the presence of I_2. Essentially no deuterium incorporation was observed in the generated methyl bisulfate product, when reaction was carried out using same experimental conditions as described above, but in D_2SO_4 instead of H_2SO_4. This led to the suggestion that CH_5^+ is an unlikely intermediate in the catalytic cycle. Based on these results authors proposed $I_2^+HS_2O_7^-$ as the active catalyst resulting from the combination of I_2, SO_3, and H_2SO_4 (Scheme 12.20). The proposed active catalyst was supported by the stoichiometric reaction of methane with independently synthesized $I_2^+[Sb_2F_{11}]^-$, in oleum, to give 30% methyl bisulfate at 50°C. In the absence of SO_3 no products were observed. Other possible cationic iodine generating species $IOHSO_4$, IO_2HSO_4, and $I(HSO_4)_3$ did not yield any products under these mild reaction conditions. This observation strongly suggests the existence of I_2^+ species as the active catalyst in the catalytic species. The proposed catalytic cycle in the Scheme 12.20 was further supported by the fact that, at 150°C, methyl iodide in oleum quantitatively generates methyl bisulfate and a characteristic blue color species that was attributed to I_2^+. No significant effect on the reaction rate and selectivity was observed when the reactions were carried out in the presence of oxygen and $K_2S_2O_8$, indicating the absence of possible radical pathways in the proposed catalytic cycle. But, when the reactions were performed either with Br_2 or Cl_2 instead of I_2, signs of radical processes are observed, as polyhalogenated methane derivatives are generated in the reaction mixture (Gillespie and Morton 1971; O'Donnel 1992).

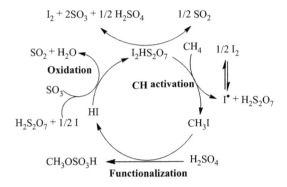

SCHEME 12.20 Proposed catalytic cycle for CH_4 oxidation by iodine dissolved in H_2SO_4/SO_3 solvent system.

Gang et al. (2004) and Michalkiewicz et al. (2009) further extended the work on these systems by replacing I_2 with KI, CH_3I, I_2O_5, KIO_3, and KIO_4. Their findings indicated that these other iodine sources could also catalyze the reaction to make methyl bisulfate from methane in H_2SO_4/SO_3 medium with similar yields and selectivity. This system was identified to yield better results at ~180°C and yielded significant over oxidation products at higher temperatures. Slightly higher methanol yields of 57% and a final methanol concentration of up to 6 M was obtained with total consumption of the added SO_3 in this system under optimized conditions.

Michalkiewicz and Balcer (2012) studied Br^- salts in oleum for methane oxidation to $MeOSO_3H$. The authors reported that 0.017 M KBr dissolved in sulfuric acid containing SO_3 (6–25 wt%) catalyzes the functionalization of methane to methyl bisulfate between 110–180°C. The study indicated that reaction proceeds only above 6.6 wt% SO_3 and 500 psi of methane. The authors reported no observation of polyhalogenated methane derivatives and only traces of CO_2 were detected in the gas phase.

In an effort to identify the true active catalyst in these I_2 catalyzed methane functionalization reactions, Davico (2005) performed gas-phase reactions and *ab initio* calculations on this system. Gas-phase reactions showed that I_2^+ did not react with methane at a measurable rate and *ab initio* calculations also indicated no favorable energetics for the reaction of I_2^+ with methane, but I^+ was found to react with methane quite rapidly both by gas-phase reactions and by *ab initio* calculations. Based on these observations Davico proposed I^+ as the active catalyst. Activation of the C–H bond was proposed to proceed by direct insertion into the C–H bond to give CH_3IH^+ after which H_2SO_4 could react with CH_3IH^+ (or its conjugated base CH_3I) to yield methyl bisulfate through a displacement reaction. The rest of the catalytic cycle and the generation of I^+ is analogous that suggested in Scheme 12.20. Davico's observations were made using gas-phase reactions and did not fully consider the solvent effects of highly polar sulfuric acid medium in his calculations which could explain the mechanistic discrepancies. Davico's observations are also in disagreement with the successful reaction of $I_2^+[Sb_2F_{11}]^-$ with methane in sulfuric acid and oleum to generate methyl bisulfate.

The reaction rates and product selectivities in the initial report of Periana et al. (2002) suggested a first-order dependence on both methane and iodine, but Jarosińska et al. (2008) reported the reaction to have a 0.5 order dependence on iodine. This indicates the requirement to carry out more theoretical and experimental studies to identify the actual catalytic species in this I_2 catalyzed reaction for functionalization of methane to methyl bisulfate in H_2SO_4/SO_3 solvent medium.

12.5 CHA2 STOICHIOMETRIC SYSTEMS THAT UTILIZE O_2/ O_2-REGENERABLE OXIDANTS AND OPERATE BY NONCHAIN, CH ACTIVATION REACTIONS

12.5.1 MERCURY SYSTEMS

In this classification, the stoichiometric oxidants used for the CHA of methane are O_2-regenerable. Very few systems that can be classified under this category have been reported. The stoichiometric reaction of methane with $Hg^{II}(CF_3SO_3)_2$, mercuric triflate, in nonoxidizing triflic acid to generate near quantitative yields of $CF_3SO_3CH_3$, methyl triflate, and the reduced mercurous triflate, $Hg_2(CF_3SO_3)_2$ falls in this category (Periana et al.1993). This is an O_2-regenerable system as the $Hg^{II}/$ Hg^0 couple ($E^\circ \sim 0.9$ V) (Hg^{II}/Hg^I couple ($E^\circ \sim 0.91$ V)) can be reoxidized by O_2 (E° = 1.2 V). In this stoichiometric CHA system, CH_3-Hg^{II} intermediate was generated by the direct reaction of Hg(II) with methane. The intermediate then functionalizes to yield methyl triflate. This system was also made catalytic by replacing triflic acid with oxidizing concentrated H_2SO_4. As shown in Equation (12.4), Hg(II) system has the potential to further develop into a system where the reaction can be carried out in a compatible inert, aprotic solvent medium by utilizing the acids being released during the reaction. However, the toxicity of mercury may hamper the further development of these stoichiometric systems into commercial process.

12.5.2 PALLADIUM SYSTEMS

The stoichiometric reaction of $Pd(OAc)_2$ with methane to generate $CF_3CO_2CH_3$, methyl trifluoroacetate, falls in this category. Sen and coworkers (Gretz et al. 1987, Sen et al. 1989) reported the first Pd(II) mediated stoichiometric functionalization of arenes and alkanes in trifluoroacetic acid to the corresponding alkyl and aryl trifluoroacetates, Equation (12.10), via an electrophilic CH activation mechanism This system was shown to react with various hydrocarbons (e.g. methane, adamantane, toluene, p-xylene, and p-dimethoxybenzene). Based on thermodynamic data, Table 12.2, this system can be reoxidized with O_2. Trifluoroacetic acid was used as the solvent due to the absence of CH bonds and it also facilitates the generation of highly electrophilic metal centers. Relative to added Pd(II), a 60% yield of methyl trifluoroacetate (MeTFA) was reported when $Pd(OAc)_2$ was reacted with methane (800 psig) in trifluoroacetic acid solvent at 80°C after 1 hr. Unfortunately, the high cost of Pd, and the stoichiometric reaction with methane makes it a nonpractical system:

$$Pd^{II}\text{-}X + CH_3\text{-}H \underset{+HX}{\overset{-HX}{\rightleftharpoons}} Pd\text{-}CH_3 \xrightarrow{Nu} Pd^0 + CH_3\text{-}Nu \qquad (12.10)$$

$$Nu=nucleophile$$

12.5.3 ANTIMONY SYSTEMS

More recently, Koppaka et al. (2019) reported an Sb(V) based, nonsuperacid system for the functionalization of methane to methayl trifluoroacetate (MeTFA) in trifluoroacetic acid and trifluoroacetic anhydride solvent mixture. Although authors did not demonstrate the oxygen regenerability of the system, the Sb(V)/Sb(III) couple has a reduction potential in the range of 0.8–1.0 V and can be readily reoxidized by O_2 ($E° = 1.23$ V). The antimony system was developed in an effort to move away from toxic main-group metal systems, Tl and Pb, which were found to be efficient non-O_2 regenerable systems for the functionalization of methane (Hashiguchi et al. 2014). Using this relatively less toxic Sb(V) system authors achieved ~ 6% yield (based on Sb(V)) of functionalized methane product (MeTFA) at 180°C after 3 hrs with greater than 90% selectivity. The authors performed theoretical and experimental studies and provided evidence for the metal centered CH activation pathway for the functionalization of methane to MeTFA (Scheme 12.21). This is an interesting and significant finding given that Sb(V) is a well-known, powerful coordination Lewis acid that can generate Brønsted superacids, which functionalize alkanes through proton-mediated mechanisms or through generation of carbocations as shown in Olah superacid chemistry (Olah and Prakash 1985). It is possible with these main-group cations in more coordinating, nonsuperacid media that, despite the very high metal electrophilicity, the d^{10} electronic configuration and resulting lack of ligand field stabilization minimize the barrier to alkane coordination that is required for CH activation. This allows these highly electrophilic metal centers to facilitate both CHA (by stabilizing inner-sphere covalent bonding to carbon) and facile two-electron, nonradical, reductive functionalization of metal alkyl intermediate (TFA$_4$Sb-CH$_3$) to MeTFA and reduced metal species(Sb(III)).

12.6 CHO1 CATALYTIC SYSTEMS THAT UTILIZE O$_2$/O$_2$- REGENERABLE OXIDANTS AND OPERATE BY NONCHAIN, CH OXIDATION REACTIONS

12.6.1 EUROPIUM SYSTEMS

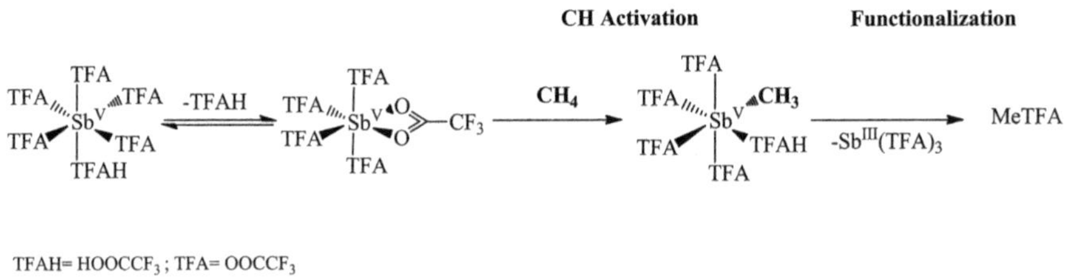

TFAH= HOOCCF$_3$; TFA= OOCCF$_3$

SCHEME 12.21 Mechanism of the functionalization of methane to MeTFA by Sb(TFA)$_5$.

Yamanaka and coworkers (1995, 2002) reported Eu^{III}/Zn^0 system for the catalytic oxidation of methane to methanol under mild reaction conditions in trifluoroacetic acid. Molecular oxygen was used as the terminal oxidant and the reaction mixtures were worked up with NaOH to obtain methanol. Authors achieved a maximum methanol TON of 5.3 in 1 hr at 50°C with 235 psig of methane pressure. This system also produced a substantial amount of CO_2 with a TON of 2 in 1 hr, which was believed to be due to the oxidation of the solvent trifluoroacetic acid by Eu^{III} as it was also observed in the absence of methane. The proposed catalytic cycle involves reduction of Eu^{III} by Zn to Eu^{II}, which then reacts with O_2 to make active oxygen-Eu^{II} intermediate. Methane or trifluoroacetic acid then reacts with this intermediate to generate methanol ester or CO_2, respectively.

The authors also improved the TON for methanol formation by introducing various metal salts as promoters (Yamanaka et al. 1996a). In the presence of bis(2,4-pentane-dionate)TiO, $(acac)_2TiO$ or TiO_2 they observed a substantial increase in the methanol TON > 10 in 1 hr along with a large increase in the TON of CO_2 as well. The increase in TON of products was believed to be due to the acceleration in the rate of formation of active oxygen-Eu^{II} species by Ti^{IV} in the catalytic system. Not surprisingly no change in the selectivity of methanol to CO_2 was observed. The reported comparative kinetic studies on the oxidation of cyclopentane and cyclohexane suggested the presence of alkyl radical intermediates in the oxidation process (Yamanaka et al. 1996b). KIE studies suggested the abstraction of proton as the key step in forming the proposed alkyl radicals and thus the oxidation of alkanes. It is further supported by the nonretention of configuration in the oxidation of *cis*- and *trans*-1,2-dimethyl-cyclohexanes. Although no insights were provided on whether the radical reaction proceeds either by chain or nonchain mechanism, based on the use of oxygen as the sole oxidant, the system is believed to be proceeds by a nonchain mechanism.

12.6.2 RUTHENIUM SYSTEMS

The binuclear oxo bridged Ru^{3+} salen dimer $[(HSalen)_2Ru_2(\mu-O)(\mu-CH_3COO)_2]$, was shown to efficiently catalyze the oxidation of methane to methanol by molecular oxygen with minimal formation of formaldehyde in a mixture of 1:1(v/v) acetone-water solvent (Khokhar et al. 2009). At 30°C a TON of 58 was achieved with mixture of 150 psig of methane and 75 psig of O_2 partial pressures. They did not report any formation of CO_2. A nonradical, ionic mechanism has been proposed by the authors for the oxidation of methane to methanol based on kinetics and radical scavenger studies. The proposed mechanism involved the activation of both oxygen and methane by ruthenium complex and transfer of oxygen atom to the carbonium ion, resulting from the abstraction of hydride from the methane, and formation of methanol as a final product.

12.7 CHO2 STOICHIOMETRIC SYSTEMS THAT UTILIZE O_2/O_2-REGENERABLE OXIDANTS AND OPERATE BY NONCHAIN, CH OXIDATION REACTIONS

Interestingly, no system that fit into this category was identified. It might be that these O_2 generable oxidants may not be sufficiently oxidizing to directly oxidize methane. This may represent an area of focus for future research.

12.8   SUMMARY AND CONCLUSIONS

The direct conversion of methane to fuels or chemicals at substantially below 800°C and without the need to generate syngas is one of the remaining "Grand Challenges" in chemistry. Given the ubiquity of vast amounts of natural gas, such a process would significantly reduce the dependence on oil and thus the ever-increasing greenhouse gas emissions, with huge environmental and economic dividends. In this chapter, we discussed various homogeneous systems for the conversion of methane to various functionalized products. Although these systems have made significant progress towards addressing the grand challenge, unfortunately, no single system has come close to disrupting the existing high-temperature, syngas-based commercial processes. Hence the need for more research in this area is highly warranted. Most importantly, need to develop methods that require the use of O_2 as the only co-reactant and generate industrially important oxygenate products such as methanol, dimethyl ether, acetic acid, and non oxygenated products such as olefins or higher hydrocarbons are highly required.

To better understand and gain more insights into the practicality we have tried to classify these systems based on reaction mechanism in to three classes, CH activation, CH oxidation and chain reactions with emphasis on the fundamental differences. CH activation and CH oxidation reactions are further divided into stoichiometric and catalytic.

Encouragingly a significant number of reported systems that can fall into O_2-regenerable category (based on the *thermodynamic potential* whether O_2 can be used as the only co-reactant or not; and in fact in most of the reported systems it has not been demonstrated) have been reported. It is also promising to find that these O_2-regenerable systems use stoichiometric oxidants that could potentially be used in a Wacker-type process. Unfortunately, most of these are H_2SO_4/SO_3–based systems and are impractical as they suffer from the disadvantage of high cost involved in separating the functionalized products, and in some cases high toxicity of the stoichiometric oxidant used. Consequently, the focus of the current research should be on identification of non-toxic systems that do not utilize strong acids.

It is interesting to mention that there are no reports that fall under the category O_2-regenerable, stoichiometric, CH oxidation (CHO2). This could be due to the higher oxidation potential limits of the oxidants and the requirement for these species to carry out elementary reactions with methane that directly generate oxidized methyl intermediates or products. Whereas CHA systems, considered as acid- base reaction doesn't generate a methyl intermediate that is formally oxidized at the carbon center through elementary reaction with methane. It is likely that CHA reactions, as is generally the case with acid-base reactions, could be more feasible than the CHO reactions, as the latter requires a CH bond cleavage with more electronic rearrangements in the transition state.

Although no emphasis was laid on any particular approach towards addressing the problem, we believe that, given the challenges involved in the direct use of oxygen, such as generation of radicals and low selectivity, development of an indirect system based on a modified Wacker process is ideal. Although significant progress has been made in this field, continued research is still required to address the "Grand Challenge."

ACKNOWLEDGMENTS

This work was supported by The Scripps Research Institute.

ASSOCIATED CONTENT

Notes: The authors declare no competing financial interest.

REFERENCES

Ahlquist, M., Periana, R. A., and W. A. Goddard. 2009. C–H activation in strongly acidic media. The co-catalytic effect of the reaction medium. *Chem. Commun.* 7:2373–2375.

An, Z. J., Pan, X. L., Liu, X. M., Han, X. W., and X. H. Bao. 2006. Combined redox couples for catalytic oxidation of methane by dioxygen at low temperatures. *J. Am. Chem. Soc.* 128:16028–16029.

Arakawa, H., Aresta, M., Armor, J. N., Barteau, M. A., Beckman, E. J., Bell, A. T., Bercaw, J. E., Creutz, C., Dinjus, E., Dixon, D. A., Domen, K., DuBois, D. L., Eckert, J., Fujita, E., Gibson, D. H., Goddard, W. A., Goodman, D. W., Keller, J., Kubas, G. J., Kung, H. H., Lyons, J. E., Manzer, L. E., Marks, T. J., Morokuma, K., Nicholas, K. M., Periana, R., Que, L., Rostrup-Nielson, J., Sachtler, W. M. H., Schmidt, L. D., Sen, A., Somorjai, G. A., Stair, P. C., Stults, B. R., and W. Tumas. 2001. Catalysis research of relevance to carbon management: progress, challenges, and opportunities. *Chem. Rev.* 101:953–996.

Basickes, N., Hogan, T. E., and A. Sen. 1996. Radical-initiated functionalization of methane and ethane in fuming sulfuric acid. *J. Am. Chem. Soc.* 118:13111–13112.

Bischof, S. M., Hashiguchi, B. G., Konnick, M. M., and Periana, R. A. 2013. Designing molecular catalysts for selective CH functionalization. In *Topics in Organometallic Chemistry: Inventing Reactions*, ed. L. J. Gooßen, 195–232. Berlin: Springer-Verlag.

Butikofer, J. L., Parson, T. G., and D. M. Roddick. 2006. Adduct studies and reactivity of trans-$[(C_2F_5)_2MeP]_2Pt(Me)X$ $(X = O_2CCF_3, OTF, OSO_2F)$. *Organometallics* 25:6108–6114.

Caballero, A., and P. J. Perez. 2013. Methane as raw material in synthetic chemistry: the final frontier. *Chem. Soc. Rev.* 42:8809–8820.

Caballero, A., Despagnet-Ayoub, E., Diaz-Requejo, M. M., Diaz-Rodriguez, A., Gonzalez-Nunez, M. E., Mello, R., Munoz, B. K., Ojo, W. S., Asensio, G., Etienne, M., and P. J. Perez. 2011. Silver-catalyzed C–C bond formation between methane and ethyl diazoacetate in supercritical CO_2. *Science* 332:835–838.

Chempath, S., and A. T. Bell. 2006. Density functional theory analysis of the reaction pathway for methane oxidation to acetic acid catalyzed by Pd^{2+} in sulfuric acid. *J. Am. Chem. Soc.* 128:4650–4657.

Chen, G. S., Labinger, J. A., and J. E. Bercaw. 2007. The role of alkane coordination in C–H bond cleavage at a Pt(II) center. *Proc. Natl. Acad. Sci. U. S. A.* 104:6915–6920.

Cheng, J. H., Li, Z. W., Haught, M., and Y. C. Tang. 2006. Direct methane conversion to methanol by ionic liquid-dissolved platinum catalysts. *Chem. Commun.* 44:4617–4619.

Chepaikin, E. G., Bezruchenko, A. P., and A. A. Leshcheva. 2002. Homogeneous rhodium-copper-halide catalytic systems for the oxidation and oxidative carbonylation of methane. *Kinet. Catal.* 43:507–513.

Chepaikin, E. G., Bezruchenko, A. P., Boiko, G. N., Gekhman, A. E., and I. I. Moiseev. 2006. Isotope effects in the oxidative functionalization of methane in the presence of rhodium-containing homogeneous catalytic systems. *Kinet. Catal.* 47:12–19.

Chepaikin, E. G., Bezruchenko, A. P., Leshcheva, A. A., Boyko, G. N., Kuzmenkov, I. V., Grigoryan, E. H., and A. E. Shilov. 2001. Functionalisation of methane under dioxygen and carbon monoxide catalyzed by rhodium complexes – oxidation and oxidative carbonylation. *J. Mol. Catal. A Chem.* 169:89–98.

Chepaikin, E. G., Bezruchenko, A. P., Menchikova, G. N., Moiseeva, N. I., and A. E. Gekhman. 2010. Homogeneous catalytic oxidation of light alkanes: C–C bond cleavage under mild conditions. *Kinet. Catal.* 51:666–671.

Chepaikin, E. G., Boyko, G. N., Bezruchenko, A. P., Leshcheva, A. A., and E. H. Grigoryan. 1998. Oxidative carbonylation of methane in the presence of Rh complexes in aqueous acetic acid. *J. Mol. Catal. A Chem.* 129:15–18.

Conley, B. L., Tenn, W. J., Young, K. J. H., Ganesh, S. K., Meier, S. K., Ziatdinov, V. R., Mironov, O., Oxgaard, J., Gonzales, J., Goddard, W. A., and R. A. Periana. 2006. Design and study of homogeneous catalysts for the selective, low temperature oxidation of hydrocarbons. *J. Mol. Catal. A Chem.* 251:8–23.

Cook, A. K., Schimler, S. D., Matzger, A. J., and M. S. Sanford. 2016. Catalyst-controlled selectivity in the C–H borylation of methane and ethane. *Science* 351:1421–1424.

Crabtree, R. H. 1985. The organometallic chemistry of alkanes. *Chem. Rev.* 85:245–269.

Crabtree, R. H. 1995. Aspects of methane chemistry. *Chem. Rev.* 95:987–1007.

Cundari, T. R., and A. Yoshikawa. 1998a. Computational study of methane activation by mercury(II) complexes. *J. Comput. Chem.* 19:902–911.

Cundari, T. R., Snyder, L. A., and A. Yoshikawa. 1998b. Ligand and substituent effects in methane activation by mercury(II) complexes. *J. Mol. Struct. Theochem.* 425:13–24.

Davico, G. E. 2005. The conversion of methane to methanol: a reaction catalyzed by I^+ or I_2^+? *J. Phys. Chem. A* 109:3433–3437.

Diaz-Urrutia, C., and T. Ott. 2019. Activation of methane to CH_3^+: a selective industrial route to methanesulfonic acid. *Science* 363:1326–1329.

Dimroth, O. 1902. Ueber die mercurirung aromatischer verbindungen. *Ber. Dtsch. Chem. Ges.* 35:2853–2873.

Eckert, M., Fleischmann, G., Jira, R., Bolt, H. M., and K. Golka. 2005. Acetaldehyde. In *Ullmann's Encyclopedia of Industrial Chemistry*, 191–207. Weinheim, Germany: Wiley-VCH.

Edwards, J. H., and N. R. Foster. 1986. The potential for methanol production from natural gas by direct catalytic partial oxidation. *Fuel Sci. Technol. Int.* 4:365–390.

Farrell, B. L., Igenegbai, V. O., and S. Linic. 2016. A viewpoint on direct methane conversion to ethane and ethylene using oxidative coupling on solid catalysts. *ACS Catal.* 6:4340–4346.

Fekl, U., and K. I. Goldberg. 2003. Homogeneous hydrocarbon C–H bond activation and functionalization with platinum. *Adv. Inorg. Chem.* 54:259–320.

Fink, J. K. 2006. *Reactive Polymers Fundamentals and Application*. Oxford: William Andrew Publishing.

Fleming, I. 1976. *Frontier Orbitals and Organic Chemical Reactions*. Chichester: John Wiley and Sons.

Forster, D. 1979. Mechanistic pathways in the catalytic carbonylation of methanol by rhodium and iridium complexes. *Adv. Organomet. Chem.* 118:4574–4580.

Fujiwara, Y., Takaki, K., and Y. Taniguchi. 1996. Exploitation of synthetic reactions via C–H bond activation by transition metal catalysts. Carboxylation and aminomethylation of alkanes or arenes. *Syn. Lett.* 27:591–599.

Fuller, J. T., Butler, S., Devarajan, D., Jacobs, A., Hashiguchi, B. G., Konnick, M. M., Goddard, W. A., Gonzales, J., Periana, R. A., and D. H. Ess. 2016. Catalytic mechanism and efficiency of methane oxidation by Hg(II) in sulfuric acid and comparison to radical initiated conditions. *ACS Catal.* 6:4312–4322.

Galadima, A., and O. Muraza. 2016. Revisiting the oxidative coupling of methane to ethylene in the golden period of shale gas: a review. *J. Ind. Eng. Chem.* 37:1–13.

Gang, X., Birch, H., Zhu, Y. M., Hjuler, H. A., and N. J. Bjerrum. 2000. Direct oxidation of methane to methanol by mercuric sulfate catalyst. *J. Catal.* 196:287–292.

Gang, X., Zhu, Y. M., Birch, H., Hjuler, H. A., and N. J. Bjerrum. 2004. Iodine as catalyst for the direct oxidation of methane to methyl sulfates in oleum. *Appl. Catal. A* 261:91–98.

Gilbert, T. M., Hristov, I., and T. Ziegler. 2001. Comparison between oxidative addition and sigma-bond metathesis as possible mechanisms for the Catalytica methane activation process by platinum(II) complexes: a density functional theory study. *Organometallics* 20:1183–1189.

Gillespie, J., and M. J. Morton. 1971. Halogen and interhalogen cations. *Q. Rev. Chem. Soc.* 25:553–570.

Gol'dshleger, N. F., Es'kova, V. V., Shteinman, A. A., and A. E. Shilov. 1972. Reactions of alkanes in solutions of chloride complexes of platinum. *Russ. J. Phys. Chem. USSR* 46:785–786.

Gretz, E., Oliver, T. F., and A. Sen. 1987. Carbon-hydrogen bond activation by electrophilic transition-metal compounds – palladium(II)-mediated oxidation of arenes and alkanes including methane. *J. Am. Chem. Soc.* 109:8109–8111.

Grice, K. A., and K. I. Goldberg. 2009. Insertion of dioxygen into a platinum(II)-methyl bond to form a platinum(II) methylperoxo complex. *Organometallics* 28:953–955.

Gunsalus, N. J., Koppaka, A., Park, S. H., Bischof, S. M., Hashiguchi, B. G., and R. A. Periana. 2017. Homogeneous functionalization of methane. *Chem. Rev.* 117:8521–8573.

Gunsalus, N. J., Park, S. H., Hashiguchi, B. G., Koppaka, A., Smith, S. J., Ess, D. H., and R. A. Periana. 2019. Selective N functionalization of methane and ethane to aminated derivatives by main-group-directed C–H activation. *Organometallics* 38:2319–2322.

Hashiguchi, B. G., Hövelmann, C. H., Bischof, S. M., Lokare, K. S., Leung, C. H., and R. A. Periana. 2010. Methane-to-methanol conversion. In *Energy Production and Storage: Inorganic Chemical Strategies for a Warming World*, ed. R. Crabtree. New York, NY: Wiley.

Hashiguchi, B. G., Konnick, M. M., Bischof, S. M., Gustafson, S. J., Devarajan, D., Gunsalus, N., Ess, D. H., and R. A. Periana. 2014. Main-group compounds selectively oxidize mixtures of methane, ethane, and propane to alcohol esters. *Science* 343:1232–1237.

Heidbrink, J. L., Ramirez-Arizmendi, L. E., Thoen, K. K., Guler, L., and H. I. Kenttamaa. 2001. Polar effects control hydrogen-abstraction reactions of charged, substituted phenyl radicals. *J. Phys. Chem. A* 105:7875–7884.

Holm, R. H., and J. P. Donahue. 1993. A thermodynamic scale for oxygen atom transfer-reactions. *Polyhedron* 12:571–589.

Hristov, I. H., and T. Ziegler. 2003a. The possible role of SO_3 as an oxidizing agent in methane functionalization by the catalytica process. A density functional theory study. *Organometallics* 22:1668–1674.

Hristov, I. H., and T. Ziegler. 2003b. Density functional theory study of the direct conversion of methane to acetic acid by RhCl3. *Organometallics* 22:3513–3525.

Jarosińska, M., Lubkowski, K., Sosnicki, J. G., and B. Michalkiewicz. 2008. Application of halogens as catalysts of CH_4 esterification. *Catal. Lett.* 126:407–412.

Jones, J. H. 2000. The cativa (TM) process for the manufacture of acetic acid iridium catalyst improves productivity in an established industrial process. *Platin. Met. Rev.* 44:94–105.

Kataja, K., Song, X. M., and M. Huuska. 1994. Modification of a mercury-catalyzed system for the oxidation of methane to methanol. *Catal. Today* 21:513–517.

Khokhar, M. D., Shukla, R. S., and R. V. Jasra. 2009. Selective oxidation of methane by molecular oxygen catalyzed by a bridged binuclear ruthenium complex at moderate pressures and ambient temperature. *J. Mol. Catal. A Chem.* 299:108–116.

Khusnutdinova, J. R., Zavalij, P. Y., and A. N. Vedernikov. 2007. C–O Coupling of LPt^{IV}Me(OH)X Complexes in Water (X = ^{18}OH, OH, OMe; L = di(2-pyridyl)methane sulfonate). *Organometallics* 26:3466–3483.

Konnick, M. M., Bischof, S. M., Yousufuddin, M., Hashiguchi, B. G., Ess, D. H., and R. A. Periana. 2014. A mechanistic change results in 100 times faster CH functionalization for ethane versus methane by a homogeneous Pt catalyst. *J. Am. Chem. Soc.* 136:10085–10094.

Koppaka, A., Park, S. H., Hashiguchi, B. G., Gunsalus, N. J., King, C. R., Konnick, M. M., Ess, D. H., and R. A. Periana. 2019. Selective C–H functionalization of methane and ethane by a molecular Sb-V complex. *Angew. Chem. Int. Ed.* 58:2241–2245.

Kua, J., Xu, X., Periana, R. A., and W. A. Goddard. 2002. Stability and thermodynamics of the PtCl$_2$ type catalyst for activating methane to methanol: a computational study. *Organometallics* 21:511–525.

Labinger, J. A. 1988. Oxidative coupling of methane: an inherent limit to selectivity? *Catal. Lett.* 1:371–375.

Labinger, J. A., and J. E. Bercaw. 2002. Understanding and exploiting C–H bond activation. *Nature* 417:507–514.

Laidler, K. J. 1996. A glossary of terms used in chemical kinetics, including reaction dynamics. *Pure Appl. Chem.* 68:149–192.

Lide, D. R. 2007. *CRC Handbook of Chemistry and Physics*, 88th ed. Boca Raton, FL: CRC Press.

Lin, M., and A. Sen. 1994. Direct catalytic conversion of methane to acetic acid in an aqueous medium. *Nature* 368:613–615.

Lin, M. R., Hogan, T. E., and A. Sen. 1996. Catalytic carbon-carbon and carbon-hydrogen bond cleavage in lower alkanes. Low-temperature hydroxylations and hydroxycarbonylations with dioxygen as the oxidant. *J. Am. Chem. Soc.* 118:4574–4580.

Lunsford, J. H. 1995. The catalytic oxidative coupling of methane. *Angew. Chem. Int. Ed.* 34:970–980.

Lunsford, J. H. 2000. Catalytic conversion of methane to more useful chemicals and fuels: a challenge for the 21st century. *Catal. Today* 63:165–174.

McCoy, M. 2016. German firm claims new route to methanesulfonic acid. *Chem. Eng. News* 94:10.

McMurry, J. 2004. *Organic Chemistry*. Belmont, CA: Brooks/Cole-Thompson Learning.

Michalkiewicz, B., and P. Kosowski. 2007. The selective catalytic oxidation of methane to methyl bisulfate at ambient pressure. *Catal. Commun.* 8:1939–1942.

Michalkiewicz, B., and S. Balcer. 2012. Bromine catalyst for the methane to methyl bisulfate reaction. *Pol. J. Chem. Technol.* 14:19–21.

Michalkiewicz, B., Jarosińska, M., and I. Lukasiewicz. 2009 Kinetic study on catalytic methane esterification in oleum catalyzed by iodine. *Chem. Eng. J.* 154:156–161.

Michalkiewicz, B., Kalucki, K., and J. G. Sošnicki. 2003. Catalytic system containing metallic palladium in the process of methane partial oxidation. *J. Catal.* 215:14–19.

Michalkiewicz, B., Ziebro, J., and M. Tomaszewska. 2006a. Preliminary investigation of low pressure membrane distillation of methyl bisulphate from its solutions in fuming sulphuric acid combined with hydrolysis to methanol. *J. Membr. Sci.* 286:223–227.

Mironov, O. A., Bischof, S. M., Konnick, M. M., Hashiguchi, B. G., Ziatdinov, V. R., Goddard, W. A. 3rd, Ahlquist, M., and R. A. Periana. 2013. Using reduced catalysts for oxidation reactions: mechanistic studies of the "Periana-Catalytica" system for CH$_4$ oxidation. *J. Am. Chem. Soc.* 135:14644–14658.

Mukhopadhyay, S., and A. T. Bell. 2003. Direct catalytic sulfonation of methane with SO$_2$ to methanesulfonic acid (MSA) in the presence of molecular O$_2$. *Chem. Commun.* 9:1590–1591.

Mukhopadhyay, S., and A. T. Bell. 2004. Catalyzed sulfonation of methane to methanesulfonic acid. *J. Mol. Catal. A Chem.* 211:59–65.

Nakata, K., Miyata, T., Jintoku, T., Kitani, A., Taniguchi, Y., Takaki, K., and Y. Fujiwara. 1993. Palladium and copper-catalyzed carboxylation of alkanes with carbon monoxide – remarkable effect of the mixed-metal salts. *Bull. Chem. Soc. Jpn.* 66:3755–3759.

Nakata, K., Yamaoka, Y., Miyata, T., Taniguchi, Y., Takaki, K., and Y. Fujiwara. 1994. Palladium(II) and/or copper(II)-catalyzed carboxylation of small alkanes such as methane and ethane with carbon monoxide. *J. Organomet. Chem.* 473:329–334.

Nishiguchi, T., Nakata, K., Takaki, K., and Y. Fujiwara. 1992. Transition-metal catalyzed acetic acid synthesis from methane and CO. *Chem. Lett.* 21:1141–1142.

O'Donnel, T. A. 1992. *Superacid and Acidic Melts as Inorganic Chemical Reaction Media.* New York, NY: VCH Publishers Inc.

Ogura, K., and K. Takamagari. 1986. Direct conversion of methane to methanol, chloromethane and dichloromethane at room-temperature. *Nature* 319:308.

Olah, G. A., and G. K. S. Prakash. 1985. *Superacids.* New York, NY: John Wiley and Sons.

Olah, G. A., and H. C. H. Lin. 1971. Electrophilic reactions at single bonds. V. Nitration and nitrolysis of alkanes and cycloalkanes with nitronioum salts. *J. Am. Chem. Soc.* 93(5):1259–1261.

Olah, G. A., Goeppert, A., and G. K. S. Prakash. 2006. *Beyond Oil and Gas: The Methane Economy.* Weinheim, Germany: Wiley-VCH.

Olsbye, U. 2016. Single-pass catalytic conversion of syngas into olefins via methanol. *Angew. Chem. Int. Ed.* 55:7294–7295.

Parsons, A. F. 2000. *An Introduction to Free-Radical Chemistry.* Oxford: Wiley-Blackwell.

Paul, A., and C. B. Musgrave. 2007. A detailed theoretical study of the mechanism and energetics of methane to methanol conversion by cisplatin and catalytica. *Organometallics* 26:793–809.

Pawlikowski, A. V., Getty, A. D., and K. I. Goldberg. 2007. Alkyl carbon–nitrogen reductive elimination from Platinum(IV) – sulfonamide complexes. *J. Am. Chem. Soc.* 129:10382–10393.

Periana, R. A., Mirinov, O., Taube, D. J., and S. Gamble. 2002. High yield conversion of methane to methyl bisulfate catalyzed by iodine cations. *Chem. Commun.* 20, 2376–2377.

Periana, R. A., Mironov, O., Taube, D., Bhalla, G., and C. J. Jones. 2003. Catalytic, oxidative condensation of CH_4 to CH3COOH in one step via CH activation. *Science* 301:814–818.

Periana, R. A., Taube, D. J., Evitt, E. R., Loffler, D. G., Wentrcek, P. R., Voss, G., and T. Masuda. 1993. A mercury-catalyzed, high-yield system for the oxidation of methane to methanol. *Science* 259:340–343.

Periana, R. A., Taube, D. J., Gamble, S., Taube, H., and H. Fuji. 1997. High yield, low temperature oxidation of methane to methanol. In *Catalytic Activation and Functionalization of Light Alkanes. Advances and Challenges*, ed. E. G. Derouane, J. Haber, F. Lemos, F. Ramôa Ribeiro, and M. Guisnet, 297–310. Dordrecht: Kluwer Academic Publishers.

Periana, R. A., Taube, D. J., Gamble, S., Taube, H., Satoh, T., and H. Fujii. 1998. Platinum catalysts for the high-yield oxidation of methane to a methanol derivative. *Science* 280:560–564.

Rostovtsev, V. V., Henling, L. M., Labinger, J. A., and J. E. Bercaw. 2002. Structural and mechanistic investigations of the oxidation of dimethylplatinum(II) complexes by dioxygen. *Inorg. Chem.* 41:3608–3619.

Rostovtsev, V. V., Labinger, J. A., Bercaw, J. E., Lasseter, T. L., and K. I. Goldberg. 1998. Oxidation of dimethylplatinum(II) complexes with dioxygen. *Organometallics* 17:4530–4531.

Russell, G. A., and H. C. Brown. 1995. The competitive halogenation of cyclohexane and aralkyl hydrocarbons; evidence as to the nature of the transition states in halogenation reactions. *J. Am. Chem. Soc.* 77:4578–4582.

Russell, G. A., and R. C. Williamson. 1964a. Nature of polar effect in reactions of atoms and radicals. II. Reactions of chlorine atoms and peroxy radicals. *J. Am. Chem. Soc.* 86:2357–2364.

Russell, G. A., and R. C. Williamson. 1964b. Directive effects in aliphatic substitutions. XXV. Reactivity of aralkanes aralkenes and benzylic ether toward peroxy radicals. *J. Am. Chem. Soc.* 86:2364–2367.

Sadow, A. D., and T. D. Tilley. 2003. Catalytic functionalization of hydrocarbons by sigma-bond-metathesis chemistry: dehydrosilylation of methane with a scandium catalyst. *Angew. Chem. Int. Ed.* 42:803–805.

Schwarz, H. 2011. Chemistry with methane: concepts rather than recipes. *Angew. Chem. Int. Ed.* 50:10096–10115.

Sen, A. 1998. Catalytic functionalization of carbon-hydrogen and carbon-carbon bonds in protic media. *Acc. Chem. Res.* 31:550–557.

Sen, A., Benvenuto, M. A., Lin, M. R., Hutson, A. C., and N. Basickes. 1994. Activation of methane and ethane and their selective oxidation to the alcohols in protic media. *J. Am. Chem. Soc.* 116:998–1003.

Sen, A., Gretz, E., Oliver, T. F., and Z. Jiang. 1989. Palladium(II) mediated oxidative functionalization of alkanes and arenes. *New J. Chem.* 13:755–760.

Shilov, A. E. 1984. *Acitvation of Saturated Hydrocarbons by Transition Metal Complexes.* Dordrecht and Boston, MA: D. Riedel Publishing Company.

Shilov, A. E., and A. A. Shteinman. 1977. Activation of saturated hydrocarbons by metal complexes in solution. *Coord. Chem. Rev.* 24:97–143.

Shilov, A. E., and G. B. Shul'pin. 1997. Activation of C–H bonds by metal complexes. *Chem. Rev.* 97:2879–2932.

Shilov, A. E., and G. B. Shul'pin. 2000. *Activation and Catalytic Reactions of Saturated Hydrocarbons in the Presence of Metal Complexes.* Dordrecht: Kluwer Academic.

Smith, K. T., Berritt, S., Gonzalez-Moreiras, M., Ahn, S., Smith, M. R., Baik, M. H., and D. J. Mindiola. 2016. Catalytic borylation of methane. *Science* 351:1424–1427.

Snyder, J. C., and A. V. Grosse. 1950. Reaction of methane with sulfurtrioxide. *U.S. Patent 2,493,038*, Jan 3, 1950.

Souma, Y., Tsumori, N., Willner, H., Xu, Q., Mori, H., and Y. Morisaki. 2002. Carbonylation of hydrocarbons and alcohols by cationic metal carbonyl catalysts. *J. Mol. Catal. A Chem.* 189:67–77.

Sunley, G. J., and D. J. Watson. 2000. High productivity methanol carbonylation catalysis using iridium – the Cativa (TM) process for the manufacture of acetic acid. *Catal. Today* 58:293–307.

Tedder, J. M. 1982. Which factors determine the reactivity and regioselectivity of free-radical substitution and addition-reactions. *Angew. Chem. Int. Ed.* 21:401–410.

Vedernikov, A. N. 2007. Recent advances in the platinum-mediated CH bond functionalization. *Curr. Org. Chem.* 11:1401–1416.

Vedernikov, A. N., Binfield, S. A., Zavalij, P. Y., and J. R. Khusnutdinova. 2006. Stoichiometric aerobic Pt^{II}-Me bond cleavage in aqueous solutions to produce methanol and a $Pt^{II}(OH)$ complex. *J. Am. Chem. Soc.* 128:82–83.

Wick, D. D., and K. I. Goldberg. 1999. Insertion of dioxygen into a platinum-hydride bond to form a novel dialkylhydroperoxo Pt(IV) complex. *J. Am. Chem. Soc.* 121:11900–11901.

Xu, X., Kua, J., Periana, R. A., and W. A. Goddard. 2003. Structure, bonding, and stability of a catalytica platinum(II) catalyst: a computational study. *Organometallics* 22:2057–2068.

Xu, Z. T., Oxgaard, J., and W. A. Goddard. 2008. The mechanism by which ionic liquids enable Shilov-type CH activation in an oxidizing medium. *Organometallics* 27:3770–3773.

Yahav-Levi, A., Goldberg, I., Vigalok, A., and A. N. Vedernikov. 2008. Competitive aryl-iodide vs aryl-aryl reductive elimination reactions in Pt(IV) complexes: experimental and theoretical studies. *J. Am. Chem. Soc.* 130:724–731.

Yamanaka, I. 2002. Reductive activation of O_2 and monooxygenation of hydrocarbons by Eu catalyst. *Catal. Surv. Asia* 6:63–72.

Yamanaka, I., Nakagaki, K., Akimoto, T., and K. Otsuka. 1996b. Reactivity of active oxygen species generated in the $EuCl_3$ catalytic system for monooxygenation of hydrocarbons. *J. Chem. Soc. Perkin Trans.* 2:2511–2517.

Yamanaka, I., Soma, M., and K. Otsuka. 1995. Oxidation of methane to methanol with oxygen catalyzed by europium trichloride at room-temperature. *J. Chem. Soc. Chem. Commun.* 21, 2235–2236.

Yamanaka, I., Soma, M., and K. Otsuka. 1996a. Enhancing effect of titanium(II) for the oxidation of methane with O_2 by an $EuCl_3$-Zn-CF_3CO_2H-catalytic system at 40 °C. *Chem. Lett.* 25:565–566.

Yang, J., Haynes, A., and P. M. Maitlis. 1999. The carbonylation of methyl iodide and methanol to methyl acetate catalysed by palladium and platinum iodides. *Chem. Commun.* 2, 179–180.

Yuan, J. L., Liu, L., Wang, L. L., and C. J. Hao. 2013a. Partial oxidation of methane with the catalysis of palladium(II) and molybdovanadophosphoric acid using molecular oxygen as the oxidant. *Catal. Lett.* 143:126–129.

Yuan, J. L., Wang, L. L., and Y. Wang. 2011. Direct oxidation of methane to a methanol derivative using molecular oxygen. *Ind. Eng. Chem. Res.* 50:6513–6516.

Yuan, J. L., Wang, Y., and C. J. Hao. 2013b. Water-promoted palladium catalysts for methane oxidation. *Catal. Lett.* 143:610–615.

Zavitsas, A. A., and J. A. Pinto. 1972. Meaning of polar-effect in hydrogen abstractions by free-radicals – reactions of tert butoxy radical. *J. Am. Chem. Soc.* 94:7390–7396.

Zerella, M., and A. T. Bell. 2006a. Pt-catalyzed oxidative carbonylation of methane to acetic acid in sulfuric acid. *J. Mol. Catal. A Chem.* 259:296–301.

Zerella, M., Kahros, A., and A. T. Bell. 2006b. Methane oxidation to acetic acid catalyzed by Pd^{2+} cations in the presence of oxygen. *J. Catal.* 237:111–117.

Zerella, M., Mukhopadhyay, S., and A. T. Bell. 2004. Direct oxidation of methane to acetic acid catalyzed by Pd^{2+} and Cu^{2+} in the presence of molecular oxygen. *Chem. Commun.* 2004:1948–1949.

Zimmermann, T., Soorholtz, M., Bilke, M., and F. Schuth. 2016. Selective methane oxidation catalyzed by platinum salts in oleum at turnover frequencies of large-scale industrial processes. *J. Am. Chem. Soc.* 138:12395–12400.

13 3D Printed Immobilized Biocatalysts for Conversion of Methane

Jennifer M. Knipe

CONTENTS

13.1 INTRODUCTION

Biocatalysts, either whole cells or enzymes isolated from bacteria and other organisms, are able to harness the elegant and efficient reactions found in nature with far less energy input or capital investment required than chemical processes (Meyer et al. 2013; Tang and Zhao 2009). Additionally, biocatalysis is considered a more green and sustainable chemistry as it avoids the use of toxic and hazardous reagents and solvents, transition metal catalysts, and costly purification steps (Sheldon and Brady 2018; Bornscheuer et al. 2012). Isolated enzymes catalyze chemical reactions with unparalleled selectivity under ambient reaction conditions and are consequently carving out a fast-growing market as industrial biocatalysts (Grunwald 2015). The predominant classes of enzyme biocatalysts are hydrolases, transferases, and oxidoreductases, and they are used to produce a wide range of products including detergents, biofuels, food and beverages, agricultural feed, biopharmaceuticals, cosmetics, and polymers (Meyer et al. 2013; Markets and Markets 2015).

The oxidoreductase methane monooxygenase (MMO), found in methanotroph bacteria, catalyzes the partial oxidation of methane to methanol at ambient conditions with the highest known selectivity of any catalyst of this reaction, biological or chemical (Hammond et al. 2012). Methanotrophic bacteria and isolated MMOs are interesting biocatalysts for methane conversion because they operate at low temperature and pressures and can withstand the presence of some contaminants such as hydrogen sulfide (Mühlemeier, Speight, and Strong 2018), unlike most inorganic or chemical catalysts. Additionally, methanotrophs have been genetically engineered to convert methane into various valuable products besides methanol, such as bioprotein (Bothe et al. 2002; Harrison and Hamer 1971), polyhydroxybutyrate (Khosravi-Darani et al. 2013), and lactate (Henard et al. 2016). Catalysis of methane gas-to-liquid products is important because U.S. emissions of methane, a greenhouse gas 70 times more potent than CO_2 over a 20 year period (Stocker et al. 2013), are expected to rise 20% from natural gas production over the next 15 years (U.S. Department of State 2014) and sources such as farming and arctic melt (Stolaroff et al. 2012) will also increase. It is essential to mitigate small and remote methane streams < 1000 ppm to control these emissions (Stolaroff et al. 2012), which are currently released or flared due to the impracticality and expense of traditional chemical processing at these remote sites with low concentrations of methane.

Biocatalysis may offer a cost-efficient solution to convert these problematic methane emissions to methanol or other liquid products that are both easy to transport and have considerable market value. Development is needed at the intersection of protein engineering, enzyme immobilization, process engineering, and life cycle analysis not only to improve the process in commercial industries that have already adopted biocatalysis, such as the pharmaceutical industry (Jiménez-González et al. 2011), but especially to drive the expansion of biocatalysis into other industries such as biofuels and methane conversion (Chapman, Ismail, and Dinu 2018). This chapter will highlight recent progress in whole-cell and isolated enzyme immobilization strategies with emphasis on those that couple advanced manufacturing techniques such as 3D printing with biocatalysts to explore reactor design for process intensification, which is essential to achieving economical, industrially relevant processes. These technologies are included in the context of their applicability to overcoming the current challenges and limitations faced by biocatalyzed reactions, in particular the biocatalyzed conversion of gas-to-liquid reactions such as methane conversion to liquid products.

13.2 CURRENT LIMITATIONS OF GAS-TO-LIQUID BIOCATALYSIS

Biocatalysis may utilize either isolated enzymes or whole cells in the conversion process. Using whole cells offers a wider range of potential products, avoids complex and expensive enzyme isolation methods that may affect the activity and stability, and may also eliminate the necessary addition of expensive cofactors for electron transfer (Britton, Majumdar, and Weiss 2018). However, this approach requires energy for upkeep and metabolism of the organisms that reduces conversion efficiency, involves mass transfer across the cell membrane, and introduces competing enzymes that may result in unwanted side reactions. Metabolic engineering and synthetic biology have enabled engineered organisms capable of overcoming some of these challenges, such as engineered methanotrophs that carry out more efficient methane activation (Haynes and Gonzalez 2014). Isolated enzymes, on the other hand, offer the promise of highly controlled reactions at ambient conditions with higher conversion efficiency and greater flexibility of reactor and process design (Bornscheuer et al. 2012), but they present challenges in terms of purification methods as well as low activity and stability. Protein engineering and directed evolution have progressed rapidly over recent decades to address these issues.

Limitations in reactor design, however, apply to both isolated enzyme and whole-cell biocatalysts and remain largely overlooked. Biocatalysis reactions have traditionally been carried out in low-throughput unit operations in which mass transfer is limiting, such as stirred-tank reactors (Haynes and Gonzalez 2014; Fei et al. 2014). Although well-studied, stirred-tank reactors have several inherent challenges, including low productivity, high operating costs, and variability in the quality of the product (Rios et al. 2004). Improving mass transfer has been identified as a major hurdle to economical bioconversion (Haynes and Gonzalez 2014), particularly in achieving efficient transfer rate of poorly soluble gases during gas fermentation at scale (Vega, Clausen, and Gaddy 1990; Munasinghe and Khanal 2010). Thus, the stirred-tank reactor is not the optimal design for gas-to-liquid reactions such as methane conversion to liquid products since it does not allow efficient delivery of reactant gases to enzymes or microbes dispersed in the bulk solution. In a stirred-tank reactor, the gas is typically delivered by sparging through the bulk phase liquid, a technique known to suffer from poor mass transfer. Further limiting the availability of gaseous reactants to enzymes or microbes in the liquid phase, methane and oxygen are only sparingly soluble in aqueous solvents: 1.5 mM/atm and 1.3 mM/atm, respectively, at 25°C (Poling et al. 2008). While reaction conditions such as temperature, headspace pressure, and media composition can be tuned to affect gas solubility, reactor design is key to increasing gas contact area and mixing to enhance gas transfer to the biocatalyst, and Stone et al. (2017) have reviewed this topic as it relates to biochemical methane conversion.

Reactor types other than stirred tank have been studied for gas fermentation, such as bubble column and airlift reactors, fluidized bed and trickle bed reactors, packed-bed reactors, and membrane reactors (Mühlemeier, Speight, and Strong 2018; Vega, Clausen, and Gaddy 1990; Munasinghe and Khanal 2010). A review of bioreactor types for gas fermentation as it relates to methanotrophy has been published by Mühlemeier, Speight, and Strong (2018) and although some reactor types improve substrate mass transfer efficiency, such as microbubble generation (Munasinghe and Khanal 2010) and membrane bioreactors (Pen et al. 2014), they still have disadvantages including high shear stress, high pressure drop, difficult scalability, and membrane fouling. Notably, a significant number of these advanced reactor designs require immobilization of the biocatalyst on a substrate or membrane within the reactor.

13.3 BIOCATALYST IMMOBILIZATION

Improved bioreactor design will be instrumental in improving the efficiency of gas-to-liquid reactions reliant upon biocatalysts, and the ability to immobilize the biocatalysts enables more sophisticated and novel designs that can mimic the biocatalyst's natural environment. Methanotrophic bacteria have evolved to use methane as their sole source of carbon and energy; thus, designing bioreactors that mimic their method of immobilizing MMO is a logical approach. In methanotrophs, particulate

MMO (pMMO) is expressed abundantly in high surface area, lipid membrane structures that can comprise a large part of the cellular volume (Choi et al. 2003). Although poorly soluble in water, methane and oxygen are much more soluble in lipids (Miller, Hammond, and Porter 1977) and some hydrophobic polymers, such as silicones (Lin and Freeman 2006, 2005). Therefore, the expression of pMMO in the lipid membrane is likely functionally significant as methane is about 10 times more soluble in the lipid membrane than in the aqueous cytosol (Miller, Hammond, and Porter 1977). Immobilizing biocatalysts that utilize gas phase reactants on or within an organic, polymeric material, for example, allows tuning of the material gas solubility, permeability, and surface area to mimic the natural immobilization of pMMO within the lipid membrane. Additionally, immobilization of biocatalysts may also significantly increase their stability, selectivity, reusability, ease of separation and recovery, and reaction rate relative to cells and enzymes in suspension (Mühlemeier, Speight, and Strong 2018; Rodrigues et al. 2013).

Immobilization of the biocatalyst on or within a support also creates the potential for a flow-through reactor design and continuous use of the biocatalyst (Eş, Vieira, and Amaral 2015). A continuous-flow reaction consists of pumping a fluid containing starting material through a reactor in a continuous manner to yield a stream of product. This is in contrast to batch reactions, in which all the reactants are combined in a reactor, and then the reaction is stopped at some point to recover the product. Separation of the biocatalyst from the product in a batch operation often requires lysing of cells and/or filtration or centrifugation of the solution, a step that usually renders the biocatalyst no longer usable.

Compared to batch operation, continuous-flow biocatalysis offers improved mixing, mass transfer, thermal control, pressurized processing, decreased variation, automation, process analytical technology, and the potential for in-line purification (Britton, Majumdar, and Weiss 2018). Continuous-flow processes have the potential to accelerate biocatalytic conversions, make large-scale production more economically feasible in significantly smaller equipment with a substantial decrease in reaction time, and improve in space-time yield (Tamborini et al. 2018). Continuous-flow biocatalysis typically results in longer biocatalyst lifetime and higher turnover number. Continuous processing and bioprocessing (including biocatalysis) have been identified by the ACS GCI Pharmaceutical Roundtable as the top two key green engineering research areas in the pharmaceutical industry (Jiménez-González et al. 2011), but these research priorities are not limited to pharmaceutical applications. A continuous process is appealing in the case of methanotrophic conversion of methane, where mass transfer is limiting, accumulation of product may inactivate the methanotrophs, and efficiency and throughput are essential to scalable production. To address scalability, the modular capability of continuous-flow reactors enables flexible production volume simply by changing the number of reactors in series or parallel (Tamborini et al. 2018).

13.3.1 SURFACE IMMOBILIZATION TECHNIQUES

For the purposes of this chapter, it is useful to include an overview of surface immobilization techniques for biocatalysts. The selection of immobilization supports is critical, and ideally they should have at least some of the following properties: large surface area, sufficient functional groups for attachment, hydrophilic character, water insolubility, chemical and thermal stability, mechanical strength, high rigidity, resistance to microbial degradation, and ease of regeneration, while also being non-toxic, biocompatible, and inexpensive (Britton, Majumdar, and Weiss 2018; Zdarta et al. 2018). Inorganic, organic, hybrid or composite supports may be used, and physical properties such as the particle size and pore structure must also be taken into account (Jesionowski, Zdarta, and Krajewska 2014). In terms of 3D printed supports, these are likely to be organic materials such as synthetic polymers and natural biopolymers like chitin, chitosan, cellulose, gelatin, and alginate.

The biocatalyst may be immobilized onto the support via adsorption, covalent immobilization, affinity immobilization and many other techniques, which have been reviewed in detail elsewhere (Sheldon and van Pelt 2013; Illanes et al. 2012; Datta, Christena, and Rajaram 2013). Briefly, adsorption immobilization involves a physical interaction such as hydrophobic, ionic, van der Waals, or hydrogen bonding between the enzyme or cell and the immobilization substrate that is relatively weak and, importantly, it does not usually change the native structure of the enzyme or cell membrane (Jesionowski, Zdarta, and Krajewska 2014). This helps to preserve the activity of the biocatalyst. Affinity immobilization is similar to adsorption immobilization in that it relies on physical interactions between the biocatalyst and support, but the affinity between the two depends on the environmental conditions, such as pH or temperature. These techniques are relatively simple, inexpensive, reversible, and allow the support material to be reused, but the interactions between the biocatalyst and support are not very stable and some biocatalyst may be lost during operation (Eş, Vieira, and Amaral 2015).

Covalent immobilization, on the other hand, involves a chemical bond between the enzyme or cell and the support. This generally requires modification of the support surface with functional groups, followed by covalent linkage to the biocatalyst that may or may not impact the active site or otherwise disrupt catalytic activity. However, covalent immobilization offers the advantage of a stronger bond that can withstand a wider range of process conditions, resulting in decreased leaching of the biocatalyst and potentially extending the biocatalyst lifetime of use (Britton, Majumdar, and Weiss 2018).

Patel et al. (2016) have completed several studies aimed at improving methane conversion by various methanotroph species via immobilization of the microbes. First, they investigated adsorption and covalent immobilization of *Methylosinus sporium* on various supports and found that cells immobilized on chitosan supports had the greatest methanol production, and in particular cells immobilized covalently outperformed adsorbed cells and were more stable over six reuse cycles; the covalently immobilized cells retained 64.3% of their methanol production efficiency whereas adsorbed cells retained 26.6% and cells in suspension retained only 17.3%, as shown in Figure 13.1. The researchers obtained similar results with *Methylocella tundrae* using simulated biohythane as a substrate (Patel et al. 2017) and with *Methylocystis bryophilla* using simulated biogas as a substrate (Patel et al. 2018); both cell strains were covalently immobilized on chitosan and generated more than double the cumulative concentration of methanol over eight cycles as compared to a free cell suspension.

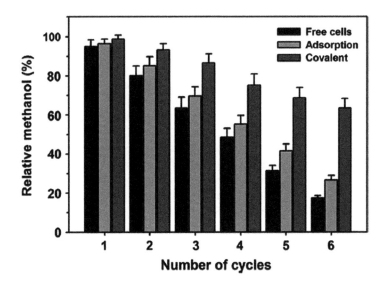

FIGURE 13.1   Reusability of free and immobilized *M. sporium* cells. Initial methanol production was considered as 100%. (Reprinted from Patel et al. (2016) © 2016 with permission from Elsevier.)

Sun et al. (2018) immobilized a consortia of methane oxidizing bacteria by cell adhesion onto the surface of different porous substrates – sponge, activated carbon, and volcanic rock – to remove methane from the exhaust stream of a biogas upgrading process. The researchers indeed saw an improvement in the methane removal by immobilized cells compared to free cell suspensions and noted that cells immobilized on the sponge performed best due to the rough surface, hydroscopicity, and large pore volume made up of pore sizes greater than 50 nm in diameter. They concluded that the roughness, hydroscopicity, pore size and particle size of the support material were critical for the bacteria immobilization and performance in methane conversion.

Thus, using 3D printing methods to fabricate a customized support structure and then using surface immobilization techniques to incorporate the biocatalyst may be a good strategy to enhance the biocatalyst efficiency. The conditions for support fabrication, functionalization, and biocatalyst immobilization will likely be specific to the support and biocatalyst system and may require extensive research to develop and optimize. While the process conditions can be controlled to avoid denaturation or degradation of the biocatalyst during immobilization via these methods, each step of the process is subject to limitations in immobilization yield and efficiency due to inaccessibility, incomplete reaction, or deactivation of the biocatalyst that may ultimately reduce loading and activity.

13.3.2   ENTRAPMENT OR ENCAPSULATION

Entrapment or encapsulation immobilization entails trapping the biocatalysts into a caged network via covalent or noncovalent interactions with an immobilization support; this generally means the network is formed in the presence of the biocatalyst and is irreversible (Britton, Majumdar, and Weiss 2018; Sheldon and van Pelt 2013). Examples of this include polymer networks, hydrogel networks, inorganic sol-gel networks, membrane or hollow fiber devices, and microcapsules. While this type of immobilization offers protection of the biocatalyst from environmental conditions and potentially higher biomass loading than surface immobilization, the maximum loading may be limited by diffusion of reactants and products through the network.

Taylor et al. (2018) immobilized *Methylosinus trichosporium* OB3b by encapsulation within alginate beads to evaluate the effect of immobilization on methane uptake and methanol yields in batch reactors and semi-continuous flow columns. They found that the rates of methane consumption were not significantly different between the cells immobilized in alginate beads of three different diameters compared to cells in suspension when the biomass concentration was relatively low for all cases, indicating that the alginate did not interfere with the transport of methane to the cells. However, when the biomass concentration within the alginate beads was increased, mass transfer limitation of methane to the cells was observed. These results were consistent with the findings of a previous study in which methanotrophs were immobilized in alginate (Uchiyama et al. 1995). This highlights a practical challenge with biocatalysis – high biomass loading is necessary to achieve an industrially relevant process, but such loading of biomass is likely to be limited by the mass transport of methane to the cells. Thus, it is critical to develop immobilization materials with small feature sizes, high gas permeability, and/or tunable porosity to maximize the transport of methane to the cells.

Fortunately, this method of immobilization may allow direct 3D printing of immobilized biocatalysts, as many of the polymers that are used for encapsulation and entrapment immobilization may also be amenable to formulation as feedstocks for various forms of 3D printing. This offers immobilization of the biocatalyst throughout the support, high biocatalyst loading, and controlled 3D printed structures with feature sizes on the order of 100 μm or less to improve mass transfer. However, in order for encapsulated biocatalyst to be directly 3D printed, it must be able to withstand the build time, conditions, and chemistry of the materials. Additionally, this immobilization technique requires that the reactants and products are able to diffuse throughout the support and if it is to be 3D printed the rheology and mechanical properties of the feedstock may need to meet certain specifications, so the encapsulation material must be appropriately selected.

13.4 ADDITIVE MANUFACTURING FOR BIOREACTOR DESIGN

Additive manufacturing (AM), which is synonymous with the term "3D printing", is the use of computer-aided design to build objects layer by layer. This is in contrast to subtractive manufacturing, in which unwanted material is removed from a solid piece of starting material. AM can be used to fabricate highly customized parts from metals, ceramics, and polymers without the need for molds or machining that is typically used for conventional formative and subtractive production (Ligon et al. 2017). This not only saves material but also could reduce production time and enable complex geometries that could not be manufactured with conventional fabrication techniques.

AM may be used to print biocompatible and biodegradable porous supports as scaffolds for subsequent biocatalyst immobilization, but it may also be used to "bio-print" viable cells and enzymes directly via the 3D printing process (Ligon et al. 2017). This offers the ability to rapidly prototype and test reactor designs and tune material architecture to optimize mass and heat transfer for a specific biocatalyzed reaction. Additionally, bioprinting cells, growth factors, and enzymes at the same time into precisely defined 3D structure with resolution on the order of 50–100 μm creates the possibility to pattern enzyme channels and engineer metabolic pathways (Pröschel et al. 2015).

Although bioprinting has been widely investigated for decades, the core of the development has been focused on regenerative medicine, tissue engineering, drug delivery, and sensor applications. Thus, the use of bioprinting for bioreactor and bio-catalyst development is largely untapped. Below, several types of 3D bioprinting and their use in developing advanced bioreactors and biocatalysis design are discussed in detail.

13.4.1 Types of 3D Bioprinting

We emphasize that this is not an exhaustive list of 3D bioprinting techniques; rather, this section highlights the types of 3D printing most commonly used in bioprint-ing and the fabrication of advanced reactors and printed reactors containing cata-lysts, or "reactionware". These three types of bioprinting are photostereolithography, extrusion-based, and droplet-based bioprinting. For a thorough review of 3D printing techniques and polymers used, we refer the reader to Ligon et al. (2017).

13.4.1.1 Photostereolithography

In photostereolithography systems, light sources (usually in the UV range) are used to induce polymerization and cross-linking of a liquid resin as the laser beam scans the surface of the resin according to the desired pattern and cures the part from the bottom-up in a layer-by-layer manner (Ligon et al. 2017). This results in an additively built part that has the advantage of high spatial resolution provided by the spot size of the focused laser beam.

More specifically, projection microstereolithography (PμSL) is a type of photoste-reolithography that is a low-cost, high-throughput AM technique capable of gen-erating three-dimensional structures from micro- to meso-scales with microscale architecture and submicron precision (Zheng et al. 2012). Rather than use a laser beam to scan the image, PμSL uses near-UV light-emitting diode (LED) arrays to project the image using a dynamic mask to initiate the photopolymerization of each layer, show schematically in Figure 13.2, making this AM process more energy effi-cient and capable of high-throughput fabrication than laser scanning-based photoste-reolithography. This type of 3D printing has been used to fabricate materials with complex geometries such as the octet-truss, a geometry that cannot be created with traditional fabrication methods and results in ultralight, ultrastiff mechanical meta-materials (Zheng et al. 2014).

In one recent example of bioprinting with PμSL, Miri et al. (2018) used digital micromirror device projection microstereolithography, a high-throughput printing

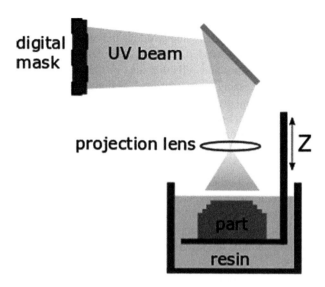

FIGURE 13.2 Schematic of projection microstereolithography (PμSL), a layer-by-layer additive manufacturing technique. To build the part, a digital mask is projected via UV light onto the liquid resin as the substrate is lowered into the resin in the Z-direction. (Adapted from Zheng et al. 2014.)

technique, to bioprint high-fidelity hydrogel constructs encapsulating mammalian cells as tissue engineering scaffolds. A microfluidic device was incorporated to enable multimaterial bioprinting of gelatin methacryloyl (GelMA) and polyethylene glycol dimethacrylate (PEGDMA), two hydrogel materials that are known to have good cell biocompatibility. The team was able to bioprint multimaterial structures resembling biological tissues with a printing resolution of ~20–30 μm. The bioinks contained various cell types including fibroblasts, skeletal muscle cells, mesenchymal stem cells, or osteoblasts, and all cell types maintained satisfactory proliferation and metabolic activity up to 7 days post bioprinting. Even though the physical size of the printed constructs was severely limited by the microfluidic chip, the ability to bioprint multimaterial constructs on the order of seconds rather than minutes is a major advantage of this bioprinting platform, as longer fabrication time can significantly decrease cell viability. Although this bioprinting platform was developed in the context of tissue engineering, one can envision its application to biocatalysis as it allows for spatial control of cells or enzymes within complex printed geometries, enabling investigation of advanced reactor design and giving rise to the potential to engineer a metabolic pathways within a reactor; however, use on an industrial scale is unlikely unless the printable area can be significantly increased.

13.4.1.2 Extrusion-Based Bioprinting

Extrusion-based bioprinting is the computer-controlled layer-by-layer pneumatically-driven deposition of polymers, pastes, polymer solutions, and dispersions containing biological components through a movable nozzle or orifice serving as the extrusion print head (Ligon et al. 2017). The extruded bioink may be cured by exposure to a heat or light source capable of inducing various physical or chemical linkages, allowing a variable processing window and enabling a broad range of possible materials that may be extruded. This is particularly advantageous for extruding biocatalysts like enzymes and cells that may be denatured, killed, or otherwise inactivated by high processing temperatures, UV light, or toxic monomers. Unlike photostereolithography and droplet-based bioprinting, extrusion-based bioprinting is well-suited for dealing with highly viscous bioinks that may contain high concentrations of biocatalysts. Additionally, multiple cartridges of bioink may be loaded into a single extrusion printer to allow for bioprinting of heterogeneous structures (Panwar and Tan 2016).

Direct ink write, shown schematically in Figure 13.3A, is one type of extrusion-based printing that employs a computer-controlled translation stage to move an ink-deposition nozzle that extrudes a continuous, filamentary element to create materials with controlled architecture and composition and structures with minimum feature sizes ranging from hundreds of microns to the submicron scale (Lewis and Gratson 2004; Lewis 2006). Deposition of concentrated colloidal gels via direct ink write was used to create 3D periodic lattices and radial arrays where both the patterned materials and pore channels are interconnected in all three dimensions, shown in Figures 13.3B and 13.3C. The feature of interconnected pores and material is attractive for reactor design. It should be noted, though, that extruding submicron strands of material may drastically increases the print time; bioprinting parameters such as

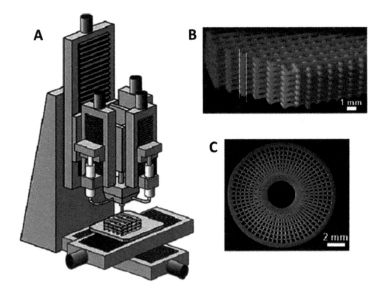

FIGURE 13.3 (A) Schematic of computer-controlled extrusion deposition apparatus; (B) optical image of 3D periodic lattice with a simple tetragonal geometry; and (C) optical image of 3D radial array assembled by robotic deposition (nozzle diameter = 200 μm, deposition speed = 6 mm/s) of a concentrated colloidal gel-based ink. (Reprinted with permission from Lewis and Gratson (2004) © 2004 Elsevier and Smay, Cesarano, and Lewis (2002) © 2002 American Chemical Society.)

printing speed, filament resolution, dispensing pressure, and movement distance are dependent upon the formulation of the bioink (Panwar and Tan 2016). These printing parameters as well as the bioink formulation must be tuned to maximize printability while maintaining viability and activity of the biological component.

13.4.1.3 Droplet-Based Bioprinting

Droplet-based bioprinting (DBB) offers several advantages over other 3D bioprinting techniques due to its simplicity, agility, versatility, and the great control over the deposition pattern (Gudapati, Dey, and Ozbolat 2016). Types of droplet-based bioprinting include inkjet, acoustic-droplet-ejection, and micro-valve printing. Inkjet printing, the original and perhaps most widely used type of droplet-based bioprinting, leverages gravity, atmospheric pressure, and fluid mechanics of the bioink to allow the high-throughput deposition of ink droplets at predefined locations on a substrate. The ability of inkjet printing to deposit droplets with picoliter volumes at micrometer-length accuracy makes it a promising method to immobilize biocatalysts and other biological agents on supporting substrates in a straightforward, specific, and systematic manner (Zhu, Chen, and Wei 2018). This technique has the potential to create spatially controlled constructs with well-defined positioning of biocatalysts.

Inkjet bioprinting can be further classified into droplet on demand (DOD), continuous inkjet, and electrohydrodynamic jet bioprinting (Gudapati, Dey, and Ozbolat 2016). We will focus here on types of DOD printing, shown in Figure 13.4, as these are typically preferred for bioprinting cells and are the most prevalent type in the literature related to biocatalysis. With thermal DOD inkjet printing, heat is applied to the liquid ink causing a bubble to form within the ink reservoir, creating pressure that propels a droplet of ink out of the print head via a microscopic orifice. Alternatively, piezoelectric actuators or electrodes can be used instead to induce a volume change and corresponding pressure wave to form the droplet, eliminating

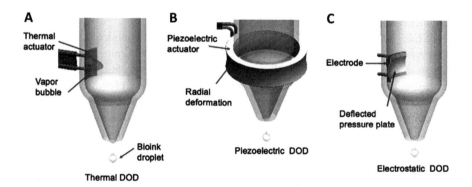

FIGURE 13.4 Mechanisms of droplet-based bioprinting. Inkjet bioprinting techniques: (A) thermal drop-on-demand bioprinting employs a thermal actuator to locally heat bioink solutions to generate droplets; (B) piezoelectric drop-on-demand bioprinting depends on radial deformation of a piezoelectric actuator to generate droplets; C) electrostatic bioprinting relies on deflection of pressure plate to generate droplets. (Reprinted from Gudapati, Dey, and Ozbolat 2016 © 2016 with permission from Elsevier.)

the need for volatile solvents used in thermal DOD, which is advantageous for bio-catalysis applications.

DOD inkjet bioprinting may be used to build freestanding objects with high build speed and large build volume, but it does not have the high spatial resolution of photostereolithography (Ligon et al. 2017). However, the two have been combined in inkjet lithographic 3D-printers, in which an inkjet head deposits a layer of droplets of both a build material and a support material, then a UV light source is used to cure the layer. At the end of the build, the support material is removed from the finished part. This technique can be used with multiple inkjet heads to build a part of mul-tiple materials but is limited by the need for a support material and strict viscosity requirements (Ligon et al. 2017).

While DOD has higher resolution than extrusion-based printing and is more ame-nable to heterogeneous bioinks (containing multiple biologics) and high-throughput printing, it suffers from clogging of the orifice and the volume and placement of the droplets is strictly controlled by the nozzle, limiting the geometry of the resulting printed part (Gudapati, Dey, and Ozbolat 2016).

13.4.2 3D PRINTING REACTORS FOR BIOCATALYSIS

3D printing simplifies and reduces the costs of iterative reactor design and prototyp-ing, allowing exploration on the effect of reactor design on reaction kinetics and pro-ductivity. A 3D printed bioreactor with immobilized biocatalyst could have similar performance advantages of a microchannel reactor, in which the micro-dimensions of the reactor walls allow for rapid heat and mass transfer, improved mixing, small solvent volumes, and large surface area to volume ratios (Sheldon and van Pelt 2013). Symes et al. (2012) demonstrated the ability to 3D print catalysts directly into the structure of a polymer reactor for chemical synthesis, dubbing the printed part "reac-tionware" and thereby creating new possibilities in the multifunctional reactor space. Since then, 3D printed reactors with catalysts have been used for the synthesis of both inorganic and organic chemicals.

While reactionware has been used for some time for chemical synthesis, the first reports of reactors 3D printed with biocatalysts for continuous-flow synthesis began appearing within the past five years. In one study, Peris et al. (2017) used an extrusion-based 3D printer to fabricate reactor devices made of nylon, then subse-quently functionalized the reactor surface with hydrochloric acid and glutaraldehyde to covalently immobilize ω-transaminase enzymes as biocatalysts in a continuous-flow reaction. The enzyme-immobilized reactionware was then used to convert (R)-methylbenzylamine into acetophenone; the researchers found that nearly full conversion of the (R)-enantiomer was achieved, and that the activity and selectivity of the reactor were maintained for 100 hours of flowtime. The researchers noted a key advantage of 3D printing reactors is the ability to rapidly screen immobilization conditions, which is often done by trial and error and can be accelerated via high-throughput methods such as microplates or 3D printed substrates.

Su et al. (2016) used an extrusion-based 3D printer and acrylonitrile butadiene styrene (ABS) thermoplastics to manufacture a flow reactor designed to operate as an on-line, flow-through biosensor to monitor metabolites in the brain extracellular fluid of rats in vivo. Flow-through biosensors are attractive for reusability, specificity, and interfacing with analytical instrumentation. The reactors were 3D printed, and then post printing either glucose oxidase or lactate oxidase was covalently immobilized on the surface of the printed part. The oxidases react with glucose and lactate in the brain extracellular fluid being analyzed to produce hydrogen peroxide, which was subsequently quantified via a fluorescent assay. The researchers found that the immobilized glucose oxidase enzymes in particular were quite stable; when stored at 4°C, the glucose oxidase-maintained activity for > 40 days, while lactate oxidase lost activity within a week. The researchers took advantage of the rapid manufacturing of the 3D printed parts to optimize system parameters such as immobilization pH, the volume of reaction chamber, and the carrier flow rate (immobilization time) to increase the reaction efficiency. The study demonstrated that not only could 3D printing allow improvements in reaction efficiency, but that it facilitated the tuning of immobilization conditions to increase the biocatalyst stability and that the 3D printed reactors were suitable for detecting metabolites in flow-through reaction conditions.

In contrast to post-printing covalent immobilization of enzymes, 3D printing hydrogels with physically entrapped biological components as reactionware can be a more flexible and faster fabrication process. Qian et al. (2019) formulated Baker's yeast as a model whole-cell microbe with polyethylene glycol diacrylate (PEGDA) and a photoinitiator into bioinks with remarkably high cell densities. The bioinks had appropriate rheological properties for extrusion by direct ink writing into various complex geometries with feature sizes as small as 200 μm and had excellent cell viability and homogeneity at all cell densities as shown in Figure 13.5. The researchers showed that 3D printing a lattice structure improved the mass transfer when the immobilized yeast was used to convert glucose to ethanol, increasing the volumetric

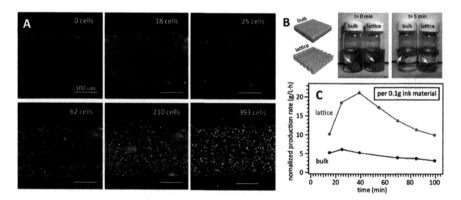

FIGURE 13.5 (A) Cell viability assays and control of cell densities; live–dead stain of a 200-μm-thick single-filament from a printed lattice. (B) Left: schematic of a bulk film and a printed lattice of living materials. Right: photographs of two vials containing bulk films and printed lattices, immediately after immersion into medium and after 5 min incubation in the medium. White arrows highlighted the location of the samples. (C) Normalized ethanol production rate of lattice (gray) and bulk hydrogels (black) as a function of time. (Adapted and reprinted with permission from Qian et al. 2019. © 2019 American Chemical Society.)

productivity of ethanol by at least three times in comparison to bulk material as shown in Figures 13.5B and 13.5C. This study demonstrated that not only can much higher cell densities be achieved with bioinks than in conventional stirred-tank reactors, but that the cells themselves act as a thickener to achieve the desired rheological properties for printing. In addition to that, the ability to 3D print the cells increases the surface area, enhances mass transport, and improves volumetric productivity. This research establishes the feasibility of directly printing cell-based bioinks, and one can envision using other types of microbes such as methanotrophs in a similar manner.

Similarly, Maier et al. (2018) encapsulated and 3D printed thermostable enzymes using an extrusion-based printing method, but their method did not require sophisticated lithography instrumentation, additional components such as photoinitiators, or post printing treatments and/or sacrificial scaffold materials. The researchers encapsulated esterase or alcohol dehydrogenase wild-type enzymes from thermophilic organisms as well as a decarboxylase from a mesophilic source that has been engineered for increased thermostability within thermoreversible agarose and extruded the bioink into various geometries about 10–20 mm in size. The enzymes maintained at least 85% of the original activity following the printing process conditions at 60°C, demonstrating the proof-of-concept of heated extrusion printing. Additionally, when operated in a flow-through reactor configuration, the team found that larger molecular weight enzymes did not elute from the agarose gel and were protected from denaturation by organic solvents, and the printed disks could be incorporated in a modular fashion to increase the concentration of the product in the outflow. This heated extrusion-based printing technique could be applicable to methanotrophs for the oxidation of methane, as thermophilic strains of the bacteria are known to exist (Islam et al. 2016; Krukenberg et al. 2018) and thermophilic bioprocessing of methane conversion may even have economical advantages (Levett et al. 2016).

Schmieg, Schimek, and Franzreb (2018) also developed an extrudable formulation of PEGDA hydrogels by incorporating additives rather than cell mass to tune the rheology, and then entrapped the enzyme β-galactosidase within the hydrogels. The researchers found that the PEGDA oligomers slightly decreased the enzyme activity, as did some of the additives, but that the shear force during extrusion had little to no impact on enzyme activity. Thus, when it was observed that the enzyme entrapped in the hydrogel was an order of magnitude less active than enzyme in solution, the researchers showed that this was due to mass transfer limitations within the hydrogel network. In order to overcome the mass transfer limitation, either hydrogel structures would have to be decreased to strands having a diameter of around 200 μm, or enzyme concentration within the hydrogel must be reduced. This demonstrates the delicate balance that must be struck between formulating a bioink with the correct mechanical properties for 3D printing while also providing adequate mass transfer of reactants and biocompatibility with the material to maintain biocatalyst activity.

A recent review of additive manufacturing used for energy applications (Zhakeyev et al. 2017) included the use of 3D printing with inorganic catalysts for methane steam reforming (Lee et al. 2013), combustion of methane gas (Stuecker et al. 2004), and oxidative coupling of methane (Michorczyk, Hędrzak, and Węgrzyniak 2016), as well as one report of the use of 3D printing with biocatalysts for methane conversion.

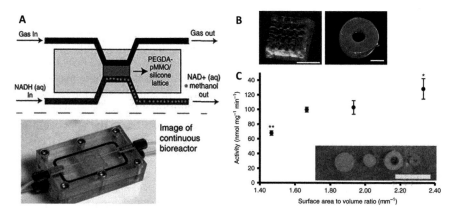

FIGURE 13.6 (A) Schematic and image of the flow-through bioreactor used with the PEG-pMMO hydrogel material. (B) Printing PEG-pMMO structures with PμSL allows high-resolution (features on the order of tens of μm) and flexibility in bioreactor component design. Printed PEG-pMMO grid structure with small feature size (left, scale bar, 500 μm). Large area PμSL was used to print cylinders with varying surface area to volume ratios on a shorter timescale with reduced resolution (right, scale bar, 1 mm). (C) The dependence of PEG-pMMO activity on surface area to volume ratio ($N = 3$, error bars represent s.d., statistical significance determined by pairwise t-test where $*P < 0.1$, $**P < 0.05$). Inset: printed cylinders with surface area to volume ratios of (left to right) 1.47, 1.67, 1.93 and 2.33 (scale bar, 10 mm). (Adapted from Blanchette et al. 2016. and reprinted under the Creative Commons CC BY license.)

Blanchette et al. (2016) developed and optimized a 3D printable biocatalytic material consisting of active pMMO enzyme entrapped within PEGDA hydrogel for the partial oxidation of methane to methanol. PEGDA was selected as the encapsulating polymer due to its decent biocompatibility and ease of photo-initiated cross-linking under mild conditions. The researchers showed that enzyme was able to maintain nearly 100% of its native activity following encapsulation within the PEG material. This material was then used to build a continuous flow-through reactor, shown in Figure 13.6A. The researchers further demonstrated that the enzyme-PEGDA bioink could be used as a substrate for 3D printing using PμSL, and demonstrated that using 3D printing to increase the surface area to volume ratio could increase the enzyme activity as shown in Figures 13.6B and 13.6C. These findings are the first to use 3D printing techniques with biocatalysis to convert methane, and they have major implications for bioreactor design and industrial methane conversion. The study demonstrated the potential for 3D printable methane bioreactors but also the need to optimize both the material and the reactor design to overcome mass transfer limitations and achieve high volumetric productivity.

13.5 CONCLUSIONS AND FUTURE PERSPECTIVES

Development of methane catalysis methods is imperative to address increasing methane production and emissions, particularly since methane is an especially deleterious greenhouse gas contributing to the rise in global temperature. Additionally, methane emissions are a potential source of energy that thus far remain underutilized due to challenges with conversion to a value-added liquid fuel or product. Biocatalytic conversion of methane by MMO enzymes or methanotroph bacteria is appealing due to is specificity, selectivity, and ambient operating conditions, but it has been limited by conventional process design and reactor equipment. In particular, the mass transfer of low-solubility methane to biocatalysts suspended in aqueous solution has severely limited the reaction kinetics, volumetric productivity, and industrial relevance.

Thus, in recent years researchers have begun to investigate the applicability of additive manufacturing, or 3D printing, to address the limitations of biocatalysis. 3D printing offers rapid prototyping, modular scalability, and micron-scale resolution; combined, these advantages have allowed researchers to more efficiently explore larger parameter spaces within biocatalyst immobilization conditions, reactor material and geometry, and continuous-flow process design. Though reports of 3D printing for biocatalysis applications have only appeared within the past few years, the results have been promising. Researchers have demonstrated that the ability to tune immobilization conditions can maximize biocatalyst activity, the ability to increase surface area can increase mass transfer and productivity, and the ability to combine immobilization with customized reactor geometry can enable efficient continuous flow-production and reuse of the biocatalyst.

While the initial results have indicated the potential of 3D printing to revolutionize biocatalysis, the area of research is quite nascent, and much development remains. The scientific value of rapidly manufacturing bioreactors with AM has already become apparent in terms of investigating reaction conditions and reactor design, but the reported results are far from optimized and the scalability to industrial relevance remains unproven. The immediate challenge is ensuring that biocatalyst activity can survive the chemical, UV, temperature and mechanical stress exposures encountered during 3D printing. The 3D printing methods themselves may need to be improved to limit these exposures, reduce print time, or achieve higher printing resolution. Next, comprehensive investigation of biocatalyst loading density, printed feature size, and reactor geometry is required to determine the optimization of mass transfer and productivity. Finally, the culmination of 3D printing bioreactors will be the demonstration of production on an industrially relevant scale.

The applicability of 3D printing with respect to biocatalytic conversion of methane remains virtually unexplored. The few reports on immobilization of MMOs and methanotrophs as well as the sole report of 3D printing pMMO indicate that this is an area worthy of additional investigation. In particular, methanotrophs and MMOs are an interesting biocatalyst to study in this context due to the numerous strains with varying tolerance to temperature and chemicals that may allow them to survive 3D printing conditions with high retention of catalytic activity. 3D printing is also perfectly suited to address the critical limitation of biocatalytic methane conversion, which is mass transfer of methane to the catalyst. The design of a continuous-flow bioreactor with optimal biocatalyst loading, feature size, and geometry to maximize volumetric productivity and allow reuse of the biocatalyst is the ultimate goal. Additionally, the biocatalytic conversion of methane to more valuable liquid products such as lactic acid may bolster the economic feasibility on an industrial scale.

ACKNOWLEDGMENTS

This work was performed under the auspices of the U.S. Department of Energy by Lawrence Livermore National Laboratory under Contract DE-AC52-07NA27344. LLNL-BOOK-775060.

REFERENCES

Blanchette, C. D., J. M. Knipe, J. K. Stolaroff, et al. 2016. Printable Enzyme-Embedded Materials for Methane to Methanol Conversion. *Nat. Commun.* 7: 11900.

Bornscheuer, U. T., G. W. Huisman, R. J. Kazlauskas, S. Lutz, J. C. Moore, and K. Robins. 2012. Engineering the Third Wave of Biocatalysis. *Nature* 485 (7397): 185–194.

Bothe, H., K. M. Jensen, A. Mergel, et al. 2002. Heterotrophic Bacteria Growing in Association with Methylococcus Capsulatus (Bath) in a Single Cell Protein Production Process. *Appl. Microbiol. Biotechnol.* 59 (1): 33–39.

Britton, J., S. Majumdar, and G. A. Weiss. 2018. Continuous Flow Biocatalysis. *Chem. Soc. Rev.* 47 (15): 5891–5918.

Chapman, J., A. E. Ismail, and C. Z. Dinu. 2018. Industrial Applications of Enzymes: Recent Advances, Techniques, and Outlooks. *Catalysts* 8 (6): 238.

Choi, D.-W., R. C. Kunz, E. S. Boyd, et al. 2003. The Membrane-Associated Methane Monooxygenase (PMMO) and PMMO-NADH: Quinone Oxidoreductase Complex from Methylococcus Capsulatus Bath. *J. Bacteriol.* 185 (19): 5755–5764.

Datta, S., L. R. Christena, and Y. R. S. Rajaram. 2013. Enzyme Immobilization: An Overview on Techniques and Support Materials. *3 Biotech* 3 (1): 1–9.

Eş, I., José D. G. Vieira, and A. C. Amaral. 2015. Principles, Techniques, and Applications of Biocatalyst Immobilization for Industrial Application. *Appl. Microbiol. Biotechnol.* 99 (5): 2065–2082.

Fei, Q., M. T. Guarnieri, L. Tao, et al. 2014. Bioconversion of Natural Gas to Liquid Fuel: Opportunities and Challenges. *Biotechnol. Adv.* 32 (3): 596–614.

Grunwald, P., ed. 2015. *Industrial Biocatalysis. Pan Stanford Series on Biocatalysis 1.* Singapore: Pan Stanford Publishing Pte. Ltd.

Gudapati, H., M. Dey, and I. Ozbolat. 2016. A Comprehensive Review on Droplet-Based Bioprinting: Past, Present and Future. *Biomaterials* 102: 20–42.

Hammond, C., M. M. Forde, M. H. Ab-Rahim, et al. 2012. Direct Catalytic Conversion of Methane to Methanol in an Aqueous Medium by Using Copper-Promoted Fe-ZSM-5. *Angew. Chem. Int. Ed.* 51 (21): 5129–5133.

Harrison, D. E. and G. Hamer. 1971. C1 Compounds as Substrates for the Production of Single-Cell Protein. *Biochem. J.* 124 (5): 78P.

Haynes, C. A. and R. Gonzalez. 2014. Rethinking Biological Activation of Methane and Conversion to Liquid Fuels. *Nat. Chem. Biol.* 10 (5): 331–339.

Henard, C. A., H. Smith, N. Dowe, M. G. Kalyuzhnaya, P. T. Pienkos, and M. T. Guarnieri. 2016. Bioconversion of Methane to Lactate by an Obligate Methanotrophic Bacterium. *Sci. Rep.* 6: 21585.

Illanes, A., A. Cauerhff, L. Wilson, and G. R. Castro. 2012. Recent Trends in Biocatalysis Engineering. *Bioresour. Technol. Biocatal.* 115: 48–57.

Islam, T., V. Torsvik, Ø. Larsen, L. Bodrossy, L. Øvreås, and N.-K. Birkeland. 2016. Acid-Tolerant Moderately Thermophilic Methanotrophs of the Class Gammaproteobacteria Isolated From Tropical Topsoil with Methane Seeps. *Front. Microbiol.* 7: 851.

Jesionowski, T., J. Zdarta, and B. Krajewska. 2014. Enzyme Immobilization by Adsorption: A Review. *Adsorption* 20 (5): 801–821.

Jiménez-González, C., P. Poechlauer, Q. B. Broxterman, et al. 2011. Key Green Engineering Research Areas for Sustainable Manufacturing: A Perspective from Pharmaceutical and Fine Chemicals Manufacturers. *Org. Process Res. Dev.* 15 (4): 900–911.

Khosravi-Darani, K., Z.-B. Mokhtari, T. Amai, and K. Tanaka. 2013. Microbial Production of Poly(Hydroxybutyrate) from C1 Carbon Sources. *Appl. Microbiol. Biotechnol.* 97 (4): 1407–1424.

Krukenberg, V., D. Riedel, H. R. Gruber-Vodicka, et al. 2018. Gene Expression and Ultrastructure of Meso- and Thermophilic Methanotrophic Consortia. *Environ. Microbiol.* 20 (5): 1651–1666.

Lee, S., T. Boeltken, A. K. Mogalicherla, U. Gerhards, P. Pfeifer, and R. Dittmeyer. 2013. Inkjet Printing of Porous Nanoparticle-Based Catalyst Layers in Microchannel Reactors. *Appl. Catal. A Gen.* 467: 69–75.

Levett, I., G. Birkett, N. Davies, et al. 2016. Techno-Economic Assessment of Poly-3-Hydroxybutyrate (PHB) Production from Methane—The Case for Thermophilic Bioprocessing. *J. Environ. Chem. Eng.* 4 (4, Part A): 3724–3733.

Lewis, J. A. 2006. Direct Ink Writing of 3D Functional Materials. *Adv. Funct. Mat.* 16 (17): 2193–2204.

Lewis, J. A. and G. M. Gratson. 2004. Direct Writing in Three Dimensions. *Mat. Today* 7 (7): 32–39.

Ligon, S. C., R. Liska, J. Stampfl, M. Gurr, and R. Mülhaupt. 2017. Polymers for 3D Printing and Customized Additive Manufacturing. *Chem. Rev.* 117 (15): 10212–10290.

Lin, H. and B. D. Freeman. 2005. Gas and Vapor Solubility in Cross-Linked Poly(Ethylene Glycol Diacrylate). *Macromolecules* 38 (20): 8394–8407.

Lin, H. and B. D. Freeman. 2006. Gas Permeation and Diffusion in Cross-Linked Poly(Ethylene Glycol Diacrylate). *Macromolecules* 39 (10): 3568–3580.

Maier, M., C. P. Radtke, J. Hubbuch, C. M. Niemeyer, and K. S. Rabe. 2018. On-Demand Production of Flow-Reactor Cartridges by 3D Printing of Thermostable Enzymes. *Angew. Chem. Int. Ed.* 57 (19): 5539–5543.

Markets and Markets. 2015. *Biocatalysis & Biocatalysts Market By Type, Application, and Geography – Forecast Till 2019*. CH 3151. marketsandmarkets.com (accessed on Jan 7, 2020).

Meyer, H.-P., E. Eichhorn, S. Hanlon, et al. 2013. The Use of Enzymes in Organic Synthesis and the Life Sciences: Perspectives from the Swiss Industrial Biocatalysis Consortium (SIBC). *Catal. Sci. Technol.* 3 (1): 29–40.

Michorczyk, P., E. Hędrzak, and A. Węgrzyniak. 2016. Preparation of Monolithic Catalysts Using 3D Printed Templates for Oxidative Coupling of Methane. *J. Mat. Chem. A* 4 (48): 18753–18756.

Miller, K. W., L. Hammond, and E. G. Porter. 1977. The Solubility of Hydrocarbon Gases in Lipid Bilayers. *Chem. Phys. Lipids* 20: 229–241.

Miri, A. K., D. Nieto, L. Iglesias, et al. 2018. Microfluidics-Enabled Multimaterial Maskless Stereolithographic Bioprinting. *Adv. Mat.* 30 (27): 1800242.

Mühlemeier, I. M., R. Speight, and P. J. Strong. 2018. Biogas, Bioreactors and Bacterial Methane Oxidation. In *Methane Biocatalysis: Paving the Way to Sustainability*, edited by M. G. Kalyuzhnaya and X.-H. Xing, 213–235. Cham: Springer International Publishing.

Munasinghe, P. C. and S. K. Khanal. 2010. Biomass-Derived Syngas Fermentation into Biofuels: Opportunities and Challenges. *Bioresour. Technol.* 101 (13): 5013–5022.

Panwar, A. and L. P. Tan. 2016. Current Status of Bioinks for Micro-Extrusion-Based 3D Bioprinting. *Molecules* (Basel, Switzerland) 21 (6): 685.

Patel, S. K. S., C. Selvaraj, P. Mardina, et al. 2016. Enhancement of Methanol Production from Synthetic Gas Mixture by Methylosinus Sporium through Covalent Immobilization. *Appl. Energy* 171: 383–391.

Patel, S. K.S., R. K. Singh, A. Kumar, et al. 2017. Biological Methanol Production by Immobilized Methylocella Tundrae Using Simulated Biohythane as a Feed. *Bioresour. Technol.* 241: 922–927.

Patel, S. K. S., S. Kondaveeti, S. V. Otari, et al. 2018. Repeated Batch Methanol Production from a Simulated Biogas Mixture Using Immobilized Methylocystis Bryophila. *Energy* 145: 477–485.

Pen, N., L. Soussan, M.-P. Belleville, J. Sanchez, C. Charmette, and D. Paolucci-Jeanjean. 2014. An Innovative Membrane Bioreactor for Methane Biohydroxylation. *Bioresour. Technol.* 174: 42–52.

Peris, E., O. Okafor, E. Kulcinskaja, et al. 2017. Tuneable 3D Printed Bioreactors for Transaminations under Continuous-Flow. *Green Chem.* 19 (22): 5345–5349.

Poling, B. E., G. H. Thomson, D. G. Friend, R. L. Rowley, and W. V. Wilding. 2008. *Perry's Chemical Engineers'* Handbook 8/E Section 2: *Physical and Chemical Data.* New York, NY: McGraw-Hill.

Pröschel, M., R. Detsch, A. R. Boccaccini, and U. Sonnewald. 2015. Engineering of Metabolic Pathways by Artificial Enzyme Channels. *Front. Bioeng. Biotechnol.* 3: 168.

Qian, F., C. Zhu, J. M. Knipe, et al. 2019. Direct Writing of Tunable Living Inks for Bioprocess Intensification. *Nano Lett.* 19 (9): 5829–5835.

Rios, G. M., M. P. Belleville, D. Paolucci, and J. Sanchez. 2004. Progress in Enzymatic Membrane Reactors – A Review. *J. Membrane Sci.* 242 (1–2): 189–196.

Rodrigues, R. C., C. Ortiz, Á. Berenguer-Murcia, R. Torres, and R. Fernández-Lafuente. 2013. Modifying Enzyme Activity and Selectivity by Immobilization. *Chem. Soc. Rev.* 42 (15): 6290–6307.

Schmieg, B., A. Schimek, and M. Franzreb. 2018. Development and Performance of a 3D-Printable Poly(Ethylene Glycol) Diacrylate Hydrogel Suitable for Enzyme Entrapment and Long-Term Biocatalytic Applications. *Eng. Life Sci.* 18 (9): 659–667.

Sheldon, R. A. and D. Brady. 2018. The Limits to Biocatalysis: Pushing the Envelope. *Chem. Commun.* 54 (48): 6088–6104.

Sheldon, R. A. and S. van Pelt. 2013. Enzyme Immobilisation in Biocatalysis: Why, What and How. *Chem. Soc. Rev.* 42 (15): 6223–6235.

Smay, J. E., J. Cesarano, and J. A. Lewis. 2002. Colloidal Inks for Directed Assembly of 3-D Periodic Structures. *Langmuir* 18 (14): 5429–5437.

Stocker, T. F., D. Qin, G.-K. Plattner, et al. eds. 2013. *Climate Change 2013: The Physical Science Basis. Contribution of Working Group I to the Fifth Assessment Report of the Intergovernmental Panel on Climate Change.* Cambridge, United Kingdom: Cambridge University Press.

Stolaroff, J. K., S. Bhattacharyya, C. A. Smith, W. L. Bourcier, P. J. Cameron-Smith, and R. D. Aines. 2012. Review of Methane Mitigation Technologies with Application to Rapid Release of Methane from the Arctic. *Environ. Sci. Technol.* 46 (12): 6455–6469.

Stone, K. A., M. V. Hilliard, Q. P. He, and J. Wang. 2017. A Mini Review on Bioreactor Configurations and Gas Transfer Enhancements for Biochemical Methane Conversion. *Biochem. Eng. J.* 128: 83–92.

Stuecker, J. N., J. E. Miller, R. E. Ferrizz, J. E. Mudd, and J. Cesarano. 2004. Advanced Support Structures for Enhanced Catalytic Activity. *Ind. Eng. Chem. Res.* 43 (1): 51–55.

Su, C.-K., S.-C. Yen, T.-W. Li, and Y.-C. Sun. 2016. Enzyme-Immobilized 3D-Printed Reactors for Online Monitoring of Rat Brain Extracellular Glucose and Lactate. *Anal. Chem.* 88 (12): 6265–6273.

Sun, M.-T., Z.-M. Yang, S.-F. Fu, X.-L. Fan, and R.-B. Guo. 2018. Improved Methane Removal in Exhaust Gas from Biogas Upgrading Process Using Immobilized Methane-Oxidizing Bacteria. *Bioresour. Technol.* 256: 201–207.

Symes, M. D., P. J. Kitson, J. Yan, et al. 2012. Integrated 3D-Printed Reactionware for Chemical Synthesis and Analysis. *Nat. Chem.* 4 (5): 349–354.

Tamborini, L., P. Fernandes, F. Paradisi, and F. Molinari. 2018. Flow Bioreactors as Complementary Tools for Biocatalytic Process Intensification. *Trends Biotechnol.* 36 (1): 73–88.

Tang, W. L. and H. Zhao. 2009. Industrial Biotechnology: Tools and Applications. *Biotechnol. J.* 4 (12): 1725–1739.

Taylor, A., P. Molzahn, T. Bushnell, et al. 2018. Immobilization of Methylosinus Trichosporium OB3b for Methanol Production. *J. Ind. Microbiol. Biotechnol.* 45 (3): 201–211.

Uchiyama, H., K. Oguri, M. Nishibayashi, E. Kokufuta, and O. Yagi. 1995. Trichloroethylene Degradation by Cells of a Methane-Utilizing Bacterium, Methylocystis Sp. M, Immobilized in Calcium Alginate. *J. Fermentation Bioeng.* 79 (6): 608–613.

U.S. Department of State. 2014. *2014 United States Climate Action Report to the UN Framework Convention on Climate Change.* http://www.state.gov/e/oes/rls/rpts/car6/index.htm (accessed on Jan 7, 2020).

Vega, J. L., E. C. Clausen, and J. L. Gaddy. 1990. Design of Bioreactors for Coal Synthesis Gas Fermentations. *Resour. Conserv. Recycl. Bioprocess. Coals* 3 (2): 149–160.

Zdarta, J., A. S. Meyer, T. Jesionowski, and M. Pinelo. 2018. A General Overview of Support Materials for Enzyme Immobilization: Characteristics, Properties, Practical Utility. *Catalysts* 8 (2): 92.

Zhakeyev, A., P. Wang, L. Zhang, W. Shu, H. Wang, and J. Xuan. 2017. Additive Manufacturing: Unlocking the Evolution of Energy Materials. *Adv. Sci.* 4 (10): 1700187.

Zheng, X., H. Lee, T. H. Weisgraber, et al. 2014. Ultralight, Ultrastiff Mechanical Metamaterials. *Science* 344 (6190): 1373–1377.

Zheng, X., J. Deotte, M. P. Alonso, et al. 2012. Design and Optimization of a Light-Emitting Diode Projection Micro-Stereolithography Three-Dimensional Manufacturing System. *Rev. Sci. Instr.* 83 (12): 125001.

Zhu, B., Y. Chen, and N. Wei. 2018. Engineering Biocatalytic and Biosorptive Materials for Environmental Applications. *Trends Biotechnol.* 37 (6): 661–676.

14 Biological Conversion of Natural Gas

*Qiang Fei, Haritha Meruvu, Hui
Wu, and Rongzhan Fu*

CONTENTS

14.1 INTRODUCTION

Natural gas exists in nature as a mixture of hydrocarbons and is chiefly composed of methane (CH_4) with variable contents of higher alkanes and minor amounts of carbon dioxide, hydrogen sulfide, nitrogen, and other compounds as shown in Table 14.1 (Viswanathan 2017). Natural gas can be procured through conventional and unconventional approaches. Conventional natural gas is formed when the methane from decayed dead plants and animals is pocket-trapped in between adjacent rocks (reservoir to be harnessed) along with water and liquid hydrocarbons based upon immiscibility and buoyancy (Benner et al. 2015). On the other hand, unconventional forms including shale gas and tight gas are mainly produced from reservoir rocks stacked together as in coal, shale or porous sandstones, which can be extracted by advanced techniques such as horizontal drilling and fracking (Tan et al. 2019, Max and Johnson 2016). The details of these techniques can be found in the other chapters of this book. Currently, within the chemical-energy industrial scenario, natural gas is produced at a surplus globally due to efficient extraction making its price hover around an all-time low in recent years (Viswanathan 2017). However, more than 150 billion m^3 of natural gas has been flared annually as a wasted resource globally (Howarth 2014, World Bank 2013). Thus, it is a critical necessity to find other options to utilize all forms of natural gas more efficiently and economically.

As an energy resource, natural gas has been mainly used via combined heat and power (CHP) for coproducing electricity or mechanical power and useful thermal energy (heating and/or cooling) (Fei and Pienkos 2018, Garg, Clomburg, and Gonzalez 2018). However, to tap the benefits and fully capture the value of this

TABLE 14.1
General Composition of Natural Gas

Compound	Symbol	Composition range (mol %)
Hydrocarbons		
Methane	CH_4	84–96
Ethane	C_2H_6	2.0–9.0
Propane	C_3H_8	0–6.0
Isobutane	C_4H_{10}	0–1.5
n-butane	C_4H_{10}	0–1.5
Isopentane	C_5H_{12}	0–0.5
n-pentane	C_5H_{12}	0–0.2
Hexane	C_6H_{14}	0–0.5
Heptane	C_7H_{16}	0–1.0
Nonhydrocarbons		
Carbon dioxide	CO_2	0–5.0
Helium	He	0–0.5
Hydrogen sulfide	H_2S	0–5.0
Nitrogen	N_2	0–10.0
Argon	A	0–0.05

incredible feedstock, it needs to be processed (fractionate alkanes and remove impurities), and either catalytically or biologically converted to mitigate storage or delivery restraints as well as improve energy contents for complying with market-specified needs and demands. Catalytic conversion of natural gas into higher hydrocarbons can be accomplished through catalysts like molybdenum, copper-loaded zeolite systems, iron–ZSM-5, nickel–ZSM-5, and H-SAPO-34, but the rate of conversion is limited due to catalytic deactivation, coke formation, thermodynamic and kinetic limitations (Park, Park, and Ahn 2019, Khodagholi and Irani 2013, Batamack, Mathew, and Prakash 2017). With the development of skills and tools of biotechnology, the manufacturing of products including fuels and chemicals via biological conversion of natural gas has been drawing more attention in recent years (Figure 14.1).

Methanotrophs are methanophilic prokaryotes which can metabolize methane for their growth and energy requirements. Methanotrophs' capacity of using CH_4 as the sole carbon and energy sources to produce platform chemicals, nutrients and biofuels, is a more favorable and ecofriendly alternative for the valorization of natural gas, compared to traditional thermo-chemical conversion methods (Henard, Smith et al. 2016, Clomburg, Crumbley, and Gonzalez 2017). Since the discovery of *Bacillus methanicus* in last century (Söhngen 1906), methanotrophs have been studied extensively due to their unique ability to oxidize CH_4 aerobically or anaerobically into various bioproducts. Hence, by harnessing the potential of methanotrophs, it is possible to mitigate the potential threats posed by greenhouse gases and contribute to the global biogeochemical carbon cycle. This chapter will only focus on aerobic methanotrophs in terms of methanotrophic biodiversity, biocatalytic properties, and cultivation modes. The potential of genetic engineering and metabolic engineering

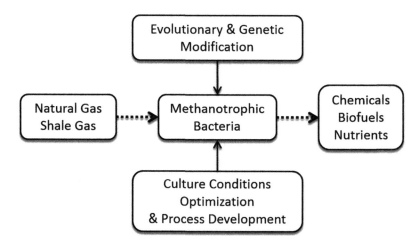

FIGURE 14.1 Scheme of biological conversion of natural gas into bio-based products using various biotechnologies.

to alter methanotrophic metabolism for improved technical performance and allied cost-cutting strategies will also be discussed. Readers who are interested in anaerobic methanotrophs may find more details in relevant literature (Cui et al. 2015, Winkel et al. 2018, Yang et al. 2018).

14.2 METHANOTROPHS AND THEIR BIOCATALYTIC PROPERTIES

Methanotrophs as a subset of methylotrophs can assimilate CH_4 using metalloenzymes called methane monooxygenases (MMOs) that oxidize CH_4 directly into methanol. Two forms of MMO have been well-studied: the soluble form (sMMO) containing iron (Fe) and the particulate form (pMMO) with copper (Cu) (Henard and Guarnieri 2018, Nichol, Murrell, and Smith 2019, Smith, Balasubramanian, and Rosenzweig 2011). Given their ability to thrive at variable temperature extremes and adapt to both acidic and alkaline environments, methanotrophs are present in diverse ecological niches such as soil sediments, agricultural farmlands, marshes, peatlands, hot acidic springs, and cold alkaline lakes (Nguyen et al. 2018, Aimen, Sohail Khan, and Kanwal 2018, Etiope and Whiticar 2019, Iguchi et al. 2019, Mateos-Rivera et al. 2018, Farhan Ul Haque et al. 2018). This adaptability allows methanotrophs to play into a significant role in global bioremediation, CH_4 sequestration, and bioconversion of natural gas.

Methanotrophs can be grouped into different categories based on taxonomic physiology and biological features, which have been reviewed before in other reports (Khmelenina et al. 2019, Kalyuzhnaya, Gomez, and Murrell 2019, Kalyuzhnaya et al. 2008, Amils 2011). All aerobic methanotrophs initially oxidize CH_4 to methanol and formaldehyde in turn, through MMO and methanol dehydrogenase (MDH), respectively. However, every methanotrophic family has a unique way to metabolize and assimilate carbon. Type I and Type X use the ribulose monophosphate (RuMP) pathway and Type II methanotrophs follow the serine pathway (Hanson and Hanson

1996, Smith, Balasubramanian, and Rosenzweig 2011). The RuMP pathway operates by fixing, cleaving and rearranging all cellular carbon directly from formaldehyde with no reduction steps. Six molecules of formaldehyde are combined with three molecules of ribulose-5-phosphate to form three molecules of hexulose-6-phosphate and a molecule of glyceraldehyde-3-phosphate, which is channeled for biosynthesis (Henard and Guarnieri 2018, Jiang et al. 2010). In the serine pathway, both formaldehyde and CO_2 are integrated using two ATP and NADH molecules each to form acetyl-CoA and CH_4 is oxidized to CO_2 via membrane-based electron transportation. A single 3-phosphoglycerate molecule is synthesized from 1.5 formaldehyde and 1.5 CO_2 molecules to produce two 2-phosphoglycerate molecules, which yield oxaloacetate and consequently malyl-CoA. Some γ-proteobacterial methanotrophs can use both pathways, as they possess genes encoding enzymes for the RuMP and serine cycles (Fei et al. 2014). However, methanotrophs like *Methylococcus capsulatus*, *Methylocaldum szegediense*, *Methylocapsa palsarum*, *Methyloferula stellata*, *Methylocella silvestris*, and *Methylocapsa acidiphila* can autotrophically fix CO_2 by Calvin Benson Bassham (CBB) cycle but cannot rely solely upon the CBB pathway for carbon assimilation (Kalyuzhnaya, Gomez, and Murrell 2019, van Teeseling et al. 2014).

Genetic engineering techniques in methanotrophs are challenging and the functional expression of MMO in heterologous hosts has been pursued for decades, with little or no success so far. Metabolic regulation can amend the metabolic pathway and carbon flux in methanotrophs to produce useful end-products like biofuels and biochemicals. Table 14.2 summarizes various genetic/metabolic techniques undertaken to manipulate methanotrophs to augment the rates of CH_4 bioconversion into industrially useful bioproducts. Modulated growth interrelated to carbon utilization has been analyzed in *Methylomicrobium alcaliphilum* using genomics and proteomics approaches (Nguyen et al. 2018, Nguyen and Lee 2019). The cultivation of *M. alcaliphilum* was manipulated by supplementing growth media with calcium or lanthanum, acting as electron-transport carriers, which can trigger the production of tricarboxylic acid (TCA) cycle metabolites and repress the formation of intermediates within the RuMP pathway (Akberdin et al. 2018). It was reported that lipid production from *Methylomicrobium buryatense* 5GB1 can be augmented threefold by limiting the glycogen biosynthesis and this bioconversion elucidates the remarkable CH_4-biofixation mechanism used by methanotrophs (Fei and Pienkos 2018). Moreover, a genetically engineered *M. buryatense* was constructed to be able to bioconvert CH_4 with an enhanced carbon flux into crotonic acid (70 mg/L) and butyric acid (40 mg/L), which are industrially useful chemicals (Garg et al. 2018).

Methanotrophs have unique, conserved sMMO regulatory systems, which complicates heterologous expression in other hosts (Conrado and Gonzalez 2014, Kalyuzhnaya, Puri, and Lidstrom 2015). Model industrial strains like *Escherichia coli* and *Saccharomyces cerevisiae* grow rapidly and can be metabolically engineered, but expression of fully active methanotrophic machinery in these hosts remains elusive. Until now, most of the works regarding sMMO expression have been done based on homologous expression (Smith and Nichol 2018), which is contrary to pMMO. Recently, the subunit B (PmoB) of pMMO was successfully expressed in *E. coli*, which triggered the enriched *E. coli* membranes to produce hydrogen

TABLE 14.2
Summary of Genetic Approaches to Use Methanotrophs as a Biosynthetic Platform

Strain	Outcome/product	Production performance	Genetic/metabolic strategy	Refs
M. capsulatus (Bath)	Methanol production in heterologous host comparable to native methanotroph	High-yield recombinant pMMO heterologously expressed in *E. coli* (with native pMMO properties conserved)	pMMO catalytic domains mimicked through ferritin scaffold by molecular assembly	(Kim et al. 2019)
Methanosarcina acetovorans	Acetate production from methane	Production of 10 mM acetate, a biofuel precursor	Cloning and insertion of methyl-coenzyme M reductase gene through reverse mutagenesis	(Soo et al. 2016)
M. alcaliphilum 20Z	Butanediol production from methane	86.2 mg/L 2,3-Butanediol per gram of CH_4 yielded at O_2 limited conditions	Expression of 2,3-Butanediol synthetic genes inserted from *Klebsiella pneumoniae* and *Bacillus subtilis* followed by in silico gene knockout mutations	(Nguyen et al. 2018)
M. alcaliphilum 20Z	Enhanced production and tolerance of Putrescine from methane	Genetically evolved strain produced 98.08 mg/L putrescine, eightfold more than actual strain	Genes from *E. coli* and *Methylosinus trichosporium* OB3b inserted; genes for spermidine synthase, acetate kinase, lactate dehydrogenase (LDH) gene knocked out; followed by adaptive evolution	(Nguyen and Lee 2019)
Methylomonas sp. strain 16a	Carotenoid synthesis was increased up to 20-fold	Production of C40 carotenoids, canthaxanthin and astaxanthin	Carotenoid gene clusters inserted into chromosome (fliCS, hsdM, ccp-3, cysH, nirS regions) insertion using random transposon mutagenesis	(Ye and Kelly 2012)
M. buryatense 5GB1	Enhanced bioconversion of methane to lactate	Heterologous expression of lactate from *Lactobacillus helveticus* 70-fold higher than wild-type strain	LDH from *Lactobacillus* cloned into vector pMS3 and inserted via conjugation with *E. coli* S17-1	(Henard, Smith, et al. 2016)

(Continued)

TABLE 14.2 (CONTINUED)

Summary of Genetic Approaches to Use Methanotrophs as a Biosynthetic Platform

Strain	Outcome/product	Production performance	Genetic/metabolic strategy	Refs
M. buryatense 5GB1	Glycogen production knocked out by vector conjugation to enhance methane metabolism	Genes toxic for *E. coli* like pMMO can be expressed in *M. buryatense*	Conjugation with native IncP plasmid using vector *E. coli* donor strain S17-1 pcm433	(Puri et al. 2015)
M. capsulatus (Bath)	Bioconversion of methane to succinate	Recombinant strain over-expressed pyruvate carboxylase yielding 70 mg/L succinate, 50% higher produce than wild-type strain	Succinate pathway genes expressed in *M. capsulatus* by recombination	(Lee et al. 2016)
M. buryatense 5GB1	Increased biomass and lipid yield from methane	2.6-fold increase in biomass yield and lipids compared to wild-type	Overexpression of phosphoketolase through cloning and plasmid propagation genes	(Henard, Smith, and Guarnieri 2017)
M. buryatense 5GB1	Growth and methane metabolism enhanced	Rapid genetic manipulation possible, suitable for industrial processes	Direct electroporation, site-specific recombination (FLP-FRT) and counterselection systems	(Yan et al. 2016)
M. capsulatus (Bath)	Bioconversion of methane to isobutanol	expected yield of isobutanol from natural gas is approximately 70% (w/w)	Insertion of alcohol dehydrogenase and pyruvate synthase genes from *Saccharomyces cerevisiae* ADH6	(Lee et al. 2016)

peroxide as a result of reaction between the dinuclear copper site of PmoB protein and dioxygen (Lu et al. 2019). While this is a step in the right direction, no one has yet been able to construct a strain of *E. coli* capable of growth on CH_4 as the sole carbon source. However, stable and high-yield pMMO mimicking catalytic protein constructs have been designed and expressed in *E. coli*, resulting in methanol production. This was achieved by encoding the assembly of pMMO catalytic domains upon a biosynthetic scaffold like apoferritin (Kim et al. 2019). Despite this progress, the industrial application of genetic engineering tools specifically for CH_4 conversion to methanol is still in progress, and success will require a better understanding of methanotrophic physiology and gene regulation.

14.3 METHANOTROPHIC RESILIENCE TO TOXIC COMPOUNDS

As listed in Table 14.1, CH_4 is the major compound of natural gas along with ethane, higher alkanes and other impurities including sulfur dioxide, hydrogen sulfide (H_2S), halogenated or aromatic hydrocarbons and traces of metal ions. It has been reported that some of these impurities could be detoxified by methanotrophs. In fact, methanotrophic communities can co-metabolize organic micropollutants like sulfamethoxazole and benzotriazole through pMMO catalysis triggered when supplemented with copper ions (to induce pMMO) and acetylene (Benner et al. 2015). The alkaliphilic *M. buryatense* was found to be possess resistance to impurities in natural gas due to its dynamic Embden–Meyerhof–Parnas (EMP) pathway (Henard, Rohrer, et al. 2016, Henard and Guarnieri 2018, Garg, Clomburg, and Gonzalez 2018). The *Methylocaldum sp.* SAD2 isolated from an anaerobic digester was claimed to be able to tolerate 1000 ppm H_2S (Zhang et al. 2016). *Methylocaldum sp.* 14B is also an obligate methanotroph tolerant to H_2S, which can be explained by its sulfur metabolism with sulfide dehydrogenase, sulfite oxidase, sulfite reductase, and sulfur transferase (Wei et al. 2017).

Methanotrophs possess methanobactins (MBs) responsible for bioremediation of organic contaminants like alkanes, alkenes, aromatics, and metal ions like copper, mercury, silver, gold, iron, manganese, nickel, lead and uranium (Semrau, DiSpirito, and Yoon 2010, Singh and Singh 2017). MBs contain two heterocyclic rings (made of imidazole/ oxazolone/ pyrazinedione) that can couple with an enethiol group to participate in metal binding, which could detoxify the aforementioned chemicals and metal ions by metabolizing them into simpler compounds (Khmelenina et al. 2019, Pandey et al. 2014). Methanotrophs of *Methylosinus* and *Methylococcus* genera can naturally detoxify methyl mercury (CH_3Hg^+) by either assimilating and/or degrading it, which is because MBs form a complex with CH_3Hg^+ to cleave the C–Hg bonds via MDH resulting in demethylation and degradation (Lu et al. 2017). Furthermore, the microbial demethylation is mediated at a biomolecular level through mer operons, which can trigger organomercurial lyases (MerB) and mercuric reductases (MerA) to cleave the Hg(II) (unbind) and then reduce it into volatile elemental Hg(0) (Lu et al. 2017, Semrau et al. 2018). Methanotrophs uptake metals using various forms of MBs with commensurate metal affinities based upon their metabolic requisites and the surrounding bioavailability. Interestingly, MBs synthesized by certain types of methanotrophs can influence the metal bioavailability of the entire community.

It has been reported that MBs from *Methylosinus trichosporium* OB3b, *Methylocystis sp.* SB2, and *Methylomicrobium album* BG8 could bind with both copper and mercury ions to reduce mercury toxicity and improve growth in associated colonies of both α-proteobacterial and γ-proteobacterial methanotrophs (Semrau, DiSpirito, and Yoon 2010). In addition, MBs can also bind with mercury in the presence or absence of copper ions to form a mercury-methanobactin complex reducing Hg (II) to Hg (0) using TonB-dependent transporters and Hg (II) reductases. It is clear that methanotrophs of *Methylocystis* genera possess diverse genes phylogenetically related to mercury and arsenic reducing bacteria (*Bradyrhizobium sp.*, *Paracoccus halophilus*, *Rhodopseudomonas* sp.) capable of bioremediating heavy metals like mercury and arsenic (Baral et al. 2014, Shi et al. 2019).

14.4 CULTURE CONDITIONS FOR NATURAL GAS BIOCONVERSION

A typical fermentation medium is usually comprised of proper nutrition ingredients to enhance cell growth and biocatalytic performance. Therefore, medium optimization is driven by the process demands based on strain development, ease of scale up, and production kinetics, all of which are critical for commercial viability. Media utilized for culturing methanotrophs are usually composed of nitrate mineral salts (NMS) along with phosphorus, potassium, copper ions and other traces (Chidambarampadmavathy et al. 2017, Chidambarampadmavathy, Obulisamy, and Heimann 2015, Xing et al. 2018). Type I methanotrophs quintessentially need ample amounts of copper to improve the pMMO activity and CH_4 oxidation rate in order to induce growth and produce extensive intra-cytoplasmic membranes (DiSpirito et al. 2016). However, the effect is contrary to methanotrophs with sMMO, which responds to limited copper availability (Semrau, DiSpirito, and Yoon 2010). The effects of chelators on CH_4 oxidation have been investigated in previous reports, which indicate that the presence of Na-EDTA favored trichloroethylene oxidation by methanotrophs. The underlying reason for this phenomenon is that EDTA is an organic ligand that can complex cations like iron through precipitation making them bioavailable and concomitantly sequester inhibitory or toxic metals rendering them unavailable (Kalyuzhnaya 2016, Nguyen et al. 2018, Henry and Grbic-Galic 1990).

An interplay of environmental factors like temperature and pH also has a profound influence upon methanotrophic growth and natural gas bioconversion. Culture temperature significantly affects CH_4 utilization or solubility and the optimal range is from 24 to 36 °C (Haroon et al. 2013, Zigah et al. 2015). Since temperature can influence abundance of most terminal restriction fragments (T-RFs) assigned to ammonium monooxygenase (amoA) genes and pMMO genes, better growth of Type II methanotrophs of rice soils was observed with temperatures higher than 15°C, compared to Type I methanotrophs (Schnell and King 1996, Mohanty, Bodelier, and Conrad 2007). However, because of their innate adaptive mechanisms, certain methanotrophs have been reported to be both cryotolerant and thermotolerant with capabilities to thrive at diverse temperature optima like *Methylovulum psychrotolerans* (4°C) (Bale et al. 2019), *Methylococcus capsulatus* (Bath) (45°C) (Strong, Xie, and Clarke 2015), *Methylocystis strain* Se48 (53°C) (Tsyrenzhapova et al. 2007), and *Methylothermus thermalis* (67°C) (Tsubota et al. 2005). A 5.0–10.0 pH range optimally affects physiological growth and activity of methanotrophic genera like *Methylobacter and Methylomicrobium*, and a circumneutral pH is commonly used for methanotrophic isolation and culturing (Kalyuzhnaya et al. 2008, Sorokin, Jones, and Gijs Kuenen 2000). However, some methanotrophs can tolerate and/or grow at pH values below 5 and above 9. For example, isolates of verrucomicrobial methanotrophs need high acidity for growth and survive temperatures up to 65°C, and isolates from hot springs and acidic swamplands can thrive under a wide range of conditions like 22.5–81.6°C and 1.8–5.0 pH (Islam et al. 2008, Sharp et al. 2014).

Oxidative phosphorylation in methanotrophs will be affected by oxygen availability, which is influenced by mass transfer and reaction kinetics. It was reported that unlike α-proteobacteria, γ-proteobacterial methanotrophs can thrive at relatively higher oxygen levels and limited CH_4 due to the different metabolism dependence in regard to the cell growth and accumulation of desired products (Aimen, Sohail Khan, and Kanwal 2018, Hanson and Hanson 1996, Frenzel, Thebrath, and Conrad 1990). An optimal O_2/CH_4 ratio (0.25 v/v) can synergistically improve removal of nitrogen in mixed cultures, where CH_4 oxidation is inhibited by aerobic methane oxidation coupled to denitrification (AME-D). This ratio causes methanotrophs to generate enough organic substrates required by denitrifiers and reduces the oxygen concentration below the level that inhibits nitrite removal. Methanol, butyrate, and formaldehyde were predicted to be the chief trophic and thermodynamic links of AME-D process of such analysis (Zhu et al. 2017). AME-D in dysoxic conditions was also demonstrated in *Methylomonas denitrificans*, where trivial expanses of O_2 are required to trigger CH_4 oxidation and oxidized nitrogen was used as the terminal electron acceptor (Kits, Klotz, and Stein 2015). Although several O_2/CH_4 ratios have been investigated in that study, the effects of CH_4 concentrations within the explosion range (5–15% of CH_4 in air) on cell growth and metabolic system have not been reported yet. Thus, more efforts are essential to resolve this issue with safety controls in the future.

14.5 CURRENT APPLICATIONS

The biocatalytic conversion of natural gas into commercial products has chiefly focused on replacing petroleum-based chemicals (Hwang et al. 2018, Versantvoort et al. 2019). Pilot scale productions of biopolymers (PHA, PHB, other copolymers) from CH_4 have been reported in methanotrophs including *M. parvus* OBBP, *M. trichosporium* OB3b, *Methylocystis sp.* WRRC, *Methylosinus* genera, and *Methylocella* genera with the biopolymer content up 20 to 50% (wt) in batch and continuous cultivations (Huang, Yu, and Chistoserdova 2018, Pérez et al. 2019, Strong et al. 2016, Yu, Beck, and Chistoserdova 2017). Valerate is commonly used as the hydroxyvalerate precursor stimulating methanotrophic metabolism to synthesize copolymers (Cal, Sikkema, et al. 2016, Colombo et al. 2017). Enhanced PHBV (poly 3-hydroxybutyrate-co-3-hydroxyvalerate) production was observed in the cultivation of Type II methanotrophs using CH_4 as the carbon source along with supplementing valerate (Myung et al. 2017, Cal, Dirk Sikkema, et al. 2016). Methanotroph *Methylocystis sp.*

WRRC1 can also produce various biopolymers with supplying valerate or n-pentanol, and the polymer production was directly proportional to valeric acid concentration supplied during the cultures (Cal, Sikkema et al. 2016). Newlight Technologies, Inc. (USA) is commercializing the development of methanotrophic techniques devoted to PHA production from natural gas. The CH_4-derived PHA has been used for packing computers by Dell Inc. and making cell phone cases by Sprint Corp, a US telecommunications company (Shogren et al. 2019). Both companies claim that the bioplastic products from Newlight Technologies are cheaper than other market available plastics (Shogren et al. 2019). Moreover, Mango Materials (USA), a start-up company is attempting to produce PHAs commercially using methane obtained from diluted emissions of industrial gases ($\leq 5\%$, v/v) or other wasted sources in order to cut down production costs even lower (Cantera et al. 2018).

Production of bioproducts such as ectione, single cell protein (SCP) for animal feed, bioplastics, biofuels and extracellular polysaccharides (EPS) mostly depended upon expensive and edible carbon sources in the past. However, this situation now can be circumvented by using natural gas or flared gas, which is a readily available and nonedible feedstock. Researches or commercial successes have been reported before for this purpose (Cantera et al. 2018, 2017, 2016, Pieja, Morse, and Cal 2017, Cal, Sikkema, et al. 2016). SCP production from methanotrophs like *Methylococcus capsulatus* is currently gaining interest in the industrial sector (Conrado and Gonzalez 2014). Companies like Calysta Inc. and Cargill Inc. are manufacturing natural gas-derived proteins for animal or fish feeds (FeedKind®). UniBio A/S has reported a pilot scale production with a productivity of 4 kg biomass/m³/h for feed protein using the patented U-loop fermenter, which is an innovative bioreactor with gas recirculation internally designed for methane oxidation and production of SCP (Petersen et al. 2017, Cantera et al. 2018). Calysta, Inc. has also claimed that the natural gas-derived biodiesel costs only half as much as conventional diesel (Petersen et al. 2017, Ritala et al. 2017).

14.6   FINAL REMARKS

Due to the development of fracking technologies, natural gas (mostly shale gas) has been becoming an abundantly available feedstock. Methanotrophs capable of using CH_4 as the sole substrate could serve as novel biocatalysts to use diffuse and remote sources of natural gas on a large- or small-scale basis. The bioconversion technologies discussed in this chapter offer a promising strategy to convert natural gas into liquid fuels or chemicals for simplified handling and transportation. Currently, SCP, biofuels, biopolymers, and other bioproducts have been produced from natural gas in pilot or commercial scales. In addition to CH_4, other components in natural gas can also be co-oxidized and further metabolized by methanotrophs for different applications.

To understand the socioeconomic benefits and industrial appraisal of natural gas and methanotrophic capabilities, it is imperative to search for and employ enhanced technologies to harness the full potential of this novel biology. The future prospects in methanotrophic biotechnology also rely upon securing taxonomically diverse methanotrophs that can thrive at faster rates. Therefore, unconventional and innovative isolation techniques could be employed for novel methanotroph isolation using approaches such as microfluidics, single cell techniques, and high throughput screening skills (Zhu et al 2017; Zigah et al 2015). In the near future, more advanced techniques like Raman activated cell sorting could be coupled with Raman activated cell ejection to discreetly detect and quantify cells by sensing their presence through biological activities. Overall, unleashing the potential of utilizing natural gas with the aid of methanotrophic bacteria could make an augmented impact on our current market economy and greenhouse gas reduction.

ACKNOWLEDGMENTS

Authors would like to express their gratitude to Dr. Philip Pienkos for reviewing this manuscript. This work is supported by the National Key R&D Programs of China (2018YFA0901500), National Natural Science Foundation of China (21878241), the Key Research and Development Program of Shaanxi Province (2017GY-146), and Open Funding Project of the State Key Laboratory of Bioreactor Engineering.

REFERENCES

Aimen, Hadiqa, Areej Sohail Khan, and Nayab Kanwal. 2018. "Methanotrophs: the natural way to tackle greenhouse effect." *Journal of Bioremediation & Biodegradation* 9:432.

Akberdin, Ilya R., David A. Collins, Richard Hamilton, Dmitry Y. Oshchepkov, Anil K. Shukla, Carrie D. Nicora, Ernesto S. Nakayasu, Joshua N. Adkins, and Marina G. Kalyuzhnaya. 2018. "Rare earth elements alter redox balance in methylomicrobium alcaliphilum 20ZR." *Frontiers in Microbiology* 9 (2735). doi:10.3389/fmicb.2018.02735.

Amils, Ricardo. 2011. "Methanotroph." In *Encyclopedia of Astrobiology*, edited by Muriel Gargaud, Ricardo Amils, José Cernicharo Quintanilla, Henderson James Cleaves, William M. Irvine, Daniele L. Pinti, and Michel Viso, 1039–1039. Berlin, Heidelberg: Springer Berlin Heidelberg.

Bale, Nicole, Irene Rijpstra, Diana Sahonero, Igor Oshkin, Svetlana E. Belova, Svetlana Dedysh, and J. Sinninghe-Damste. 2019. "Fatty acid and hopanoid adaption to cold in the methanotroph Methylovulum psychrotolerans." *Frontiers in Microbiology* 10:589.

Baral, Bipin S., Nathan L. Bandow, Alexy Vorobev, Brittani C. Freemeier, Brandt H. Bergman, Timothy J. Herdendorf, Nathalie Fuentes, Luke Ellias, Erick Turpin, Jeremy D. Semrau, and Alan A. DiSpirito. 2014. "Mercury binding by methanobactin from Methylocystis strain SB2." *Journal of Inorganic Biochemistry* 141:161–169. doi:10.1016/j.jinorgbio.2014.09.004.

Batamack, Patrice T. D., Thomas Mathew, and G. K. Surya Prakash. 2017. "One-pot conversion of methane to light olefins or higher hydrocarbons through H-SAPO-34-catalyzed in situ halogenation." *Journal of the American Chemical Society* 139 (49):18078–18083. doi:10.1021/jacs.7b10725.

Benner, Jessica, Delfien De Smet, Adrian Ho, Frederiek-Maarten Kerckhof, Lynn Vanhaecke, Kim Heylen, and Nico Boon. 2015. "Exploring methane-oxidizing communities for the co-metabolic degradation of organic micropollutants." *Applied Microbiology and Biotechnology* 99 (8):3609–3618. doi:10.1007/s00253-014-6226-1.

Cal, Andrew J., W. Dirk Sikkema, Maria I. Ponce, Diana Franqui-Villanueva, Timothy J. Riiff, William J. Orts, Allison J. Pieja, and Charles C. Lee. 2016. "Methanotrophic production of polyhydroxybutyrate-co-hydroxyvalerate with high hydroxyvalerate content." *International Journal of Biological Macromolecules* 87:302–307. doi:10.1016/j.ijbiomac.2016.02.056.

Cantera, Sara, José M. Estrada, Raquel Lebrero, Pedro A. García-Encina, and Raúl Muñoz. 2016. "Comparative performance evaluation of conventional and two-phase hydrophobic stirred tank reactors for methane abatement: mass transfer and biological considerations." *Biotechnology and Bioengineering* 113 (6):1203–1212. doi:10.1002/bit.25897.

Cantera, Sara, Osvaldo D. Frutos, Juan Carlos López, Raquel Lebrero, and Raúl Muñoz Torre. 2017. "Technologies for the bio-conversion of GHGs into high added value products: current state and future prospects." In *Carbon Footprint and the Industrial Life Cycle: From Urban Planning to Recycling*, edited by Roberto Álvarez Fernández, Sergio Zubelzu, and Rodrigo Martínez, 359–388. Cham: Springer International Publishing.

Cantera, Sara, Raúl Muñoz, Raquel Lebrero, Juan Carlos López, Yadira Rodríguez, and Pedro Antonio García-Encina. 2018. "Technologies for the bioconversion of methane into more valuable products." *Current Opinion in Biotechnology* 50:128–135. doi:10.1016/j. copbio.2017.12.021.

Chidambarampadmavathy, K., K. P. Obulisamy, and K. Heimann. 2015. "Role of copper and iron in methane oxidation and bacterial biopolymer accumulation." *Engineering in Life Sciences* 15 (4):387–399. doi:10.1002/elsc.201400127.

Chidambarampadmavathy, Karthigeyan, Obulisamy Parthiba Karthikeyan, Roger Huerlimann, Gregory E. Maes, and Kirsten Heimann. 2017. "Responses of mixed methanotrophic consortia to variable Cu2+/Fe2+ ratios." *Journal of Environmental Management* 197:159–166. doi:10.1016/j.jenvman.2017.03.063.

Clomburg, James M., Anna M. Crumbley, and Ramon Gonzalez. 2017. "Industrial biomanufacturing: the future of chemical production." *Science* 355 (6320):aag0804. doi:10.1126/ science.aag0804.

Colombo, Bianca, Francesca Favini, Barbara Scaglia, Tommy Pepè Sciarria, Giuliana D'Imporzano, Michele Pognani, Anna Alekseeva, Giorgio Eisele, Cesare Cosentino, and Fabrizio Adani. 2017. "Enhanced polyhydroxyalkanoate (PHA) production from the organic fraction of municipal solid waste by using mixed microbial culture." *Biotechnology for Biofuels* 10 (1):201. doi:10.1186/s13068-017-0888-8.

Conrado, R. J., and R. Gonzalez. 2014. "Envisioning the bioconversion of methane to liquid fuels." *Science* 343 (6171):621–623. doi:10.1126/science.1246929.

Cui, Mengmeng, Anzhou Ma, Hongyan Qi, Xuliang Zhuang, and Guoqiang Zhuang. 2015. "Anaerobic oxidation of methane: an 'active' microbial process." *MicrobiologyOpen* 4 (1):1–11. doi:10.1002/mbo3.232.

DiSpirito, Alan A., Jeremy D. Semrau, J. Colin Murrell, Warren H. Gallagher, Christopher Dennison, and Stéphane Vuilleumier. 2016. "Methanobactin and the link between copper and bacterial methane oxidation." *Microbiology and Molecular Biology Reviews* 80 (2):387. doi:10.1128/MMBR.00058-15.

Etiope, G., and M. J. Whiticar. 2019. "Abiotic methane in continental ultramafic rock systems: towards a genetic model." *Applied Geochemistry* 102:139–152. doi:10.1016/j. apgeochem.2019.01.012.

Farhan Ul Haque, Muhammad, Andrew Crombie, Scott A. Ensminger, Calin Baciu, and J. Colin Murrell. 2018. "Facultative methanotrophs are abundant at terrestrial natural gas seeps." *Microbiome* 6:118.

Fei, Qiang, and Philip T. Pienkos. 2018. "Bioconversion of methane for value-added products." In *Extremophilic Microbial Processing of Lignocellulosic Feedstocks to Biofuels, Value-Added Products, and Usable Power*, edited by Rajesh K. Sani, and Navanietha Krishnaraj Rathinam, 145–162. Cham: Springer International Publishing.

Fei, Qiang, Michael T. Guarnieri, Ling Tao, Lieve M. L. Laurens, Nancy Dowe, and Philip T. Pienkos. 2014. "Bioconversion of natural gas to liquid fuel: opportunities and challenges." *Biotechnology Advances* 32 (3):596–614. doi:10.1016/j. biotechadv.2014.03.011.

Frenzel, Peter, Bernward Thebrath, and Ralf Conrad. 1990. "Oxidation of methane in the oxic surface layer of a deep lake sediment (Lake Constance)." *FEMS Microbiology Letters* 73 (2):149–158. doi:10.1111/j.1574-6968.1990.tb03935.x.

Garg, Shivani, Hao Wu, James M. Clomburg, and George N. Bennett. 2018. "Bioconversion of methane to C-4 carboxylic acids using carbon flux through acetyl-CoA in engineered Methylomicrobium buryatense 5GB1C." *Metabolic Engineering* 48:175–183. doi:10.1016/j.ymben.2018.06.001.

Garg, Shivani, James M. Clomburg, and Ramon Gonzalez. 2018. "A modular approach for high-flux lactic acid production from methane in an industrial medium using engineered Methylomicrobium buryatense 5GB1." *Journal of Industrial Microbiology & Biotechnology* 45 (6):379–391. doi:10.1007/s10295-018-2035-3.

Hanson, R. S., and T. E. Hanson. 1996. "Methanotrophic bacteria." *Microbiological Reviews* 60 (2):439–471.

Haroon, Mohamed F., Shihu Hu, Ying Shi, Michael Imelfort, Jurg Keller, Philip Hugenholtz, Zhiguo Yuan, and Gene W. Tyson. 2013. "Anaerobic oxidation of methane coupled to nitrate reduction in a novel archaeal lineage." *Nature* 500:567. doi:10.1038/nature12375.

Henard, Calvin, Holly Rohrer, Nancy Dowe, Marina Kalyuzhnaya, Philip Pienkos, and Michael Guarnieri. 2016. "Bioconversion of methane to lactate by an obligate methanotrophic bacterium." *Scientific Reports* 6:21585.

Henard, Calvin A., and Michael T. Guarnieri. 2018. "Metabolic engineering of methanotrophic bacteria for industrial biomanufacturing." In *Methane Biocatalysis: Paving the Way to Sustainability*, edited by Marina G. Kalyuzhnaya, and Xin-Hui Xing, 117–132. Cham: Springer International Publishing.

Henard, Calvin A., Holly K. Smith, and Michael T. Guarnieri. 2017. "Phosphoketolase overexpression increases biomass and lipid yield from methane in an obligate methanotrophic biocatalyst." *Metabolic Engineering* 41:152–158. doi:10.1016/j.ymben.2017.03.007.

Henard, Calvin A., Holly Smith, Nancy Dowe, Marina G. Kalyuzhnaya, Philip T. Pienkos, and Michael T. Guarnieri. 2016. "Bioconversion of methane to lactate by an obligate methanotrophic bacterium." *Scientific Reports* 6:21585. doi:10.1038/srep21585.

Henry, Susan M., and Dunja Grbic-Galic. 1990. "Effect of mineral media on trichloroethylene oxidation by aquifer methanotrophs." *Microbial Ecology* 20 (1):151–169. doi:10.1007/BF02543874.

Howarth, Robert W. 2014. "A bridge to nowhere: methane emissions and the greenhouse gas footprint of natural gas." *Energy Science & Engineering* 2 (2):47–60.

https://www.nature.com/articles/nature12375#supplementary-information.

https://www.nature.com/articles/srep21585#supplementary-information.

Huang, Jing, Zheng Yu, and Ludmila Chistoserdova. 2018. "Lanthanide-dependent methanol dehydrogenases of XoxF4 and XoxF5 clades are differentially distributed among methylotrophic bacteria and they reveal different biochemical properties." *Frontiers in Microbiology* 9 (1366). doi:10.3389/fmicb.2018.01366.

Hwang, In Yeub, Anh Duc Nguyen, Thu Thi Nguyen, Linh Thanh Nguyen, Ok Kyung Lee, and Eun Yeol Lee. 2018. "Biological conversion of methane to chemicals and fuels: technical challenges and issues." *Applied Microbiology and Biotechnology* 102 (7):3071–3080. doi:10.1007/s00253-018-8842-7.

Iguchi, Hiroyuki, Ryohei Umeda, Hiroki Taga, Tokitaka Oyama, Hiroya Yurimoto, and Yasuyoshi Sakai. 2019. "Community composition and methane oxidation activity of methanotrophs associated with duckweeds in a fresh water lake." *Journal of Bioscience and Bioengineering*. doi:10.1016/j.jbiosc.2019.04.009.

Islam, Tajul, Sigmund Jensen, Laila Johanne Reigstad, Øivind Larsen, and Nils-Kåre Birkeland. 2008. "Methane oxidation at 55°C and pH 2 by a thermoacidophilic bacterium belonging to the Verrucomicrobia phylum." *Proceedings of the National Academy of Sciences* 105 (1):300. doi:10.1073/pnas.0704162105.

Jiang, Hao, Yin Chen, Peixia Jiang, Chong Zhang, Thomas J. Smith, J. Colin Murrell, and Xin-Hui Xing. 2010. "Methanotrophs: multifunctional bacteria with promising applications in environmental bioengineering." *Biochemical Engineering Journal* 49 (3):277–288. doi:10.1016/j.bej.2010.01.003.

Kalyuzhnaya, M. G. 2016. "Chapter 13 – methane biocatalysis: selecting the right microbe." In *Biotechnology for Biofuel Production and Optimization*, edited by Carrie A. Eckert, and Cong T. Trinh, 353–383. Amsterdam: Elsevier.

Kalyuzhnaya, Marina G., Aaron W. Puri, and Mary E. Lidstrom. 2015. "Metabolic engineering in methanotrophic bacteria." *Metabolic Engineering* 29:142–152. doi:10.1016/j.ymben.2015.03.010.

Kalyuzhnaya, Marina G., Oscar A. Gomez, and J. Colin Murrell. 2019. "The methane-oxidizing bacteria (methanotrophs)." In *Taxonomy, Genomics and Ecophysiology of Hydrocarbon-Degrading Microbes*, edited by Terry J. McGenity, 1–34. Cham: Springer International Publishing.

Kalyuzhnaya, Marina G., Valentina Khmelenina, Bulat Eshinimaev, Dimitry Sorokin, Hiroyuki Fuse, Mary Lidstrom, and Yuri Trotsenko. 2008. "Classification of halo(alkali)philic and halo(alkali)tolerant methanotrophs provisionally assigned to the genera Methylomicrobium and Methylobacter and emended description of the genus Methylomicrobium." *International Journal of Systematic and Evolutionary Microbiology* 58 (3):591–596. doi:10.1099/ijs.0.65317-0.

Khmelenina, Valentina N., J. Colin Murrell, Thomas J. Smith, and Yuri A. Trotsenko. 2019. "Physiology and biochemistry of the aerobic methanotrophs." In *Aerobic Utilization of Hydrocarbons, Oils, and Lipids*, edited by Fernando Rojo, 73–97. Cham: Springer International Publishing.

Khodagholi, Mohammad Ali, and Mohammad Irani. 2013. "Catalytic and noncatalytic conversion of methane to olefins and synthesis gas in an AC parallel plate discharge reactor." *Journal of Chemistry* 2013:7. doi:10.1155/2013/676901.

Kim, Hyun Jin, June Huh, Young Wan Kwon, Donghyun Park, Yeonhwa Yu, Young Eun Jang, Bo-Ram Lee, Eunji Jo, Eun Jung Lee, Yunseok Heo, Weontae Lee, and Jeewon Lee. 2019. "Biological conversion of methane to methanol through genetic reassembly of native catalytic domains." *Nature Catalysis* 2 (4):342–353. doi:10.1038/s41929-019-0255-1.

Kits, K. Dimitri, Martin G. Klotz, and Lisa Y. Stein. 2015. "Methane oxidation coupled to nitrate reduction under hypoxia by the Gammaproteobacterium Methylomonas denitrificans, sp. nov. type strain FJG1." *Environmental Microbiology* 17 (9):3219–3232. doi:10.1111/1462-2920.12772.

Lee, Ok Kyung, Dong Hoon Hur, Diep Thi Ngoc Nguyen, and Eun Yeol Lee. 2016. "Metabolic engineering of methanotrophs and its application to production of chemicals and biofuels from methane." *Biofuels, Bioproducts and Biorefining* 10 (6):848–863. doi:10.1002/bbb.1678.

Li, Qianwen, Xiongqi Pang, Ling Tang, Wei Li, Kun Zhang, Tianyu Zheng, and Xue Zhang. 2019. "Insights into the origin of natural gas reservoirs in the devonian system of the marsel block, Kazakhstan." *Journal of Earth Science*. doi:10.1007/s12583-019-1016-4.

Lu, Xia, Wenyu Gu, Linduo Zhao, Muhammad Farhan Ul Haque, Alan A. DiSpirito, Jeremy D. Semrau, and Baohua Gu. 2017. "Methylmercury uptake and degradation by methanotrophs." *Science Advances* 3 (5):e1700041. doi:10.1126/sciadv.1700041.

Lu, Yu-Jhang, Mu-Cheng Hung, Brian T. A. Chang, Tsu-Lin Lee, Zhi-Han Lin, I. Kuen Tsai, Yao-Sheng Chen, Chin-Shuo Chang, Yi-Fang Tsai, Kelvin H. C. Chen, Sunney I. Chan, and Steve S. F. Yu. 2019. "The PmoB subunit of particulate methane monooxygenase (pMMO) in Methylococcus capsulatus (Bath): the CuI sponge and its function." *Journal of Inorganic Biochemistry* 196:110691. doi:10.1016/j.jinorgbio.2019.04.005.

Mateos-Rivera, Alejandro, Tajul Islam, Ian P. G. Marshall, Lars Schreiber, and Lise Øvreås. 2018. "High-quality draft genome of the methanotroph Methylovulum psychrotolerans Str. HV10-M2 isolated from plant material at a high-altitude environment." *Standards in Genomic Sciences* 13 (1):10. doi:10.1186/s40793-018-0314-2.

Max, Michael D., and Arthur H. Johnson. 2016. "Energy overview: prospects for natural gas." In *Exploration and Production of Oceanic Natural Gas Hydrate: Critical Factors for Commercialization*, 1–38. Cham: Springer International Publishing.

Mohanty, Santosh R., Paul L. E. Bodelier, and Ralf Conrad. 2007. "Effect of temperature on composition of the methanotrophic community in rice field and forest soil." *FEMS Microbiology Ecology* 62 (1):24–31.

Myung, Jaewook, James C. A. Flanagan, Robert M. Waymouth, and Craig S. Criddle. 2017. "Expanding the range of polyhydroxyalkanoates synthesized by methanotrophic bacteria through the utilization of omega-hydroxyalkanoate co-substrates." *AMB Express* 7 (1):118. doi:10.1186/s13568-017-0417-y.

Nguyen, Anh Duc, In Yeub Hwang, Ok Kyung Lee, Donghyuk Kim, Marina G. Kalyuzhnaya, Rina Mariyana, Susila Hadiyati, Min Sik Kim, and Eun Yeol Lee. 2018. "Systematic metabolic engineering of Methylomicrobium alcaliphilum 20Z for 2,3-butanediol production from methane." *Metabolic Engineering* 47:323–333. doi:10.1016/j.ymben.2018.04.010.

Nguyen, Linh Thanh, and Eun Yeol Lee. 2019. "Biological conversion of methane to putrescine using genome-scale model-guided metabolic engineering of a methanotrophic bacterium Methylomicrobium alcaliphilum 20Z." *Biotechnology for Biofuels* 12 (1):147. doi:10.1186/s13068-019-1490-z.

Nichol, Tim, J. Colin Murrell, and Thomas J. Smith. 2019. "Biochemistry and molecular biology of methane monooxygenase." In *Aerobic Utilization of Hydrocarbons, Oils, and Lipids*, edited by Fernando Rojo, 99–115. Cham: Springer International Publishing.

Pandey, V. C., J. S. Singh, D. P. Singh, and R. P. Singh. 2014. "Methanotrophs: promising bacteria for environmental remediation." *International Journal of Environmental Science and Technology* 11 (1):241–250. doi:10.1007/s13762-013-0387-9.

Park, Min Bum, Eun Duck Park, and Wha-Seung Ahn. 2019. "Recent progress in direct conversion of methane to methanol over copper-exchanged zeolites." *Frontiers in Chemistry* 7 (514). doi:10.3389/fchem.2019.00514.

Pérez, Rebeca, Sara Cantera, Sergio Bordel, Pedro A. García-Encina, and Raúl Muñoz. 2019. "The effect of temperature during culture enrichment on methanotrophic polyhydroxyalkanoate production." *International Biodeterioration & Biodegradation* 140:144–151. doi:10.1016/j.ibiod.2019.04.004.

Petersen, Leander A. H., John Villadsen, Sten B. Jørgensen, and Krist V. Gernaey. 2017. "Mixing and mass transfer in a pilot scale U-loop bioreactor." *Biotechnology and Bioengineering* 114 (2):344–354. doi:10.1002/bit.26084.

Pieja, Allison J., Molly C. Morse, and Andrew J. Cal. 2017. "Methane to bioproducts: the future of the bioeconomy?" *Current Opinion in Chemical Biology* 41:123–131. doi:10.1016/j.cbpa.2017.10.024.

Puri, Aaron W., Sarah Owen, Frances Chu, Ted Chavkin, David A. C. Beck, Marina G. Kalyuzhnaya, and Mary E. Lidstrom. 2015. "Genetic tools for the industrially promising methanotroph Methylomicrobium buryatense." *Applied and Environmental Microbiology* 81 (5):1775–1781. doi:10.1128/AEM.03795-14.

Ritala, Anneli, Suvi T. Häkkinen, Mervi Toivari, and Marilyn G. Wiebe. 2017. "Single cell protein—state-of-the-art, industrial landscape and patents 2001–2016." *Frontiers in Microbiology* 8 (2009). doi:10.3389/fmicb.2017.02009.

Schnell, S., and G. M. King. 1996. "Responses of methanotrophic activity in soils and cultures to water stress." *Applied and Environmental Microbiology* 62 (9):3203–3209.

Semrau, Jeremy D., Alan A. DiSpirito, and Sukhwan Yoon. 2010. "Methanotrophs and copper." *FEMS Microbiology Reviews* 34 (4):496–531. doi:10.1111/j.1574-6976.2010.00212.x.

Semrau, Jeremy D., Alan A. DiSpirito, Wenyu Gu, and Sukhwan Yoon. 2018. "Metals and methanotrophy." *Applied and Environmental Microbiology* 84 (6):e02289–e02317. doi:10.1128/aem.02289-17.

Sharp, Christine, Angela Smirnova, Jaime M. Graham, Matthew Stott, Roshan Kdh, Tim Moore, Stephen Grasby, Maria Strack, and Peter Dunfield. 2014. "Distribution and diversity of Verrucomicrobia methanotrophs in geothermal and acidic environments." *Environmental Microbiology* 16:1867–1878.

Shi, Ling-Dong, Yu-Shi Chen, Jia-Jie Du, Yi-Qing Hu, James P. Shapleigh, and He-Ping Zhao. 2019. "Metagenomic evidence for a methylocystis species capable of bioremediation of diverse heavy metals." *Frontiers in Microbiology* 9 (3297). doi:10.3389/fmicb.2018.03297.

Shogren, Randal, Delilah Wood, William Orts, and Gregory Glenn. 2019. "Plant-based materials and transitioning to a circular economy." *Sustainable Production and Consumption* 19:194–215. doi:10.1016/j.spc.2019.04.007.

Singh, Jay Shankar, and D. P. Singh. 2017. "Methanotrophs: an emerging bioremediation tool with unique broad spectrum methane monooxygenase (MMO) enzyme." In *Agro-Environmental Sustainability: Volume 2: Managing Environmental Pollution*, edited by Jay Shankar Singh, and Gamini Seneviratne, 1–18. Cham: Springer International Publishing.

Smith, Stephen M., Ramakrishnan Balasubramanian, and Amy C. Rosenzweig. 2011. "Metal reconstitution of particulate methane monooxygenase and heterologous expression of the pmoB subunit." *Methods in Enzymology* 495:195–210. doi:10.1016/B978-0-12-386905-0.00013-9.

Smith, Thomas J., and Tim Nichol. 2018. "Engineering soluble methane monooxygenase for biocatalysis." In *Methane Biocatalysis: Paving the Way to Sustainability*, edited by Marina G. Kalyuzhnaya, and Xin-Hui Xing, 153–168. Cham: Springer International Publishing.

Söhngen, N. L. 1906. "Über bakterien, welche methan als kohlenstoffnahrung und energiequelle gebrauchen." *Zentrabl Bakteriol Parasitenk Infektionskr* 15:513–517.

Soo, Valerie W. C., Michael J. McAnulty, Arti Tripathi, Fayin Zhu, Limin Zhang, Emmanuel Hatzakis, Philip B. Smith, Saumya Agrawal, Hadi Nazem-Bokaee, Saratram Gopalakrishnan, Howard M. Salis, James G. Ferry, Costas D. Maranas, Andrew D. Patterson, and Thomas K. Wood. 2016. "Reversing methanogenesis to capture methane for liquid biofuel precursors." *Microbial Cell Factories* 15 (1):11. doi:10.1186/s12934-015-0397-z.

Sorokin, Dimitry Yu, B. E. Jones, and J. Gijs Kuenen. 2000. "An obligate methylotrophic, methane-oxidizing Methylomicrobium species from a highly alkaline environment." *Extremophiles* 4 (3):145–155. doi:10.1007/s007920070029.

Strong, Peter James, Bronwyn Laycock, Syarifah Nuraqmar Syed Mahamud, Paul Douglas Jensen, Paul Andrew Lant, Gene Tyson, and Steven Pratt. 2016. "The opportunity for high-performance biomaterials from methane." *Microorganisms* 4 (1):11.

Strong, P. J., S. Xie, and W. P. Clarke. 2015. "Methane as a resource: can the methanotrophs add value?" *Environmental Science & Technology* 49 (7):4001–4018. doi:10.1021/es504242n.

Tan, Peng, Yan Jin, Bin Hou, Liang Yuan, and Zhenyu Xiong. 2019. "Experimental investigation of hydraulic fracturing for multi-type unconventional gas co-exploitation in Ordos basin." *Arabian Journal for Science and Engineering*. doi:10.1007/s13369-019-03974-9.

Tsubota, Jun, Bulat Ts Eshinimaev, Valentina N. Khmelenina, and Yuri A. Trotsenko. 2005. "Methylothermus thermalis gen. nov., sp. nov., a novel moderately thermophilic obligate methanotroph from a hot spring in Japan." *International Journal of Systematic and Evolutionary Microbiology* 55:1877–1884.

Tsyrenzhapova, I. S., B. Ts Eshinimaev, V. N. Khmelenina, George Osipov, and Yu A. Trotsenko. 2007. "A new thermotolerant aerobic methanotroph from a thermal spring in Buryatia." *Microbiology* 76:118–121.

van Teeseling, Muriel C. F., Arjan Pol, Harry R. Harhangi, Sietse van der Zwart, Mike S. M. Jetten, Huub J. M. Op den Camp, and Laura van Niftrik. 2014. "Expanding the verrucomicrobial methanotrophic world: description of three novel species of Methylacidimicrobium gen. nov." *Applied and Environmental Microbiology* 80 (21):6782–6791. doi:10.1128/AEM.01838-14.

Versantvoort, Wouter, Arjan Pol, Lena J. Daumann, James A. Larrabee, Aidan H. Strayer, Mike S. M. Jetten, Laura van Niftrik, Joachim Reimann, and Huub J. M. Op den Camp. 2019. "Characterization of a novel cytochrome cGJ as the electron acceptor of XoxF-MDH in the thermoacidophilic methanotroph Methylacidiphilum fumariolicum SolV." *Biochimica et Biophysica Acta (BBA) – Proteins and Proteomics* 1867 (6):595–603. doi:10.1016/j.bbapap.2019.04.001.

Viswanathan, Balasubramanian. 2017. "Chapter 3 – natural gas." In *Energy Sources*, edited by Balasubramanian Viswanathan, 59–79. Amsterdam: Elsevier.

Wei, Xiangdong, Xumen Ge, Yebo Li, and Zhongtang Yu. 2017. "Draft genome sequence of Methylocaldum sp. SAD2, a methanotrophic strain that can convert raw biogas to methanol in the presence of hydrogen sulfide." *Genome Announcements* 5 (32):e00716–e00717. doi:10.1128/genomeA.00716-17.

Winkel, Matthias, Julia Mitzscherling, Pier P. Overduin, Fabian Horn, Maria Winterfeld, Ruud Rijkers, Mikhail N. Grigoriev, Christian Knoblauch, Kai Mangelsdorf, Dirk Wagner, and Susanne Liebner. 2018. "Anaerobic methanotrophic communities thrive in deep submarine permafrost." *Scientific Reports* 8 (1):1291. doi:10.1038/s41598-018-19505-9.

World_Bank. 2013. *Global Gas Flaring Reduction Partnership*. Washington, DC, World Bank.

Xing, Zhilin, Tiantao Zhao, Lijie Zhang, Yanhui Gao, Shuai Liu, and Xu Yang. 2018. "Effects of copper on expression of methane monooxygenases, trichloroethylene degradation, and community structure in methanotrophic consortia." *Engineering in Life Sciences* 18 (4):236–243. doi:10.1002/elsc.201700153.

Yan, Xin, Frances Chu, Aaron W. Puri, Yanfen Fu, and Mary E. Lidstrom. 2016. "Electroporation-based genetic manipulation in type I methanotrophs." *Applied and Environmental Microbiology* 82 (7):2062. doi:10.1128/AEM.03724-15.

Yang, Yuyin, Jianfei Chen, Baoqin Li, Yong Liu, and Shuguang Xie. 2018. "Anaerobic methane oxidation potential and bacteria in freshwater lakes: seasonal changes and the influence of trophic status." *Systematic and Applied Microbiology* 41 (6):650–657. doi:10.1016/j.syapm.2018.08.002.

Ye, Rick W., and Kristen Kelly. 2012. "Construction of carotenoid biosynthetic pathways through chromosomal integration in methane-utilizing bacterium Methylomonas sp. strain 16a." In *Microbial Carotenoids from Bacteria and Microalgae: Methods and Protocols*, edited by José-Luis Barredo, 185–195. Totowa, NJ: Humana Press.

Yu, Zheng, David A. C. Beck, and Ludmila Chistoserdova. 2017. "Natural selection in synthetic communities highlights the roles of methylococcaceae and methylophilaceae and suggests differential roles for alternative methanol dehydrogenases in methane consumption." *Frontiers in Microbiology* 8 (2392). doi:10.3389/fmicb.2017.02392.

Zhang, Wenxian, Xumeng Ge, Yueh-Fen Li, Zhongtang Yu, and Yebo Li. 2016. "Isolation of a methanotroph from a hydrogen sulfide-rich anaerobic digester for methanol production from biogas." *Process Biochemistry* 51 (7):838–844. doi:10.1016/j.procbio.2016.04.003.

Zhu, Jing, Xingkun Xu, Mengdong Yuan, Hanghang Wu, Zhuang Ma, and Weixiang Wu. 2017. "Optimum O2:CH4 ratio promotes the synergy between aerobic methanotrophs and denitrifiers to enhance nitrogen removal." *Frontiers in Microbiology* 8 (1112). doi:10.3389/fmicb.2017.01112.

Zigah, Prosper K., Kirsten Oswald, Andreas Brand, Christian Dinkel, Bernhard Wehrli, and Carsten J. Schubert. 2015. "Methane oxidation pathways and associated methanotrophic communities in the water column of a tropical lake." *Limnology and Oceanography* 60 (2):553–572. doi:10.1002/lno.10035.

15 System Integration Approaches in Natural Gas Conversion

Hisham Bamufleh and Mahmoud M. El-Halwagi

CONTENTS

15.1 INTRODUCTION

The discovery and production of substantial amounts of shale gas have spurred significant interest in monetizing shale gas into value-added products (e.g., Elbashir et al., 2019; Al-Douri et al., 2017; Bamufleh et al., 2016; Ehlinger et al., 2014; Siirola, 2014). The successful implementation of monetization initiatives requires a systematic approach to overcome the following challenges and questions:

- What products should be obtained from shale gas?
- What chemical pathways should be adopted?
- How to synthesize the process flowsheets and screen the various alternatives?
- How to predict the performance of the emerging gas conversion technologies with limited experimental/pilot data or industrial precedents?
- How to integrate the monetization processes with existing infrastructures?

Because of the complexity of the aforementioned challenges and the possibility of numerous alternatives, a systematic approach is needed to guide the decision makers. In this context, System integration offers a highly effective framework that is enabled with power tools and techniques. System integration is aimed at the synergistic aggregation of various elements to address a collective objective or set of objectives. A hallmark of system integration is that the functionality of the aggregated system is superior to the mere addition of individual functionalities. Over the

415

past three decades, substantial progress has been made in the area of system integration for the design and operation of industrial processes. For comprehensive coverage of the field of integrated process design and operation, the reader is referred to several textbooks (Biegler et al., 1997; El-Halwagi, 1997, 2006, 2017a; Foo, 2012; Majozi, 2010; Seider et al., 2017; Smith, 2016; Towler and Sinnott, 2013; Turton et al., 2018). The adaptation of system integration techniques to monetizing shale gas offers a powerful framework and a unique opportunity towards innovation, research, and technology development. This chapter provides an overview of important system integration concepts and techniques that can be adapted and applied for shale-gas monetization. Specifically, the following topics will be covered:

- Flowsheet synthesis and analysis
- Process integration and multi-objective decision making
- Modular manufacturing
- Industrial symbiosis
- Coupling of systems integration and experimental work

Relevant case studies will also be discussed to illustrate the merits and applicability of system integration tools in developing and optimizing shale-gas monetization processes.

15.2 FLOWSHEET SYNTHESIS AND ANALYSIS

There are numerous alternatives to convert shale gas into value-added products. It is, therefore, critically important to be able to efficiently generate and screen the promising pathways. Several techniques have been proposed for reaction pathway synthesis. One such approach is based on species generation, sorting, and reaction via a superstructure that embeds potential configurations (e.g., Topolski et al., 2019; Noureldin et al., 2014; Ponce-Ortega et al., 2012; Pham and El-Halwagi, 2012; Bao et al., 2011). Figure 15.1 shows the superstructure for this approach. It involves multiple stages. Within each stage, a set of possible intermediates is generated and sorted then stoichiometrically valid reactions are created to connect them. The constituents of shale gas as well as external species (e.g., other reactants, flare gases, etc.) constitute the initial set of species. Chemical conversion is used to create other species. The generated pathways may be tested using experimental work and the experimental results can be used to refine the theoretical predictions (Alsuhaibani et al., 2019). Prescreening rules are used to eliminate non-promising pathways. An example of preliminary screening tools is the use of the Metric for Inspecting Sales and Reactants (MISR), which is defined by (El-Halwagi, 2017b):

$$\text{MISR} = \frac{\sum_{p=1}^{N_{\text{products}}} P_p \times S_p}{\sum_{r=1}^{N_{\text{reactants}}} R_r \times C_r} \tag{15.1}$$

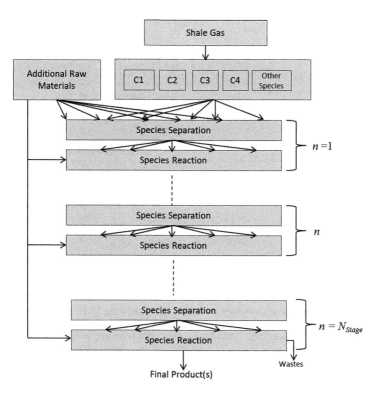

FIGURE 15.1 A superstructure for embedding monetization pathways.

where P_p is the annual production rate of product p, S_p is the selling price of product p, R_r is the annual consumption of reactant r and C_r is the purchased price of reactant r. A pathway having an MISR value less than one is removed from further consideration.

There are several examples in literature that illustrate the generation and reduction of reaction-pathways superstructures. For instance, Panjapakul and El-Halwagi (2018) considered 75 reaction-pathways for the production of isopropanol. Using the aforementioned screening tools (such as MISR and the maximum number of processing steps), the superstructure was reduced to only six pathways.

Next, process synthesis techniques are used to generate process flow diagrams (PFDs). Data from similar processes, experiments, or literature may be used to provide critical information for the process flow diagrams. Next, simulation is used to analyze the performance of the devised configurations. A high-level techno-economic analysis is utilized to select a preferred base case. Figure 15.2 shows the sequence of activities.

15.3 PROCESS INTEGRATION AND MULTICRITERIA DECISION MAKING

Process integration is a holistic approach to design and operation that emphasizes the unity of the process (El-Halwagi, 1997). It provides a powerful framework for

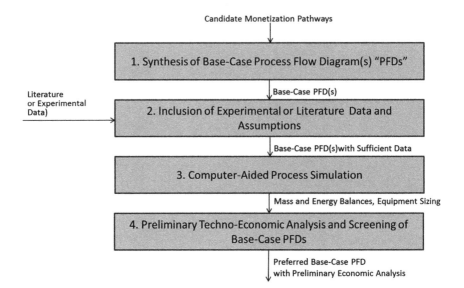

Candidate Monetization Pathways

1. Synthesis of Base-Case Process Flow Diagram(s) "PFDs"

Base-Case PFD(s)

Literature
or Experimental
Data)

2. Inclusion of Experimental or Literature Data and
Assumptions

Base-Case PFD(s)with Sufficient Data

3. Computer-Aided Process Simulation

Mass and Energy Balances, Equipment Sizing

4. Preliminary Techno-Economic Analysis and Screening of
Base-Case PFDs

Preferred Base-Case PFD
with Preliminary Economic Analysis

FIGURE 15.2 Sequence of activities for selecting a base-case PFD.

conserving natural resources, enhancing yield, mitigating pollution, and debottle-necking. There are three primary categories of process integrations: energy, mass, and property integration (El-Halwagi, 2017a; Smith, 2016; Towler and Sinnott, 2013). Graphical techniques such as "pinch analysis" have been used to construct composite diagrams for energy, mass, and property flows, allocation, and conversion for the whole process and to determine performance benchmarks (targets). Algebraic and optimization techniques have also been used. Graphical techniques offer valuable visualization insights but become cumbersome when complex problems are encountered (in which case, algebraic and optimization techniques become the preferred method of solution). Next, it is necessary to assess and improve several objectives for the process such as operability, controllability, sustainability, reliability, and safety. Systematic techniques may be used to address these objectives and to generate improved PFDs (e.g., Ortiz-Espinoza et al., 2019; Ade et al., 2018; Thiruvenkataswamy et al., 2016). Figure 15.3 shows the incorporation of process integration and multicriteria decision making in the design.

In addition to the use of multi-objective optimization to reconcile and trade off the various objectives, it is also possible to incorporate several objectives into a financial framework. An example is the sustainability weighted return-on-investment metric (El-Halwagi, 2017b). It extends the return-on-investment (ROI) concept by using targets from process integration and indicators representing sustainability. Suppose there are several PFD alternatives: $p = 1,2,..., N_{Projects}$. For the pth project, the annual sustainability profit "ASP" may be calculated according to the following definition:

$$ASP_p = AEP_P \left[1 + \sum_{i=1}^{N_{indicators}} w_i \left(\frac{Indicator_{p,i}}{Indicator_i^{Target}} \right) \right] \qquad (15.2)$$

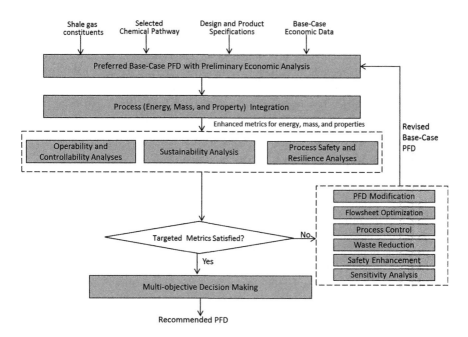

FIGURE 15.3 Incorporation of process integration and multicriteria decision making in PFD selection and optimization.

where i is an index for the sustainability indicators. Each weight, w_i, represents the relative value of the ith sustainability indicator compared to the economic profit. The term $Indicator_{p,i}$ represents the value of the ith sustainability indicator associated with the pth PFD and the term $Indicator_i^{Target}$ is determined via the target value of the ith sustainability indicator as determined from the process-integration benchmarking studies.

The sustainability weighted return-on-investment metric "SWROIM" of the pth PFDis defined as:

$$SWROIM_p = \frac{ASP_p}{TCI_p} \tag{15.3}$$

Conventional rules for the acceptability of ROI are applied to SWROIM while including the multiple sustainability objectives. A similar metric can be used to include safety and sustainability indicators in addition to the profitability criteria (Guillen-Cuevas et al., 2018). Agrawal et al. (2018) illustrated the use of sustainability- and safety-based return-on-investment for the creation, screening, and selection of a propane-to-propylene process. In addition to safety and sustainability, resilience can also be included as one of the objectives. Resilience may be defined as the ability of the manufacturing process (e.g., gas monetization plant), its value chains, and its impacted communities to predict, prepare for, mitigate, withstand, adapt to, and recover from the impact of external stressors such as natural disasters. This issue is becoming increasingly important in light of the catastrophic effects of recent natural disasters (such as hurricanes, floods, earthquakes, and fires) on the gas monetization plants and supply chains. For instance, in 2017, Hurricane Harvey caused a major disruption in the shale-gas monetization supply chains including the shutting down of about 80% of the ethylene production in Texas (DeRosa et al., 2019). Moreno-Sader et al. (2019) included resilience, reliability, safety, and sustainability in the economic screening and evaluation of process designs through an extended return-on-investment metric. The use of the devised metric was illustrated using a case study on designing and optimizing a hydrocracking system.

15.4   MODULAR PROCESS INTEGRATION

Because of the distributed nature of shale-gas supply and product demand as well as the dynamic variability of gas production, there is a growing need to integrate modular systems into the infrastructure of shale-gas monetization. As an illustration, consider the problem of designing a gas-processing facility for a shale-gas source that features a steep decline in production (Allen et al., 2019). If the facility is designed for the initial production capacity, it will be oversized and the capital productivity will be poor. On the other hand, designing the facility for a mid-production level will entail flaring the unprocessed gas during the initial production phase and under-utilization of the process capacity during the latter production stages. To address such a dilemma, the integration of modular units into the facility design is particularly effectives. Gas-processing modules are added and transported on a need basis. This approach requires the simultaneous optimization of design, operation, transportation, and scheduling. Allen et al. (2019) generated scheduling schemes of skid-mounted modular gas-processing units to address variable input flows of shale gas. A similar approach can be adopted for treating shale-gas wastewater and flared gases with modular and transportable units to address the temporal and spatial variabilities (Thompson, 2019; Al-Aboosi and El-Halwagi, 2019; Lira-Barragán et al., 2016; El-Halwagi et al., 2009).

15.5   INTEGRATION OF SHALE-GAS MONETIZATION
INTO INDUSTRIAL SYMBIOSIS

Industrial symbiosis involves the integration of material, energy, and infrastructure among multiple facilities that are typically within the same region (Gibbs and Deutz, 2007; Chertow, 2007; Lowe, 2001). Shale-gas monetization is an excellent candidate for incorporation within an industrial symbiosis framework. In this regard, a particularly attractive conceptual framework is the design of a carbon-hydrogen-oxygen symbiosis network (CHOSYN). The main idea is to develop multiscale integration opportunities between shale and flared gases with existing and planned infrastructures to induce synergistic effects (Noureldin and El-Halwagi, 2015). Benchmarking and pathway synthesis are developed at the atomic, unit, process, and multi-process levels using a consistent and systematic framework. Several recent research contributions have been made in the area of design and system integration of shale and flared gases into CHOSYNs. El-Halwagi (2017a) introduced a multiscale benchmarking technique for the generation of reaction-pathways and design of CHOSYNs using algebraic techniques. Mukherjee and El-Halwagi (2018) addressed the problem of designing

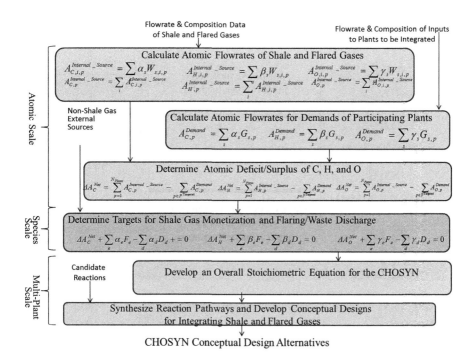

FIGURE 15.4 A multiscale approach to the integration of shale and flared gases into a CHOSYN. (Adapted, with revisions, from El-Halwagi, 2017b.)

CHOSYNs under uncertainty in flowrate and composition of shale and flared gases. The incorporation of shale gas into CHOSYNs can lead to substantial reduction in the carbon footprint (Panu et al., 2019) and higher intensification in using mass and energy (Topolski et al., 2019, Al-Fadhli et al., 2019). These integration opportunities can be created for existing and planned infrastructures (Topolski et al., 2018). Figure 15.4 shows a typical approach for the multiscale integration of shale and flared gases into a CHOSYN. First, the flowrate and composition data of available gas are used to evaluate the content of carbon, hydrogen, and oxygen atoms which are subsequently integrated with existing infrastructures and candidate non–shale gas external resources. The result is the calculation of atomic surplus or deficit targets for carbon, hydrogen, and oxygen. Next, economic factors are used to evaluate targets for product and intermediate species that can be used to construct overall reaction stoichiometry. Reaction pathway synthesis techniques are then used to create reaction alternatives that are included in a conceptual design framework. Techno-economic analysis is used to screen the alternatives and generate base-case designs.

15.6 SYNERGISM BETWEEN SYSTEMS INTEGRATION AND EXPERIMENTAL WORK

Although this chapter has focused on systems integration, it is important to note the imperative need and value of coupling process systems engineering with experimental

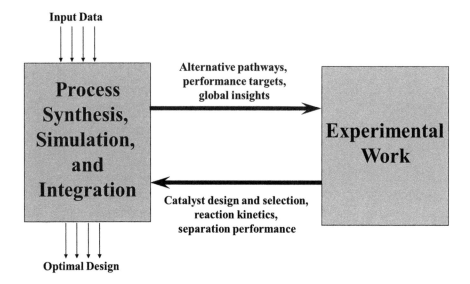

Input Data

Process Synthesis, Simulation, and Integration

Alternative pathways,
performance targets,
global insights

Experimental Work

Catalyst design and selection,
reaction kinetics,
separation performance

Optimal Design

FIGURE 15.5 Synergistic coupling of process systems engineering with experimental work.

work. Process synthesis can be used to quickly generate candidate monetization pathways and process analysis and simulation can be used to screen the alternatives. For the promising candidates, there is a need to use actual data that may not be predictable via simulation. These data are best obtained through experimental work. Therefore, process synthesis and simulation can direct the scope of experimental work and focus the needs for certain results. Process integration can provide performance benchmarks for the experimental work. On the other hand, the experimental work leads to the design and selection of the catalysts, identification of reaction kinetics and implementation hurdles, separation performance, and verifiable data that can be used to enhance the design and aid in optimization. This synergistic effect is shown by Fig. 5. Alsuhaibani et al. (2019) illustrated the value of coupling systems integration with experimental work via a case study for the production of methanol from shale gas, which involved the accelerated development of catalyst and reaction system while optimizing the overall profitability of the monetization process (Figure 15.5).

15.7 CONCLUSIONS

System integration frameworks and enabling techniques have evolved over the past three decades as the definitive framework for the design and optimization of industrial processes. Their use for the emerging field of natural gas monetization offers remarkable advantages for the creation of plausible pathways, the quick screening of alternatives, the selection of promising configurations, the benchmarking of potential performance, and the tradeoff among multiple decision-making criteria. Specific adaptations and development activities are needed to address the particular characteristics of shale-gas conversion and to handle temporal and spatial variabilities.

In this regard, modular system design provides new directions and challenges in the value-added conversion of stranded and flared gases. Consistent frameworks can be used to carry out integration at multiple scales (e.g., atomic, species, unit, subprocess, process, and a region). Integration of shale-gas conversion into an industrial symbiosis framework creates synergistic opportunities for coupling with existing and planned infrastructures. The net result is the ability to identify win-win situations and to develop technological breakthroughs that lead to enhanced profitability, improved sustainability, better resilience, and increased safety.

REFERENCES

Ade, N., G. Liu, A. F. Al-Douri, M. M. El-Halwagi, and M. S. Mannan. 2018. Investigating the Effect of Inherent Safety Principles on System Reliability in Process Design. *Process Safety and Environmental Protection*, 117, 100–110. doi:10.1016/j.psep.2018.04.011.

Agrawal, A., D. Sengupta, and M. M. El-Halwagi. 2018. A Sustainable Process Design Approach for On-Purpose Propylene Production and Intensification. *ACS Sustainable Chemistry & Engineering*, 6, 2407–2421. doi:10.1021/acssuschemeng.8b00235.

Al-Aboosi, F. Y. and M. M. El-Halwagi. 2019. A Stochastic Optimization Approach to the Design of Shale Gas/Oil Wastewater Treatment Systems with Multiple Energy Sources under Uncertainty. *Sustainability*, 11(18), 4865–4906. doi:10.3390/su11184865.

Al-Douri, A., D. Sengupta, and M. M. El-Halwagi. 2017. Shale Gas Monetization – A Review of Downstream Processing to Chemicals and Fuels. *Journal of Natural Gas Science & Engineering*, 45, 436–455. doi:10.1016/j.jngse.2017.05.016.

Al-Fadhli, F. M., H. Baaqeel, and M. M. El-Halwagi. 2019. Modular Design of Carbon-Hydrogen-Oxygen Symbiosis Networks over a Time Horizon with Limited Natural Resources. *Chemical Engineering and Processing – Process Intensification*, 141, 107535. doi:10.1016/j.cep.2019.107535.

Allen, R. C., D. Allaire, and M. M. El-Halwagi. 2019. Capacity Planning for Modular and Transportable Infrastructure for Shale Gas Production and Processing. *Industrial & Engineering Chemistry Research*, 58, 5887–5897. doi:10.1021/acs.iecr.8b04255.

Alsuhaibani, A. S., S. Afzal, M. Challiwala, N. O. Elbashir, and M. M. El-Halwagi. 2019. The Impact of the Development of Catalyst and Reaction System of the Methanol Synthesis Stage on the Overall Profitability of the Entire Plant: A Techno-Economic Study. *Catalysis Today*, 343, 191–198. doi:10.1016/j.cattod.2019.03.070.

Bamufleh, H. S., M. M. B. Noureldin, and M. M. El-Halwagi. 2016. Sustainable Process Integration in the Petrochemical Industries, Chapter 6, in *Petrochemical Catalyst Materials, Processes, and Emerging Technologies*, 155–168, H. Al-Megren and T. Xiao, Editors. *A Volume in the Advances in Chemical and Materials Engineering (ACME) Book Series*. IGI Global, Hershey, PA. doi:10.4018/978-1-4666-9975-5.

Bao, B., D. K. S. Ng, D. H. S. Tay, A. Jiménez-Gutiérrez, and M. M. El-Halwagi. 2011. A Shortcut Method for the Preliminary Synthesis of Process-Technology Pathways: An Optimization Approach and Application for the Conceptual Design of Integrated Biorefineries. *Computers and Chemical Engineering*, 35(8), 1374–1383. doi:10.1016/j.compchemeng.2011.04.013.

Biegler, L. T., I. E. Grossmann, and A. W. Westerberg. 1997. *Systematic Methods of Chemical Process Design*. Prentice Hall, NJ.

Chertow, M. R. 2007. Uncovering Industrial Symbiosis. *Journal of Industrial Ecology*, 11(1), 11–30.

DeRosa, S. E., Y. Kimura, M. A. Stadtherr, G. McGaughey, E. McDonald-Buller, and D. T. Allen. 2019. Network Modeling of the US Petrochemical Industry under Raw Material and Hurricane Harvey Disruptions. *Industrial & Engineering Chemistry Research*, 58(28), 12801–12815.

Ehlinger, V. M., K. J. Gabriel, M. M. B. Noureldin, and M. M. El-Halwagi. 2014. Process Design and Integration of Shale Gas to Methanol. *ACS Sustainable Chemistry & Engineering*, 2(1), 30–37. doi:10.1021/sc400185b.

Elbashir, N. O., M. M. El-Halwagi, K. R. Hall, and I. Economou (Editors). 2019. *Natural Gas Processing from Midstream to Downstream*. Wiley, New Jersey.

El-Halwagi, M. M. 1997. *Pollution Prevention through Process Integration: Systematic Design Tools*. Academic Press, San Diego, CA.

El-Halwagi, M. M. 2006. *Process Integration*. Elsevier, Amsterdam.

El-Halwagi, M. M. 2017a. A Return on Investment Metric for Incorporating Sustainability in Process Integration and Improvement Projects. *Clean Technologies and Environmental Policy*, 19, 611–617. doi:10.1007/s10098-016-1280-2.

El-Halwagi, M. M. 2017b. *Sustainable Design through Process Integration: Fundamentals and Applications to Industrial Pollution Prevention, Resource Conservation, and Profitability Enhancement*, 2nd Edition. IChemE/Elsevier, the Netherlands.

El-Halwagi, M. M., D. Harell, and H. D. Spriggs. 2009. Targeting Cogeneration and Waste Utilization through Process Integration. *Applied Energy*, 86(6), 880–887. doi:10.1016/j.apenergy.2008.08.011.

Foo, D. C. Y. 2012. *Process Integration for Resource Conservation*. CRC Press, Boca Raton, FL, US.

Gibbs, D. and P. Deutz. 2007. Reflections on Implementing Industrial Ecology Through Eco-Industrial Park Development. *Journal of Cleaner Production*, 15, 1683–1695. doi:10.1016/j.jclepro.2007.02.003.

Guillen-Cuevas, K., A. P. Ortiz-Espinoza, E. Ozinan, A. Jiménez-Gutiérrez, N. K. Kazantzis, and M. M. El-Halwagi. 2018. Incorporation of Safety and Sustainability in Conceptual Design via A Return on Investment Metric. *ACS Sustainable Chemistry and Engineering*, 6, 1411–1416. doi:10.1021/acssuschemeng.7b03802.

Lira-Barragán, L., J. M. Ponce-Ortega, G. Guillén-Gosálbez, and M. M. El-Halwagi. 2016. Optimal Water Management under Uncertainty for Shale Gas Production. *Industrial & Engineering Chemistry Research*, 55(5), 1322–1335. doi:10.1021/acs.iecr.5b02748.

Lowe, E. A. 2001. *Eco-Industrial Park Handbook for Asian Developing Countries Report to Asian Development Bank*. Asian Development Bank, Metro Manila, Philippines.

Majozi, T. 2010. *Batch Chemical Process Integration: Analysis, Synthesis, and Optimization*. Springer, Heidelberg.

Moreno-Sader, K., P. Jain, L. C. B. Tenorio, M. S. Mannan, and M. M. El-Halwagi. 2019. Integrated Approach of Safety, Sustainability, Reliability, and Resilience Analysis via a Return on Investment Metric. *ACS Sustainable Chemistry & Engineering*, 7, 19522–19536. doi:10.1021/acssuschemeng.9b04608.

Mukherjee, R. and M. M. El-Halwagi. 2018. Reliability of C-H-O Symbiosis Networks under Source Streams Uncertainty. *Smart and Sustainable Manufacturing Systems*, 2(2), 132–153. doi:10.1520/SSMS20180022.

Noureldin, M. M. B. and M. M. El-Halwagi. 2015. Synthesis of C-H-O Symbiosis Networks. *AIChE Journal*, 61(4), 1242–1262. doi:10.1002/aic.14714.

Noureldin, M. M. B., N. O. Elbashir, and M. M. El-Halwagi. 2014. Optimization and Selection of Reforming Approaches for Syngas Generation from Natural/Shale Gas. *Industrial & Engineering Chemistry Research*, 53(5), 1841–1855. doi:10.1021/ie402382w.

Ortiz-Espinoza, A. P., V. Kazantzi, F. T. Eljack, A. Jimenez-Gutierrez, M. M. El-Halwagi, and N. K. Kazantzis. 2019. A Framework for Design Under Uncertainty Including Inherent Safety, Environmental Assessment and Economic Performance of Chemical Processes. *Industrial & Engineering Chemistry Research*, 58(29), 13239–13248.

Panjapakkul, W. and M. M. El-Halwagi. 2018. Techno-Economic Analysis of Alternative Pathways of Isopropanol Production. *ACS Sustainable Chemistry and Engineering*, 6, 10260–10272. doi:10.1021/acssuschemeng.8b01606.

Panu, M., K. Topolski, S. Abrash, and M. M. El-Halwagi. 2019. CO_2 Footprint Reduction via the Optimal Design of Carbon-Hydrogen-Oxygen SYmbiosis Networks (CHOSYNs). *Chemical Engineering Science*, 203, 1–11. doi:10.1016/j.ces.2019.03.066.

Pham, V. and M. M. El-Halwagi. 2012. Process Synthesis and Optimization of Biorefinery Configurations. *AIChE Journal*, 58(4), 1212–1221. doi:10.1002/aic.12640.

Ponce-Ortega, J. M., V. Pham, M. M. El-Halwagi, and A. A. El-Baz. 2012. A Disjunctive Programming Formulation for the Optimal Design of Biorefinery Configurations. *Industrial & Engineering Chemistry Research*, 51, 3381–3400. doi:10.1021/ie201599m.

Seider, W. D., D. R. Lewin, J. D. Seader, S. Widagdo, R. Gani, and K. M. Ng. 2017. *Product and Process Design Principles: Synthesis, Analysis, and Design*, 4th Edition. Wiley, New York, NY.

Siirola, J. J. 2014. The impact of shale gas in the chemical industry. *AIChE Journal*, 60(3), 810–819.

Smith, R. 2016. *Chemical Process Design and Integration*, 2nd Edition. Wiley, New York, NY.

Thiruvenkataswamy, P., F. T. Eljack, N. Roy, M. S. Mannan, and M. M. El-Halwagi. 2016. Safety and Techno-Economic Analysis of Ethylene Technologies. *Journal of Loss Prevention in the Process Industries*, 39, 74–84. doi:10.1016/j.jlp.2015.11.019.

Thompson, D. 2019. *Enhancing Sustainability of Natural Gas Extraction via Technological Innovations in Wastewater Management*. https://engineering.tamu.edu/news/2019/08/enhancing-sustainability-of-natural-gas-extraction-via-technological-innovations-in-wastewater-management.html (accessed on October 6, 2019).

Topolski, K., L. F. Lira-Barragán, M. Panu, J. M. Ponce-Ortega, and M. M. El-Halwagi. 2019. Integrating Mass and Energy through the Anchor-Tenant Approach for the Synthesis of Carbon-Hydrogen-Oxygen Symbiosis Networks. *Industrial & Engineering Chemistry Research*, 58(36), 16761–16776. doi:10.1021/acs.iecr.9b02622.

Topolski, K., M. M. Noureldin, F. T. Eljack, and M. M. El-Halwagi. 2018. An Anchor-Tenant Approach to the Synthesis of Carbon-Hydrogen-Oxygen Symbiosis Networks. *Computers & Chemical Engineering*, 116(4), 80–90. doi:10.1016/j.compchemeng.2018.02.024.

Towler, G. and R. Sinnott. 2013. *Chemical Engineering Design: Principles, Practice and Economics of Plant and Process Design*, 2nd Edition. Elsevier, Amsterdam.

Turton, R., J. A. Shaeiwitz, D. Bhattacharyya, and W. B. Whiting. 2018. *Analysis, Synthesis, and Design of Chemical Processes*, 5th Edition. Prentice Hall, New Jersey.

16 Techno-Economic Analysis of Microwave-Assisted Conversion Processes
Application to a Direct Natural Gas-to-Aromatics Process

Chirag Mevawala and Debangsu Bhattacharyya

CONTENTS

16.1 INTRODUCTION

Microwave (MW) is an electromagnetic radiation with a frequency range between 300 MHz to 300 GHz (Durka 2013). When a material is irradiated with MWs, it reflects, transmits, and/or absorbs the MW energy. Materials that mostly reflect MW are called conductive materials such as gold, copper, aluminum, silver. Materials that allow MW to pass through them with negligible interactions or attenuation are called insulating materials. Lastly, materials that absorb the MW radiations are called absorbing materials. Such materials are also called dielectrics, and they generate heat due to their ability to get polarized by MW (Stankiewicz et al. 2019). The dielectric properties of a material play a key role since the main mechanisms that promotes heating due to MW radiation are ionic conduction and dipolar polarization (Gabriel et al. 1998). The dielectric response i.e. the interaction of the material with the electric field is commonly considered as a standard measure of the susceptibility of the material to MWs (Harrison and Whittaker 2003):

$$\varepsilon^* = \varepsilon' - i\varepsilon'' \tag{16.1}$$

The dielectric response of a material depends on two main parameters as shown in Equation 16.1. The dielectric constant (ε') represents the ability of the material to be polarized by the electric field, and the dielectric loss factor (ε'') represents the ability of the material to convert the absorbed electromagnetic energy into heat (Harrison and Whittaker 2003). The ratio of the dielectric constant and the dielectric loss factor is called the loss tangent or loss factor, which is a good indicator of the ability of the material to be heated by the electromagnetic field. If a material has a high loss factor, then that material can be heated rapidly using MW and vice versa.

Microwave heating has been commercially used for many years with applications in wood drying, pharmaceuticals, and food industry for cooking, thawing, pasteurization and preservation of food materials. Over the past several years, it is being increasingly investigated for assisting chemical reactions both for the homogenous and heterogeneous reactive systems (Lidström et al. 2001). Enhancement of reaction rate and improvement of product yields in MW-assisted reactors have been the two main reasons for the increased interest in application of MW for reaction engineering. Thermal and nonthermal effects are two main factors that have been attributed to the improvement in reaction mechanisms and higher productivity. Dominating mechanisms for a particular reactive system are still debatable. However, it is well agreed that MW leads to rapid heating of the reaction sites thus enhancing the reaction rates. The solid–gas catalytic systems that can benefit from the MW are those that have reasonable concentration of MW-absorbing materials, which, in turn, helps in rapid heating of the entire bed as well as possible local hotspots that can selectively catalyze desired reactions improving selectivity. For example, in methane dehydroaromatization, the reactor is filled with molybdenum (Mo)/HZSM-5 catalyst. The Mo metal particles absorb the electromagnetic waves and dissipate it in the form of heat to the rest of bed (Bai et al. 2019). For more details on the MW technology, especially from the perspective of assisting chemical reactions, interested readers are referred to several good resources in this area (Metaxas and Meredith 2008; Loupy 2008; Collin 2000).

MW-assisted reactors provide distinct advantages over conventional reactors. First, since MW can increase the reaction rate (De Souza 2015), it can help to achieve higher single-pass conversion leading to lower flowrate of the recycle streams, which, in turn, reduces the capital and operating costs of practically all equipment items that are involved in recycling. Second, since MW can help to improve yield in comparison to the conventional processes (De Souza 2015), the improved yield leads to increased production of value-added products thus generating more revenue for a given feed flow rate of the raw materials. Third, MW can help to selectively heat up the catalysts locally. Conventional heating of catalysts happens due to surface heating, where the heat is supplied to the catalysts mainly via conduction/convection through the external surface of the catalyst. In contrast, MW heating is volumetric heating where the MW-sensitive material directly absorbs the radiation and converts it into heat locally, which then gets conducted to the entire catalyst matrix. Selective heating of the catalyst not only helps to improve energy efficiency by avoiding the need to heat the entire reactive system including the reaction vessel, but also helps to improve the selectivity by designing catalysts to enable thermal gradients to reduce undesired reactions while improving the rate of desired reactions. Thus MW-assisted reactors offer great flexibility to obtain high selectivity without compromising conversion. Due to increased reaction rates and higher selectivity, MW-assisted reactors are also expected to be smaller than their conventional counterparts. However, these advantages need to be weighed with respect to several other cost factors. First, the cost of the MW generator and the associated equipment items need to be taken into account. Second, cost of the modified reactor internals due to MW integration need to be considered. Third, cost of the electricity for generating the electrogmantic radiation should be accounted for. Furthermore, while rich experience exists in scaling up conventional reactors, there is still a lack of experience in developing, installing, and operating large-scale MW-assisted reactor for various reactive processes. Therefore, the risk of scaling MW-assisted reactors may be higher compared to conventional processes. Due to the tradeoffs noted above, techno-economic analysis (TEA) is critical for evaluating the commercialization potential of MW-assisted processes.

16.2   DESIGN OF MW REACTORS

There are several aspects that has to be taken into consideration while designing/selecting a MW reactor: penetration depth (PD), MW frequency, mode of the reactor, and material of construction (Priecel and Lopez-Sanchez 2019). All these aspects are inter-dependent on each other. PD is defined as the depth at which the intensity of the electromagnetic waves drops to 37% of the original intensity. The PD of an electromagnetic wave depends on the wavelength and the properties of the propagation medium, as given by Equation 16.2 (Metaxas and Meredith 2008):

$$PD = \frac{\lambda\sqrt{\varepsilon'}}{2\pi\varepsilon''} \qquad (16.2)$$

In Equation 16.2, λ is wavelength; ε' is dielectric constant; ε'' is loss factor

The homogeneity of the reactor medium heating will depend on the penetration depth (PD) of the MW. If the PD is too short, then the reaction medium may not be heated evenly possibly leading to hotspots. The PD of the MW also depends on the frequency of the MW used. Higher frequency wave such as 2.45 GHz has a wavelength of 12.2 cm. As the frequency is reduced, the wavelength increases. For example, at 915 MHz, the wavelength increases to 33.3 cm. Selection of the MW frequency also depends on the mode of the reactor. There are mainly two basic types of MW reactor modes: multimode and mono-mode (Priecel and Lopez-Sanchez 2019; Stankiewicz et al. 2019; Durka 2013). In the multimode reactor, the MW enters a rectangular cavity through a waveguide in which the reactor is integrated. Distribution of the MW is nonhomogenous due to the reflections on the cavity walls and the catalyst/reaction sample. Due to this nonhomogenous field, heating of the catalyst/reaction sample is also nonhomogenous. To have homogenous heating of the catalyst/reaction sample, a rotating disk or stirrer is used. The advantage of having a multimode MW cavity is that several reactors can be irradiated at the same time. The disadvantage of the multimode cavity is that the electromagnetic field loses its energy due to several incidence/reflection on the cavity wall and as a result has a lower energy efficiency. In contrast, in a mono-mode cavity, the electromagnetic wave has a well-defined phase and generates a standing wave inside the cavity. The reactor is placed directly inside the waveguide where the electric field is at its maximum. Even though the characteristics of the electromagnetic wave is well understood in an empty mono-mode cavity, the characteristics of the electromagnetic wave and heating pattern in presence of catalyst/reaction sample is not well understood (Durka 2013). The advantage of mono-mode MW cavity is that high intensity electromagnetic field can be produced with less energy consumption. In addition, it is more targeted. The disadvantage of the mono-mode cavity is that multiple reactors cannot be irradiated at the same time. Both batch and flow type reactors can be of mono or multimode cavity. On laboratory scale, mostly, batch reactors with multimode cavity is used for liquid phase synthesis, while a flow system with mono-mode cavity is typically used for solid-gas reaction chemistry. Besides these factors, the material of construction is also an important design parameter for MW reactors. The MW reactor, including the waveguide, has to be constructed with such a material that the energy of electromagnetic radiation does not reduce much when incident on the waveguide or reactor wall, and the loss of the MW radiation through the reactor wall is negligible. Typical materials of construction are metals such as steel that have high resistance to chemicals, and also reflect the electromagnetic waves with minimum loss of energy. MW reactors typically have variable frequency and variable power output as opposed to the MW ovens used for cooking. The design and the scale of the MW reactors depend on the application. For gas–solid catalysis, mainly continuous flow mono-mode cavity reactors are used, while for heating food, organic liquid phase reactions, and synthesis of catalytic materials, mostly batch-type reactors are used. More information on type of reactor and their application can be found elsewhere (Nüchter et al. 2004; Bagley et al. 2005; Baxendale, Hayward, and Ley 2007; Buttress et al. 2019; Priecel and Lopez-Sanchez 2019; Durka 2013).

16.3 MW SCALE-UP CHALLENGES AND ALTERNATIVES

On a laboratory scale, MW-assisted synthesis is mainly focused on catalyst and reactor design, method development, and optimization of the operating parameters. However, when scale-up to industrial scale is considered, the focus is on the process throughput, reproducibility of the results, and process economics. Plant capacity depends on the specific product and therefore varies considerably. For example, producing multi-gram of active ingredient for a drug might be sufficient for a pharmaceutical industry. However, the same scale is not appropriate for a chemical plant producing aromatics or liquid fuels due to the large demand of these products. Therefore, while the larger-scale plants can economically benefit due to the economy of scale (Turton et al. 2018), capacity of MW reactors can widely vary. Irrespective of the capacity of the MW reactor(s), they need to be suitably designed with due consideration of scale-up uncertainties. There are several commercially available designs for batch (Moseley et al. 2008) and flow type (Dąbrowska et al. 2018) MW reactors. The commercially available reactors are still significantly smaller than what would be desired for large-scale processes like production of aromatics or fuels from natural gas.

During scale-up of MW reactors, the spatial variation in the temperature needs to be carefully accounted for with due consideration of the dimensions of the reactors and reaction energetics. A suboptimal temperature distribution can lead to undesired concentration profile, undesired side reactions like coke formation, and possible material degradation due to local hotspots. The spatial temperature profile, both locally and in the bulk, can be affected due to the PD of the MW field, as mentioned before. The PD can be increased by lowering the frequency. For example, by shifting the frequency from 2.45 GHz to 915 MHz, the PD can be increased approximately by three times (Priecel and Lopez-Sanchez 2019). Moreover, 915 MHz magnetron can provide up to 100 kW of power, compared to just 30 kW of power by 2.45 GHz (Priecel and Lopez-Sanchez 2019). This can be useful in heating larger reactors. However, other aspects must be considered while shifting the frequency such as material-MW interaction, reaction chemistry characteristics, and reactor performance. If a higher frequency is desired, then one way of addressing the limitations of the PD especially for batch-type reactors would be to have several reactors in parallel with one MW energy distributor as shown in Figure 16.1. However, the time consumed to load/unload the parallel reactors needs to be taken into considers. In addition, the advantages due to the economy of scale may not be fully realized. For continuous flow type reactors, the size of the MW irradiation zone can be designed to reduce the undesired temperature gradients.

In addition, orientation of the MW applicators can be designed to facilitate the desired design criteria. One option for the orientation of the MW applicator would be

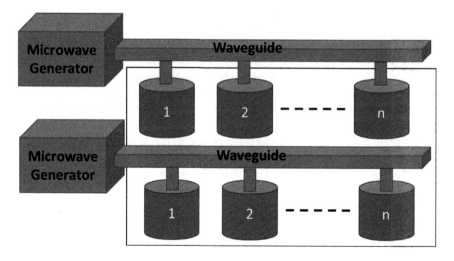

FIGURE 16.1   Parallel batch-type reactor scale-up.

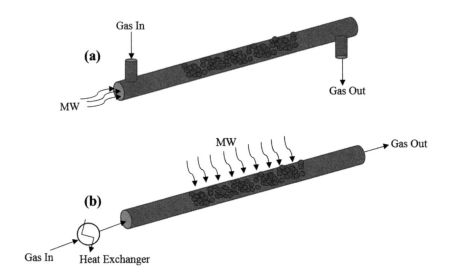

FIGURE 16.2 Fixed bed configuration for flow type reactor with MW propagating (a) parallel and (b) perpendicular to the axis of the reactor.

such that the electromagnetic waves are transmitted parallel to the axis of the reactor as shown in Figure 16.2a. However, this configuration would require transmitting the electromagnetic waves over the entire length of the reactor and/or restricting the reactor length. Obviously, the diameter of the reactor is also constrained depending on the frequency of the MW. Another possible orientation can be where the electromagnetic waves are transmitted perpendicular to the axis of the reactor as shown in Figure 16.2a. Obviously for both types, the incoming gas may be heated in conventional heat exchanger(s) before being introduced to the reactor. Such a hybrid concept is shown in Figure 16.2b. One potential concept is to accomplish bulk heating by using a furnace or a heat exchanger while the MW irradiation could be used to assist the reactions through thermal and nonthermal mechanisms.

For large-scale fixed bed reactors, one potential approach is to embed the transmission lines inside the reactor (Durka 2013) as shown in Figure 16.3. The transmission lines can be strategically placed inside the reactor such that the entire volume of the reactor is heated evenly. A potential advantage with this approach is to strategically transmit the electromagnetic waves in certain area of the reactor to obtain the desired conversion and selectivity.

A precise control of the temperature profile in the MW reactor can be highly advantageous for certain reactive systems. For example, in a direct nonoxidative conversion of methane to aromatics, coking is a major issue. While both the single-pass conversion and coking are favored at a higher temperature, selectivity towards aromatics is favored at a lower temperature. So, it will be advantageous to have a reactor temperature profile such that the reactor temperature towards the inlet can be higher for initiating the methane dimerization reaction while the later sections can be at a lower temperature to increase the selectivity of the aromatization reactions thus reducing coking of the catalyst. While the fixed bed MW reactor

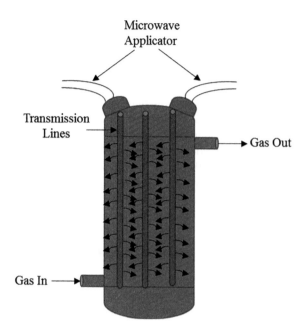

FIGURE 16.3   Fixed bed reactor with MW transmission line embedded.

technology is reasonably matured, another technology that can address the issue with the undesired temperature profile is the fluidized bed reactor (Durka 2013). Circulation of the solid bed improves the uniformity of exposure for the active sites to the irradiation by the electromagnetic waves, hence providing a considerably more uniform temperature distribution. In a fluidized bed system, the magnetrons can be mounted on the side of the reactor, so the electromagnetic waves propagate in the radial direction. The number of magnetrons needed will depend upon the size of the reactor. There are other type of reactor technologies, e.g., microchannels with travelling waves, coaxial travelling MW reactor, and so on (Durka 2013; Stefanidis et al. 2014).

One of the major challenges with MW catalysis especially with the large-scale reactors is the accuracy of the temperature measurement. In a gas–solid system, the solid metal particles that couple with the electromagnetic field can be at a significantly higher temperature than the bulk. Thermocouples that are used in conventionally-heated reactors cannot be used for MW system because of their possibility to couple with the MW field and getting heated up. Recently, infrared (IR) sensors have been utilized in temperature measurement of MW reactors. However, the IR sensors may only measure the bulk temperature which can be considerably lower than the core of the catalyst bed. Fiber optic (FO) sensors can be a good option as of now for temperature measurement because it is MW transparent. The FO sensor has to be inserted inside the catalyst bed for temperature measurement, but one has to be careful with the homogeneity of the reaction medium. More information about different types of temperature measurement techniques and their advantages/disadvantages can be found elsewhere (Priecel and Lopez-Sanchez 2019).

While scaling up large-scale MW reactors is still a challenge that needs considerably more experimental study, computational models can certainly help to provide more understanding in the design of the large-scale reactors. Developing accurate computational models of MW reactors is challenging due to various complicated and not-so-well-understood phenomena that take place in these reactors. Interaction of electromagnetic field with the catalyst, distribution of the field within the reactor, and the effect of reactor conditions on the behavior of electromagnetic waves are highly specific to a given reactive system and reactor design. Furthermore, spatial variation of material properties, concentration of the reactants, temperature, and pressure can be considerable. Therefore, a spatially distributed model where the electromagnetic field equations are coupled with heat and mass transfer equations and the model of reaction chemistry is desired. Computational models can be leveraged for optimization of the reactor design and operating variables with due consideration of the plant-wide system.

16.4   NATURAL GAS CONVERSION TO HYDROCARBONS

The existing routes for converting natural gas to aromatics and olefins involves a multistep process (indirect route) in which natural gas is first converted to syngas (a mixture of H_2 and CO) typically through steam reforming, partial oxidation, or autothermal reforming. In steam reforming, steam reacts with methane at high temperature (800–1000°C) and pressure (up to 30 bar) generating CO and H_2 (Rostrup-Nielsen and Rostrup-Nielsen 2002). This reaction is highly endothermic, so a large amount of energy has to be provided to sustain the reaction(s). In a partial oxidation reaction, fuel-air mixture reacts with methane to produce CO and H_2 at high temperatures. This reaction route has been reported to form soot thus makes this route unattractive (Nourbakhsh et al. 2018). Autothermal reforming is a combination of steam reforming and partial oxidation where the energy required by the endothermic steam reforming reaction is provided by the exothermic partial oxidation reaction. However, this reaction route is still energy intensive with the reactor operating at temperatures > 600°C, and high pressure (Mevawala, Jiang, and Bhattacharyya 2017; Aasberg-Petersen et al. 2011). After syngas formation, the H_2/CO ratio in the syngas can be adjusted to produce various liquid fuels using the Fischer–Tropsch process (Dry 2002). Alternately, aromatics or olefins can be produced via two routes, such as syngas-to-methanol and methanol-to-olefins/aromatics (Tian et al. 2015; Olsbye et al. 2012).The main issues with the multistep process are that the syngas generation step is highly energy intensive and that either CO or H_2 needs to be added to remove the oxygen when producing olefins/aromatics. As a result, there is low carbon utilization efficiency (< 50%). Furthermore, valuable hydrogen product is used for removing oxygen for hydrocarbon production. Moreover, in the multistep process, CO_2 is also generated increasing the energy penalty and capital cost due to separation and disposal of CO_2. Therefore, a route that does not require syngas generation is preferred.

An alternative process is to convert natural gas directly to hydrocarbons. However, direct conversion of methane, the main component of natural gas, is challenging because it is a very stable molecule and it cannot be easily activated. Methane has a strong C–H bond with a first bond dissociation energy of 439.3 kJ/mol (Schwach, Pan, and Bao 2017), which indicates that high amount of energy would be needed in order to break the bond. Moreover, the C–H bond of methane is stronger than higher hydrocarbons products and as a result these products are more reactive than methane (Holmen 2009). Existing approaches for direct conversion of methane to higher hydrocarbons are oxidative coupling of methane (OCM), nonoxidative methane dehydroaromatization (MDA), and methane conversion to aromatics, olefins, and hydrogen (MTOAH) (Alvarez-Galvan et al. 2011). In OCM, gas phase reaction takes place between oxygen and methane typically around 625–971°C temperature, and 1–41 bar pressure to produce CO, CO_2, $\geq C_2$ hydrocarbons. The maximum methane conversion that can be currently obtained is about 20% with a maximum hydrocarbons selectivity of 50% (Zanthoff and Baerns 1990; Chen, Couwenberg, and Marin 1994). In MDA, the reactions take place in presence of a metal-supported catalyst around 700–850°C temperature and atmospheric pressure (Ismagilov, Matus, and Tsikoza 2008; Zeng et al. 1998). The maximum methane conversion that can be obtained is limited to 22%, with selectivity to aromatics reaching > 90% (Zeng et al. 1998). Recently, MTOAH reactions were carried out in presence of $FeSiO_2$ catalyst to obtain a methane conversion of 48% and selectivity to C_2 and aromatics of approximately 70% at 1090°C and atmospheric pressure (Guo et al. 2014). Some of these reaction routes for direct methane conversion looks promising, however practically all of them suffer from low methane conversion, harsher reaction conditions, low product selectivity, and high coke formation (Alvarez-Galvan et al. 2011). An alternative is to use MW catalysis to directly convert the methane in natural gas to aromatics in presence of metal-supported catalyst. In this route, the methane molecule gets activated, i.e., the C–H bond breaks in presence of MW at much lower temperature than the current reaction routes and produces aromatics. One study has shown that MW catalysis can achieve the same methane conversion at a significantly lower temperature than the current direct methane conversion routes (Will, Scholz, and Ondruschka 2004). Operating the reactor at a lower temperature can assist in reducing the operating cost of the process. Moreover, coke formation, which is favored at higher temperature, can also be limited and this can increase the lifetime of the catalyst. Furthermore, considerably higher methane conversion than the non-MW routes has been reported (Zhang, Hayward, and Mingos 1999; Marún 1999). Having a higher methane conversion can reduce the capital cost of the process because less methane needs to be recycled and this reduces the equipment size. To evaluate the impact of the MW catalytic process for methane conversion, in this chapter, process modeling and techno-economic analysis are performed and compared with the conventional methane to aromatics process via methanol conversion, which is a matured technology. Evaluation of separation technologies, impact of market conditions, and effect of coke formation on the plant economics are also analyzed.

16.5   DIRECT NONOXIDATIVE CONVERSION OF NATURAL GAS TO AROMATICS: PLANT-WIDE MODELING AND TECHNO-ECONOMIC ANALYSIS

The process and economic model for the direct nonoxidative conversion of methane to aromatics (DHA) via MW catalysis is developed using Aspen Plus, and Aspen Process

TABLE 16.1

Product Distribution of Methane DHA MW Reactor for 10% and 37% Coke Selectivity

	Product Distribution (mass %)	
Components	37% Coke selectivity	10% Coke selectivity
H2	4.75%	4.44%
CH_4	77.48%	77.48%
C_2H_4	0.06%	0.06%
C_2H_6	0.51%	0.51%
C_6H_6	6.40%	6.40%
C_7H_8	0.38%	0.38%
C_8H_{10}	0.19%	0.30%
$C_{10}H_8$	4.00%	8.74%
C	6.23%	1.69%

Economic Analyzer (APEA). The economic study is carried out by using a yield reactor that is simulated based on the product distribution obtained from the experimental studies performed at National Energy Technology Laboratory (NETL) and West Virginia University (WVU). The plant performance and economics depend on the reactor product yield and the technologies used for purifying the products. The product distribution is likely to change depending on reactor operating conditions, reactor setup, and the catalyst. For this study, two reactor yield values are considered, as shown in Table 16.1. The coke selectivity is assumed to be 10%. This is similar to the literature. For example, Huang et al. have assumed that a maximum of 20% converted methane goes to coke (Huang et al. 2018). The case with low coke selectivity reflects the best-case scenario. The methane conversion for both the cases are about 22%.

In this reactor, a large quantity of hydrogen is produced. The hydrogen is a valuable product and therefore needs to be separated. Two technologies are considered for separating hydrogen–cryogenic separation, and pressure swing adsorption (PSA). A total of four processes cases are analyzed as listed here:

Case 1 – 37% of converted carbon goes to coke and cryogenic hydrogen separation

Case 2 – 10% of converted carbon goes to coke and cryogenic hydrogen separation

Case 3 – 37% of converted carbon goes to coke and pressure swing adsorption for hydrogen separation

Case 4 – 10% of converted carbon goes to coke and pressure swing adsorption for hydrogen separation

The economics of these process cases is compared with the conventional natural gas-to-aromatics (NG-to-A) process developed by Niziolek, Onel, and Floudas (2016), where the aromatics production section of the process was developed using the National Renewable Energy Laboratory (NREL) study on gasoline from wood via the integrated gasification technology (Phillips et al. 2011). The block flow diagram of the methane DHA process via MW catalysis with cryogenic hydrogen separation is shown in Figure 16.4. In this process, pure methane, stream S1, is preheated to the desired reactor temperature using the outlet stream (S3) of the MW reactor and a furnace, and fed to the MW reactor. After heating the stream (S1), stream (S3) is then used to heat the unconverted methane recycle stream (S2). The unconverted methane recycle stream (S2) is further heated to the desired reactor temperature using a furnace and fed to the MW reactor. Product from the MW reactor (S3) is then cooled down to separate out naphthalene and coke from rest of the products. The stream (S4) containing aromatics, C2, and light gases is compressed using a multistage intercooled compressor, and then fed to a flash separator (F1) for separation of aromatics from unconverted methane, C2, and light gases.

The liquid stream S5 from F1 is processed through a second flash separator (F2) and benzene distillation column to obtain > 99.8 mass% pure benzene (stream S7). The liquid stream from the benzene distillation column is fed to the toluene distillation column to obtain 99.4 mass% pure toluene (stream S8), and 99.8 mass% pure O-xylene (stream S9). The vapor stream S6 from the flash separator (F1) is further cooled to the desired temperature using a cooler. A four-stage refrigeration cycle is considered to provide additional cooling in the process. The cold stream, S6, is then sent to a flash separator (F3) to separate hydrogen from unconverted methane and C2 gases. Stream S10 has 99.1 mol% pure hydrogen and stream S11 has > 98 mass% methane, which is recycled to the reactor. Design specifications are used in Aspen Plus to meet the product specifications and to ensure that there is no temperature cross in the heat exchangers. It is ensured that the refrigeration duties available in the outgoing cold streams are efficiently recovered.

The process configuration for Case 3 and 4 is shown in Figure 16.5. The upstream portion of the process, i.e., until the compression of the stream (S4) is similar to Case 1 and 2. In these cases, the outlet stream (S4) is cooled to the desired temperature using a process heat exchanger and a utility heat exchanger. A single stage ethylene refrigeration cycle is used to provide the cooling duty. The vapor stream (S6) from flash separator (F1) is cooled down to 30°C and is fed to a PSA unit for hydrogen separation where 85% of hydrogen is recovered with a purity of 99.1 mol%. Product purity specifications for the aromatics are similar to Case 1 and 2.

The conventional NG-to-aromatics process is shown in Figure 16.6. The conventional process is not modeled in Aspen Plus; rather, mass and energy balance calculations are performed for this process using the information given in the literature (Niziolek, Onel, and Floudas 2016; Phillips et al. 2011). The detailed description of the methanol to aromatics process including the product distribution and plant configuration can be found elsewhere (Niziolek, Onel, and Floudas 2016; Phillips et al. 2011). For both conventional and MW-assisted processes, benzene, toluene, O-xylene, and naphthalene are considered to be the primary products. Credit for all other secondary products are included in the economic calculations.

For the MW catalysis process, the cost of all the equipment items except the MW reactor and the PSA unit is estimated using APEA. The cost of the MW reactor is

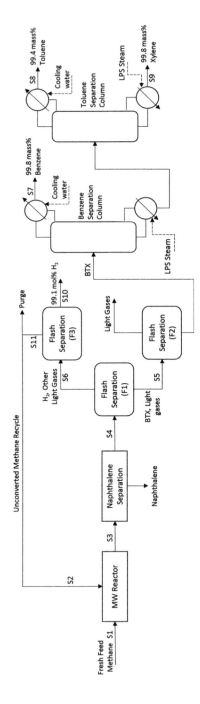

FIGURE 16.4 Block flow diagram of methane DHA process via MW catalysis with cryogenic hydrogen separation (Cases 1 and 2).

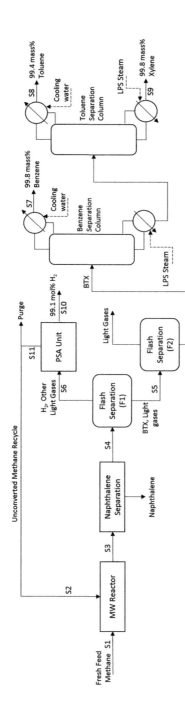

FIGURE 16.5 Block flow diagram of methane DHA process via MW catalysis with PSA separation technology (Cases 3 and 4).

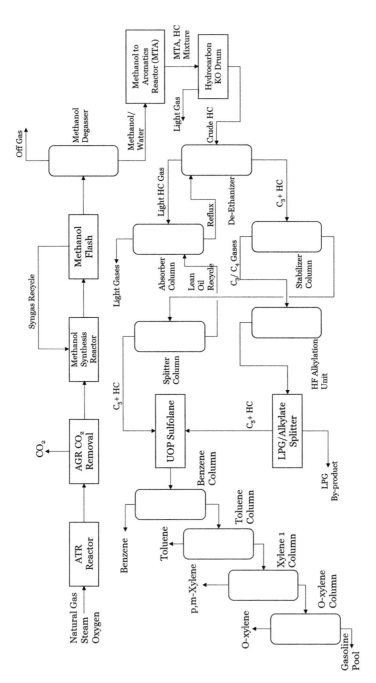

FIGURE 16.6   Conventional natural gas-to-aromatics process via methanol synthesis.

based on the internal information from vendor and WVU. The cost of the PSA unit is calculated using cost-correlation from the literature (Niziolek, Onel, and Floudas 2016; Mevawala, Jiang, and Bhattacharyya 2019). It should be noted that the cost of the raw material, products, and byproducts is based on the literature data and current market trend, and is shown in Table 16.2. An efficiency factor of 0.33 is used to convert the energy generated by burning coke into medium-pressure steam. In the conventional natural-to-aromatics plant, a methanol synthesis reaction and MTA reactor are highly exothermic. The energy generated by these reactors are used to generate medium-pressure and high-pressure steams, respectively. The price for both type of steams is taken from Turton et al. (2018). An efficiency factor of 0.45 is used to calculate the electricity equivalent of purge gases. To calculate electricity equivalent, the LHV value is first calculated, and then the efficiency factor is used to convert it into electricity.

APEA is then used to perform economic analysis and estimate key economic drivers like net present value (NPV) and internal rate of return (IRR). The economic assumptions are shown in Table 16.3.

For conventional processes, the equipment cost is estimated using cost correlations given by Niziolek, Onel, and Floudas (2016). The comparison between the MW and conventional process for 8000 kg/h methane feed is shown in Table 16.4.

As seen in Table 16.4, the selectivity of methane to aromatics for the MW-assisted processes, i.e., Cases 1–4, is higher compared to the conventional process. Moreover, in the MW process, hydrogen, a valuable byproduct, is also produced which can be either utilized in the process or sold over the fence. There are fewer products generated in the MW process compared to the conventional process thus separation process is considerably simpler for the MW process. In the conventional process, methanol converts to aromatics (MTA) to form a range of hydrocarbons and water. Very little amount of hydrogen is generated. Moreover, in the conventional NG-to-aromatics process, considerable amount of carbon is lost in form of carbon dioxide

TABLE 16.2
Prices for Products, Byproducts, Credits Given to the Process

Raw material/products/byproducts	Price	Unit
Methane	2.54	$/MMBTU
Electricity	17.5	$/GJ
Cooling Water	0.354	$/GJ
Benzene	0.867	$/kg
Toluene	0.898	$/kg
O-Xylene	0.763	$/kg
Naphthalene	0.4921	$/kg
Hydrogen	1.55	$/kg
P,M-Xylene	0.839	$/kg
LPG	0.235	$/kg
Gasoline pool	0.688	$/kg

TABLE 16.3
Parameters for Investment Analysis

Investment Parameter	Value	Investment parameter	Value
Contingency	30%	Working capital percentage	12% per year
Tax rate	40% per year	Plant overhead	50%
Desired internal rate of return	20% per year	G & A expenses	8% per year
Salvage value	20%	O & M escalation	3% per year
Project capital escalation	1% per year	No. of periods for analysis	20 year
Products and raw material escalation	1% per year each	Operating hours per year	8000 h
Utility escalation	1% per year	Length of start-up period	30 week

TABLE 16.4
Performance Comparison of MW Process Cases and Conventional NG-to-Aromatics Process

Parameter	Case 1	Case 2	Case 3	Case 4	Conventional NG-to-A
Relative CAPEX	−29.3%	−26.6%	−35.3%	−32.04%	-
Relative OPEX	2.75%	9.02%	−10.1%	−13.4%	-
Syngas-to-Methanol (yield mass %)	-	-	-	-	61.0%
Methanol/Methane to Aromatics (selectivity mass %)	53.7%	76.3%	54.5%	77.2%	41.8%
Overall Process Efficiency	54.9%	63.4%	57.7%	68.4%	57.4%
Benzene + Toluene Production Cost ($/kg)	0.800	0.735	0.728	0.570	1.78

thus lowering the carbon efficiency. As shown, the overall process efficiency of the MW catalysis process (Case 2 and 4) is higher than the conventional NG-to-aromatics process. Case 1 and 3 have lower efficiency due to the requirement of extreme low temperatures for hydrogen separation. The overall process efficiency is calculated by taking the ratio of the energy output from the process to energy input. Lower heating value (LHV) of products and raw materials is considered in the efficiency calculation. Relative capital cost (CAPEX) and operating cost (OPEX) for all the MW cases are also compared with the conventional process. If the relative CAPEX/OPEX of a MW-assisted case is negative, it indicates that the CAPEX/OPEX of that case is lower than the conventional process. The CAPEX of all the MW process cases is lower than the conventional process. The OPEX of Case 1 and 2 is higher than the conventional process, while Case 3 and 4 have lower OPEX. It should be noted that

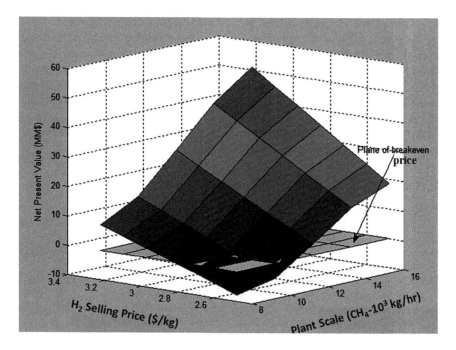

FIGURE 16.7   Sensitivity of plant scale and H2 selling price on NPV (36% coke with PSA separation).

the MW-assisted cases are not optimized, while the conventional process is highly optimized. The CAPEX and OPEX of MW-assisted cases are expected to improve further if techno-economic optimization is undertaken. The benzene plus toluene product cost for all the MW-assisted processes is much lower than the conventional process. Profitability of the MW process is affected by the plant scale like most other conventional processes as well as the prices of hydrogen. Hydrogen is mostly sold over the fence or within limited distance and its market price can vary significantly depending on the feedstock used and the location. A sensitivity analysis is performed to investigate the effect of plant scale and hydrogen price on the NPV. Figure 16.7 shows that when the plant scale is below 12,000 kg/h of methane feed, i.e., 360 tons/day of products produced and the hydrogen price is below about $2.8/kg, the process is no more profitable.

16.6   CONCLUSIONS

MW technology is a very promising technology for chemical synthesis. For commercial success of this technology, various issues such as the challenges of temperature measurement, uniform heating, and reaction configuration needs to be addressed especially when large-scale systems are considered. As an example, the direct nonoxidative MW-assisted conversion of methane to aromatics is compared with the conventional oxidative conversion of methane to aromatics via methanol. It was observed that the higher selectivity of the MW-assisted process also makes separation easier. The MW-assisted process also earns considerable revenue by selling hydrogen, a valuable byproduct, which is not produced much in the conventional route. Carbon utilization of the MW-assisted route is much superior. The study shows that the required price of benzene plus toluene for the MW-assisted process to break even is about one-third of that required for the conventional process. Overall, the study reflects the promise of the MW-assisted reactional pathways. The authors believe that large class of chemical reactive systems can significantly benefit from the MW technology through innovative reactor and catalyst design, optimization of operating conditions and process hardware, and optimal operations.

REFERENCES

Aasberg-Petersen, K., I. Dybkjaer, C. V. Ovesen, N. C. Schjodt, J. Sehested, and S. G. Thomsen. 2011. Natural gas to synthesis gas – catalysts and catalytic processes. *J. Nat. Gas Sci. Eng.* 3(2): 423–459.

Alvarez-Galvan, M. C., N. Mota, M. Ojeda, et al. 2011. Direct methane conversion routes to chemicals and fuels. *Catal. Today* 171(1): 15–23.

Bagley, M. C., R. L. Jenkins, M. C. Lubinu, C. Mason, and R. Wood. 2005. A simple continuous flow microwave reactor. *J. Org. Chem.* 70(17): 7003–7006.

Bai, X., B. Robinson, C. Killmer, Y. Wang, L. Li, and J. Hu. 2019. Microwave catalytic reactor for upgrading stranded shale gas to aromatics. *Fuel* 243: 485–492.

Baxendale, I., J. Hayward, and S. Ley. 2007. Microwave reactions under continuous flow conditions. *Comb. Chem. High Throughput Screen.* 10(10): 802–836.

Buttress, A. J., G. Hargreaves, A. Ilchev, et al. 2019. Design and optimisation of a mw reactor for kilo-scale polymer synthesis. *Chem. Eng. Sci. X* 2: 100022. doi:10.1016/j.cesx.2019.100022.

Chen, Q., P. M. Couwenberg, and G. B. Marin. 1994. Effect of pressure on the oxidative coupling of methane in the absence of catalyst. *AIChE J.* 40(3): 521–535.

Collin, R. E. 2000. *Fundamentals of Microwave Engineering.* New York, NY: Wiley-IEEE Press.

Dąbrowska, S., T. Chudoba, J. Wojnarowicz, and W. Łojkowski. 2018. Current trends in the development of microwave reactors for the synthesis of nanomaterials in laboratories and industries: a review. *Crystals.* doi:10.3390/cryst8100379.

De Souza, R. O. M. A. 2015. Theoretical aspects of microwave irradiation practices. In: Fang, Z., Smith, Jr. R., Qi, X. (eds.), *Production of Biofuels and Chemicals with Microwave. Biofuels and Biorefineries*, vol. 3. Dordrecht: Springer.

Dry, M. E. 2002. The Fischer-Tropsch process: 1950–2000. *Catal. Today* 71(3–4): 227–241.

Durka, T. 2013. *Microwave Effects in Heterogeneous Catalysis : Application to Gas-Solid Reactions for Hydrogen Production.* PhD Thesis. Warsaw University of Technology, Poland.

Gabriel, C., S. Gabriel, E. H. Grant, B. S. J. Halstead, and D. M. P. Mingos. 1998. Dielectric parameters relevant to microwave dielectric heating. *Chem. Soc. Rev.* 27: 213–224.

Guo, X., G. Fang, G. Li, et al. 2014. Direct, nonoxidative conversion of methane to ethylene, aromatics, and hydrogen. *Science* 344(6184): 616–619.

Harrison, A., and A. G. Whittaker. 2003. Microwave heating. *Compr. Coord. Chem. II* 1(January). Pergamon: 741–745. doi:10.1016/B0-08-043748-6/01162-2.

Holmen, A. 2009. Direct conversion of methane to fuels and chemicals. *Catal. Today* 142(1–2): 2–8.

Huang, K., J. B. Miller, G. W. Huber, J. A. Dumesic, and C. T. Maravelias. 2018. A general framework for the evaluation of direct nonoxidative methane conversion strategies. *Joule* 2(2): 349–365.

Ismagilov, Z. R., E. V. Matus, and L. T. Tsikoza. 2008. Direct conversion of methane on Mo/ZSM-5 catalysts to produce benzene and hydrogen: achievements and perspectives. *Energy Environ. Sci.* 1: 526–541.

Lidström, P., J. Tierney, B. Wathey, and J. Westman. 2001. Microwave assisted organic synthesis – a review. *Tetrahedron* 57: 9225–9283.

Loupy, A. 2008. *Microwaves in Organic Synthesis*, 2nd Edition. doi:10.1002/9783527619559.

Marún, C. 1999. Catalytic oligomerization of methane via microwave heating. *J. Phys. Chem. A* 103(22): 4332–4340.

Metaxas, A. C., and R. J. Meredith. 2008. *Industrial Microwave Heating*, 1st Edition. Milton Keynes: Lightning Source UK Ltd.

Mevawala, C., Y. Jiang, and D. Bhattacharyya. 2017. Plant-wide modeling and analysis of the shale gas to dimethyl ether (DME) process via direct and indirect synthesis routes. *Appl. Energy* 204: 163–180.

Mevawala, C., Y. Jiang, and D. Bhattacharyya. 2019. Techno-economic optimization of shale gas to dimethyl ether production processes via direct and indirect synthesis routes. *Appl. Energy* 238: 119–134.

Moseley, J. D., P. Lenden, M. Lockwood, et al. 2008. A comparison of commercial microwave reactors for scale-up within process chemistry. *Org. Proc. Res. Dev.* 12(1): 30–40.

Niziolek, A. M., O. Onel, and C. A. Floudas. 2016. Production of benzene, toluene, and xylenes from natural gas via methanol: process synthesis and global optimization. *AIChE J.* 62(5): 1531–1556.

Nourbakhsh, H., J. R. Shahrouzi, A. Zamaniyan, H. Ebrahimi, and M. R. J. Nasr. 2018. A thermodynamic analysis of biogas partial oxidation to synthesis gas with emphasis on soot formation. *Int. J. Hydrogen Energy* 43(33): 15703–15719.

Nüchter, M., B. Ondruschka, W. Bonrath, and A. Gum. 2004. Microwave assisted synthesis – a critical technology overview. *Green Chem.* 6: 128–141.

Olsbye, U., S. Svelle, M. Bjrgen, et al. 2012. Conversion of methanol to hydrocarbons: how zeolite cavity and pore size controls product selectivity. *Angew. Chem. Int. Ed.* 51: 5810–5831.

Phillips, S. D., J. K. Tarud, M. J. Biddy, and A. Dutta. 2011. Gasoline from wood via integrated gasification, synthesis, and methanol-to-gasoline technologies. doi:10.2172/1004790.

Priecel, P., and J. A. Lopez-Sanchez. 2019. Advantages and limitations of mw reactors: from chemical synthesis to the catalytic valorization of biobased chemicals. *ACS Sust. Chem. Eng.* 7(1): 3–21.

Rostrup-Nielsen, J. R., and T. Rostrup-Nielsen. 2002. Large-scale hydrogen production. *CATTECH* 6: 150–159.

Schwach, P., X. Pan, and X. Bao. 2017. Direct conversion of methane to value-added chemicals over heterogeneous catalysts: challenges and prospects. *Chem. Rev.* 117(13): 8497–8520.

Stankiewicz, A., F. E. Sarabi, A. Baubaid, P. Yan, and H. Nigar. 2019. Perspectives of microwaves-enhanced heterogeneous catalytic gas-phase processes in flow systems. *Chem. Rec.* 19: 40–50.

Stefanidis, G. D., A. N. Muñoz, G. S. J. Sturm, and A. Stankiewicz. 2014. A helicopter view of microwave application to chemical processes: reactions, separations, and equipment concepts. *Rev. Chem. Eng.* 30(3). De Gruyter: 233–259.

Tian, P., Y. Wei, M. Ye, and Z. Liu. 2015. Methanol to olefins (MTO): from fundamentals to commercialization. *ACS Catal.* 5(3): 1922–1938.

Turton, R., J. A. Shaeiwitz, D. Bhattacharyya, and W. B. Whiting. 2018. *Analysis, Synthesis, and Design of Chemical Processes*, 5th Edition. Upper Saddle River, NJ: Prentice Hall.

Will, H., P. Scholz, and B. Ondruschka. 2004. Microwave-assisted heterogeneous gas-phase catalysis. *Chem. Eng. Technol.* 27(2): 113–122.

Zanthoff, H., and M. Baerns. 1990. Oxidative coupling of methane in the gas phase: kinetic simulation and experimental verification. *Ind. Eng. Chem. Res.* 29: 2–10.

Zeng, J. L., Z. T. Xiong, H. B. Zhang, G. D. Lin, and K. R. Tsai. 1998. Nonoxidative dehydrogenation and aromatization of methane over W/HZSM-5-based catalysts. *Catal. Lett.* 53: 119–124.

Zhang, X., D. O. Hayward, and D. M. P. Mingos. 1999. Apparent equilibrium shifts and hotspot formation for catalytic reactions induced by microwave dielectric heating. *Chem. Commun.* 11(11): 975–976.

Index